DISTRIBUTED ANTENNA SYSTEMS

WIRELESS NETWORKS AND MOBILE COMMUNICATIONS

Series Editor: Yan Zhang

Millimeter Wave Technology in Wireless PAN, LAN, and MAN
Shao-Qiu Xiao, Ming-Tuo Zhou and Yan Zhang
ISBN: 0-8493-8227-0

Security in Wireless Mesh Networks
Yan Zhang, Jun Zheng and Honglin Hu
ISBN: 0-8493-8250-5

Resource, Mobility and Security Management in Wireless Networks and Mobile Communications
Yan Zhang, Honglin Hu, and Masayuki Fujise
ISBN: 0-8493-8036-7

Wireless Mesh Networking: Architectures, Protocols and Standards
Yan Zhang, Jijun Luo and Honglin Hu
ISBN: 0-8493-7399-9

Mobile WIMAX: Toward Broadband Wireless Metropolitan Area Networks
Yan Zhang and Hsiao-Hwa Chen
ISBN: 0-8493-2624-9

Distributed Antenna Systems: Open Architecture for Future Wireless Communications
Honglin Hu, Yan Zhang and Jijun Luo
ISBN:1-4200-4288-2

AUERBACH PUBLICATIONS

www.auerbach-publications.com
To Order Call: 1-800-272-7737 • Fax: 1-800-374-3401
E-mail: orders@crcpress.com

DISTRIBUTED ANTENNA SYSTEMS

Open Architecture for Future Wireless Communications

Edited by
Honglin Hu ◆ Yan Zhang ◆ Jijun Luo

Auerbach Publications
Taylor & Francis Group
Boca Raton   New York

Auerbach Publications is an imprint of the
Taylor & Francis Group, an **informa** business

Auerbach Publications
Taylor & Francis Group
6000 Broken Sound Parkway NW, Suite 300
Boca Raton, FL 33487-2742

© 2007 by Taylor & Francis Group, LLC
Auerbach is an imprint of Taylor & Francis Group, an Informa business

No claim to original U.S. Government works
Printed in the United States of America on acid-free paper
10 9 8 7 6 5 4 3 2

International Standard Book Number-10: 1-4200-4288-2 (Hardcover)
International Standard Book Number-13: 978-1-4200-4288-7 (Hardcover)

Library of Congress Cataloging-in-Publication Data

Distributed antenna systems : open architecture for future wireless communications / editors Honglin Hu, Yan Zhang, Jijun Luo.
    p. cm.
Includes bibliographical references and index.
ISBN-13: 978-1-4200-4288-7 (alk. paper)
ISBN-10: 1-4200-4288-2 (alk. paper)
   1. Wireless communication systems. 2. Antenna arrays. I. Hu, Honglin, 1975- II. Zhang, Yan, 1977- III. Luo, Jijun.

TK6565.A6D57 2007
621.382'4--dc22
    2006101714

Visit the Taylor & Francis Web site at
http://www.taylorandfrancis.com

and the Auerbach Web site at
http://www.auerbach-publications.com

CONTENTS

PREFACE

The rapid growth in mobile communications has led to an increasing demand for wide-band high data rate communications services. In recent years, Distributed Antenna Systems (DAS) has emerged as a promising candidate for future (beyond 3G or 4G) mobile communications, as illustrated by projects such as FRAMES and FuTURE. The architecture of DAS inherits and develops the concepts of pico- or micro-cell systems, where multiple distributed antennas or access points (AP) are connected to and controlled by a central unit. DAS owns the property of open wireless architecture (OWA), which is one of the key features of future wireless communication systems. Due to its open architecture, DAS can be assigned new and more flexible radio resource management, and thus outperforms conventional centralized wireless communication systems. These distributed antenna techniques are being studied intensively for cellular systems, and are especially suitable for ad hoc and mesh networks due to their inherent distributed characteristics. However, many topics involving the physical layer and (especially) high-efficiency protocols for DAS still need further investigation.

Distributed Antenna Systems: Open Architecture for Future Wireless Communications is the first book to provide readers with a comprehensive technical guide to the fundamental concepts, recent advances, and open issues of DAS. The subject is explored via various key challenges in numerous diverse scenarios including architecture, capacity, connectivity, scalability, medium access control, scheduling, dynamic channel assignment, and cross-layer optimization. The primary focus of this book is on concept introduction, effective protocol proposal, system integration, performance analysis techniques, simulation, experimentation, and, more importantly, future directions in DAS research. The objective of the book is to serve as a valuable reference for scientists, faculty members, researchers, students, engineers, and research strategists in this rapidly evolving field.

This book is organized in three parts:

- Part I: Channel and Theoretical Issues
- Part II: MAC and Protocols
- Part III: Case Studies and Applications

In Part I, DAS fundamentals, including channel models for DAS and theoretical issues for DAS are introduced, which enable the readers to understand the capacity of DAS with different structures. Part II concentrates on the MAC and protocols for DAS, including distributed signal processing, optimal resource allocation, cooperative MAC protocols, cross-layer design, and distributed organization. Part III illustrates case studies and applications of DAS, including experiments, RF engineering, and applications.

This book has the following salient features:

- Provides a comprehensive reference to state-of-the-art DAS technology, including concepts, protocols, architecture, and system implementation
- Identifies advanced DAS research topics and future research directions
- Provides an easy-to-understand introduction to DAS via the use of illustrative figures
- Provides a complete cross-reference to the different layers of DAS protocol stacks
- Details techniques that can be used to efficiently improve the performance of the DAS
- Explores emerging DAS standardization activities and specifications

The book can serve as a useful reference for students, educators, faculties, telecom service providers, research strategists, scientists, researchers, and engineers in the field of wireless networks and mobile communications.

ACKNOWLEDGMENTS

We would like to acknowledge the time and effort invested by all of the contributors and to thank them for their excellent work. They were all extremely professional and cooperative. Special thanks go to Richard O'Hanley, Jessica Vakili, Ari Silver, Jennifer Strong, and many other colleagues of the Taylor & Francis Group for their support, patience, and professionalism from the beginning until the final stage. We are also grateful to Charles Devaux for his help in typesetting. Last, but not least, a special thank you to our families and friends for their constant encouragement, patience, and understanding throughout this project.

Honglin Hu, Yan Zhang, and Jijun Luo

ABOUT THE EDITORS

Honglin Hu received his Ph.D. in communications and information systems in 2004 from the University of Science and Technology of China (USTC) in Hefei, China. From July 2004 to January 2006, he worked for Future Radio, Siemens AG Communications in Munich, Germany. In January 2006, he joined the Shanghai Research Center for Wireless Communications (SHRCWC), which is also known as the International Center for Wireless Collaborative Research (WirelessCoRe). Dr. Hu also serves as an associate professor at the Shanghai Institute of Microsystem and Information Technology (SIMIT) in the Chinese Academy of Science (CAS), where his main focus is on international standardization and other collaborative activities. Moreover, he is a member of IEEE, IEEE ComSoc, and IEEE TCPC. He also serves as a member of Technical Program Committee for IEEE WirelessCom 2005, IEEE ICC 2006, IEEE IWCMC 2006, IEEE/ACM Q2SWinet 2006, IEEE ICC 2007, IEEE WCNC 2007. Since June 2006, he has served on the editorial board of *Wireless Communications and Mobile Computing*, John Wiley & Sons. Email: hlhu@ieee.org

Yan Zhang received his Ph.D. from the School of Electrical & Electronics Engineering, Nanyang Technological University, Singapore. From August 2004 to May 2006, he worked with NICT Singapore, a branch of the National Institute of Information and Communications Technology (NICT). Since August 2006, he has worked for Simula Research Laboratory, Norway (http://www.simula.no/). He is on the editorial board of the *International Journal of Network Security*. He is also currently serving as a book series editor for the "Wireless Networks and Mobile Communications" series (Auerbach Publications, CRC Press, Taylor & Francis Group) and has also served as co-editor for several books: *Resource, Mobility and Security Management in Wireless Networks and Mobile Communications; Wireless Mesh Networking: Architectures, Protocols and Standards; Millimeter-Wave Technology in Wireless PAN, LAN and MAN; Distributed Antenna Systems: Open Architecture for Future Wireless Communications; Security in Wireless Mesh Networks; Wireless Metropolitan Area Networks: WiMAX and Beyond; Wireless Quality-of-Service: Techniques, Standards and Applications; Broadband Mobile Multimedia: Techniques and Applications; Internet of Things: From RFID to the Next-Generation Pervasive Networked Systems* and *Handbook of Research on Wireless Security*. Dr. Zhang served as program co-chair for IEEE PCAC'07, special track co-chair for "Mobility and Resource Management in Wireless/Mobile Networks" in ITNG 2007, special session co-chair for "Wireless Mesh Networks" in PDCS 2006, and is a member of Technical Program Committee for IEEE AINA 2007, IEEE CCNC 2007, WASA'06, IEEE GLOBECOM'2006, IEEE WoNGeN'06, IEEE IWCMC 2006, IEEE IWCMC 2005, ITST 2006, and ITST 2005. His research interests include resource, mobility, energy, and security management in wireless networks and mobile computing. He is a member of IEEE and IEEE ComSoc. Email: yanzhang@ieee.org

Jijun Luo received his M. Eng. from Shandong University, China 1999 and M. Sc. from Munich University of Technology, Germany in 2000, respectively. He joined Siemens in 2000 and received his Doktor-Ingenieur (Dr. Ing.) degree in 2006 from RWTH Aachen University, Germany. He has published more than a hundred technical papers, co-edited three books and holds many patents. His technical contributions mainly cover his findings on wireless communication system design, radio protocol, system architecture, radio resource management, signal processing, coding, and modulation technologies. He is active in international academy and research activities. Dr. Luo has been nominated as the session chair of many high level technical conferences organized by IEEE and European research organizations. He leads research activities in several European research projects and is active in international industrial standardization bodies. His main interests are transmission technologies, radio resource management, reconfigurability (software defined radio), and radio system design. He is also a member of IEEE. Email: jesse.luo@ieee.org

CONTRIBUTORS

Fumiyuki Adachi
Tohoku University
Sendai, Japan

Andreas Ahrens
University of Rostock
Rostock, Germany

Jeffrey G. Andrews
The University of Texas
Austin, Texas

Yeheskel Bar-Ness
New Jersey Institute of Technology
Newark, New Jersey

Min Chen
University of British Columbia
Vancouver, Canada

Yifan Chen
Nanyang Technological University
Singapore

Wan Choi
Information and Communication
University
Daejeon, South Korea

Lin Dai
University of Delaware
Newark, Delaware

I.A. Glover
University of Strathclyde
Glasgow, Scotland

Dennis L. Goeckel
University of Massachusetts
Amherst, Massachusetts

Martin Haardt
Ilmenau University of Technology
Ilmenau, Germany

Alexander M. Haimovich
New Jersey Institute of Technology
Newark, New Jersey

Honglin Hu
Shanghai Research Center for Wireless
    Communications
Shanghai, China

Sudharman K. Jayaweera
University of New Mexico
Albuquerque, New Mexico

Volker Jungnickel
Fraunhofer Institute for
    Telecommunications
Heinrich Hertz Institut
Berlin, Germany

Jee Hyun Kim
Siemens AG
Munich, Germany

Eisuke Kudoh
Tohoku University
Sendai, Japan

J. Nicholas Laneman
University of Notre Dame
Notre Dame, Indiana

Jijun Luo
Siemens AG Communications
Munich, Germany

Boris Rankov
ETH Zurich
Zurich, Switzerland

Anna Scaglione
Cornell University
Ithaca, New York

Martin Schubert
Fraunhofer German - Sino Lab for
    Mobile Communications MCI
Berlin, Germany

Shlomo (Shitz) Shamai
Israel Institute of Technology
Haifa, Israel

D.J. Shyy
The MITRE Corporation
McLean, Virginia

Osvaldo Simeone
New Jersey Institute of Technology
Newark, New Jersey

Oren Somekh
New Jersey Institute of Technology
Newark, New Jersey

T.B. Sørensen
Aalborg University
Aalborg, Denmark

Umberto Spagnolini
Politecnico di Milano
Milan, Italy

Craig J. Stanziano
Distributed Wireless Group
Newbury Park, California

John A. Stine
The MITRE Corporation
McLean, Virginia

Fei Tong
Motorola Inc.
Swindon, UK

Ramanarayanan Viswanathan
Southern Illinois University
Carbondale, Illinois

Jörg Wagner
ETH Zurich
Zurich, Switzerland

Tobias Weber
University of Rostock
Rostock, Germany

Armin Wittneben
ETH Zurich
Zurich, Switzerland

Xiaodong Yang
Göettingen University
Göettingen, Germany

Yong Yuan
Huazhong University of Science
    and Technology
Wuhan, China

Chau Yuen
Institute for Infocomm Research
Singapore

Yan Zhang
Simula Research Laboratory
Lysaker, Norway

Zhenrong Zhang
Guangxi University
Nanning, Guangxi, China

Wolfgang Zirwas
Siemens AG
Munich, Germany

PART I

CHANNEL AND THEORETICAL ISSUES

1

DIVERSITY AND MULTIPLEXING FOR DAS: CHANNEL MODELING PERSPECTIVE

Yifan Chen, Yan Zhang, Chau Yuen and Zhenrong Zhang

Contents

Distributed antenna system (DAS) is a new architecture for future public wireless access, which refers to a generalized multiple-input multiple-output (MIMO) system comprising an antenna array at one side of the link and several largely separated antenna arrays at the other side. With DASs, both diversity and multiplexing gains can be achieved. This chapter presents formulation of diversity (DIV) and degrees-of-freedom (DOF) for a system-dependent wireless channel by taking a microscopic view of the two measures. As an initial study, we will consider a simple network topology and DAS architecture. Starting with a degenerate single-cluster single-bounce (SCSB) channel and perfect receiving conditions, the analysis is extended to more general scattering and receiver-operating scenarios. The utility of the proposed methodology is demonstrated by presenting various numerical examples to provide insight into the impact of the signal power spectrum on the DIV and DOF estimates. The chapter is concluded by identifying some important open issues in DAS channel measurement and modeling, DIV-DOF tradeoff, and design of an advanced signaling scheme under the rubric of "system-dependent channel modeling."

1.1 INTRODUCTION

In recent years, considerable attention has been drawn to multiple-input multiple-output (MIMO) communication techniques due to the prospect of significant improvements of system performance [1,2]. This idea has also been extended to a distributed antenna system (DAS), which refers to a MIMO system comprising multiple antennas co-located at one end of the radio link and several geographically scattered access points (APs), each with multiple antennas co-located within an AP at the other end [3,4]. The DAS has the advantage of macrodiversity that is inherent to the widely spaced antenna and, therefore, offers the capability to enhance signal quality, increase system capacity, and improve coverage. Relevant literature has discussed the performance improvement from two different directions. One set of techniques, called space-time coding, improves the link quality by utilizing the diversity gains offered by multiple antenna arrays [5,6], whereas the other scheme, called spatial multiplexing, utilizes the parallel spatial subchannels by exploiting channel information at the receiver [7]. Tradeoff between the two approaches has been analyzed from both information-theoretic [8–10] and system performance [11] viewpoints recently. In the existing literature, diversity (DIV) [6] and degrees-of-freedom (DOF) [8–10] have been widely used as the simple measures of performance gains obtained from using MIMO systems. DIV is commonly defined as the negative exponent of signal-to-noise ratio (SNR) in the probability of error expression and DOF is referred to as the coefficient of log(SNR) occurring in the expression for capacity, both in the high SNR region. To shed a light on the nature of the two measures, DIV can be thought of as the redundancy of the transmitted signal in a particular communication system, whereas DOF are considered as the number of independent spatial-temporal channels available for communication.

Nonetheless, to the best of our knowledge, little work has been reported to incorporate the theoretical framework with real-life channel conditions to obtain a mathematical model that allows system designers to develop a deeper understanding of the performance bound of these two schemes. In the past, a few models have been proposed to relate the spatial DOF to the array geometry and the scattering field patterns [12–16]. As the channel response is sandwiched between the antenna responses, a versatile practice is to apply the separation hypothesis by assuming the separability between *radio channel* and *propagation channel* [17–19], where the former embraces the transmit block, the propagation channel, and the receive block. This hypothesis is imposed in most of the existing channel models, which, however, may become inaccurate for some realistic environments. One example is the indoor ultra-wideband measurement conducted in the

Intel research laboratory [20]. It is shown in the experiment that the number of clusters and their subtended angular intervals decreased with increasing operating frequency, which demonstrated the dependency of multipath structures on signal frequency. Furthermore, it is also plausible that the channel structure is dependent on antenna gain patterns due to the finite sensitivity of measurement devices [21–23]. Motivated by these facts, we introduce the paradigm of DIV and DOF densities, which facilitate a microscopic examination of the diversity and multiplexing levels, thereby capturing the system-dependent nature of multipath channels. Subsequently, the DIV and DOF can be calculated as an integral of the two densities over the signal space. Thus, the main contribution of this work is to integrate the intuitive explanations of the two measures with physical channel modeling to obtain a more fundamental bound to the available DIV and DOF, given a constraint on the signal space and the multipath structure.

In wireless communication channels, scatterers are not distributed uniformly throughout the whole coverage area, but rather occur in clusters [24–27]. The scattering of each cluster in physical environments will be characterized by a set of aperture frequency dependent parameters, including the subtended angular and delay intervals and the cluster radius and keyhole radius, where the first two serve to characterize the cluster boundaries and the last two reflect its internal radio propagation conditions. Starting with a simple single-cluster single-bounce (SCSB) channel and perfect receiving conditions (see e.g., [23]), the mathematical framework is then extended to more general situations (channels with multiple reradiation processes and scattering clusters). By this methodology, we provide insight into the fundamental bound of diversity and capacity gains in the generalized DAS scenario from the actual wave-propagation viewpoint. Furthermore, we will use numerical examples to show that it is important for system designers to prefilter the signal power spectrum before applying any conventional transmitter optimization techniques.

This chapter is organized as follows. Section 1.2 states our viewpoint on the system-independent and system-dependent modeling perspectives. Then, a comprehensive discussion on system-dependent channel characterization is given in Section 1.3. In Section 1.4, the DAS topology and channel models will be presented. Sections 1.5 and 1.6 discuss the analytical approaches for DIV and DOF computation. The utility of the proposed methodology is demonstrated in Section 1.7 by presenting various numerical examples to provide insight into the impact of the signal power spectrum on the DIV and DOF estimates. Finally, some concluding remarks are drawn in Section 1.8.

1.2 COMPARISON BETWEEN SYSTEM-INDEPENDENT AND SYSTEM-DEPENDENT CHANNEL MODELS

Multipath channel modeling is the most important and fundamental amongst all the research activities on wireless systems that have taken place in the past years. The term "channel" takes on different meanings in the scientific literature depending on the relevant technical issues [19]. For example, the physical propagation paths that result from multipath environments are called "propagation channels." The signal received by a loaded antenna system is referred to as the output from the "signal channel," which comprises the functional blocks of the propagation channel, the effects of the interaction of the antenna and the field, and the antenna itself. Throughout the stages of downconversion, demodulation, and decoding (baseband processing), different types of channels can be defined as "baseband radio channel," "digital channel," "raw data channel," etc. [19]. In this chapter, we will consider the antenna and its associated signal processing and front-end circuitry as "system," which can effectively alter the antenna interaction with the free space waves.

For a system-independent characterization of mobile radio channels, the so-called double-directional channel concept is proposed [17,18]. To separate the influence of transmit/receive (Tx/Rx) sites and the propagation channel itself, it is a versatile practice to distinguish between the *radio channel* and the *propagation channel*. The former is described by the nondirectional channel response, whereas the latter by the double-directional channel response excluding both the Tx and Rx antennas (Figure 1.1(a)). In between is the so-called single-directional channel with the introduction of directionality at the Rx site. With the arbitrary inclusion of system-specific information, such as antenna beam patterns and operating frequency, the system-independent approach features an improved scatterer identification that offers the following advantages [17]:

■ Insertion of any desired system configuration (e.g., antenna patterns).
■ Evaluation of path dispersion and dissimilarities [28].
■ A generic channel model for MIMO systems that enables the signal-processing optimization process at both ends of the radio link.

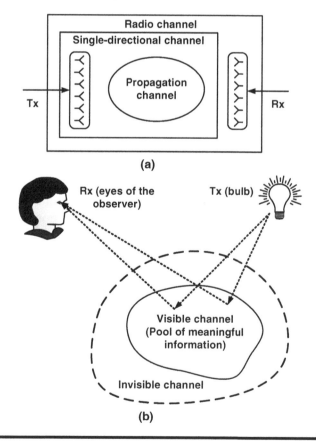

Figure 1.1 Illustration of (a) System-Independent and (b) System-Dependent Channel Modeling Concepts.

In general, the double-directional channel models separate three functional blocks in the radio transmission chain:

- Transmitting block, which operates at a specific frequency band and distributes the signal into the desired angle-of-departure (AOD). These AODs can be selected from the group of input directions provided by the double-directional channel measurements.
- Propagation channel, which includes all the resolvable propagation paths between the transmit and receive sites. Each path is fully characterized by its impulse response including the excess delay, the complex amplitude, and the associated angle-of-arrival (AOA) and AOD. In general, all the parameters will also depend on the absolute time, and the set of multipath components (MPCs) may vary with time as well.
- Receiving block, which collects the signal from the various AOAs by weighted combination and filters the signal by a bandpass filter.

The perspective of system-dependent channel characterization is completely different from the aforementioned philosophy. Referring to Figure 1.1(b), the Tx behaves like a bulb which probes the environment by illuminating the lights. The Rx behaves like the eyes of an observer who is trying to obtain any useful information from the Tx by receiving the lights. Depending on the brightness of the source and the eyesight, only certain environmental details are discernible to the eyes. Therefore, the observer will interpret the environment as what he has seen, which also conveys the meaningful information to him. In a similar manner, in system-dependent channel characterization, we distinguish between the visible channel (VC) and the invisible channel (IVC). The former represents a pool of meaningful information, from which the useful signal can be extracted for further processing given the specified system sensitivity and the radio channel structure. Surrounding the VC is a large invisible region that extends to infinity. Apparently, the boundary line of the VC is dependent on the received signal properties such as signal strength, which in turn is jointly determined by the *system properties* and the *multipath compositions*. Also note that it is usually unfeasible to define the boundary line of the VC in a deterministic manner. A more realistic approach would be defining the VC boundary probabilistically.

In general, a thorough description of the propagation environment is almost impossible due to the sophisticated channel structure. There are, in effect, an infinite number of paths that could be enumerated in an ideal double-directional channel, which should be system-independent according to its original definition. However, certain system-dependency has been introduced in the actual implementation of double-directional channel models by extracting the so-called "effective" multipath [29]. A path is considered to be effective only if it carries a significant percentage of the total power and it is separable in the angle and delay domains, both of which are determined by the system sensitivity and the useful information that is required for a certain application (e.g., MIMO capacity computation, position location, etc.). From this viewpoint, channel shall not be defined before the properties of the transmit and receive systems have been specified. We will illustrate how the system properties will alter the multipath structure in the following section.

1.3 SYSTEM-DEPENDENT CHANNEL CHARACTERIZATION

1.3.1 General Description of Multipath Structures

Consider a two-dimensional, single-bounce environment with scattering objects spanning over a single solid angle as depicted in Figure 1.2, where Ω_T (Ω_R) is obtained by first projecting the scattering cluster onto the unit circle enclosing the transmit (receive) array and then projecting the arc onto the arrays. This domain will be called the array-constrained angular (ACA) domain in the following discussion. For instance, in the ideal Clarke's scattering field, the channel solid angle in the ACA domain is 2. Without loss of generality, we represent the scattering cluster as a circular area, $\mathcal{C}$, with radius R (see Figure 1.3(a)), where there are total N scatterers, each of which is randomly located in this area independent of other scatterers. We further define a keyhole, $\mathcal{K}$, as the circular sub-area with radius r within which the orthogonality of the received waves at any two points, $\mathbf{a_{C1}}$ and $\mathbf{a_{C2}}$, is below a prespecified critical value ς regardless of the monochromatic transmitting sources. Stated alternatively,

$$\int_{\forall \mathcal{A}_T} G^\dagger(\mathbf{a_T}, \mathbf{a_{C2}}, \lambda) G(\mathbf{a_T}, \mathbf{a_{C1}}, \lambda) \mathbf{da_T} \leq \varsigma \tag{1.1}$$

where $G(\mathbf{a_T}, \mathbf{a_C}, \lambda)$ is the Green's function, or the possible waves at position $\mathbf{a_C}$ resulting from a point source at position $\mathbf{a_T}$ within any transmit space $\mathcal{A}_T$. $(\cdot)^\dagger$ denotes the Hermitian operator. Note that the keyhole defined here is slightly different from the original definition in [30]. We refer to keyhole as any scenario that causes a significant reduction of the rank of the channel transfer matrix $\mathbf{H}$ as compared to the idealized infinite-rank case. In this work we consider continuous one-dimensional (1D) arrays that are composed of an infinite number of antennas separated by infinitesimal distances. This eliminates the need to specify *a priori* the number of antennas and their relative positions on the arrays.

The channel response from the transmitter to the receiver is

$$\mathbf{H}(\kappa_T, \kappa_R) = \sum_{n=1}^{N} \alpha_n \delta(\kappa_T - \kappa_{T,n}) \delta(\kappa_R - \kappa_{R,n}) \tag{1.2}$$

where α_n is the attenuation and polarization between the Tx direction κ_T and Rx direction

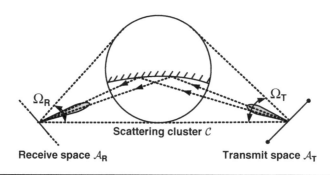

Figure 1.2   Illustration of a Single-Cluster Environment.

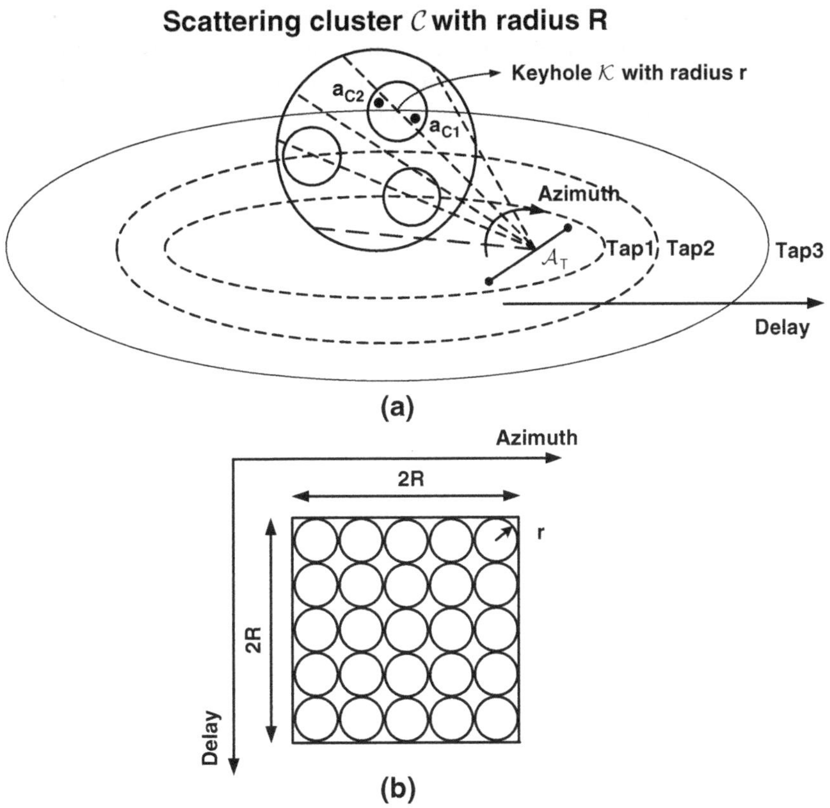

Scattering cluster $\mathcal{C}$ with radius R

(a)

(b)

Figure 1.3   Multipath Structure of a Scattering Cluster in the Azimuth-Delay Domain: (a) Natural Coordinate and (b) Cartesian Coordinate Representations.

$\kappa_{\mathbf{R}}$. $\mathbf{H}(\cdot, \cdot)$ can be approximated as

$$\mathbf{H}(\kappa_{\mathbf{T}}, \kappa_{\mathbf{R}}) \approx \sum_{i=1}^{M} \alpha'_i \sum_{j \in \mathrm{P}_{\mathcal{K},i}} \delta(\kappa_{\mathbf{T}} - \kappa_{\mathbf{T},j}) \delta(\kappa_{\mathbf{R}} - \kappa_{\mathbf{R},j}) \qquad (1.3)$$

where α'_i is the average attenuation of the ith keyhole and $\mathrm{P}_{\mathcal{K},i}$ is the set of propagation paths in the ith keyhole.

The keyhole radius describes the correlation scale of MPCs within a scattering cluster. Subsequently, we introduce another two parameters to characterize the cluster boundaries, namely, the path delay interval T and the subtended angular interval Ω_{T} (Ω_{R}). Then, to ensure the well-conditionedness of $\mathbf{H}(\cdot, \cdot)$, the channel response satisfies

$$\mathbf{H}(\kappa_{\mathbf{T}}, \kappa_{\mathbf{R}}) \neq 0 \quad \text{if and only if} \quad (\kappa_{\mathbf{T}}, \kappa_{\mathbf{R}}) \in \Omega_{\mathrm{T}} \times \Omega_{\mathrm{R}} \qquad (1.4)$$

We further assume that it is possible to identify a set of joint azimuth-delay taps as depicted in Figure 1.3(a). Each tap (scattering center) is enclosed by its corresponding keyhole $\mathcal{K}$ and all the $\mathcal{K}$s will cover the whole cluster without overlaps or gaps. It has been found in [24] that the average delay-tap angular interval is just a few degrees less than the angular interval of that cluster. Therefore, the reference system in Figure 1.3(a) that describes the location and size of each tap can be converted to the equivalent Cartesian coordinate in Figure 1.3(b). The point spectrum of $\mathbf{H}(\cdot, \cdot)$ ranges between 1 and infinity as r varies between R and 0. For example, a horizontal roof-edge diffraction brings the rank of $\mathbf{H}(\cdot, \cdot)$ to unity because all MPCs go through a single point and, hence, $\frac{R}{r} \to 1$. Another example of a degenerate channel is when the scatterers have smooth, well-reflecting surfaces, such as pure glass fronts.

1.3.2 Impact of Antenna Directivity and System-Operating Frequency on Channel Characteristics

We will now present system-dependent channel models to describe the effect of system properties on channel characteristics. We first characterize the relationship for either Ω_T or T versus the array length L. Consider the situation depicted in Figure 1.4(a), where a large number of multipaths emanating from a finite region are impinging on a continuous linear array $\mathcal{A}$ of length L placed symmetrically along the a-axis. Ω_T is defined as

$$\Omega_T \doteq \int_{\theta_p - \Theta}^{\theta_p + \Theta} \cos \theta d\theta \tag{1.5}$$

where θ denotes the AOA, θ_p is the mean AOA, and Θ is the angular spread (AS) of the scattering cluster. All of the angles are measured relative to the broadside of the continuous linear array $\mathcal{A}$. Note that in the case when the mean AOA is aligned with the broadside axis ($\theta_p = 0°$), (1.5) reduces to $\Omega_T = 2 \sin \Theta$. In the subsequent discussion, it is assumed that the diversity levels in the angular domain and the delay domain are approximately uncorrelated. The weak correlation is justified by observing that the AS of the signal arrivals within a cluster does not vary depending on the delay time [24–27].

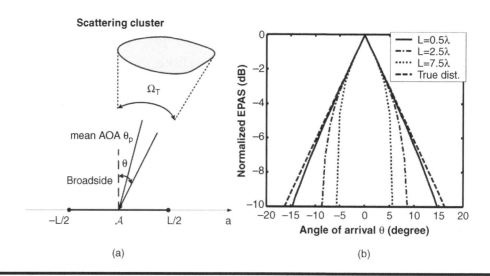

(a)   (b)

Figure 1.4 (a) Multiple Rays Impinging on a Continuous Linear Array and (b) Normalized EPAS Viewed from $\mathcal{A}$ for Different Array Lengths. The System Dynamic Range is Set to $\zeta = 10$ dB.

Therefore, the channel power delay-azimuth density spectrum as a function of delay and angle is separable in these two domains, from which independent descriptions of the multipath time-of-flight and AOA can be developed [25–27].

Firstly, the relationship for Ω_T versus the array length L will be characterized. In Figure 1.4(a), the space factor is given by [31]

$$F_A(\theta) = \int_{-L/2}^{L/2} I(a) e^{j[ka\sin\theta + \varphi(a)]} da \qquad (1.6)$$

$k = 2\pi/\lambda$ is the free space wavenumber where λ is the wavelength of propagation. $I(a)$ and $\varphi(a)$ represent, respectively, the amplitude and phase distributions along the source. For brevity, we assume a normalized uniform current distribution of the form $I(a) = I_0/L$ and a constant phase distribution $\varphi(a) = 0$. In such a case, (1.6) reduces to

$$F_A(\theta) = \frac{I_0 \sin(kL\sin\theta/2)}{kL\sin\theta/2} \qquad (1.7)$$

We further assume that the true distribution of angles has a Laplacian density [20, 25–27]:

$$f_{tr}(\theta) = C_1 \exp\left(-\frac{|\theta - \theta_p|}{\sigma_\theta}\right), \quad \theta \in (-\pi, \pi) \qquad (1.8)$$

where $\sqrt{2}\sigma_\theta$ is the standard deviation of the distribution and C_1 is a normalization factor to ensure that f_{tr} is a probability density function (pdf).

At each θ, the received signal will be the summation of numerous incident rays along that direction weighted by $F_A(\theta)$. The sum will vary randomly due to the random amplitude and phase of each incident ray. Assuming that the magnitude of the received complex envelope $|H(\theta)|$ has a Rayleigh fading, $|H(\theta)|^2$ would be an exponential variate with expectation

$$\mathbb{E}\left\{|H(\theta)^2|\right\} = f_{tr}(\theta)\left[F_A(\theta)\right]^2 \qquad (1.9)$$

where $\mathbb{E}\{\cdot\}$ denotes the expectation operator. We can interpret (1.9) as representing the *mean* power azimuthal profile. Because the usual way to eliminate the effect of instrumental noise is setting to zero the values of multipaths whose power lies below the noise threshold [22], the effective mean power azimuthal profile viewed from $\mathcal{A}$ is given by

$$p_{eff}(\theta) = \int_{\mathcal{N}}^{+\infty} p \cdot f_{p|\theta}(p|\theta) dp$$

$$= \int_{\mathcal{N}}^{+\infty} \frac{p}{\mathbb{E}\{|H(\theta)|^2\}} \exp\left\{-\frac{p}{\mathbb{E}\{|H(\theta)|^2\}}\right\} dp \qquad (1.10)$$

where $f_{p|\theta}$ is the probability of received power conditioned on angle. $\mathcal{N}$ denotes the noise threshold, which is quantified through the system dynamic range ζ, defined as the number of decibels (dB) below the maximum value. It is of special interest to find the effective power azimuthal spectrum (EPAS), $f_{eff}(\theta)$, referring to the actual distribution of angles after the antenna effects are deconvolved from the measured data. The EPAS is derived from (1.10) as

$$f_{eff}(\theta) = C_2 \cdot \frac{p_{eff}(\theta)}{\left[F_A(\theta)\right]^2} \qquad (1.11)$$

with C_2 being a normalization factor to ensure that f_{eff} is a pdf. It is worth mentioning that as $\mathcal{N}$ approaches 0, $f_{eff}(\theta)$ approaches $f_{tr}(\theta)$ as expected.

Figure 1.4(b) illustrates the plots of EPAS for $\zeta = 10$ dB, $\theta_p = 0°$, and $\sigma_\theta = 10°$. The maximum values of all the curves have been normalized to 0 dB. Notably, the 10-dB angular interval, Θ, of the EPAS decreases from around 15° to 5° as the array length increases from 0.5λ to 7.5λ. This is due to the fact that the distribution of angles at a small array is spread over a wider angular range and, thus, the AOA statistics would be less distorted by truncation of the tails that are below the noise cutoff. Figure 1.5(a) captures the relationship for Θ versus L/λ for different noise thresholds, which can be fitted by an exponential regression curve as follows

$$\Theta(L/\lambda) = \beta_1 \exp(\gamma_1 L/\lambda) \quad \text{(degree)} \qquad (1.12)$$

where β_1 and γ_1 are the curve-fit parameters. In general, larger β_1 and γ_1 result from a larger value of ζ, leading to a wider angular interval. It is worth emphasizing here that other line-source amplitude and phase distributions will result in different space factor

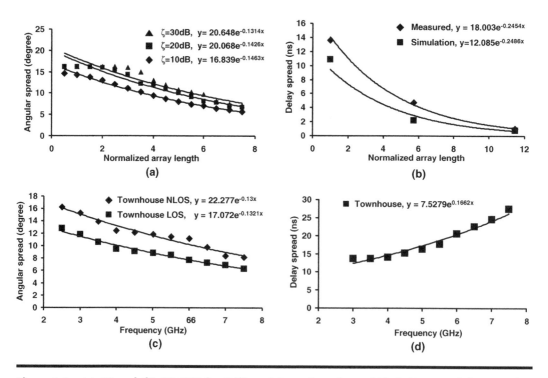

Figure 1.5 (a) AS and (b) DS [32] for Different Array Lengths and Their Exponential Regression Curves; (c) AS and (d) DS in the Townhouse Environment [20] and Their Exponential Regression Curves.

patterns [31]. A similar procedure is used to analyze the relationship for AS versus the array length when other distributions are considered. In such a case, it is possible to postulate different regression models and curve-fit parameters that provide better fits. Nevertheless, the general trend as predicted in Figure 1.5(a) is still expected because the underlying idea remains the same.

Secondly, we consider the relationship for T versus L. It has been shown in extensive measurement campaigns that the use of directive antennas at the remote terminal can reduce the multipath delay interval [32–34]. The work reported in [32] explicitly related the channel delay spread (DS) to the antenna directivities through indoor measurements and, thus, will be used as the basis for quantitative analysis. It was demonstrated that the DS decreased from 13.59 ns to 1.05 ns as the antenna's 3-dB beamwidth decreased from 60° to 5° (i.e., the corresponding array length will increase from λ to 11.46λ). The empirical observation was also verified by ray-tracing simulation. It was observed that the estimated DS decreased from 10.91 ns to 0.79 ns as the beamwidth of the receive antenna decreased, which is consistent with the empirical observation. For brevity, we propose to describe the relationship as follows

$$T(L/\lambda) = \beta_2 \exp(\gamma_2 L/\lambda) \quad \text{(ns)} \tag{1.13}$$

β_2 and γ_2 are the parameters that yield the curve closest to the measurement data. Equation (1.13) is shown to have a fairly good agreement with the results reported in [32] as illustrated in Figure 1.5(b). However, the regression in (1.13) is deduced from a very small data set. It is possible that other models could obtain better curve fits when a larger measurement sample is available.

The next step is to investigate the relationship for Θ versus the frequency f. In [20], Poon and Ho presented measurement data from 2 to 8 GHz at residential environments. The resulting plot of average AS versus frequency, shown in Figure 1.5(c), can be matched in the exponential form

$$\Theta(f) = \beta_3 \exp(\gamma_3 f) \quad \text{(degree)} \tag{1.14}$$

where β_3 and γ_3 are the regression coefficients. Note that (1.14) can also be deduced from Figure 1.4(b) by keeping the array length fixed and allowing the frequency to vary.

Finally, the average delay interval T versus f may be best expressed in terms of a negative exponential function following from the experimental results obtained in [20], as follows,

$$T(f) = \beta_4 \exp(\gamma_4 f) \quad \text{(ns)} \tag{1.15}$$

where β_4 and γ_4 are the curve-fit parameters. This observation has been shown in Figure 1.5(d).

The combination of (1.12) to (1.15) forms a useful model for estimating DIV and DOF in a system-dependent wideband directional channel. Due to the limited amount of data available in the literature, this can only be an approximation. Nevertheless, these assumptions are mathematically expedient and it is also believed that they can serve as a basis in the planning and evaluation of future measurement programs.

1.4 DAS TOPOLOGY AND CHANNEL MODELS

In the previous section, the system-dependent channel model for a single array with co-located (infinitesimal) antenna elements is presented, which deals with the *local* propagation phenomenon. In this section, the *global* DAS network topology and channel models will be investigated. The original form of the DAS has a bus-type antenna configuration and consists of omnidirectional Tx/Rx antennas with minimum complexity that are scattered around the service region [35,36]. This simple DAS, however, shows limited efficiency in fading and interference reduction, because all the antenna nodes share a single connecting link. In order to overcome the shortcoming with the common feeder, the sectorized DAS, where each node has a separate feeder to a central base station (BS), has been proposed [37]. This architecture ensures that various benefits of a MIMO system, such as diversity combining and interference reduction, can be realized.

In the current work, we consider a generalized version of a DAS [38] as illustrated in Figure 1.6(a). Each deployment cell (DC) has a number of distributed APs spaced apart by a large distance. Each AP itself is an antenna array and is connected to the central BS through either optical fibers, coaxial cable, or radio link. It is further assumed that these APs are uniformly distributed in a circular DC with radius R_c. This topology model is similar to the random antenna layout in [39] and owing to the complex landform of real environments, some sort of randomness in the AP placement is expected. Nevertheless, different from the model in [39], the user terminal (UT) may not locate at the center of the circular area. In real-life deployment, there could be multiple DCs to serve the region of interest; however, we focus on a single-DC scenario in this initial study.

As the system DIV and DOF are dependent on the fading correlation of the channel matrix in the generalized DAS, it is of special interest to investigate the cross correlation between two communication links from two APs to the common UT. Consider the system geometry illustrated in Figure 1.6(a). We are interested in finding the mean value of the cross-correlation coefficient. Past work shows the correlation coefficient between shadow fading signals from two different APs has been viewed as a function of the angle-of-arrival difference (AAD) or the relative distance difference. In most cases, the correlation is considered as a function of the AAD only (e.g., [40,41]) because the AAD

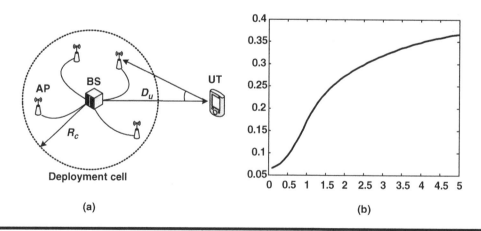

(a)　(b)

Figure 1.6 (a) Pictorial Illustration of a Generalized DAS, and (b) Mean Value of Cross-Correlation Against the Ratio of D_u to R_c.

has the strongest influence on the results. Nevertheless, models considering the effects of both the AAD and relative distance difference also exist [42]. In the present work, we will apply the piecewise model proposed by Sørensen for the cross correlation ρ [40]

$$
\rho(\Delta\phi) = \begin{cases} 0.78 - 0.0056|\Delta\phi|, & \text{if } 0° \leq |\Delta\phi| \leq 15° \\ 0.48 - 0.0056|\Delta\phi|, & \text{if } 15° \leq |\Delta\phi| \leq 60° \\ 0, & \text{if } |\Delta\phi| \geq 60° \end{cases} \tag{1.16}
$$

where $\Delta\phi$ is the AAD. Note that although the actual ρ against the AAD is likely to change for a different environment, the general methodology presented below is still applicable to other shapes of $\rho(\Delta\phi)$.

As the APs are uniformly distributed within a circle, the pdf of the AOA viewed from the UT for $\frac{D_u}{R_c} \geq 1$ (i.e., the UT locates outside the DC) is given by [43]

$$
f_\phi(\phi) = \begin{cases} \frac{2D_u\cos\phi\sqrt{D_u^2\cos^2\phi - D_u^2 + R_c^2}}{\pi R_c^2}, & -\sin^{-1}\left(\frac{R_c}{D_u}\right) \leq \phi \leq \sin^{-1}\left(\frac{R_c}{D_u}\right) \\ 0, & \text{otherwise} \end{cases} \tag{1.17}
$$

where D_u is the distance of separation between the central BS and the UT as depicted in Figure 1.6(a). When the UT lies within the DC, $\frac{D_u}{R_c} < 1$. The pdf of the AOA can be derived as [44]

$$
f_\phi(\phi) = \frac{R_c^2 + D_u^2 + 2R_c D_u \cos\left[\sin^{-1}\frac{D_u\sin\phi}{R_c} + \phi\right]}{2\pi R_c^2}, \quad 0 \leq \phi \leq 2\pi \tag{1.18}
$$

Subsequently, the pdf of the AAD, $\Delta\phi$, is obtained as [45]

$$
f_{\Delta\phi}(\Delta\phi) = \begin{cases} \int_0^\infty f_\phi(\Delta\phi + \phi) f_\phi(\phi) d\phi, & \Delta\phi \geq 0 \\ \int_{-\Delta\phi}^\infty f_\phi(\Delta\phi + \phi) f_\phi(\phi) d\phi, & \Delta\phi < 0 \end{cases} \tag{1.19}
$$

and the mean value of cross correlation is

$$
\mathbb{E}(\rho) = \int \rho(\Delta\phi) f_{\Delta\phi}(\Delta\phi) d(\Delta\phi) \tag{1.20}
$$

Figure 1.6(b) plots the mean cross correlation against $\frac{D_u}{R_c}$. As $\frac{D_u}{R_c}$ increases, the UT moves away from the center of the cell. The angular range of the AOA at the UT decreases. Therefore, ρ increases as expected. However, as $\frac{D_u}{R_c}$ increases further, the mean value of cross correlation will eventually approach an upper bound.

Armed with the model preliminaries in Section 1.3 and Section 1.4, both DIV and DOF in a generalized DAS can be estimated, as will be discussed in the following sections.

1.5 DIV AND DOF IN DEGENERATE SCSB CHANNELS

1.5.1 Diversity Gain Calculation

It is conceptually well-known that the channel AS and DS determine the correlation level of the receiving process in the space and frequency domains, respectively [46–50]. Let B_c represent either the coherence distance or the coherence bandwidth for a critical correlation level c, and ϵ represent either AS or DS. A number of different relationships to describe the dependence of B_c and ϵ have been reported in the literature [46,49,50]. A current practice assumes that $B_c = 1/\epsilon$ [12,51], which is a heuristic statement of Heisenberg's uncertainty relations [46]. In effect, B_c defines a forbidden zone of correlation in the signal space. The antenna elements (or frequency carriers) must be placed mutually outside one another's forbidden zone to ensure that the envelope correlation is lower than c. Subsequently, the maximum number of forbidden zones obtainable from the system will be $W \times T \times (L/\lambda_0) \times \Omega_T$, where W is the system bandwidth and λ_0 is the wavelength corresponding to the center frequency f_0 [12,51]. This also gives us the total number of independently faded replicas of the data symbol that can be obtained at the receiver end, and therefore characterizes the inherent diversity afforded by a wide-sense stationary uncorrelated scattering (WSSUS) channel. In particular, the following two conditions have to be fulfilled:

1. The wireless transmission scheme occupies a small fractional bandwidth W/f_0, such that L/λ_0 is a good approximation to the normalized array length across the entire frequency band.
2. T and Ω_T are fixed parameters independent of the signal power spectrum density (PSD) functions.

However, the deployment of ultra-wideband technology [52], which utilizes large bandwidths to achieve data rates of the order of gigabits per second (Gbps), has nullified the first condition. Furthermore, in real-life channels, both T and Ω_T will exhibit array length and frequency dependencies as discussed in Section 1.3.2. Therefore, we examine the DIV on a microscopic scale to better reveal the system-dependent nature of propagation channels. First, we represent each signal point as a vector $\mu(y, f)$ on the signal space $\mathcal{S}$. The y-domain can be explained using Figure 1.7(a), which illustrates a nonuniform power distribution $\mathcal{D}$ over the transmitting aperture a_T. $y(a_T) = [I(a_T)]^2$ is the line-source power. Interestingly, $\mathcal{D}$ can be thought of as the combination of many differential arrays each having a uniform distribution (see also Figure 1.7(a)):

$$\mathcal{D} \equiv \sum_i 2a_T(y_i) \cdot \delta y_i \tag{1.21}$$

which is applicable for any symmetric $\mathcal{D}$ peaked at $a_T = 0$. Let $\delta\mu_{i,j}$ be the area of a small neighborhood centered on $\mu(y_i, f_j)$. δy_i and δf_j are the differential intervals in the respective domains. As illustrated in Figure 1.7(b), the average diversity gains on $\delta\mu_{i,j}$ can be thought of as the volume of each box scaled by $\delta y_i/y(0)$, which is the probability of occurrence when the differential array length is $2a_T(y_i)$. The volume may be expressed as

$$\delta\text{DIV}_{i,j} = \left\{ \delta f_j \times [\text{T}(2a_T(y_i), f_j)] \times \frac{2a_T(y_i)}{\lambda_j} \right.$$

$$\left. \times [\Omega_T(2a_T(y_i), f_j)] \right\} \times \frac{\delta y_i}{y(0)} \tag{1.22}$$

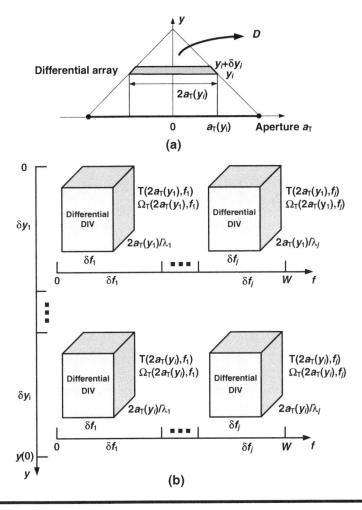

Figure 1.7 (a) A Non-Uniform Line-Source Distribution $\mathcal{D}$ Where $y(a_T) = [I(a_T)]^2$ and (b) A Microscopic View of DIV. The Volume of Each Box Scaled by $\delta y_i/y(0)$ Represents the Diversity Gains on the Signal Space $\delta\mu_{i,j}$.

Subsequently, the total number of DIV is computed as

$$\text{DIV} = \sum_j \sum_i \delta\text{DIV}_{i,j}$$

$$= \iint_S \frac{2a_T(y)\, f\, \Omega_T(2a_T(y),\, f)\text{T}(2a_T(y),\, f)}{v_c y(0)}\, dy\, df \qquad (1.23)$$

where v_c is the speed of electromagnetic waves. The integral kernel in (1.23) is referred to as the DIV density.

Although the representation in (1.23) is quite general, it provides a robust and parsimonious estimation of the number of independent fading coefficients that can be averaged over to detect the symbol. Intuitively, it corresponds to the SNR exponent of the error probability at high SNR [8,51]. Hence, it can be used as a theoretical benchmark to access the performances of various diversity-based schemes that have been extensively pursued in recent years.

If the multipath channel is ill-conditioned by propagation effects like diffraction or waveguiding, which lead to a reduction of the rank of $\mathbf{H}(\cdot, \cdot)$, the size of $\mathcal{K}$ could be comparable to the cluster size. In such a case, the keyhole population (KP) may not be sufficient to support the propagation of all diversity paths. Since the propagation mechanisms that cause channel rank deficiency are usually sensitive to the system-operating conditions [30], we can similarly take a microscopic view of the KP on $\delta\mu_{i,j}$ as

$$\delta\text{KP}_{i,j} = \left[\frac{R\left(2a_{\text{T}}(y_i),\, f_j\right)}{r\left(2a_{\text{T}}(y_i),\, f_j\right)} \right]^2 \frac{\delta y_i \cdot \delta f_j}{y(0) \cdot W} \tag{1.24}$$

Summation of (1.24) yields

$$\text{KP} = \sum_j \sum_i \delta\text{KP}_{i,j} = \iint_{\mathcal{S}} \left[\frac{R\left(2a_{\text{T}}(y),\, f\right)}{r\left(2a_{\text{T}}(y),\, f\right)} \right]^2 \frac{1}{W y(0)}\, dy df \tag{1.25}$$

with the integrand being referred to as the KP density. Subsequently, the overall diversity estimate is given by

$$\overline{\text{DIV}} = \min\{\text{DIV}, \text{KP}\} \tag{1.26}$$

The above analysis is applicable to the conventional MIMO system where there is one UT and one AP. We will now extend the analysis to the generalized DAS, for which there is one UT and multiple spatially distributed APs. Consider the simplest DAS architecture with one UT and two APs. Let us assume that the cross-correlation of the lognormal macroscopic fading between the two APs is ρ, and the diversity gains of the two radio links connecting the UT to APs are $\overline{\text{DIV}}_{\text{AP1}}$ and $\overline{\text{DIV}}_{\text{AP2}}$, respectively. It is further assumed that the number of diversity paths that are common to AP1 and AP2 is z.

When $\rho = 0$, the slow fading signals received from noncollocated APs are independent and the overall diversity gain can be obtained as

$$\widetilde{\text{DIV}} = \overline{\text{DIV}}_{\text{AP1}} + \overline{\text{DIV}}_{\text{AP2}} \tag{1.27}$$

On the other hand, when $\rho = 1$, the signals received from the two APs are fully correlated. In such a case, $\overline{\text{DIV}}_{\text{AP1}} = \overline{\text{DIV}}_{\text{AP2}}$, and the overall diversity gain is given by

$$\widetilde{\text{DIV}} = \overline{\text{DIV}}_{\text{AP1}} = \overline{\text{DIV}}_{\text{AP2}} \tag{1.28}$$

When $0 < \rho < 1$, the number of paths that are common to both links is $z = \rho\sqrt{\overline{\text{DIV}}_{\text{AP1}}\overline{\text{DIV}}_{\text{AP2}}}$ [53]. The overall diversity level for the generalized DAS can be approximated by

$$\widetilde{\text{DIV}} = \overline{\text{DIV}}_{\text{AP1}} + \overline{\text{DIV}}_{\text{AP2}} - \rho\sqrt{\overline{\text{DIV}}_{\text{AP1}}\overline{\text{DIV}}_{\text{AP2}}} \tag{1.29}$$

The above discussion can be extended to more general scenarios of U APs ($U \geq 2$). The DIV is computed as the average of diversity gains obtainable from any two APs, u_1 and

u_2, belonging to the U APs:

$$\widetilde{\text{DIV}} = \frac{1}{U-1} \cdot \sum_{u_1} \sum_{u_2} \left[\overline{\text{DIV}}_{\text{AP}\,u_1} + \overline{\text{DIV}}_{\text{AP}\,u_2} - \rho_{u_1,u_2} \sqrt{\overline{\text{DIV}}_{\text{AP}\,u_1} \overline{\text{DIV}}_{\text{AP}\,u_2}} \right] \qquad (1.30)$$

where ρ_{u_1,u_2} is the cross-correlation between AP u_1 and AP u_2. $\overline{\text{DIV}}_{\text{AP}\,u_1}$ and $\overline{\text{DIV}}_{\text{AP}\,u_2}$ are the diversity estimates of the two links connecting the UT to AP u_1 and AP u_2, respectively.

1.5.2   Degrees-of-Freedom (DOF) Calculation

DOF are essentially the number of pairs of functions (communication modes), with each pair consisting of one function for the transmit space and one for the receive space. The different communication modes are mutually orthogonal and define the two sets of functions with the strongest possible couplings between the spaces. We first consider the DOF in the time domain. In waveform channels, if a transmitted signal is first frequency limited to $[-\frac{W}{2}, \frac{W}{2}]$ and then approximately time limited to $[-\frac{\tilde{T}}{2}, \frac{\tilde{T}}{2}]$, the dimension of the subspace satisfying these two physical constraints will give the number of significant singular values [54]:

$$W\tilde{T} + \chi \ln(W\tilde{T}) + o(\ln(W\tilde{T})) \qquad (1.31)$$

for $W\tilde{T} \gg 1$ and χ is a constant. Theoretically, the multipath propagation channel increases the signal duration from $1/W$ to the channel delay interval T. However, the signal bandwidth is also decreased from W to $1/T$ after being convolved with the channel impulse response. Consequently, no extra DOF can be achieved in the time domain. Stated another way, it is impossible to utilize different delay taps as parallel data pipes because all the taps will be excited once the signal is transmitted. Therefore, nature forces the entire signal space to be utilized only for DIV gains.

The next step is to solve for the spatial DOF based on a heuristic approach which is similar to the rationale presented in [55]. We first consider the monochromatic wave communication between a well-conditioned cluster $\mathcal{C}$ and the Tx array $\mathcal{A}_\text{T}$. A key step in calculating the DOF lies in understanding the degree to which the waves in the scattering cluster, being generated from different points in $\mathcal{A}_\text{T}$, are orthogonal to one another. Consider two source points, $\mathbf{a}_{\text{T1}}$ and $\mathbf{a}_{\text{T2}}$, in $\mathcal{A}_\text{T}$ as shown in Figure 1.8. If the points are close, the waves generated by them in $\mathcal{C}$ will be almost identical and will not be orthogonal. As the points move farther apart, a rapidly varying phase results, which causes the waves to undergo alternating constructive and destructive interference. Assuming that the best possible choice of relative phase between the two sources is used to eliminate this particular interference effect, as points $\mathbf{a}_{\text{T1}}$ and $\mathbf{a}_{\text{T2}}$ move farther apart, we can expect the waves to become progressively different from one another. To assess the orthogonality of the waves, we can evaluate their overlap integral in $\mathcal{C}$, which is given by

$$\kappa(\mathbf{a}_{\text{T2}}, \mathbf{a}_{\text{T1}}, \lambda) = \int_{\mathcal{C}} G^{\dagger}(\mathbf{a}_{\text{T2}}, \mathbf{a}_{\mathbf{C}}, \lambda) G(\mathbf{a}_{\text{T1}}, \mathbf{a}_{\mathbf{C}}, \lambda) d\mathbf{a}_{\mathbf{C}} \qquad (1.32)$$

We now use a simplistic model to obtain some approximate results. As mentioned above, the overlap is expected to be large initially and decrease as the two source points are separated. It is presumed that for any point $\mathbf{a}_{\text{T1}}$, there is a finite length Δl for which

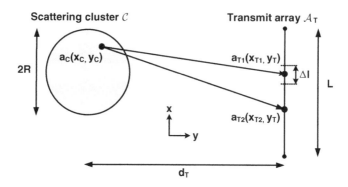

Figure 1.8 Illustration of the Source Points a_{T1} and a_{T2} in the Transmit Array and the Scattering Cluster C. Also Shown Is an Array Segment Δl Near a_{T1}. Other Sources in Δl Are Assumed to Produce Waves in C that Are Largely Similar to the Waves from the Source Point a_{T1}, Whereas Sources Outside Δl Are Assumed to Produce Waves in C that Are Substantially Orthogonal to Those from a_{T1}.

$\kappa(\mathbf{a_{T2}}, \mathbf{a_{T1}}, \lambda)$ is finite and approximately constant. In addition, $\kappa(\mathbf{a_{T2}}, \mathbf{a_{T1}}, \lambda)$ is approximately zero for $\mathbf{a_{T2}}$ outside this interval. To estimate the dimension of Δl, only $\mathcal{A}_T$ and C, which are far apart compared with their linear dimensions, are considered, such that paraxial approximations can be made. Subsequently, we have

$$G^{\dagger}(\mathbf{a_{T2}}, \mathbf{a_C}, \lambda) G(\mathbf{a_{T1}}, \mathbf{a_C}, \lambda)$$

$$= \frac{1}{(4\pi d_T)^2} \exp\left[-j\frac{2\pi}{\lambda}\left(|\mathbf{a_C} - \mathbf{a_{T1}}| - |\mathbf{a_C} - \mathbf{a_{T2}}|\right)\right]$$

$$\approx \frac{1}{(4\pi d_T)^2} \exp\left\{-j\frac{2\pi}{\lambda}\left[(y_C - y_T) + \frac{(x_C - x_{T1})^2}{2d_T}\right.\right.$$

$$\left.\left. -(y_C - y_T) - \frac{(x_C - x_{T2})^2}{2d_T}\right]\right\}$$

$$= \frac{1}{(4\pi d_T)^2} \exp\left\{-j\frac{2\pi}{\lambda}\left\{-\frac{x_{T2}^2}{2d_T} + \frac{x_{T1}^2}{2d_T} + \frac{1}{d_T}\left[x_C(x_{T2} - x_{T1})\right]\right\}\right\}$$

$$= \frac{1}{(4\pi d_T)^2} \mathcal{F}_T(\mathbf{a_{T2}}, \lambda)\mathcal{F}_T^{\dagger}(\mathbf{a_{T1}}, \lambda) \exp\left\{-j\frac{2\pi x_C(x_{T2} - x_{T1})}{\lambda d_T}\right\} \tag{1.33}$$

where $\mathcal{F}_T(\mathbf{a_T}, \lambda)$ is the focusing function corresponding to a spherical wave that is centered on the receive area and is separated to extract the underlying spherical focusing of the source [55]. Also refer to Figure 1.8 for definitions of all the other variables. The remaining exponential expression is unity for small arguments. However, after $(x_{T2} - x_{T1})$ becomes sufficiently large, the function overall becomes oscillatory as we move through C, for which the integral will tend to average to zero. A characteristic size of $(x_{T2} - x_{T1})$ for which this integral becomes negligible occurs when

$$\left|\frac{2\pi \cdot 2R(x_{T2} - x_{T1})}{\lambda d_T}\right| \approx 2\pi \tag{1.34}$$

That is, when

$$|(x_{T2} - x_{T1})| \approx \frac{\lambda d_T}{2R} \tag{1.35}$$

In this case, as we integrate from one end of $\mathcal{C}$ to another extreme in the x direction, the phase of the integrand will change by 2π. Next, we select a basis which is essentially uniform within Δl near the point $\mathbf{a_{T1}}$ and is zero elsewhere [55], i.e., within Δl, we have the eigenfunction

$$\Psi_1(\mathbf{a_{T1}}, \lambda) = \frac{1}{\sqrt{\Delta l}} \mathcal{F}_T(\mathbf{a_{T1}}, \lambda) \tag{1.36}$$

To form another function orthogonal to that of (1.36), we can simply move sideways to an adjacent array segment of the same length and construct a similar second eigenfunction in that segment. The second eigenfunction would be orthogonal to the first because no overlap occurs between the functions. This process continues until all the aperture is used, following which we have a number $O(\lambda)$ of orthogonal functions

$$O(\lambda) = \frac{2RL}{\lambda d_T} \tag{1.37}$$

all having the same eigenvalues, where L is the transmit array length.

The simplistic approximate eigenfunctions are different from the prolate spheroidal functions derived from rigorous mathematical approaches because they fill the entire aperture rather than our solutions, which are localized each within a subdivision Δl. However, we can take orthogonal linear combinations of the approximate functions. And it is easy to construct sets which fill the whole aperture with the number of orthogonal functions remaining unchanged by such linear combinations [55].

By applying the same arguments presented in Section 1.5.1, there is a need to take into account the system dependency of R for DOF computation. Following from (1.37), the differential DOF on $\delta\mu_{i,j}$ is essentially the area of each rectangle scaled by $\delta y_i/y(0) \cdot \delta f_j/W$ as illustrated in Figure 1.9, which can be obtained as

$$\delta\mathrm{DOF}_{i,j} = \frac{4R\left(2a_T(y_i), f_j\right) a_T(y_i)}{\lambda d_T} \times \frac{\delta y_i \cdot \delta f_j}{y(0) \cdot W} \tag{1.38}$$

and the total available communication modes are simply

$$\mathrm{DOF} = \sum_j \sum_i \delta\mathrm{DOF}_{i,j} = \iint_S \frac{4fR\left(2a_T(y), f\right) a_T(y)}{v_c d_T y(0) W} dy df \tag{1.39}$$

If the scattering is highly correlated, $\mathbf{H}(\cdot, \cdot)$ will be rank deficient. In such a scenario, the effective DOF are limited by the KP derived from (1.25) with the overall DOF estimate

$$\overline{\mathrm{DOF}} = \min\{\mathrm{DOF}, \mathrm{KP}\} \tag{1.40}$$

It is worth mentioning that (1.40) is a more optimistic upper bound for the DOF measure compared to the upper bound for the DIV estimate in (1.26). It has been assumed

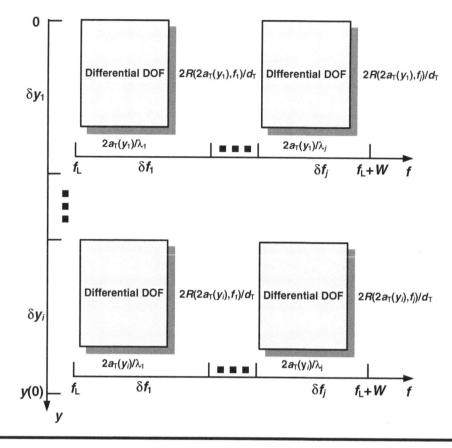

Figure 1.9 A Microscopic View of DOF; the Area of Each Rectangle Scaled by $\delta y_i / y(0) \cdot \delta f_j / W$ Represents the Degrees-of-Freedom on the Signal Space $\delta\mu_{i,j}$.

that a *complete* picture of the scattering environment can be obtained in order to fully explore the DOF offered by the channel, whereas this requirement is not enforced for the calculation of diversity gains in the previous section.

In a similar approach with (1.30), the DOF for a generalized DAS is calculated as the average of multiplexing gains obtainable from any two APs, u_1 and u_2, belonging to the U APs:

$$\widetilde{\text{DOF}} = \frac{1}{U-1} \cdot \sum_{u_1} \sum_{u_2} \left[\overline{\text{DOF}}_{\text{AP}\,u_1} + \overline{\text{DOF}}_{\text{AP}\,u_2} \right.$$

$$\left. -\rho_{u_1,u_2} \sqrt{\overline{\text{DOF}}_{\text{AP}\,u_1} \overline{\text{DOF}}_{\text{AP}\,u_2}} \right] \tag{1.41}$$

where $\overline{\text{DOF}}_{\text{AP}\,u_1}$ and $\overline{\text{DOF}}_{\text{AP}\,u_2}$ are the DOF estimates of the two communication links connecting the UT to AP u_1 and AP u_2, respectively.

1.6 COMPLETE MODEL: MULTIPLE-CLUSTER MULTIPLE-BOUNCE (MCMB) CHANNELS WITH IMPERFECT RECEIVER CONDITIONS

1.6.1 Diversity Gain Calculation

The wireless channel measurements have shown that physical paths are clustered both in the temporal and spatial domains [20, 24–27]. To apply the methodology presented in the previous section to more general situations, we first identify three classes of clusters as depicted in Figure 1.10.

1. Class-One (C-I) clusters represent the scattering objects in the vicinity of the transmit site. This group of clusters can only be observed from the transmit array and has the subtended angular interval $\Omega^{I}_{T,q}$, cluster radius R^{I}_{q}, and keyhole radius r^{I}_{q}, where $q = 1, 2, 3, \ldots, Q$. The superscript I indicates the class of a cluster.

2. Class-Two (C-II) clusters correspond to the scattering elements close to the receive site. This group of clusters can only be observed from the receive array and has the subtended angular interval $\Omega^{II}_{R,s}$, delay interval T^{II}_{s}, cluster radius R^{II}_{s}, and keyhole radius r^{II}_{s}, where $s = 1, 2, 3, \ldots, S$.

3. Class-Three (C-III) clusters have line-of-sight connections to both transmit and receive arrays. Therefore, each transmit interval $\Omega^{III}_{T,p}$ ($p = 1, 2, 3, \ldots, P$) is one-to-one mapped to a corresponding receive interval $\Omega^{III}_{R,p}$ such that the joint spectra cannot be described by a product of separate AOA and AOD spectra. This group of clusters has the delay interval of T^{III}_{p}, cluster radius R^{III}_{p}, and keyhole radius r^{III}_{p}.

We will model the joint AOA/AOD spectrum as a concatenation of a *diagonal submatrix* in which each receive angle couples with only one corresponding transmit angle, and a *generalized coupling submatrix* in which each receive angle couples with K

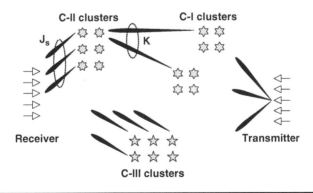

Figure 1.10 Propagation Processes in an MCMB Channel. Three Classes of Clusters Have Been Identified: C-I Clusters Can Only Be Observed from the Transmit Site, C-II Clusters Can Only Be Observed from the Receive Site, and C-III Clusters Have Line-of-Sight Connections to Both Transmit and Receive Sites. J_s Is the Number of Resolvable AOAs from the sth C-II Cluster. Furthermore, Each AOA from C-II Clusters Couples with K AODs from C-I Clusters on Average.

transmit angles. The value of K determines the channel connectivity between the C-I and C-II clusters, thereby characterizing the correlation level of the channel.

The diagonal submatrix captures the number of independent channels available through C-III clusters. The diversity measure for a *system-independent*, wideband, directional channel can be obtained by (see also Section 1.5.1)

$$\mathrm{DIV}^{\mathrm{III}} = \sum_p W \cdot \mathrm{T}_p^{\mathrm{III}} \cdot (L/\lambda_0) \cdot \Omega_{\mathrm{T},p}^{\mathrm{III}} \tag{1.42}$$

Subsequently, the generalized coupling submatrix captures the available channel diversity linked to the C-I and C-II clusters. Assuming that each AOA from C-II clusters couples with K AODs from C-I clusters, the DIV can be expressed as

$$\mathrm{DIV}^{\mathrm{I}\to\mathrm{II}} = \sum_s W \cdot \mathrm{T}_s^{\mathrm{II}} \cdot J_s \cdot K, \quad K \le \sum_q (L/\lambda_0)\Omega_{\mathrm{T},q}^{\mathrm{I}} \tag{1.43}$$

where J_s is the average number of AOAs from the sth C-II cluster impinging on the receiver. K is upper-bounded by the number of resolvable AODs from C-I clusters. The overall DIV, thus, is given by

$$\mathrm{DIV} = \mathrm{DIV}^{\mathrm{I}\to\mathrm{II}} + \mathrm{DIV}^{\mathrm{III}} \tag{1.44}$$

A pictorial illustration of the above mentioned propagation process is depicted in Figure 1.10.

Along the lines of derivation for (1.23) and noting that both the delay interval T and the angular interval Ω_{T} are system-dependent parameters, a more accurate estimate of the diversity level ought to be

$$\mathrm{DIV} = \mathrm{DIV}^{\mathrm{I}\to\mathrm{II}} + \mathrm{DIV}^{\mathrm{III}}$$

$$= \sum_s \iint_{\mathcal{S}} J_s K \frac{\mathrm{T}_s^{\mathrm{II}}(2a_{\mathrm{T}}(y), f)}{y(0)} dy df$$

$$+ \sum_p \iint_{\mathcal{S}} \frac{2a_{\mathrm{T}}(y) f \Omega_{\mathrm{T},p}^{\mathrm{III}}(2a_{\mathrm{T}}(y), f) \mathrm{T}_p^{\mathrm{III}}(2a_{\mathrm{T}}(y), f)}{v_c y(0)} dy df,$$

$$K \le \sum_q \frac{2a_{\mathrm{T}}(y) f \Omega_{\mathrm{T},q}^{\mathrm{I}}(2a_{\mathrm{T}}(y), f)}{v_c} \quad \forall \mu(y, f) \in \mathcal{S} \tag{1.45}$$

Equation (1.45) calculates the DIV for a receiver with perfect beamforming capabilities and an infinite bandwidth. In practical situations, however, the receivers will have a finite aperture window L' (henceforth, a finite AOA resolution) and filtering bandwidth W'. In such a case, J_s will be upper-bounded by the number of resolvable AOAs at the receive array, as follows,

$$J_s \le \frac{L' f \Omega_{\mathrm{R},s}^{\mathrm{II}}(2a_{\mathrm{T}}(y), f)}{v_c}, \quad \forall \mu(y, f) \in \mathcal{S}' \tag{1.46}$$

where $\mathcal{S}'$ is the windowed signal space. For diversity gains due to C-III clusters, $\mathrm{DIV}^{\mathrm{III}}$ in (1.45) has to be reformulated as

$$
\mathrm{DIV}^{\mathrm{III}} = \sum_p \iint_{\mathcal{S}'} \min \left\{ \frac{2a_{\mathrm{T}}(y) f \Omega^{\mathrm{III}}_{\mathrm{T},p}(2a_{\mathrm{T}}(y), f)}{v_c}, \right.
$$

$$
\left. \frac{L' f \Omega^{\mathrm{III}}_{\mathrm{R},p}(2a_{\mathrm{T}}(y), f)}{v_c} \right\} \cdot \frac{\mathrm{T}^{\mathrm{III}}_p(2a_{\mathrm{T}}(y), f)}{y(0)} dy df \tag{1.47}
$$

To derive the general formulation of DIV for scattering clusters with nonvanishing keyholes, we simply compare the DIV density of *each* cluster to its KP density function and substitute the smaller one to either $\mathrm{DIV}^{\mathrm{I} \rightarrow \mathrm{II}}$ or $\mathrm{DIV}^{\mathrm{III}}$. For example, consider the ℓth C-III cluster, if

$$
\min \left\{ \frac{2a_{\mathrm{T}}(y) f \Omega^{\mathrm{III}}_{\mathrm{T},\ell}(2a_{\mathrm{T}}(y), f)}{v_c}, \frac{L' f \Omega^{\mathrm{III}}_{\mathrm{R},\ell}(2a_{\mathrm{T}}(y), f)}{v_c} \right\}
$$

$$
\times \frac{\mathrm{T}^{\mathrm{III}}_\ell(2a_{\mathrm{T}}(y), f)}{y(0)} > \left[\frac{R^{\mathrm{III}}_\ell(2a_{\mathrm{T}}(y), f)}{r^{\mathrm{III}}_\ell(2a_{\mathrm{T}}(y), f)} \right]^2 \frac{1}{|\mathcal{S}'|} \tag{1.48}
$$

(1.47) is modified as

$$
\mathrm{DIV}^{\mathrm{III}} = \sum_{p \neq \ell} \iint_{\mathcal{S}'} \min \left\{ \frac{2a_{\mathrm{T}}(y) f \Omega^{\mathrm{III}}_{\mathrm{T},p}(2a_{\mathrm{T}}(y), f)}{v_c}, \right.
$$

$$
\left. \frac{L' f \Omega^{\mathrm{III}}_{\mathrm{R},p}(2a_{\mathrm{T}}(y), f)}{v_c} \right\} \times \frac{\mathrm{T}^{\mathrm{III}}_p(2a_{\mathrm{T}}(y), f)}{y(0)} dy df
$$

$$
+ \iint_{\mathcal{S}'} \frac{1}{|\mathcal{S}'|} \left[\frac{R^{\mathrm{III}}_\ell(2a_{\mathrm{T}}(y), f)}{r^{\mathrm{III}}_\ell(2a_{\mathrm{T}}(y), f)} \right]^2 dy df \tag{1.49}
$$

Finally, the overall diversity gains for the DAS can be derived by substituting (1.45) into (1.30).

1.6.2 Degrees-of-Freedom Calculation

For C-I and C-II clusters, the spatial channels involve the communicating with waves between two scattering areas in the intermediate scattering process (ISP). Consider a multi-hop propagation path ϖ from the Tx to Rx via the qth C-I cluster and the sth C-II cluster as depicted in Figure 1.11. We denote the largest free-run-length of waves as d_ϖ, which is the maximum distance traveled by a plane wave before it intercedes with the next scattering cluster. We further define the cluster radius of the two clusters connected by d_ϖ as $\overline{R}_\varpi$ and $\hat{R}_\varpi$, and their keyhole radius as $\overline{r}_\varpi$ and $\hat{r}_\varpi$. If the clusters and arrays are far apart from one another compared with their linear dimensions, we can safely infer that the spatial DOF associated with ϖ is constrained by the free path of waves.

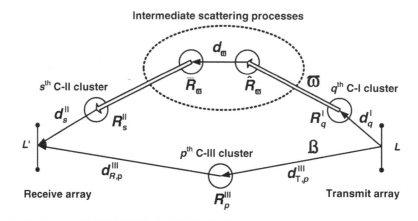

Figure 1.11   DOF Evaluation for Two Typical Propagation Paths: ϖ and ß. The Former Involves a Multiple-Scattering Process Whereas the Latter Signifies a Single-Bounce Channel.

Following from (1.39), the overall DOF for path ϖ (see also Figure 1.11) is

$$\mathrm{DOF}_{\varpi} = \min\{\mathrm{DOF}_{\varpi 1}, \mathrm{DOF}_{\varpi 2}\} \tag{1.50}$$

where

$$\mathrm{DOF}_{\varpi 1} = \iint_{\mathcal{S}'} \min\left\{ \frac{2 f R_s^{\mathrm{II}}(2a_{\mathrm{T}}(y), f) L'}{v_c d_s^{\mathrm{II}}}, \right.$$

$$\left. \frac{4 f R_q^{\mathrm{I}}(2a_{\mathrm{T}}(y), f) a_{\mathrm{T}}(y)}{v_c d_q^{\mathrm{I}}} \right\} \frac{1}{y(0) W} dy df \tag{1.51}$$

$$\mathrm{DOF}_{\varpi 2} = \iint_{\mathcal{S}'} \frac{4 f \overline{R}_{\varpi}(2a_{\mathrm{T}}(y), f) \widehat{R}_{\varpi}(2a_{\mathrm{T}}(y), f)}{v_c d_{\varpi} y(0) W} dy df \tag{1.52}$$

d_q^{I} and d_s^{II} represent the distances between the Tx and the qth C-I cluster, and the Rx and the sth C-II cluster, respectively. Equations (1.51) and (1.52) provide the DOF estimate for the signal flow at two ends of the path link and the ISP, respectively.

The general expression of $\overline{\mathrm{DOF}}_{\varpi}$ for realistic environments with nonvanishing keyholes can be obtained by selecting the smaller value of DOF_{ϖ} and the minimum KP among all the clusters associated with this path link, i.e.,

$$\mathrm{KP}_{\varpi} = \min_{\eta \in \Xi_{c,\varpi}} \left\{ \iint_{\mathcal{S}'} \frac{1}{|\mathcal{S}'|} \frac{R_{\eta}^2(2a_{\mathrm{T}}(y), f)}{r_{\eta}^2(2a_{\mathrm{T}}(y), f)} dy df \right\} \tag{1.53}$$

where $\Xi_{c,\varpi}$ is the set of all clusters along ϖ.

Now we consider the orthogonal communication modes related to the wave propagation via the pth C-III cluster. We denote this path link as ß as depicted in Figure 1.11. Similarly, by assuming a free-run-length-limited capacity gain, the DOF estimate can be easily obtained as

$$\overline{\mathrm{DOF}}_{\text{ß}} = \min\{\mathrm{DOF}_{\text{ß}}, \mathrm{KP}_{\text{ß}}\}$$

$$= \min\left\{ \iint_{\mathcal{S}'} \min\left\{ \frac{2fL'R_p^{\mathrm{III}}(2a_{\mathrm{T}}(y), f)}{v_c d_{\mathrm{R},p}^{\mathrm{III}}}, \right.\right.$$

$$\left.\frac{4fa_{\mathrm{T}}(y)R_p^{\mathrm{III}}(2a_{\mathrm{T}}(y), f)}{v_c d_{\mathrm{T},p}^{\mathrm{III}}}\right\} \frac{1}{y(0)W} dy df,$$

$$\left.\min_{\gamma \in \Xi_{C,\text{ß}}} \left\{ \iint_{\mathcal{S}'} \frac{1}{|\mathcal{S}'|} \frac{R_\gamma^2(2a_{\mathrm{T}}(y), f)}{r_\gamma^2(2a_{\mathrm{T}}(y), f)} dy df \right\} \right\} \tag{1.54}$$

$d_{\mathrm{T},p}^{\mathrm{III}}$ and $d_{\mathrm{R},p}^{\mathrm{III}}$ correspond to the distances of separation between the pth C-III cluster and the Tx and Rx arrays, respectively. $\Xi_{C,\text{ß}}$ stands for the set of all clusters associated with ß. Subsequently, the overall DOF is simply the summation of all parallel channels in the wireless link, and reads

$$\overline{\mathrm{DOF}} = \sum_{\varpi} \overline{\mathrm{DOF}}_{\varpi} + \sum_{\text{ß}} \overline{\mathrm{DOF}}_{\text{ß}} \tag{1.55}$$

Finally, the overall DOF for the DAS can be derived by substituting (1.55) into (1.41).

1.7 NUMERICAL EXAMPLES

1.7.1 SCSB Channels with Perfect Receiver Conditions

In the following discussion, it is assumed that there are five APs scattered in the circular cell. Furthermore, the ratio of D_u to R_c is assumed to be 2.5. Therefore, the average cross correlation can be computed as 0.3 following from the analysis presented in Section 1.4 (see also Figure 1.6(b)). For simplicity, the same set of channel model parameters is used for each AP.

We first probe into the diversity gains of an SCSB channel, which correspond to a maximum correlation and a minimum diversity. For brevity, it is hypothesized that the dependencies of T, Ω, and R are separable in the aperture and frequency domains. Applying the system-dependent channel models presented in (1.12) to (1.15) yields the following

$$\Omega(a_{\mathrm{T}}, f) = \sqrt{\Omega_a(a_{\mathrm{T}})\Omega_f(f)}$$

$$= \sqrt{2\sin\left[\beta_1 \exp(2\gamma_1 a_{\mathrm{T}}/\lambda)\right] \times 2\sin\left[\beta_2 \exp(\gamma_2 f)\right]} \tag{1.56}$$

$$\mathrm{T}(a_{\mathrm{T}}, f) = \sqrt{\mathrm{T}_a(a_{\mathrm{T}})\mathrm{T}_f(f)}$$

$$= \sqrt{\beta_3 \exp(2\gamma_3 a_{\mathrm{T}}/\lambda) \times \beta_4 \exp(\gamma_4 f)} \tag{1.57}$$

and

$$R(a_T, f) = \sqrt{R_a(a_T) R_f(f)}$$

$$= \sqrt{\beta_5 \exp(2\gamma_5 a_T/\lambda) \times \beta_6 \exp(\gamma_6 f)} \tag{1.58}$$

Subsequently, we study five types of typical PSD functions, where the separability of the signal spectrum is assumed.

$$f_p(a_T, f) =$$

$$\begin{cases} C_3 \exp\left(-\frac{2|a_T|}{\sigma_a} - \frac{|f-f_0|}{\sigma_f}\right), & |a_T| \leq \frac{L_0}{2}, |f - f_0| \leq \frac{W_0}{2} & (1.59.1) \\ \frac{1}{L_0 W_0}, & |a_T| \leq \frac{L_0}{2}, |f - f_0| \leq \frac{W_0}{2} & (1.59.2) \\ \frac{1}{L_0 W_0}, & |a_T| \leq \frac{L_0}{2}, |f - f_1| \leq \frac{W_0}{2}, f_1 > f_0 & (1.59.3) \\ \frac{1}{L_1 W_0}, & |a_T| \leq \frac{L_1}{2}, |f - f_0| \leq \frac{W_0}{2}, L_1 > L_0 & (1.59.4) \\ \frac{1}{L_0 W_1}, & |a_T| \leq \frac{L_0}{2}, |f - f_0| \leq \frac{W_1}{2}, W_1 > W_0 & (1.59.5) \end{cases}$$

Equation (1.59.1) is a multiplication of two truncated Laplacian distributions with C_3 being a normalization factor. It has a sharp peak at $(0, f_0)$ and characterizes signals of narrow power spectrums, while (1.59.2) represents signals of flat spectrums. For (1.59.3), the system operates at a higher frequency as compared to (1.59.2). Finally, the transmit power for (1.59.4) and (1.59.5) is distributed over an array of larger aperture length and wider frequency interval, respectively, both of which are compared to (1.59.2). The parameters used in the examples are listed in Table 1.1. After substituting (1.58) and (1.59) into (1.30), the numbers of DIV for the five PSDs are calculated as 98, 132, 140, 123, and 255. Following from (1.41) yields the DOF: 8.3, 13.4, 14.3, 19.9, and 26.1. Therefore, the following observations can be made for this numerical example: (1) both DIV and DOF will increase if the transmission scheme has a flat power spectrum, operates at a higher frequency, or occupies a broader bandwidth; (2) both DIV and DOF will not be scaled linearly with respect to the array length as usually believed (comparing the DIV and DOF for (1.59.2) and (1.59.4)); and (3) if the multipath channel is defined system-independently, the diversity and capacity gains would be no different for the transmission schemes depicted in (1.59.1), (1.59.2), and (1.59.3).

Table 1.1 Parameters Used in Numerical Examples

Parameters in (1.56) – (1.58)				Parameters in (1.59)		Other Parameters	
β_1	20	β_2	20	L_0	30 cm	L'	25 cm
γ_1	−0.14	γ_2	−0.13	W_0	2 GHz	W'	2 GHz
β_3	18	β_4	7.5	L_1	60 cm	f'	3 GHz
γ_3	−0.25	γ_4	0.16	W_1	4 GHz	d_T'''	50 m
β_5	16	β_6	25	f_0	4 GHz	d_R'''	50 m
γ_5	−0.15	γ_6	−0.2	f_1	6 GHz		
				σ_a	5 cm		
				σ_f	1 GHz		

1.7.2 SCSB Channels with Imperfect Receiver Conditions

If the receiver operates at imperfect conditions (i.e., with a finite receive aperture L' and frequency band $[f' - W'/2, f' + W'/2]$, see Table 1.1), the numbers of DIV and DOF can be obtained as 55 and 5.4 by substituting (1.58) and (1.59.2) into (1.47) and (1.54), assuming the same channel conditions as in Section 1.7.1 and the transmit signals in the form of (1.59.2). Evidently, both DIV and DOF are greatly reduced by more than half in this case, which certainly holds an intuitive appeal.

1.7.3 MCMB Channels with Imperfect Receiver Conditions

In this section, we will briefly investigate the effect of multiple scattering and multiple clusters on the system diversity estimate. Consider an abstract channel with one C-I and one C-II cluster. In order to make a fair comparison, we further assume that the system operating conditions and cluster properties are similar to the examples in Section 1.7.1, with the only exception being that the original AOA AOD characteristics of the single C-III cluster are now decoupled into two clusters. Following from (1.45), the number of DIV is calculated as 81, which is larger than the figure obtained in the previous case. Thus, in terms of the maximum diversity gain, MCMB channels are better than SCSB channels. However, the attainable DOF would remain the same, as might be expected by intuition. Therefore, characterizing the bouncing frequency in different physical environments helps system designers to deploy more efficient space-time diversity and multiplexing schemes. As a final remark of this section, our arguments have resulted from an oversimplification in the model, but they do correctly count the available DIV and DOF and yield a substantially correct intuitive picture. Nonetheless, the implementation of the general framework to represent the real-world situation would have to be justified by further measurements.

1.8 SUMMARY AND OPEN ISSUES

The DIV and DOF calculation for system-dependent wideband directional channels in a generalized DAS has been presented in this chapter. Starting with a degenerate SCSB channel and perfect receiving conditions, the analysis is extended to more general scattering and receiver-operating scenarios. The utility of the proposed methodology is demonstrated by presenting various numerical examples to provide insight into the impact of signal power spectra on the DIV and DOF estimates. It has been shown that the possible diversity and multiplexing gains are greatly dependent on the power spectrum density of the transmitted waveforms, which is based on the fact that the wireless channel could be perturbed to a certain extent when signals are shaped either in the azimuth or frequency domain, thereby allowing for a more flexible signaling strategy optimization.

There are a number of open issues that need to be addressed in future work:

■ In the current work, the APs are uniformly distributed in a circular cell. It is of interest to study the macroscopic cross-correlation of the DAS with more complicated antenna layouts and network architecture. For example, there could be multiple cells due to the specific system configurations and environments. Moreover, it is useful to study the properties of other DAS channel parameters, such as the average path gains and the standard deviation of the path gains, which provide important insight into the DAS performance in terms of the DIV and DOF values.

■ It is also of interest to investigate other DAS architectures, e.g., noncentralized UTs and APs, multiple noncollocated UTs and APs, etc.

■ We have derived the DIV and DOF as the average of diversity and capacity gains for any two APs belonging to the set of all APs. This, nevertheless, is an approximation approach. More rigorous analysis is expected to provide better estimates of these two measures.

■ A large amount of measurement data is required to support the system-dependent channel models and to provide better model parameterizations. More empirical results are expected to describe slow fading cross-correlation against azimuth separation of APs in various service environments.

■ It is also crucial to provide insight into the fundamental tradeoff between diversity and capacity gains in the DAS scenario from the actual wave-propagation viewpoint. To the best of our knowledge, little work has been done on this important aspect.

■ It would be very challenging to design advanced signaling schemes under the rubric of "system-dependent channel modeling," which poses a nonlinear optimization problem.

REFERENCES

[1] A. J. Paulraj, R. Nabar, and D. Gore, *Introduction to Space-Time Wireless Communications*, Cambridge, U.K.: Cambridge University Press, 2003.

[2] E. Larsson and P. Stoica, *Space-Time Block Coding for Wireless Communications*, Cambridge, U.K.: Cambridge University Press, 2003.

[3] N. H. Dawod, I.D. Marsland, and R.H.M. Hafez, Improved transmit null steering for MIMO-OFDM downlinks with distributed base station antenna arrays, *IEEE J. Select. Areas Commun.*, vol. 24, no. 3, pp. 419–426, Mar. 2006.

[4] W. Roh and A. Paulraj, Outage performance of the distributed antenna systems in a composite fading channel, in *Proc. IEEE Veh. Technol. Conf. 2002 Fall*, vol. 3, pp. 1520–1524, Sept. 2002.

[5] S. M. Alamouti, A simple transmit diversity technique for wireless communications, *IEEE J. Select. Areas Commun.*, vol. 16, pp. 1451–1458, Oct. 1998.

[6] V. Tarokh, N. Seshadri, and A. R. Calderbank, Space-time codes for high data rate wireless communication: performance criterion and code construction, *IEEE Trans. Inform. Theory*, vol. 44, pp. 744–765, Mar. 1998.

[7] G. J. Foschini and M. J. Gans, On limits of wireless communications in a fading environment when using multiple antennas, *Wireless Pers. Commun.*, vol. 6, no. 3, pp. 311–335, Mar. 1998.

[8] L. Zheng and D. N. C. Tse, Diversity and multiplexing: a fundamental tradeoff in multiple antennas channels, *IEEE Trans. Inform. Theory*, vol. 49, pp. 1073–1096, May 2003.

[9] D. Tse, P. Viswanath, and L. Zheng, Diversity-multiplexing tradeoff in multiple access channels, *IEEE Trans. Inform. Theory*, vol. 50, no. 9, pp. 1859–1874, Sept. 2004.

[10] M. Godavarti and A. O. Hero III, Diversity and degrees of freedom in wireless communications, in *Proc. IEEE ICASSP 2002*, vol. 3, pp. 2861–2864, May 13–17, 2002.

[11] S. Catreux, L. J. Greenstein, and V. Erceg, Some results and insights on the performance gains of MIMO systems, *IEEE J. Select. Areas Commun.*, vol. 21, pp. 839–847, June 2003.

[12] A. M. Sayeed, Deconstructing multi-antenna fading channels, *IEEE Trans. Signal Process.*, vol. 50, pp. 2563–2579, Oct. 2002.

[13] G. G. Raleigh and J. M. Cioffi, Spatio-temporal coding for wireless communications, *IEEE Trans. Commun.*, vol. 46, pp. 357–366, Mar. 1998.

[14] D. Gesbert, H. Bölcskei, D. A. Gore, and A. J. Paulraj, Outdoor MIMO wireless channels: models and performance prediction, *IEEE Trans. Commun.*, vol. 50, pp. 1926–1934, Dec. 2002.

[15] H. Bölcskei, D. Gesbert, and A. J. Paulraj, On the capacity of OFDM-based spatial multiplexing systems, *IEEE Trans. Commun.*, vol. 50, pp. 225–234, Feb. 2002.

[16] A. G. Burr, Capacity bounds and estimates for the finite scatterers MIMO wireless channel, *IEEE J. Select. Areas Commun.*, vol. 21, pp. 812–818, June 2003.

[17] M. Steinbauer, A. F. Molisch, and E. Bonek, The double-directional mobile radio channel, *IEEE Antennas Propagat. Mag.*, vol. 53, pp. 51–63, Aug. 2001.

[18] T. Zwick, D. Hampicke, A. Richter, G. Sommerkorn, R. Thomä, and W. Wiesbeck, A novel antenna concept for double-directional channel measurements, *IEEE Trans. Veh. Technol.*, vol. 53, no. 2, pp. 527–537, Mar. 2004.

[19] R. Vaughan and J. B. Andersen, *Channels, Propagation and Antennas for Mobile Communications*, The Institution of Electrical Engineers, 2004.

[20] A. S. Y. Poon and M. Ho, Indoor multiple-antenna channel characterization from 2 to 8 GHz, in *Proc. of IEEE ICC*, vol. 5, pp. 3519–3523, May 2003.

[21] Y. Chen and V. K. Dubey, A general analysis on diversity estimate in a system-dependent wideband directional channel, *IEEE Trans. Wireless Commun.*, vol. 5, no. 12, pp. 3644–3650, Dec. 2006.

[22] J. P. Rossi, Influence of measurement conditions on the evaluation of some radio channel parameters, *IEEE Trans. Veh. Technol.*, vol. 48, pp. 1304–1316, July 1999.

[23] Y. Chen and V. K. Dubey, Spatio-diversity gain of a system-dependent wideband directional channel, *IEEE Commun. Lett.*, vol. 9, no. 4, pp. 319–321, Apr. 2005.

[24] K. Yu, Q. Li, and M. Ho, Measurement investigation of tap and cluster angular spreads at 5.2 GHz, *IEEE Trans. Antennas Propagat.*, vol. 53, pp. 2156–2160, July 2005.

[25] Q. H. Spencer, B. D. Jeffs, M. A. Jensen, and A. L. Swindlehurst, Modeling the statistical time and angle of arrival characteristics of an indoor multipath channel, *IEEE J. Sel. Areas Commun.*, vol. 18, pp. 347–359, Mar. 2000.

[26] C. C. Chong, C. M. Tan, D. I. Laurenson, S. McLaughlin, M. A. Beach, and A. R. Nix, A new statistical wideband spatio-temporal channel model for 5-GHz band WLAN systems, *IEEE J. Sel. Areas Commun.*, vol. 21, pp. 139–150, Feb. 2003.

[27] R. J. Cramer, R. A. Scholtz, and M. Z. Win, Evaluation of an ultra-wide-band propagation channel, *IEEE Trans. Antennas Propagat.*, vol. 50, pp. 561–570, May 2002.

[28] M. Steinbauer, H. Özcelik, H. Hofstetter, C. F. Mecklenbräuker, and E. Bonek, How to quantify multipath separation, *IEICE Trans. Electron.*, vol. E85-C, no. 3, pp. 552–557, Mar. 2002.

[29] M. Steinbauer, *The Radio Propagation Channel - A Non-Directional, Directional, and Double-Directional Point-of-View*, Ph.D. dissertation, Technische Universität Wien, Austria, Nov. 2001.

[30] D. Chizhik, G. J. Foschini, M. J. Gans, and R. A. Valenzuela, Keyholes, correlations, and capacities of multielement transmit and receive antennas, *IEEE Trans. Wireless Commun.*, vol. 1, pp. 361–368, Apr. 2002.

[31] C. A. Balanis, *Antenna Theory: Analysis and Design*, New York: John Wiley & Sons, 1997.

[32] T. Manabe, Y. Miura, and T. Ihara, Effects of antenna directivity and polarization on indoor multipath propagation characteristics at 60 GHz, *IEEE J. Select. Areas Commun.*, vol. 14, no. 3, pp. 441–448, Apr. 1996.

[33] A. Kajiwara, Effects of polarization, antenna directivity, and room size on delay spread in LOS indoor radio channel, *IEEE Trans. Veh. Technol.*, vol. 46, no. 1, pp. 169–175, 1997.

[34] T. S. Rappaport and D. A. Hawbaker, Wide-band microwave propagation parameters using circular and linear polarized antennas for indoor wireless channels, *IEEE Trans. Commun.*, vol. 40, no. 2, pp. 240–245, 1992.

[35] H. H. Xia, A. Herrera, S. Kim, F. Rico, and B. Bettencourt, Measurements and modeling of CDMA PCS in-building systems with distributed antenna, in *Proc. IEEE Veh. Technol. Conf. 1994*, vol. 2, pp. 733–737, 1994.

[36] V. Roy and C. L. Despins, Planning of GSM-based wireless local access via simulcast distributed antennas over hybrid fiber-coax, in *Proc. IEEE GLOBECOM 1999*, vol. 2, pp. 1116–1120, 1999.

[37] H. Yanikomeroglu and E. S. Sousa, Power control and number of antenna elements in CDMA distributed antenna systems, in *Proc. IEEE ICC 1998*, vol. 2, pp. 1040–1045, 1998.

[38] W. Roh and A. Paulraj, MIMO channel capacity for the distributed antenna systems, in *Proc. IEEE Veh. Technol. Conf. 2002*, vol. 2, pp. 706–709, Sept. 2002.

[39] H. Zhuang, L. Dai, L. Xiao, and Y. Yan, Spectral efficiency of distributed antenna system with random antenna layout, *IEE Electron. Lett.*, vol. 39, no. 6, pp. 495–496, Mar. 2003.

[40] T. B. Sørensen, Slow fading cross-correlation against azimuth separation of base stations, *IEE Electron. Lett.*, vol. 35, no. 2, pp. 127–129, Jan. 1999.

[41] V. Graziano, Propagation correlations at 900 MHz, *IEEE Trans. Veh. Technol.*, vol. VT-27, no. 4, pp. 182–189, Nov. 1978.

[42] K. Zayana and B. Guisnet, Measurements and modelisation of shadowing cross-correlations between two base-stations, in *Proc. IEEE ICUPC 1998*, vol. 1, pp. 101–105, Oct. 1998.

[43] P. Petrus, J. H. Reed, and T. S. Rappaport, Geometrical-based statistical macrocell channel model for mobile environments, *IEEE Trans. Commun.*, vol. 50, pp. 495–502, Mar. 2002.

[44] Y. Chen, C. Yuen, Y. Zhang, and Z. Zhang, Cross-correlation analysis of generalized distributed antenna systems with cooperative diversity, accepted for presentation in *IEEE Veh. Technol. Conf. 2007*, Dublin, Ireland, April 2006.

[45] A. Papoulis and S. U. Pillai, *Probability, Random Variables and Stochastic Processes*, 4th Ed., New York: McGraw-Hill, 2002.

[46] B. H. Fleury, An uncertainty relation for WSS process and its application to WSSUS systems, *IEEE Trans. Commun.*, vol. 44, no. 12, pp. 1632–1634, 1996.

[47] G. D. Durgin, *Space-Time Wireless Channels*, Upper Saddle River, NJ: Prentice Hall, 2003.

[48] W. G. Newhall, R. Mostafa, K. Dietze, J. H. Reed, and W. L. Stutzman, Measurement of multipath signal component amplitude correlation coefficients versus propagation delay, in *Proc. IEEE RAWCON 2002*, Boston, MA, 2002, pp. 133–136.

[49] G. J. M. Janssen, P. A. Stigter, and R. Prasad, Wideband indoor channel measurements and BER analysis of frequency selective multipath channels at 2.4, 4.75, and 11.5 GHz, *IEEE Trans. Commun.*, vol. 44, no. 10, pp. 1272–1288, 1996.

[50] M. S. Varela and M. C. Sanchez, RMS delay and coherence bandwidth measurements in indoor radio channels in the UHF band, *IEEE Trans. Veh. Technol.*, vol. 50, no. 2, pp. 515–525, 2001.

[51] R. S. Kennedy, *Fading Dispersive Communication Channels*, New York: John Wiley & Sons, 1968.

[52] S. Roy, J. R. Foerster, V. S. Somayazulu, and D. G. Leeper, Ultrawideband radio design: the promise of high-speed, short-range wireless connectivity, *Proc. IEEE*, vol. 92, no. 2, pp. 295–311, 2004.

[53] Y. Chen, C. Yuen, Y. Zhang, and Z. Zhang, Diversity gains of generalized distributed antenna systems with cooperateive diversity, accepted for presentation in *IEEE WCNC 2007*, Hong Kong, China, March 2006.

[54] D. Slepian and H. O. Pollak, Prolate spheroidal wave functions, Fourier analysis and uncertainty - I (II), *Bell Syst. Tech. J.*, vol. 40, pp. 43–84, Jan. 1961.

[55] D. A. B. Miller, Communicating with waves between volumes: evaluating orthogonal spatial channels and limits on coupling strengths, *Appl. Opt.*, vol. 39, pp. 1681–1699, Apr. 2000.

2

AN INFORMATION THEORETIC
VIEW OF DISTRIBUTED ANTENNA
PROCESSING IN CELLULAR
SYSTEMS

**Oren Somekh, Osvaldo Simeone, Yeheskel Bar-Ness,
Alexander M. Haimovich, Umberto Spagnolini and
Shlomo Shamai (Shitz)**

Contents

This chapter presents a survey of information theoretic results available on DAS in cellular systems. The treatment focuses on the derivation of the per-cell ergodic sum-rate of different intercell and intracell communications strategies for both uplink and downlink. Only a simple symmetric family of cellular models in which the intercell interferences are emerging from the adjacent cells is considered. Although hardly realistic, this family of models accounts for essential parameters of cellular systems such as intercell interference and fading. Whenever computation of the sum-rate is intractable or yields little insight into the problem, asymptotic performance criteria (e.g., extreme-SNR parameters) are evaluated. Emphasis is placed on the assessment of benefits of co-operation among access points (AP) (i.e., joint detection/precoding). Mathematical tools underlying the results are introduced briefly where necessary. Finally, some advanced topics, such as cooperation among mobile terminals, are discussed as well.

2.1 INTRODUCTION

Information theory provides a solid reference framework for a thorough understanding of communication systems. An accurate information theoretic enables one to assess the limiting performance expected by a given technology and to yield insight into practical design solutions.

From an information theoretical standpoint, distributed antenna systems (DAS) qualify as either a vector multiple access channel (MAC) or a broadcast channel (BC), according to whether the uplink or downlink is considered. The limiting performance (capacity region) and optimal transmission/reception schemes for such scenarios have been widely studied under both Gaussian and fading channels (e.g., [1], [2], [3]). However, the distributed nature of DAS provides the analysis of this system with a unique structure that deserves a separate treatment. In particular, the main differences between the *microdiversity* provided by the use of conventional collocated antenna arrays and the *macrodiversity* provided by DAS reduces to: (1) the power profile of the resulting channels transfer matrices; (2) the power constraints; and (3) possible limitations on the capacity of links (i.e., the *backbone*) connecting (AP) to the *central processing unit*. These points are elaborated upon in the following.

1. The channel transfer matrix describes the complex channel propagation gains between each pair of transmit and receive antennas. With antenna array systems (microdiversity), the entries of the channel matrix can be modeled as identically distributed, because the elements of the antenna arrays are collocated, and thus experience similar propagation conditions. This enables the use of standard analytical tools from random matrix theory [4]. In contrast, with DAS (macrodiversity), the channel gains between any terminal and the distributed antennas of the APs may have different statistics, as dictated by the characteristics of propagation, such as path loss, shadowing, line-of-sight components, and the system topology. In general, it follows that the power profile of the corresponding channel matrix is nonuniform, preventing a simple application of known tools to the scenario at hand.

2. For collocated antenna arrays, power constraints are generally imposed on the sum-power radiated by all the elements of the array. Conversely, for distributed APs, because each element is provided with a separate power amplifier, the constraints have to be enforced on a per-AP basis. This creates specific challenges in pursuing an analytical treatment of DAS.

3. Different APs in a DAS are connected through high capacity links (referred to as the backbone) to a central processing unit, which jointly processes the signals transmitted/received by the APs. Though the backbone is generally composed of highly reliable wireline links (e.g., optical fibers), delay or locality constraints may have to be considered, especially in large DAS. This feature is radically different from the centralized processing available at little cost in collocated antenna arrays.

The system layout in DAS is likely to present distributed and regularly displaced APs according to a cellular structure. This structure can be effectively captured by a somewhat simplistic model, referred to in the literature as Wyner's model, and first presented in the context of multi-cell networks in [5,6]. In this model, intercell interference is limited to adjacent cells and is measured by a single parameter $\alpha \in [0, 1]$ (see Figures 2.1 and 2.2). To a certain extent, the model is simple enough to allow analytical treatment, but also sufficiently rich to incorporate the main features of propagation in DAS.

This chapter presents a survey of the information theoretic results available on the uplink and downlink of DAS under the Wyner model for Gaussian and fading channels. The treatment aims at presenting the main analytical results concerning both the ultimate performance of the system (under optimal transmission/reception schemes) and limitations of suboptimal solutions. Furthermore, recent results that attest to the advantages of employing cooperative technology among terminals are briefly discussed.

The rest of the chapter is organized as follows. Section 2.2 introduces the Wyner cellular model along with the transmission/reception schemes of interest and the performance criteria employed in the analysis throughout the chapter. The uplink of Wyner's model is investigated in Section 2.3, while the corresponding study of the downlink is covered in Section 2.4. The advantages of cooperation among terminals are studied in Section 2.5. Concluding remarks and related open problems are included in Section 2.6. Finally, bibliographical notes are presented in Section 2.7.

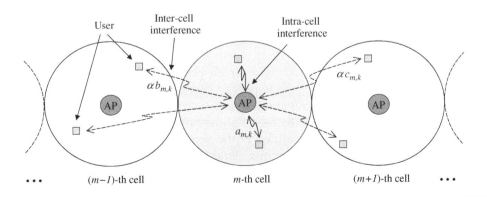

Figure 2.1 Wyner's Linear Model.

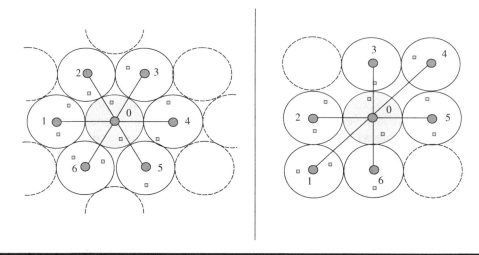

Figure 2.2 Wyner's Planar Model.

2.2 SYSTEM MODEL

In this section, we present the framework under which the results of this chapter are formulated. In particular, Section 2.2.1 introduces the physical model describing the relationship between the signal transmitted by the mobiles and received by the base stations (uplink) or, dually, transmitted by the base stations and received by the mobiles (downlink). Notice that we will use the terms base station and access point interchangeably throughout this chapter. Section 2.2.2 introduces the main transmission/reception schemes of interest, whose performance is evaluated and compared throughout the chapter. Finally, performance criteria used throughout this chapter are briefly described in Section 2.2.2.

2.2.1 Wyner's Model

A family of physical models that prescribes that each cell senses only signals radiated from a limited number of neighboring cells has been introduced by Hanly and Whiting in [5] and by Wyner in [6]. For instance, the basic *linear* model, depicted in Figure 2.1, consists of a linear array of M identical cells with single antenna base stations. The K single antenna users contained in each cell, "see" only their own cell-site antenna and the two adjacent cell-site antennas. The flat fading experienced by the users is assumed independent and identically distributed (i.i.d.) among different users and ergodic in the time index. As a further simplification, the path loss (i.e., the average fading power) toward adjacent cells is represented by a single parameter $\alpha \in [0, 1]$. Although this model is simplistic, it encompasses the impact of central factors of a cellular system and accounts for phenomena like fading and intercell interference.

As a result of the assumptions discussed above, the overall channel transfer matrix of these models (to be rigorously defined in the sequel) consists of a finite number of nonzero diagonals regardless of the matrix dimensions. This fact precludes the use of results in the field of random matrices, which were recently used successfully for calculating the sum-rates of several interesting communication setups [4]. Nevertheless, the simplicity of the Wyner model family renders analytical treatment feasible, which in turn provides much insight and understanding of DAS.

Wyner's basic linear model in Figure 2.1 will be considered in the rest of the chapter for its simplicity, in order to get insight into the role of different system and channel parameters in the system overall performance. However, extension to more complex (and realistic) scenarios is conceptually straightforward, and results will be mentioned in the following, whenever available. Among interesting generalizations of the model that have been considered in the literature, we recall:

- *Circular model*: In several cases of interest it is convenient to assume that the M cells are arranged on a circle. In this setup, introduced in [5], the first and last indexed cells are adjacent, resulting in a perfectly symmetric and homogenous model with no boundary effects. It is noted that in the limit where the number of cells goes to infinity the boundary effects are negligible, both linear array and circular array models are "equivalent" under the performance merits of interest.
- *Planar model*: A finite two-dimensional hexagonal array is considered in [6]. In this case each cell "sees" the signal radiated in the six adjacent cell sites. Because it is assumed that the path loss is independent of the user's position, by scaling the vertical axis by a factor of $1/\sqrt{3}$, rotating the plane by $45°$, and scaling it by a factor of $1/\sqrt{2}$, the hexagonal array is transformed into a rectangular $M \times M$ cell array, as depicted in Figure 2.2.
- *Cell clusters*: Cells are divided into nonoverlapping independent clusters with separate backbone networks and central processing units. This model is a first step toward introducing practical constraints on the backbone connecting different APs. In particular, only signals relative to APs within a cluster are jointly processed by the corresponding central processing unit, which is unaware of outer users' codebooks [7,8].

2.2.2 Transmission/Reception Schemes of Interest

DAS inherently refers to schemes where signals either transmitted or received by the APs are jointly processed in order to enhance the system performance. We refer to this scenario as multi-cell processing (MCP), and distinguish it from single-cell processing (SCP). With SCP, each AP processes the signal (transmitted or received) independently from other, interfering, APs or users. Throughout the chapter, it is assumed that all APs and users are fully synchronized. Being a defining feature of DAS, this chapter will be devoted mostly to the study of MCP, even though, for reference, SCP will be considered as well. The taxonomy of schemes of interest further divides the transmission/reception techniques according to the strategy employed at both an intercell and intracell level. This point is elaborated upon in the following.

2.2.2.1 *Single-Cell Processing (SCP)*

In SCP, signals radiated within other cells are considered as interference by any given AP. As far as *intracell* strategies are concerned, possible alternatives include:

- Orthogonal intracell medium access: In this case, only one user per cell is active in the considered time-frequency resource. Therefore, no intracell interference is produced and single-user transmission/reception strategies are used at each AP. In accordance with the literature on the subject, this solution will be referred to in the following as *intracell* time division multiple access (TDMA);

■ Nonorthogonal intracell medium access: Here, different users of the same cell access the time-frequency resource simultaneously, thus causing interference on the concurrent transmissions. Each AP employs multi-user processing techniques (e.g., in uplink, multi-user detection (MUD)) in order to cope with intracell interference. Following the literature, we will refer to this scenario as *intracell wideband* (WB);

■ Collaborative transmission: Same-cell users cooperate for the transmission/reception of each other's data through, e.g., relaying or more sophisticated forms of cooperative coding [9]. This situation can coexist with either orthogonal or nonorthogonal intracell medium access. We will consider this advanced intracell strategy in some detail in Section 2.5.

A simple solution commonly employed in cellular systems to reduce interference is that of using the same channel (i.e., frequency or time resource) only in cells sufficiently separated. This *intercell* strategy is usually referred to as frequency reuse, and in Wyner's model allows a total elimination of intercell interference. For instance, in the linear model, one activates odd cells in orthogonal channels (in either time or frequency) with respect to even cells, to have intercell interference-free communications. This scenario is referred to in the literature as intercell time-sharing (ICTS).

2.2.2.2 Multi-Cell Processing (MCP)

As explained above, MCP is a defining feature of the DAS, and is based on joint processing of all the signals transmitted or received by the APs. To simplify the analysis and focus on ultimate bounds, an ideal delayless, infinite-capacity, "backbone" network is assumed to connect all APs to a central unit, which jointly processes all the signals. Alleviating this assumption is the subject of a number of recent publications [10,11]. The central unit is aware of all the users' codebooks and their channel state information (CSI). Finally, for the downlink, the users are assumed to be aware of only their own CSI and codebooks. Intracell strategies are defined as above. Notice that APs here always employ multi-user transmission/reception strategies.

2.2.2.3 Performance Criteria

Throughout the chapter, the performance of different schemes and scenarios is evaluated in terms of *per-cell ergodic sum-rate* (measured in *nats/s/Hz* or *bit/s/Hz*).[1] This performance criterion is well-suited for fading channels that vary fast enough to allow each transmitted codeword to experience a large number of fading states, while it is not appropriate for delay sensitive applications. For the uplink, because each user has a separate power amplifier, a per-user power constraint of P is enforced, which amounts to a per-cell power constraint of $\bar{P} = KP$, where K is the number of users. For the downlink, because the APs serve all the users through a single power amplifier, a per-cell power constraint of $\bar{P}$ is considered.

In cases where the ergodic sum-rate provides little insight into the performance of the system due to cumbersome analytical expressions, asymptotic measures will be considered as follows.

---

[1] Due to the symmetry of the system model, the per-cell sum-rate divided by the number of users per cell is a reliable measure of the rate achieved by each user.

Extreme SNR Analysis: For low signal-to-noise ratios (SNRs), the sum-rate R of a given scheme is described by the minimum transmitted energy per bit required for reliable communication (normalized to the background noise power spectral density N_0) $E_b/N_{0\min}$ and by the slope S_0 at $E_b/N_{0\min}$ (measured in [bit/s/Hz/(3 dB)]) [12]. In particular, R can be approximated by the *affine* function of E_b/N_0 measured in [dB]:

$$R \simeq \frac{S_0}{3[\mathrm{dB}]} \left(\frac{E_b}{N_0}[\mathrm{dB}] - \frac{E_b}{N_0}{}_{\min}[\mathrm{dB}] \right). \tag{2.1}$$

The low SNR characterization, in terms of parameters $E_b/N_{0\min}$ and S_0, turns out to depend on both the first ($\dot{R}(\mathrm{SNR})$) and second ($\ddot{R}(\mathrm{SNR})$) derivatives of the achievable rate $R(\mathrm{SNR})$ when measured in [nats/s/Hz] [12]. In fact, it can be shown that:

$$\frac{E_b}{N_0}{}_{\min} = \frac{\log 2}{\dot{R}(0)} \tag{2.2a}$$

$$S_0 = \frac{2[\dot{R}(0)]^2}{-\ddot{R}(0)}. \tag{2.2b}$$

In the high SNR regime, the sum-rate of a given scheme can be expanded as an *affine* function of the SNR (measured in [dB]) [13]:

$$R \simeq S_\infty \left(\frac{\mathrm{SNR}[\mathrm{dB}]}{3[\mathrm{dB}]} - \mathcal{L}_\infty \right), \tag{2.3}$$

where the high-SNR parameters S_∞ (slope or multiplexing gain measured in [bits/s/Hz/ (3 dB)]) and $\mathcal{L}_\infty$ (power offset measured in 3 [dB] units with respect to a reference channel having the same slope, but with zero power offset), read:

$$S_\infty = \lim_{\mathrm{SNR} \to \infty} \frac{R}{\log_2 SNR} \tag{2.4}$$

$$\mathcal{L}_\infty = \lim_{\mathrm{SNR} \to \infty} \left(\log_2 \mathrm{SNR} - \frac{R}{S_\infty} \right), \tag{2.5}$$

where R is measured in [nats/s/Hz] and in [bits/s/Hz], in (2.4) and (2.5), respectively. If the multiplexing gain is zero, then a given scenario is said to be *interference limited.*

Scaling Law with the Number of Users: To study the ability of a multi-user system such as a DAS to profit from different fading conditions across the users of the system (multi-user diversity), it is customary to evaluate the scaling law of the sum-rate with respect to the number of users K. In particular for a broadcast channel with Rayleigh fading, optimality with respect to this criterion is achieved if

$$\lim_{K \to \infty} \frac{R}{\log \log K} = 1, \tag{2.6}$$

indicating that the scaling law is $\log \log K$ [14].

Finally, it is noted that a *natural logarithm* base is used throughout this chapter unless specified differently.

Spectral Efficiency: Most of the curves included in this chapter represent the spectral efficiency versus the transmitted E_b/N_0. This representation is considered to be more informative than the rate versus SNR representation, especially in the low SNR regime. The spectral efficiency $C(E_b/N_0)$ is defined through the following relations: $C(E_b/N_0) = C(\text{SNR})$ and $\text{SNR} = C(\text{SNR})E_b/N_0$.

2.3 THE UPLINK CHANNEL

Consider the uplink of Wyner's linear model with M cells as in Figure 2.1. The $M \times 1$ vector baseband representation of the signals received at all cell sites is given for an arbitrary time index by

$$y = Hx + z, \tag{2.7}$$

where x is the $MK \times 1$ complex Gaussian symbols vector $x \sim \mathcal{CN}(0, P I_{MK})$, ($I_N$ is an $N \times N$ unity matrix), z is the $M \times 1$ complex Gaussian additive noise vector $z \sim \mathcal{CN}(0, I_M)$, and H is the $M \times MK$ channel transfer matrix, given by

$$
H = \begin{pmatrix}
a_0 & \alpha c_0 & 0 & \cdots & & 0 & 0 \\
\alpha b_1 & a_1 & \alpha c_1 & 0 & & \cdots & 0 \\
0 & \alpha b_2 & a_2 & \alpha c_2 & & \ddots & \vdots \\
\vdots & 0 & \alpha b_3 & \ddots & & \ddots & 0 \\
0 & \vdots & \ddots & \ddots & & a_{M-2} & \alpha c_{M-2} \\
0 & 0 & \cdots & 0 & & \alpha b_{M-1} & a_{M-1}
\end{pmatrix}, \tag{2.8}
$$

where a_m, b_m, and c_m are $1 \times K$ row vectors denoting the fading channel coefficients (i.i.d. complex Gaussian with zero mean and unit power), experienced by the K users of the m-th, $(m-1)$-th, and $(m+1)$-th cells, respectively, when received by the m-th cell-site antenna.

2.3.1 Single-Cell Processing

In this section, single-cell processing is considered. Starting with the Gaussian (i.e., nonfading) scenario and assuming $M \gg 1$ in order to make border effects negligible, it is easily shown that the per-cell sum-rate is given for the linear model by

$$R_{\text{scp}} = \log\left(1 + \frac{\bar{P}}{1 + 2\alpha^2 \bar{P}}\right) \tag{2.9}$$

and for the planar model by

$$\hat{R}_{\text{scp}} = \log\left(1 + \frac{\bar{P}}{1 + 6\alpha^2\bar{P}}\right), \tag{2.10}$$

where $\bar{P} \triangleq KP$ is the average per-cell transmitted power (since a unit noise power is assumed, $\bar{P}$ is also the total cell SNR). These rates are achieved (not uniquely) by intracell TDMA protocols, where each user is transmitting $1/K$ of the time with power $\bar{P}$, or by the WB protocol with MUD, according to which all users are transmitting all the time with average power P. As expected, the rates demonstrate an interference limited behavior as $\bar{P}$ increases.

For reference, the extreme SNR characterization of single-cell processing over Gaussian channels is summarized for the linear and planar models as follows:

$$S_0 = \frac{2}{1+4\alpha^2}; \quad \frac{E_b}{N_0}_{\min} = \log 2; \quad S_\infty = 0$$

$$\hat{S}_0 = \frac{2}{1+12\alpha^2}; \quad \frac{\hat{E}_b}{N_0}_{\min} = \log 2; \quad \hat{S}_\infty = 0, \tag{2.11}$$

showing the deleterious effects of intercell interference in both regimes.

Introducing fading, and assuming that each cell site (say the m-th) is aware of the instantaneous channel gains a_k in vector $\boldsymbol{a}_m$ and the instantaneous interference power, the per-cell ergodic sum-rate for the WB protocol deployed in the linear model is given by

$$R_{\text{scp}}^* = E\left\{\log\left(1 + \frac{\bar{P}\sum_{k=1}^{K}|a_k|^2}{K + \alpha^2\bar{P}\sum_{k=1}^{K}(|b_k|^2 + |c_k|^2)}\right)\right\}$$

$$= E\left\{\log\left(K + \frac{1}{2}\bar{P}(S + \alpha^2 T)\right) - \log\left(K + \frac{1}{2}\bar{P}\alpha^2 T\right)\right\}, \tag{2.12}$$

where the expectation is taken over the independent random variables S and T, defined by

$$S = 2\sum_{k=1}^{K}|a_k|^2; \quad T = 2\sum_{k=1}^{K}(|b_k|^2 + |c_k|^2). \tag{2.13}$$

For the special case of Rayleigh fading, $S \sim \chi^2_{2K}$ and $T \sim \chi^2_{4K}$, where χ^2_n denotes a central chi-square distribution with n degrees of freedom. In [15] it is shown that the random variable $W = S + \alpha^2 T$ has a density distribution function given by

$$f_W(x) = \frac{x^{3K-1}\exp\left(-\frac{x}{2\alpha^2}\right)}{2^{3K}(\alpha^2)^{2K}\Gamma(3K)}\,{}_1F_1\left(K, 3K; \frac{1}{2}\left(\frac{1}{\alpha^2} - 1\right)x\right); \quad x \geq 0, \tag{2.14}$$

where $_1\mathcal{F}_1(\cdot, \cdot; \cdot)$ is the hypergeometric function of the first kind and $\Gamma(\cdot)$ is the Gamma function. Hence, (2.12) becomes

$$
R_{\text{scp}}^* = \frac{1}{\epsilon_1} \int_0^\infty \log\left(K + \frac{1}{2}\bar{P}x\right) x^{3K-1} \exp\left(-\frac{x}{2\alpha^2}\right) {}_1\mathcal{F}_1
$$

$$
\times \left(K, 3K; \frac{1}{2}\left(\frac{1}{\alpha^2} - 1\right)x\right) dx - \frac{1}{\epsilon_2} \int_0^\infty \log\left(K + \frac{1}{2}\alpha^2\bar{P}x\right) x^{2K-1}
$$

$$
\times \exp\left(-\frac{x}{2}\right) dx, \tag{2.15}
$$

where $\epsilon_1 = 2^{3K}(\alpha^2)^{2K}\Gamma(3K)$ and $\epsilon_2 = 2^{2K}\Gamma(2K)$. It is noted that the sum rate of the intracell TDMA is achieved by setting $K = 1$ in (2.15). To get the SCP sum rate for the planar model, replace $2K$ with $3K$, and $3K$ with $4K$ in (2.15).

Focusing on the case where $K \gg 1$ while the total transmit power $\bar{P}$ is kept constant, the strong law of large numbers (SLLN) can be employed to the normalized sums of (2.12). This shows that for $K \to \infty$, the rate (2.12) boils down to (2.9). Hence, for large numbers of users, the WB protocol demonstrates no penalty in the presence of fading. This result reflects the well-known phenomenon of channel "hardening" by multi-user diversity. It is noted that the same result applies for the planar model and to a general zero-mean unit-power fading distribution.

Extreme SNR characterization of SCP with Rayleigh fading for the linear and planar models reads

$$
S_0^* = \frac{2}{1 + 4\alpha^2 + \frac{1}{2K}}; \quad \frac{E_b}{N_0}^*_{\min} = \log 2; \quad S_\infty^* = 0
$$

$$
\hat{S}_0^* = \frac{2}{1 + 24\alpha^2 + \frac{1}{2K}}; \quad \frac{\hat{E}_b}{N_0}^*_{\min} = \log 2; \quad \hat{S}_\infty^* = 0, \tag{2.16}
$$

showing that, for increasing K, the performance reduces to the nonfading setup (2.11).

In Figure 2.3 the interference limited behavior of the linear model SCP spectral efficiency is demonstrated for $\alpha = 0.4$ (which represents a situation where the overall intercell interference power is about a third of the received useful signal power) and several values of K. The dependency of this sum rate on α is demonstrated in Figure 2.4 for $\bar{P} = 10$ [dB] and several values of K. In both figures the curves drawn for Rayleigh fading approach the respective nonfading related curves as K increases. It is noted that due to the numerical instability of (2.15), Monte-Carlo simulation is used to produce the curves.

Next, the analysis of the ICTS protocol is considered, according to which, in the linear Wyner model, odd and even cells are active alternatively in orthogonal channels. Hence, intercell interference is avoided (thus, it is clearly noninterference limited). Starting with the nonfading setup, the achievable sum-rate is easily shown to be

$$
R_{\text{icts}} = \frac{1}{2}\log(1 + 2\bar{P}), \tag{2.17}
$$

where the pre-log factor of $1/2$ and the power factor of 2 comes from the fact that each cell user transmits only half the time using twice the available average power. It is noted that this rate is achieved (not uniquely) by intracell TDMA and WB schemes. Deploying

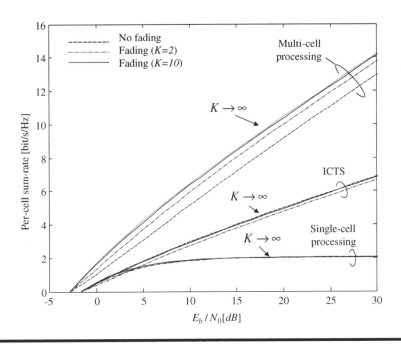

Figure 2.3   Spectral Efficiencies Versus E_b/N_0 in Wyner's Uplink Channel ($\alpha = 0.4$).

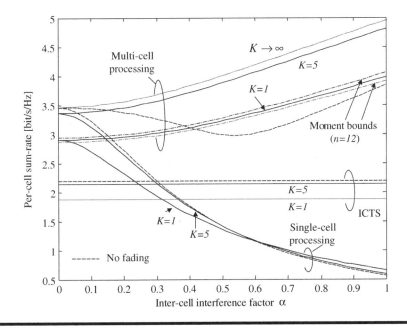

Figure 2.4   Per-Cell Sum-Rates Versus Intercell Interference Factor α in Wyner's Uplink Channel ($\bar{P} = 10$ dB).

a similar protocol for the planar model where each cell is active only 1/6 of the time using $6\bar{P}$ power results in

$$\hat{R}_{\text{icts}} = \frac{1}{6} \log(1 + 6\bar{P}). \tag{2.18}$$

Comparing the rate (2.9) with the performance of ICTS (2.17) for the linear model reveals that $R_{\text{icts}} > R_{\text{scp}}$ if the average power (or total cell SNR) $\bar{P}$ is above a certain threshold, which is given by

$$\bar{P}_{\text{th}}(\alpha) = \frac{1 - 4\alpha^2}{8\alpha^4}. \tag{2.19}$$

It is observed that for $\alpha > 1/2$, $R_{\text{icts}} > R_{\text{scp}}$ for any value of $\bar{P}$.

The extreme SNR analysis of ICTS in a nonfading scenario leads to

$$S_0 = 1; \quad \frac{E_b}{N_0}_{\min} = \log 2; \quad S_\infty = \tfrac{1}{2}; \quad \mathcal{L}_\infty = -1$$

$$\hat{S}_0 = \tfrac{1}{3}; \quad \frac{\hat{E}_b}{N_0}_{\min} = \log 2; \quad \hat{S}_\infty = \tfrac{1}{6}; \quad \hat{\mathcal{L}}_\infty = -(1 + \log_2 3), \tag{2.20}$$

proving that, as expected, ICTS is not interference limited.

Introducing flat fading, it is easily verified that the rate of the linear model is given by

$$R^*_{\text{icts}} = \frac{1}{2} E \left\{ \log \left(1 + \frac{2\bar{P}}{K} \sum_{k=1}^{K} |a_k|^2 \right) \right\} = \frac{1}{2} E \left\{ \log \left(1 + \frac{\bar{P}}{K} S \right) \right\}, \tag{2.21}$$

where the rate for the intracell TDMA is achieved by setting $K = 1$. For the special case of Rayleigh fading, where $S = 2 \sum_{k=1}^{K} |a_k|^2$ is distributed as $S \sim \chi^2_{2K}$, the rate is shown in [15] to be given by

$$R^*_{\text{icts}} = \frac{(-1)^K}{2\Gamma(K)} \frac{\partial^{K-1}}{\partial \mu^{K-1}} \left[\frac{-1}{\mu} \exp \left(\frac{K\mu}{2\bar{P}} \right) \text{Ei} \left(\frac{K\mu}{2\bar{P}} \right) \right]_{\mu=1}, \tag{2.22}$$

where $\text{Ei}(\cdot)$ is the exponential integral function. It is noted that the rate of the planar model is achieved by dividing (2.22) by 3 and replacing $\bar{P}$ with $3\bar{P}$.

Applying Jensen's inequality to (2.21), it is easily verified that the presence of fading decreases the sum-rate $R^*_{\text{icts}} \le R_{\text{icts}}$. Furthermore, increasing the number of users while keeping $\bar{P}$ constant, and applying the SLLN to (2.21), the rate boils down to (2.17). Hence, as discussed above, the effect of fading is vanishing for the WB protocol with increasing K. Finally, applying Jensen's inequality to (2.21) in a reverse manner reveals that for the ICTS protocol, the WB scheme outperforms the intracell TDMA scheme in the presence of fading. It is noted that the results mentioned in this paragraph are valid for a general zero-mean fading distribution with unit power and also for the planar model.

In the extreme SNR scenarios, and in the presence of Rayleigh fading, characterization of the performance of ICTS reads

$$S_0^* = \frac{1}{1 + \frac{1}{2K}}; \quad \frac{E_b}{N_0}^*_{\min} = \log 2; \quad S_\infty^* = \frac{1}{2};$$

$$\hat{S}_0^* = \frac{1}{3(1 + \frac{1}{2K})}; \quad \frac{\hat{E}_b}{N_0}^*_{\min} = \log 2; \quad \hat{S}_\infty^* = \frac{1}{6};$$

$$\mathcal{L}_\infty^* = \log_2 K + \left(\gamma - \sum_{\ell=1}^{K-1} \frac{1}{\ell}\right) \log_2 e - 1$$

$$\hat{\mathcal{L}}_\infty^* = \log_2 K + \left(\gamma - \sum_{\ell=1}^{K-1} \frac{1}{\ell}\right) \log_2 e - 1 - \log_2 3, \tag{2.23}$$

where γ is the Euler–Mascheroni constant

$$\gamma = \lim_{n \to \infty} \left(\sum_{\ell=1}^{n} \frac{1}{\ell} - \log n\right) \approx 0.5772. \tag{2.24}$$

As discussed above, parameters (2.23) coincide with the ones of a Gaussian scenario (2.20) asymptotically with an increasing number of users K.

In Figure 2.3 the noninterference limited behavior of the linear model ICTS spectral efficiency is demonstrated for $\alpha = 0.4$ and several values of K. The superiority of ICTS over SCP for a fixed transmission power and increasing interference power is demonstrated in Figure 2.4 for $\bar{P} = 10$ [dB] and several values of K. In both figures the curves drawn for Rayleigh fading approach the respective nonfading related curves as K increases. As with SCP protocol, Monte-Carlo simulation is used to produce the curves.

The reader is referred to [15] for a comprehensive study of single-cell processing in the uplink of Wyner's linear and planar models.

2.3.2 Multi-Cell Processing

In this section, joint processing of all signals received by all cell sites is considered. As mentioned earlier, the central receiver is aware of all the users' codebooks and CSI, and the users are not allowed to cooperate. Accounting for the underlying assumptions, the overall channel is a Gaussian MAC with KM single antenna users and M distributed antenna receivers. Assuming an optimal decoding the per-cell sum-rate capacity of the linear model is given by

$$C_{\text{mcp}}^* = \frac{1}{M} E\left\{\log \det\left(I + \frac{\bar{P}}{K} HH^\dagger\right)\right\}$$

$$= E\left\{\frac{1}{M} \sum_{m=1}^{M} \log\left(1 + \frac{\bar{P}}{K} \lambda_m(HH^\dagger)\right)\right\} \tag{2.25}$$

$$= E\left\{\int_0^\infty \log\left(1 + \frac{\bar{P}}{K} x\right) dF_{HH^\dagger}^M(x)\right\},$$

where $\boldsymbol{H}$ is the channel transfer matrix defined in (2.8), $\{\lambda_m(\boldsymbol{HH}^\dagger)\}_{m=1}^M$ are the non-negative eigenvalues of the semipositive definite (SPD) matrix $\boldsymbol{HH}^\dagger$, and $F_{\boldsymbol{HH}^\dagger}^M(x)$ denotes the empirical cumulative distribution function (or *spectrum*) of $\{\lambda_m(\boldsymbol{HH}^\dagger)\}_{m=1}^M$, defined as

$$F_{\boldsymbol{HH}^\dagger}^M(x) = \frac{1}{M} \sum_{m=1}^M 1_{\left\{\lambda_m(\boldsymbol{HH}^\dagger) \le x\right\}}, \qquad (2.26)$$

and $1_{\{\cdot\}}$ is the indicator function. The spectrum of the eigenvalues of $\boldsymbol{HH}^\dagger$ plays a key role in the sum-rate capacity calculation. Unfortunately, the power profile of $\boldsymbol{H}$ for finite K does not converge uniformly as the matrix dimensions increase [4]. Hence, *Girko*'s law for the eigenvalues of large random matrices cannot be applied to evaluate the sum-rate of (2.25) [4].

The system topology, according to which interferences arise only from adjacent cells, is well-reflected in the five-diagonal structure of $\boldsymbol{HH}^\dagger$, which is explicitly given by

$$\left[\boldsymbol{HH}^\dagger\right]_{(m,n)} = \begin{cases} \sum_{k=1}^K \left(|a_k^n|^2 + \alpha^2 |b_k^n|^2 + \alpha^2 |c_k^n|^2\right) & (n, n) \\ \alpha \sum_{k=1}^K \left(a_k^n(b_k^{n+1})^\dagger + c_k^n(a_k^{n+1})^\dagger\right) & (n, n+1) \\ \alpha \sum_{k=1}^K \left(a_k^n(c_k^{n-1})^\dagger + b_k^n(a_k^{n-1})^\dagger\right) & (n, n-1) \\ \alpha^2 \sum_{k=1}^K c_k^n(b_k^{n+2})^\dagger & (n, n+2) \\ \alpha^2 \sum_{k=1}^K b_k^n(c_k^{n-2})^\dagger & (n, n-2) \\ 0 & \text{otherwise} \end{cases}, \qquad (2.27)$$

where out-of-range indices should be ignored.

The rest of this section is dedicated to the evaluation of (2.25) under various conditions. Starting with the nonfading setup, already treated by Wyner in [6], it is easily verified that $\frac{1}{K}\boldsymbol{HH}^\dagger$ becomes a five-diagonal Toeplitz matrix

$$\frac{1}{K}\left[\boldsymbol{HH}^\dagger\right]_{(m,n)} = \begin{cases} (1 + 2\alpha^2) & (n, n) \\ 2\alpha & (n, n\pm 1) \\ \alpha^2 & (n, n\pm 2) \\ 0 & \text{otherwise} \end{cases}. \qquad (2.28)$$

Applying *Szego*'s theorem regarding the limit spectrum of infinite Toeplitz matrices [16], the per-cell sum-rate capacity for the infinite linear array is given by

$$R_{\text{mcp}} \underset{M\to\infty}{=} \int_0^1 \log\left\{1 + \bar{P}(1 + 2\alpha\cos(2\pi\theta))^2\right\} d\theta. \qquad (2.29)$$

Because the entries of $\frac{1}{K}\boldsymbol{HH}^\dagger$ are independent of K for a fixed $\bar{P}$, this rate is achievable (not uniquely) by intracell TDMA and WB protocols. Applying the two-dimensional version of *Szego*'s theorem, the per-cell sum-rate capacity of the planar model is given by

$$\hat{R}_{\text{mcp}} \underset{M\to\infty}{=} \int_0^1 \int_0^1 \log\left\{1 + \bar{P}(1 + 2\alpha F(\theta_1, \theta_2))^2\right\} d\theta_1 d\theta_2, \qquad (2.30)$$

where

$$F(\theta_1, \theta_2) \triangleq \cos(2\pi\theta_1) + \cos(2\pi\theta_2) + \cos(2\pi(\theta_1 + \theta_2)). \tag{2.31}$$

Extreme SNR characterization of MCP over Gaussian channels is summarized for the linear and planar models for $M \to \infty$, as follows

$$S_0 = \frac{2(1+2\alpha^2)^2}{1+12\alpha^2+6\alpha^4}; \qquad \frac{E_b}{N_0}_{\min} = \frac{\log 2}{1+2\alpha^2}$$

$$S_\infty = 1; \qquad \mathcal{L}_\infty = \begin{cases} -2\log_2\left(\frac{1+\sqrt{1-4\alpha^2}}{2}\right) & 0 \le \alpha \le \frac{1}{2} \\ -2\log_2 \alpha & \alpha > \frac{1}{2} \end{cases} \tag{2.32}$$

$$\hat{S}_0 = \frac{2(1+6\alpha^2)^2}{1+36\alpha^2+48\alpha^3+90\alpha^4}; \qquad \frac{\hat{E}_b}{N_0}_{\min} = \frac{\log 2}{1+6\alpha^2}$$

$$\hat{S}_\infty = 1; \qquad \hat{\mathcal{L}}_\infty = -2\int_0^1\int_0^1 \log_2|1+2\alpha F(\theta_1, \theta_2)|\, d\theta_1 d\theta_2.$$

Examining the power offset of the linear model, it is concluded that in the high-SNR regime and in the absence of fading, high intercell interference ($\alpha \approx 1$) or low intercell interference ($\alpha \approx 0$) is favorable, while intermediate intercell interference ($\alpha \approx 1/2$) provides the worst case scenario in terms of the spectral efficiency. This conclusion is well-demonstrated in Figure 2.4 for $\bar{P} = 10$ [dB]. Numerical calculations (not presented here) show that for $\bar{P} < 1$ (or 0 [dB]), the sum rate increases with α. Because MCP is the capacity achieving strategy, its rates are superior to both SCP and ICTS protocols under any condition, as is demonstrated in Figures 2.3 and 2.4.

The analysis of the fading case for finite K is impaired by the fact that the spectrum of the resulting finite band matrix $\boldsymbol{HH}^\dagger$ is currently not known, even for increasing matrix dimensions $M \to \infty$. Therefore, techniques such as bounding, asymptotics, and extreme SNR analysis are used in order to get further insight into the performance merits of interest. These alternatives are explored below. Nevertheless, a simple observation can be achieved by applying Jensen's inequality in a reverse manner to the first equality of (2.25). Accordingly, it is concluded that in the presence of fading, the MCP WB protocol ($K > 1$) outperforms the MCP intracell TDMA protocol ($K = 1$) [17]. This observation is demonstrated for Rayleigh fading in Figures 2.3 and 2.4 where the curves of the MCB WB protocol surpass the the curves of the MCP intracell TDMA protocol. It is noted that both curves are produced by Monte-Carlo simulations of a system with $M = 40$ cells.

2.3.2.1 Moment Bounds

Turning to the flat fading setup, it is easily verified that the n-th moment of the spectrum of $\boldsymbol{HH}^\dagger$ is given by

$$\mathcal{M}_n = \lim_{M\to\infty} \frac{1}{M} \text{tr}\{(\boldsymbol{HH}^\dagger)^n\}, \tag{2.33}$$

where the multiplication of matrices $(\boldsymbol{HH}^\dagger)^n$ may be interpreted as the weight summation of paths over a restricted grid (see [17] for more details). Where the amplitude of an individual fading coefficient is statistically independent of its uniformly distributed phase (e.g., *Rayleigh* fading) and intracell TDMA protocol is deployed ($K = 1$), a symbolic mathematics software has been reported in [17] to obtain the limit moments. For example, listed below are the first three limit moments:

$$\mathcal{M}_1 = m_2 + 2m_2\alpha^2$$

$$\mathcal{M}_2 = m_4 + 8m_2^2\alpha^2 + (4m_2^2 + 2m_4)\alpha^4$$

$$\mathcal{M}_3 = m_6 + (6m_2^3 + 12m_2m_4)\alpha^2 + (36m_2^3 + 12m_2m_4)\alpha^4$$

$$+ (6m_2^3 + 12m_2m_4 + 2m_6)\alpha^6, \tag{2.34}$$

where m_i is the i-th moment of the amplitude of an individual fading coefficient. When the limit moments do not grow too fast $\sqrt[2n]{\mathcal{M}_{2n}} = O(n)$, then, according to theorem 3.12 of [18], the spectrum of (2.26) converges weakly to a *unique* limit spectrum (defined by the limit moments)

$$F^M_{\boldsymbol{HH}^\dagger}(x) \xrightarrow[M\to\infty]{d} F_{\boldsymbol{HH}^\dagger}(x). \tag{2.35}$$

Rayleigh fading satisfies this last condition. Identifying that the triplex, {$\log(1 + \bar{P}x)$, $\{x^l\}_{l=1}^n$, $F_{\boldsymbol{HH}^\dagger}(x)$}, forms a T^0_+ *Tchebychev* system, the upper and lower principle representations (see [19]) of $F_{\boldsymbol{HH}^\dagger}(x)$ can be used to produce an analytical lower and upper bound to the per-cell sum-rate capacity. For example, listed below are the lower and upper bound of order $n = 2$ derived by this method

$$\frac{(\mathcal{M}_1)^2}{\mathcal{M}_2} \log\left(1 + \bar{P}\frac{\mathcal{M}_2}{\mathcal{M}_1}\right) \leq R^*_{\mathrm{msp}} \leq \log\left(1 + \bar{P}\mathcal{M}_1\right), \tag{2.36}$$

where $\mathcal{M}_1$ and $\mathcal{M}_2$ are the first and second limit moments of $F_{\boldsymbol{HH}^\dagger}(x)$ given in (2.34). Applying *Jensen's* inequality to the first equality of (2.25) yields the upper bound of (2.36). For higher orders (higher values of n), it is necessary to use symbolic mathematics software in order to find the probability masses of the upper and lower principal representation and their locations [17]. Examining the moment bounds (orders $n = 8$, 10) calculated for Rayleigh fading and presented in Figure 2.4 reveals that the bounds are uniform in the interference factor α and get tighter as the order n increases. Additional calculations [17] (not presented here) show that the bounds are tighter when $\bar{P}$ decreases. In addition, in contrast to the SCP and ICTS protocols, the rate of MCP intracell TDMA ($K = 1$) increases in the presence of Rayleigh fading for certain values of α and $\bar{P}$. It is noted that the moment bounding technique is applicable (although in a tedious manner) for any finite K and also for the planar model.

2.3.2.2 Asymptotes with the Number of Users K

Focusing on the case where the number of users is large while $\bar{P}$ is kept constant, and applying the SLLN, the entries of $\frac{1}{K}\boldsymbol{HH}^\dagger$ consolidate almost surely to their mean values

$$\frac{1}{K}\left[\boldsymbol{HH}^\dagger\right]_{(m,n)} \xrightarrow[K\to\infty]{a.s.} \begin{cases} 1 + 2\alpha^2 & (n, n) \\ 2\alpha(1 - \sigma_a^2) & (n, n\pm 1) \\ \alpha^2(1 - \sigma_a^2) & (n, n\pm 2) \\ 0 & \text{otherwise} \end{cases}, \tag{2.37}$$

where $0 \leq \sigma_a^2 \leq 1$ is the variance of an individual unit gain fading coefficient. Hence, in the limit where K is large, $\frac{1}{K}\boldsymbol{HH}^{\dagger}$ becomes Toeplitz and by applying Szego's theorem, the per-cell sum-rate capacity of the infinite linear model is approximated by

$$R_{\mathrm{msp}}^{*} \underset{M, K \to \infty}{=} \int_0^1 \log\left\{1 + \bar{P}\left[\sigma_a^2(1 + 2\alpha^2)\right.\right.$$

$$\left.\left. + (1 - \sigma_a^2)(1 + 2\alpha\cos(2\pi\theta))^2\right]\right\} d\theta. \tag{2.38}$$

Adhering to similar argumentation and applying the two-dimensional Szego's theorem, the per-cell sum-rate capacity of the infinite planar model is approximated by

$$\hat{R}_{\mathrm{mcp}}^{*} \underset{M, K \to \infty}{=} \int_0^1 \int_0^1 \log\left\{1 + \bar{P}\left[\sigma_a^2(1 + 6\alpha^2)\right.\right.$$

$$\left.\left. + (1 - \sigma_a^2)(1 + 2\alpha F(\theta_1, \theta_2))^2\right]\right\} d\theta_1 d\theta_2. \tag{2.39}$$

Because the same results, derived for a general fading distribution, are achieved by applying the Jensen's inequality to (2.25), it is concluded that the asymptotic expressions upper bound the sum rates achieved for any finite number of users K. Another observation is that both expressions *increase* with σ_a^2 [17]; it is concluded that for MCP WB protocol and large numbers of users per cell $K \gg 1$, the presence of fading is beneficial under any condition. This performance enhancement is due to the independence of the three fading processes affecting the signal of each user, as observed by the three receiving cell sites, which explains why mimicking artificial fading at the users' transmitters fails to produce the same impact [17]. The sum rates (2.38) and (2.39) are maximized for $\sigma_a^2 = 1$, which corresponds to a zero-mean unit gain fading distribution (e.g., Rayleigh fading) and is given by

$$R_{\mathrm{msp}}^{*} \underset{\substack{M, K \to \infty \\ \sigma_a^2 = 1}}{=} \log\left(1 + (1 + 2\alpha^2)\bar{P}\right), \tag{2.40}$$

for the infinite linear model, and by

$$\hat{R}_{\mathrm{msp}}^{*} \underset{\substack{M, K \to \infty \\ \sigma_a^2 = 1}}{=} \log\left(1 + (1 + 6\alpha^2)\bar{P}\right), \tag{2.41}$$

for the planar model. Examining (2.40), a *resource pooling* effect is revealed, as this rate coincides with the rate of an equivalent single-user nonfading link, but with a channel gain of $(1 + 2\alpha^2)$. The latter gain reflects the array power gain of the linear system $((1 + 6\alpha^2)$ for the planar model).

Figures 2.3 and 2.4 demonstrate the tightness of the asymptotic expression (and upper bound) (2.40) for a moderate number of users K and also that the presence of fading is beneficial for the MCP WB protocol, with large K for all values of $\bar{P}$ and α where the asymptotic curves surpass the nonfading corresponding curves.

2.3.2.3 Extreme SNR Analysis

Extreme SNR characterization of the MCP protocol per-cell sum-rate is summarized for the linear model in the presence of Rayleigh fading and $M \to \infty$, as follows:

$$S_0^* = \frac{2K}{1+K}; \quad \frac{E_b^*}{N_0}_{\min} = \frac{\log 2}{1+2\alpha^2};$$

$$\frac{1}{2} \leq S_\infty^* \leq 1; \quad -1 - \log_2(1+2\alpha^2) \leq \mathcal{L}_\infty^* \leq -\log_2(1+2\alpha^2). \tag{2.42}$$

Surprisingly, the low-SNR slope S_0^* is independent of α which comes in contrast to the low-SNR slope of the corresponding nonfading setup (2.54). Comparing the parameters of the MCP for the fading and nonfading linear setups (given in (2.42) and (2.54), respectively), reveals that in the low-SNR regime the presence of Rayleigh fading is already beneficial for K values above a threshold given by

$$K_t(\alpha) = \left\lceil \frac{(1+2\alpha^2)^2}{2\alpha^2(4+\alpha^2)} \right\rceil, \tag{2.43}$$

which is a decreasing function of α. Accordingly, in the low SNR regime the presence of Rayleigh fading is beneficial for the MCP intracell TDMA protocol ($K = 1$), for $\alpha \gtrsim 0.54$. It is noted that the bounds in (2.42) for the high SNR regime parameters are derived by the fact that the MCP WB protocol sum-rate increases with K in the presence of fading and is upper bounded for $K \to \infty$ by (2.40). Hence, the second-order moment bounds (derived for $K = 1$) (2.36) and (2.40) are a lower and an upper bound on the sum-rate, respectively. However, a closed form expression for the MCP protocol high-SNR parameter in the presence of fading for finite K is still an open problem.

2.4   THE DOWNLINK CHANNEL

Collecting the baseband signals received by all the terminals in the system at a given time instant, in a $MK \times 1$ vector $\boldsymbol{y}$, the input-output equation for the downlink of the linear Wyner model with M cells, reads

$$\boldsymbol{y} = \boldsymbol{H}^\dagger \boldsymbol{x} + \boldsymbol{z}. \tag{2.44}$$

The channel transfer matrix $\boldsymbol{H}$ is defined as in (2.7), with $\boldsymbol{a}_m$, $\boldsymbol{b}_m$, and $\boldsymbol{c}_m$ defining the $1 \times K$ row vectors denoting the channel complex fading coefficients, experienced by the K users of the m-th, $(m-1)$-th, and $(m+1)$-th cells, respectively, when receiving the transmissions of the m-th cell-site antenna. Full CSI is assumed available to the joint multi-cell transmitter. The latter assumption implies the availability of a feedback channel from the users to the APs. On the other hand, the mobile receivers are assumed to be cognizant of their own CSI, and of the employed transmission strategy. In addition, $\boldsymbol{x}$ is the $M \times 1$ complex Gaussian vector of signals transmitted by the M cell-sites $\boldsymbol{x} \sim \mathcal{CN}(0, \boldsymbol{P})$. As explained in Section 2.2, an *equal individual per-cell power constraint* of $[\boldsymbol{P}]_{(m,m)} \leq \bar{P} \; \forall m$ is assumed. Lastly, $\boldsymbol{z}$ is the $MK \times 1$ complex Gaussian additive noise vector $\boldsymbol{z} \sim \mathcal{CN}(0, \boldsymbol{I}_{MK})$.

2.4.1 Single-Cell Processing

Starting with the Gaussian case, the per-cell sum-rates for the linear and planar models are equal to the respective uplink rates, (2.9) and (2.10). In the presence of Rayleigh fading, and assuming each cell site is aware of its users' instantaneous channel gain and instantaneous interference power, a suboptimal scheme that schedules the "best" user for transmission in each cell is considered. The following rate is achieved for the linear model:

$$R^*_{\text{scp}} = E\{\log(1 + \max_k \text{SNR}_k)\}, \tag{2.45}$$

where the i.i.d. random variable SNRs are given by

$$\text{SNR}_k = \frac{\bar{P}|a_k|^2}{1 + \alpha^2 \bar{P}(|b_k|^2 + |c_k|^2)}. \tag{2.46}$$

Following [14], the probability density function (PDF) and cumulative distribution function (CDF) of an arbitrary SNR_k (2.46) are given by

$$f_s(x) = \frac{\alpha^2 e^{-x/\bar{P}}}{(1 + \alpha^2 x)^3}\left(\frac{1}{\alpha^2 \bar{P}}(1 + \alpha^2 x) + 2\right); \quad x \geq 0, \tag{2.47}$$

and

$$F_s(x) = 1 - \frac{e^{-x/\bar{P}}}{(1 + \alpha^2 x)^2}; \quad x \geq 0. \tag{2.48}$$

Because $\max_k \text{SNR}_k$ is distributed according to $f_{\hat{s}}(x) = K f_s(x)(F_s(x))^{K-1}$, the per-cell sum-rate (2.45) is given by

$$R^*_{\text{scp}} = K \int_0^\infty \log(1 + x)\left(1 - \frac{e^{-x/\bar{P}}}{(1 + \alpha^2 x)^2}\right)^{K-1}$$

$$\times \frac{\alpha^2 e^{-x/\bar{P}}}{(1 + \alpha^2 x)^3}\left(\frac{1}{\alpha^2 \bar{P}}(1 + \alpha^2 x) + 2\right) dx. \tag{2.49}$$

Adhering to similar argumentation, the per-cell sum-rate of the planar model is given by

$$\hat{R}^*_{\text{scp}} = K \int_0^\infty \log(1 + x)\left(1 - \frac{e^{-x/\bar{P}}}{(1 + \alpha^2 x)^4}\right)^{K-1}$$

$$\times \frac{\alpha^2 e^{-x/\bar{P}}}{(1 + \alpha^2 x)^5}\left(\frac{1}{\alpha^2 \bar{P}}(1 + \alpha^2 x) + 4\right) dx. \tag{2.50}$$

For an increasing number of users K and fixed $\bar{P}$, the maximum SNR in the linear and planar models behaves with high probability like $\bar{P}\log K + O(\log\log K)$ [14]. Hence, for fixed $\bar{P}$, the SCP sum-rate of both models behaves like

$$R^*_{\text{scp}} = \hat{R}^*_{\text{scp}} \underset{K \gg 1}{\cong} \log(1 + \bar{P}\log K), \tag{2.51}$$

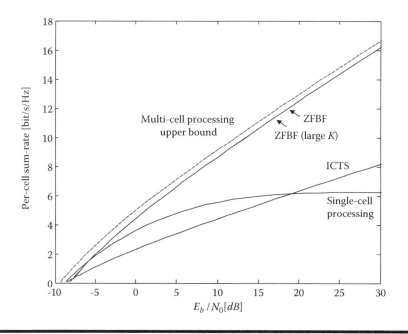

Figure 2.5 **Per-Cell Sum-Rates Versus E_b/N_0 in Wyner's Downlink Channel ($K = 100$, $\alpha = 0.4$).**

and a scaling law of $\log \log K$ with an increasing number of users per cell is revealed. On the other hand, for any finite K, the rates demonstrate an interference limited behavior. Hence, $S_\infty^* = \hat{S}_\infty^* = 0$. The interference limited behavior of the SCP protocol (linear model) in the presence of Rayleigh fading is demonstrated in Figure 2.5 using Monte-Carlo simulations for $K = 100$ users and $\alpha = 0.4$. As with the SCP protocol, the per-cell sum-rates of the ICTS protocol for the linear and planar models are equal to the respective uplink rates, (2.17) and (2.18). In the presence of Rayleigh fading, the following sum-rate is achieved by scheduling the "best" user for transmission [20]

$$R_{\text{icts}}^* = \frac{K}{2} \int_0^\infty \log(1 + 2\bar{P}x)e^{-x}\left(1 - e^{-x}\right)^{K-1} dx. \tag{2.52}$$

Because $\max_{1 \le k \le K} |a_k|^2$ (where $|a_k|^2 \sim \chi_2^2 \ \forall k$ and i.i.d.) behaves with high probability like $\log K + O(\log \log K)$ for $K \gg 1$ [14], (2.52) is well-approximated for a large number of users per cell, by

$$R_{\text{icts}}^* \underset{K \gg 1}{\cong} \frac{1}{2} \log(1 + 2\bar{P} \log K). \tag{2.53}$$

The respective rates for the planar model are obtained by replacing the constant 2 by 6, which reflects the fact that each cell is active $1/6$ of the time. It is concluded that for fixed $\bar{P}$, the ICTS per-cell sum-rates of the linear and planar models scale as $\frac{1}{2} \log \log K$ and $\frac{1}{6} \log \log K$, respectively. The high SNR regime of the ICTS protocol sum-rate per cell is characterized for large K by

$$\begin{aligned} S_\infty^* &= \tfrac{1}{2}; \quad \mathcal{L}_\infty^* = -\log_2 \log K - 1 \\ \hat{S}_\infty^* &= \tfrac{1}{6}; \quad \hat{\mathcal{L}}_\infty^* = -\log_2 \log K - 1 - \log_2 3. \end{aligned} \tag{2.54}$$

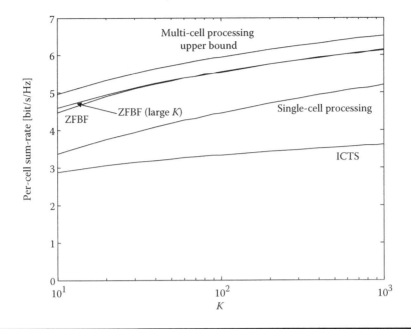

Figure 2.6   Per-Cell Sum-Rates Versus Number of Users Per-Cell K in Wyner's Downlink Channel ($\alpha = 0.4$, $\bar{P} = 10$ **dB**).

The inherent noninterference behavior and high SNR slope of the ICTS protocol are demonstrated in the presence of Rayleigh fading in Figure 2.5, using Monte Carlo simulations for $K = 100$ and $\alpha = 0.4$. The inferior scaling law of the ICTS protocol, in comparison to the SCP protocol (linear model), is demonstrated in Figure 2.6, using Monte Carlo simulations for $\alpha = 0.4$ and $\bar{P} = 10$ [dB].

2.4.2   Multi-Cell Processing

The key tool, used in the following analysis, is a recent result by Yu and Lan [21], who established a connection between the uplink–downlink duality of the Gaussian vector MAC and BC, and the Lagrangian duality in minimax optimization. Accordingly, the sumrate capacity of Wyner's downlink channel with per-cell power constraint equals the sum-rate capacity of its dual uplink channel, subject to an overall sum power constraint determining the level of cooperation between the users, and a noise power constraint capturing the equal per-cell power constraints of the original downlink channel. Hence, the per-cell sum-rate capacity of the downlink channel is given by

$$R_{\text{msp}}^* = \frac{1}{M} E_H \left\{ \min_{\boldsymbol{\Lambda}} \max_{\boldsymbol{D}} \log \frac{\det\left(\boldsymbol{HDH}^\dagger + \boldsymbol{\Lambda}\right)}{\det\left(\boldsymbol{\Lambda}\right)} \right\}, \tag{2.55}$$

where the optimization is over all nonnegative $MK \times MK$ and $M \times M$ *diagonal* matrices, $\boldsymbol{D}$ and $\boldsymbol{\Lambda}$, satisfying $\text{tr}(\boldsymbol{D}) \leq M\bar{P}$ and $\text{tr}(\boldsymbol{\Lambda}) \leq M$, respectively.

This optimization problem was recently studied in [22,23] for the *circular* Wyner model. For the nonfading setup, it is proved that the rate (2.55) is equal to the per-cell sum-rate capacity of the uplink channel (2.29) [23]:

$$R_{mcp} \underset{M \to \infty}{=} \int_0^1 \log\left(1 + \bar{P}(1 + 2\alpha\cos(2\pi\theta))^2\right) d\theta. \qquad (2.56)$$

This rate is achieved by *dirty paper coding* (DPC) techniques [2,3]. The equivalence between the downlink and uplink channels is shown to hold for *any* circulant channel transfer matrix [23]. The proof of these results is based on the symmetry induced by the nonfading circulant channel transfer matrix, and on a result by [24], in which the inner term of (2.55) is convex in Λ and concave in D (for more details see [23]).

For Rayleigh flat fading channels, upper and lower (achievable rate) bounds are derived, while focusing on the asymptotic regime in terms of the number of users per cell. The achievable rate is obtained by employing a power control scheme in the *dual uplink channel*, according to which only users received at their local cell-site with fade power levels exceeding some threshold L, are active. As $K \to \infty$, the number of active users per cell crystallizes to $K_0 = Ke^{-L}$, and it is assumed that all active users transmit at equal powers $\bar{P}/K_0$. The constant L should be chosen, so that $K_0 \to \infty$ as $K \to \infty$ (and thus, the SLLN can be applied). In particular, for $K_0 = Ke^{-L} = K^\epsilon$, the threshold is set to $L = (1 - \epsilon)\log K$, where $0 < \epsilon < 1$. The resulting large K achievable rate can be shown to constitute an upper bound for *any finite* K. The threshold-crossing scheduling scheme used in the virtual dual uplink channel is an analysis tool and has little in common with the actual DPC used to achieve the sum-rate capacity.

The capacity upper bound is derived by bounding the channel fades by the strongest fading gain (over all intracell users) received at each cell site, and observing that the maximum of K i.i.d. χ_2^2 distributed random variables behaves like $\log K + O(\log\log K)$ for $K \gg 1$ [14]. Combining the two bounds, for $K \gg 1$, the downlink per-cell sum-rate capacity satisfies,

$$\log\left(1 + \bar{P}\left((1 - \epsilon)\log K + 1 + 2\alpha^2\right)\right) \le R^*_{mcp}$$

$$\le \log(1 + (1 + 2\alpha^2)\bar{P}\log K), \qquad (2.57)$$

for some $\epsilon \underset{K \to \infty}{\to} 0$. Examining (2.57), it is observed that in addition to a noninterference limited behavior, both bounds increase with α. Moreover, both bounds demonstrate multi-user diversity gains of $\log K$ and scale like $\log\log K$ as K increases. In addition, the upper bound predicts an array power gain of $(1 + 2\alpha^2)$ in addition to the multi-user diversity gain. It is noted that the gap between the two bounds converges to $\log(1 + 2\alpha^2)$ [nat/channel use] as the number of users per cell increases. Finally, the high SNR regime characterization of R^*_{mcp} for large K is summarized as follows:

$$\mathcal{S}^*_\infty = 1 - \log_2\left(1 + 2\alpha^2\right) - \log_2\log K \le \mathcal{L}^*_\infty \le -\log_2\left((1 - \epsilon)\log K + 1 + 2\alpha^2\right). \quad (2.58)$$

In comparison to the respective nonfading setup parameters of (2.54), the presence of Rayleigh fading does not change the high SNR slope, but a multi-user diversity gain of $\log K$ is observed in the power offset.

It is noted that finding an exact expression for per-cell sum-rate capacity of the Wyner downlink channel is still an open problem.

2.4.3 Distributed Zero Forcing Beamforming

In the previous section, bounds on the ultimate per-cell sum-rate under per-cell equal power constraint have been presented. In this section, as a practical alternative to optimal DPC schemes, zero forcing beamforming (ZFBF) with a simple scheduling is considered. ZFBF is an attractive scheduling scheme because it demonstrates noninterference limited behavior and requires a single user coding–decoding scheme. In addition, with N transmit antennas and K users, the sum-rate achieved by ZFBF to a "semiorthogonal" subset of N users, demonstrates the same scaling law as the sum-rate achieved by the optimal DPC scheme for an increasing number of users [25].

Consider Wyner's circular model, where $M > 2$ cells with K users each are arranged on a circle. Assuming an intracell TDMA scheme, in which only one user is selected for transmission per cell, the $M \times 1$ vector baseband representation of the signals received by the *selected* users is given for an arbitrary time index by

$$y = H^\dagger B u + z, \tag{2.59}$$

where u is the $M \times 1$ complex Gaussian symbols vector $u \sim \mathcal{CN}(0, I_M)$, B is the beamforming $M \times M$ matrix:

$$B = \sqrt{\frac{M\bar{P}}{\mathrm{tr}\left((HH^\dagger)^{-1}\right)}} \left(H^\dagger\right)^{-1}, \tag{2.60}$$

with $M\bar{P}$ being the overall average transmit power constraint, which is ensured by definition,[2] and z is the $M \times 1$ complex Gaussian additive noise vector $z \sim \mathcal{CN}(0, I_M)$. Substituting (2.60) into (2.59), the received signal vector reduces to

$$y = \sqrt{\frac{M\bar{P}}{\mathrm{tr}\left((HH^\dagger)^{-1}\right)}} \, u + z. \tag{2.61}$$

Because (2.61) can be interpreted as a set of M identical independent parallel single-user channels, its ergodic achievable sum-rate per channel (or cell) is given by

$$R^*_{\mathrm{zfbf}} = E\left\{\log\left(1 + \frac{M\bar{P}}{\mathrm{tr}\left((HH^\dagger)^{-1}\right)}\right)\right\}. \tag{2.62}$$

Although an overall power constraint is assumed, a more natural choice for a cellular system is to maintain per-cell power constraints. Hence, we are interested in the transmitted power of an arbitrary cell, which is averaged over the TDMA time slot duration (many symbols) and is a function of the realization of H,

$$\bar{P}_m = [BB^\dagger]_{m,m} = \frac{M\bar{P}\left[(HH^\dagger)^{-1}\right]_{m,m}}{\mathrm{tr}\left((HH^\dagger)^{-1}\right)}. \tag{2.63}$$

---

[2] Later on it is argued that, under certain conditions, this scheme satisfies a per-cell average power constraint as well.

For nonfading channels, round-robin scheduling is deployed and there is no need to feed back the channel coefficients. In addition, for each time slot the channel transfer matrix becomes circulant, with $(1, \alpha, 0, \dots 0, \alpha)$ as a first row. Applying Szego's theorem regarding the spectrum of Toeplitz matrices [16], the per-cell sum-rate of the ZFBF scheme is given for $\alpha < 1/2$, by

$$R_{\text{zfbf}} \underset{M \to \infty}{=} \log\left(1 + \mathcal{F}(\alpha)\,\bar{P}\right), \tag{2.64}$$

where

$$\mathcal{F}(\alpha) \overset{\Delta}{=} \frac{1}{\int_0^1 (1 + 2\alpha\cos(2\pi\theta))^{-2}\,d\theta}. \tag{2.65}$$

This rate holds for an overall power constraint $M\bar{P}$, and for an *equal per-cell power constraint* $\bar{P}$. It is easily verified that $0 < \mathcal{F}(\alpha) \leq 1$ and that it is a decreasing function of the interference factor α. Comparing (2.64) to (2.17), it is clear that the ZFBF scheme is superior to the ICTS scheme when the SNR $\bar{P}$ is above a certain threshold

$$\bar{P}_t(\alpha) = \frac{2(1 - \mathcal{F}(\alpha))}{(\mathcal{F}(\alpha))^2}, \tag{2.66}$$

which is an increasing function of α. It is noted that for $\alpha = 1/2$ the circulant channel transfer matrix $\boldsymbol{H}$ is singular and channel inversion methods such as ZFBF are not applicable. Moreover, $\boldsymbol{H}$ is not guaranteed to be nonsingular for $\alpha > 1/2$ and any finite number of cells M. The extreme SNR characterization of the ZFBF scheme is summarized by

$$S_0 = 2; \quad \frac{E_b}{N_0}_{\min} = \frac{\log 2}{\mathcal{F}(\alpha)}; \quad S_\infty = 1; \quad \mathcal{L}_\infty = -\log_2 \mathcal{F}(\alpha), \tag{2.67}$$

proving a two-fold rate gain in the high SNR regime when compared to the ICTS protocol (2.20).

Turning to the Rayleigh fading setup, a simple scheduling algorithm selects, for each fading block (or TDMA slot), the user with the "best" local channel for transmission in each cell. In other words, the selected user in the m-th cell is

$$\tilde{k}(m) = \underset{k}{\text{argmax}}\{|a_{m,k}|^2\}, \tag{2.68}$$

where $\{a_{m,k}\}_{k=1}^K$ are the fading coefficients of the m-th cell transmitted signals as they are received by the m-th cell users. The resulting channel transfer matrix $\boldsymbol{H}^\dagger$ of this suboptimal scheduling consists of diagonal entries $a_m = a_{m,\tilde{k}(m)}$, of which their powers are the *maximum* of K i.i.d. χ_2^2 distributed random variables. The other two diagonals entries of $\boldsymbol{H}^\dagger$ are zero mean unit variance complex Gaussian distributed random variables multiplied by α.

When $\boldsymbol{H}$ is ill-conditioned, the joint beamformer can start replacing the "best" users by their second "best" users until the resulting $\boldsymbol{H}$ is well-behaved. Because we assume that $K \gg 1$, the overall statistics are not expected to change by using this replacement procedure.

The special structure of the channel transfer matrix $\boldsymbol{H}^\dagger$ resulting from the setup topology and scheduling procedure plays a key role in understanding the asymptotic scaling law of the scheme's per-cell sum-rate R^*_{zfbf} (expression (2.62)), which is asymptotically optimal with an increasing number of users per cell [26]:

$$\frac{R^*_{\text{zfbf}}}{R^*_{\text{msp}}} \xrightarrow[K \to \infty]{} 1, \tag{2.69}$$

where R^*_{msp}, which scales like $\log \log K$, is the per-cell sum-rate capacity of the channel and is bounded for large K in (2.57). Due to the scheduling process, this result can be intuitively explained by the fact that $(\boldsymbol{HH}^{\dagger})$ "becomes" diagonal $(\log K \boldsymbol{I}_M)$ when K increases. Accordingly, for large K, $(\boldsymbol{HH}^{\dagger})^{-1}$ "behaves" like $(\boldsymbol{I}_M/\log K)$, and R^*_{zfbf} (expression (2.62)) is approximated by

$$R^*_{\text{zfbf}} \underset{K \gg 1}{\cong} \log(1 + \bar{P}\log K). \tag{2.70}$$

It is also concluded that, in the presence of Rayleigh fading, the ZFBF scheme provides a twofold scaling law rather than the per-cell sum-rate scaling law of the ICTS scheme (2.53). Moreover, by definition the sum-rate of the ZFBF scheme ensures a noninterference limited behavior for any number of users K (not necessarily large) in contrast to the rate achieved by the SCP protocol. The high SNR characterization of the ZFBF scheme is summarized by

$$S^*_{\infty} = 1; \quad \mathcal{L}^*_{\infty} = -\log_2 \log K, \tag{2.71}$$

Finally, the ZFBF scheme that maintains an overall power constraint of $M\bar{P}$, is shown in [26] to ensure, *in probability*, an equal per-cell power constraint of $\bar{P}$ asymptotically with an increasing number of users per cell. Hence,

$$Pr\left(\bar{P}_m \leq \bar{P} + \epsilon\right) \xrightarrow[K \to \infty]{} 1, \tag{2.72}$$

for an arbitrary small $\epsilon > 0$. As mentioned earlier, for cellular systems, an individual per-cell power constraint is a more reasonable choice than a sum-power constraint, which is more suitable for collocated antenna arrays.

In Figure 2.5, the spectral efficiencies of the ZFBF scheme (Monte Carlo simulation for $M = 40$), along with its asymptotic expression (2.70) and the sum-rate capacity upper bound (2.57), are plotted in the presence of Rayleigh fading for $K = 100$ and $\alpha = 0.4$. A good match between the asymptotic expression and the Monte Carlo simulation resulting curve is observed. It is noted that this match degrades for lower K or larger values of α (not presented here). The gap between the ZFBF curves and the per-cell sum-rate capacity upper bound is clearly explained by the fact that the ZFBF scheme does not use the antenna array to enhance the reception power but to eliminate intercell interferences. Hence, the additional array power gain of $(1 + 2\alpha^2)$ predicted by the upper bound can not be achieved. In addition, the performance gain of the ZFBF scheme over the SCP and ICTS protocols is clearly evident. In Figure 2.6, the scaling law of the scheme with K is presented by Monte Carlo simulations, and is a good match to the asymptotic expression; however, the match degrades when the number of users decreases.

The reader is referred to [26] for more details on the analysis and derivations.

2.5 COOPERATION AMONG TERMINALS

In this section, the benefits of cooperative transmission among terminals in DAS is considered. A simple extension of the Wyner model that allows intracell terminal-to-terminal cooperation (recall Section 2.2.2) is analyzed in Section 2.5.1. Then, the per-cell throughput (sum-rate) of SCP and MCP with terminal cooperation is evaluated in Section 2.5.2 and Section 2.5.3, respectively, with special emphasis on the low SNR regime.

2.5.1 Wyner's Model with Cooperation among Terminals

The system layout is illustrated in Figure 2.7. For simplicity, the uplink channel and intracell TDMA scheme are considered. In each of the M cells, every active terminal has a relay terminal available for cooperation. As compared to the standard linear Wyner model presented in Section 2.2.1, two new parameters, β^2 and γ^2, are introduced that model the average channel gain power between the source terminal and its relay, and between the relay and the corresponding AP, respectively. The channel gains, relative to the signal received by APs of adjacent cells, from source, terminal and relay equal the square of Wyner's intercell interference factor α^2. It is assumed that a relay receives, with negligible power, the signal transmitted by terminals belonging to adjacent cells. This assumption is reasonable if the relays are terminals, but it may be questionable if the relays are fixed wireless stations with antennas placed at heights comparable to the APs. Extension to a more general model can be derived from the treatment presented below, but will not be further illustrated for the sake of simplicity.

Cooperation between terminals is assumed to follow the decode and forward (DF) protocol that is illustrated in Figure 2.7 and discussed in [9]. In the first time slot, each active terminal broadcasts to both its relay and AP. In the second time slot, the relay forwards the decoded signal to the AP. Finally, the AP decodes by considering the signal received in both time slots.

2.5.2 Single-Cell Processing and DF Cooperation between Terminals

In this section, the scenario in Figure 2.7 is investigated with SCP. According to the DF protocol, the code word transmitted by the source terminal in the first slot must be

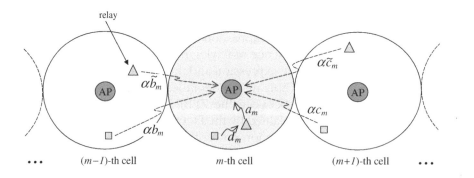

Figure 2.7 Wyner's Linear Model With Cooperation Among Terminals.

decoded by the relay. Therefore, assuming that the relay is aware of the realization of the channel gain d_m (see Figure 2.7), the achievable rate is limited by

$$R_{\text{scp+coop}} \leq R_{\text{relay}}(\bar{P}, \beta) = \frac{1}{2} E_d \left\{ \log \left(1 + \bar{P} \beta^2 |d_m|^2 \right) \right\}, \tag{2.73}$$

where the prelog scaling factor $1/2$ accounts for the two-slot transmission structure of DF. Moreover, conditioned on (2.73), the achievable rate at the AP, taking into account the signals received in both the first timeslot (directly from the source) and in the second (forwarded by the relay) is $R_{\text{scp+coop}} \leq R_d(\bar{P}, \alpha, \gamma)$, where [27]

$$R_d(\bar{P}, \alpha, \gamma) = \frac{1}{2} E \left\{ \log \left(1 + \bar{P} \left[a_m^* \; \gamma \tilde{a}_m^* \right] \mathbf{Q}(\bar{P}, \alpha)^{-1} \begin{bmatrix} a_m \\ \gamma \tilde{a}_m \end{bmatrix} \right) \right\}, \tag{2.74}$$

with $\tilde{a}_m$ denoting the channel gain between relay and corresponding AP ($\tilde{b}_m$ and $\tilde{c}_m$ are similarly defined in accordance with the notation used throughout the chapter, see Figure 2.7), and matrix $\mathbf{Q}(\bar{P}, \alpha)$ accounting for correlation of the intercell interference in the two time-slots:

$$\mathbf{Q}(\bar{P}, \alpha) = \begin{bmatrix} 1 + \alpha^2 \bar{P}(|b_m|^2 + |c_m|^2) & \alpha^2 \bar{P}(b_m \tilde{b}_m^* + c_m \tilde{c}_m^*) \\ \alpha^2 \bar{P}(b_m \tilde{b}_m^* + c_m \tilde{c}_m^*) & 1 + \alpha^2 \bar{P}(|\tilde{b}_m|^2 + |\tilde{c}_m|^2) \end{bmatrix}. \tag{2.75}$$

From (2.73) and (2.74), we finally get the ergodic per-cell achievable sum-rate:

$$R_{\text{scp+coop}} = \min\{ R_{\text{relay}}(\bar{P}, \beta), R_d(\bar{P}, \alpha, \gamma) \}. \tag{2.76}$$

2.5.2.1 Low SNR Analysis

The low SNR characterization of the achievable rate (2.76) can be shown as in [27] to read:

$$\frac{E_b}{N_0}_{\min} = \max \left\{ \frac{2 \log 2}{\beta^2}, \frac{2 \log 2}{1 + \gamma^2} \right\};$$

$$S_0 = \frac{1}{2} \min \left\{ 1, \frac{1 + 2\gamma^2 + \gamma^4}{2 + \gamma^2 + \gamma^4 + 6\alpha^2(1 + \gamma^2)} \right\}. \tag{2.77}$$

In Figure 2.8, the low SNR approximation is compared with the exact throughput (2.76) for $\alpha^2 = -3$ [dB], $\beta^2 = 20$ [dB], and $\gamma^2 = 10$ [dB], showing that the approximation holds for spectral efficiencies as large as 0.4 [bit/s/Hz]. From inspection of (2.77), it is clear that, if the average channel gains between relay and both the active terminal and AP are larger than the average channel gain of the direct link between the terminal and AP, or more precisely, if $\beta^2 > 2$ and $\gamma^2 > 1$, then relevant gains in terms of minimum energy per bit can be obtained. On the other hand, if $\beta^2 \leq 2$ or $\gamma^2 \leq 1$, cooperation between

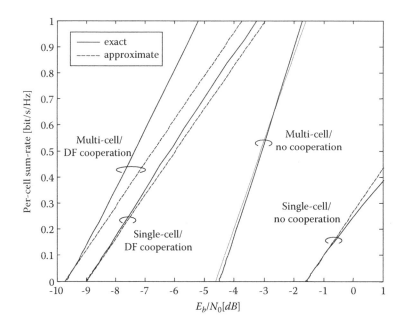

Figure 2.8 Per-Cell Ergodic Rates of Different Schemes With or Without Cooperation Between Either BSs or MTs versus E_b/N_0 ($\alpha^2 = -3$ [dB], $\beta^2 = 20$ [dB], $\gamma^2 = 10$ [dB]). The Plot Compares the Exact Achievable Rates With the Low-SNR approximation (2.1).

terminals yields a power loss as compared to the noncooperative case. However, the slope S_0 is at most $1/2$ (for the example $S_0 = 0.4172$). This reduction in the low SNR slope is immaterial if $E_b/N_{0\min}$ is sufficiently small, as for the case in Figure 2.8.

2.5.3 Multi-Cell Processing and DF Collaboration between Terminals

In this section, multi-cell processing such as in Section 2.3.2 is assumed. As explained above, the achievable rate, due to the DF protocol, is limited by the maximum rate at which the relay is able to correctly decode the transmitted signal, i.e., $R_{\text{mcp+coop}} \leq R_{\text{relay}}(\bar{P}, \beta)$ (recall (2.73)). In the first time slot, the signal received by the APs is (2.7) with $K = 1$, whereas in the second, it has the same matricial formulation $\hat{y} = \hat{H}x + \tilde{z}$, but the channel matrix $\tilde{H}$ contains the fading channels from the relays (see (2.8)). It follows that the achievable per-cell throughput satisfies the inequality $R_{\text{mcp+coop}} \leq R_{\text{m}}(\bar{P}, \alpha, \gamma)$:

$$R_{\text{m}}(\bar{P}, \alpha, \gamma) = \frac{1}{2M} E\{\log \det \{I + \text{SNR}(HH^\dagger + \tilde{H}\tilde{H}^\dagger)\}\}. \tag{2.78}$$

Then, similarly to (2.76),

$$R_{\text{mcp+coop}} = \min\{R_{\text{relay}}(\bar{P}, \beta), R_{\text{m}}(\bar{P}, \alpha, \gamma)\}. \tag{2.79}$$

2.5.3.1 Low SNR Analysis

The low SNR characterization of multi-cell processing with DF cooperation between terminals reads for M large enough (see [27] for proof):

$$\frac{E_b}{N_0}_{\min} = \max\left\{\frac{2\log 2}{\beta^2}, \frac{2\log 2}{1 + \gamma^2 + 4\alpha^2}\right\}$$

$$S_0 = \frac{1}{2}\min\left\{1, \frac{(1 + 4\alpha^2 + \gamma^2)^2}{2(8\alpha^4 + 4\alpha^2(1 + \gamma^2) + 1 + \gamma^4)}\right\}. \tag{2.80}$$

Comparison between the actual throughput (2.79) and the affine low SNR approximation is shown in Figure 2.8 for $\alpha^2 = -3$ [dB], $\beta^2 = 20$ [dB], $\gamma^2 = 10$ [dB], and $M = 20$. From (2.80) and (2.77), multi-cell processing proves to be beneficial in a system that employs DF cooperation at the terminals only if $\beta^2 > 1 + \gamma^2$ and, in this case, the energy gain is easily quantified as $\min\{(1 + \gamma^2 + 4\alpha^2)/(1 + \gamma^2), \beta^2/(1 + \gamma^2)\}$ (equal to 0.72 [dB] in the example). It is noted that this problem could be alleviated by implementing the selective DF protocol proposed in [9], wherein if the channel gain between active terminal and relay falls between a given threshold, then direct transmission is employed.

2.6 PERSPECTIVES AND DISCUSSION

The cell-site antennas (or APs) of a cellular system are spread over the coverage area of the network. At each time instance, the users "see" several colocated cell sites. In conventional systems, where each cell processes the signals of its own users, treating other users as noise, this phenomenon leads to an interference-limited behavior that significantly affects the system performance. This problem is mitigated by sharing and reusing the degrees of freedom available to the network among the cells. On the other hand, by deploying joint processing of signals associated with the different APs, and taking advantage of the powerful "backbone" network connecting the latter, the impact of interference can be theoretically eliminated because interferences become useful signals. Moreover, spatial diversity and beamforming advantages can be harnessed as well to increase the overall system performance. In this chapter, these principles are demonstrated for the family of Wyner cellular models, which lends itself to an analytical study, thus facilitating insight in more involved and realistic settings.

Ending this section is a brief list of open issues related to the information theoretic analysis of DAS under the Wyner model:

- Derivation of a closed-form expression for the per-cell ergodic sum-rate of both uplink and downlink for finite number of users K: Instrumental to this task is the calculation of the eigenvalues distribution of a finite band random matrix, which is to date an open problem even in the asymptotic case where the matrix dimensions are large.

- Investigation of the effects of impairments (finite capacity, delay) on the backbone connecting APs: Some preliminary results in this direction have been presented in [8, 10, and 11].

- Analysis of the effects of reduced feedback (imperfect CSI) at the APs in the downlink: To study the tradeoff between multi-user diversity and feedback overhead, performance of suboptimal schemes requiring a limited amount of feedback has been widely investigated in recent years for collocated antenna arrays. Extension to DAS within Wyner's model is a challenging task due to the unique structure of the problem.

- Study of the benefits of cooperation among mobile terminals under general cooperative protocols, see, e.g., [9].

- The problem of optimal degrees of freedom allocation in the Wyner framework. As an example, see the comprehensive discussion on ICTS protocols in [15].

2.7 BIBLIOGRAPHICAL NOTES

The general framework of the "multireceiver" network model was introduced and analyzed in [5] for Gaussian channels. The uplink channel of a simple cellular model, referred to as Wyner's linear and planar models, is introduced and analyzed in [6] for optimal and linear minimum mean square error (MMSE) MCP receivers, and Gaussian channels. It is noted that this model consists of a special case of the more general "multireceiver" network model. [28] extends the Wyner model to include fading channels and analyzes the performance of single- and two-cell processing under various setups. In [17], the results of [6] are extended to include fading channels. [29] and [30] consider SCP of randomly spread code division multiple access (CDMA) in Wyner's linear model for nonfading and flat fading, respectively (see also [31]). References [8] and [7] extend the results of [29] and [30] for MCP limiting the number of cooperating cells (clusters). Various iterative decoding schemes based on local message passing between adjacent cells are considered in [10,11,32–34] for the Wyner linear and planar models. The capacity under outage constraint for "strongest-user-only" SCP receivers in the Wyner linear uplink channel is derived in [35,36] (see [37] for a corresponding single-cell setup). [38] considers a MCP linear minimum mean square error (LMMSE) receiver for a finite number of cells version of the Wyner linear model. [22,23] (see also [39]) introduce the soft-handoff model (according to which the cells are located on a circle and the cell-site antennas are located on the cells' boundaries) and consider MCP for Gaussian and fading channels. The uplink channel of the soft-handoff model with MCP can be viewed as a tap intersymbol-interference (ISI) channel analyzed by [40]. The variant of the "multireceiver" network model which is extended to include MIMO fading channels is introduced in [41] and is analyzed under asymptotic numbers of the receive-transmit antennas' setup.

Turning to the downlink channel, [42] applies the results of [43], while focusing on Wyner's linear model. Here, the *LQ*-factorization based linear MCP scheme, combined with DPC, is analyzed for an *overall* power constraint. In [44], the problem of transmitter optimization to maximize the downlink sum-rate of a *multiple antenna* cellular system is addressed, but with a more realistic *separate* power constraint per each cell site. The MCP DPC is also used in [45] and [46], where the problem of providing the best possible service to new users joining the system without affecting existing users is addressed. The new users are required to be invisible, interference wise, to existing users, and the network is referred to as "PhantomNet." MCP for the downlink channel of a "Wyner-like" planar model in the presence of fading is considered in [47] under asymptotic numbers of the receive-transmit antennas' setup and overall power constraint. See [48] for a downlink capacity analysis where no cell-site cooperation is assumed, and multi-user detection is employed at the mobile receivers for mitigating co-channel interference, though multi-user cooperation has been considered in [9] in a similar setup. In [49] (see also [50]) a generic framework is proposed for the study of base station cooperation in the downlink, utilizing multi-cell DPC, and a cooperative base station selection procedure. Throughput outage calculations of several cooperative joint transmission schemes, including "zero forcing" and DPC, are reported in [51]. Bounds to the sum-rate capacity supported by the downlink of the soft-handoff model have been reported in [22,23] under per-cell power constraints. [52] considers multi-cell beamforming under minimum receive signal-to-interference ratio constraints, for a general cellular downlink channel model. In [53], distributed beamforming based on local message passing between neighboring base stations is considered for the Wyner linear model through an equivalent virtual LMMSE estimation problem. Multi-cell ZFBF with a simple user selection scheme is considered in [26] for the downlink channel of Wyner's linear model with Gaussian and fading channels.

A brief survey for MCP in cellular uplink and downlink channels and related issues is provided in [54]. MCP schemes are also considered as part of next-generation cellular wireless systems physical-layer designs in [55].

ACKNOWLEDGEMENTS

This research was supported by the National Science Foundation under Grant No. CNS-0626611 and by a Marie Curie Outgoing International Fellowships within the 6th European Community Framework Programme.

REFERENCES

[1] A. Goldsmith, S.A. Jafar, N. Jindal, and S. Vishwanath, "Capacity limits of MIMO channels," *IEEE Journal on Selected Areas in Communications*, pp. 684–702, June 2003.

[2] S. Vishwanath, N. Jindal, and A. Goldsmith, "Duality, achievable rates, and sum-rate capacity of Gaussian MIMO broadcast channels," *IEEE Transactions on Information Theory*, vol. 49, pp. 2658–2669, Oct. 2003.

[3] H. Weingarten, Y. Steinberg, and S. Shamai (Shitz), "The capacity region of the Gaussian MIMO broadcast channel," in *Proceedings of the 2004 IEEE International Symposium on Information Theory (ISIT'04)*, (Chicago), p. 174, June 27–July 2, 2004.

[4] A.M. Tulino and S. Verdú, "Random matrix theory and wireless communications," *Foundations and Trends in Communications and Information Theory*, vol. 1, 2004.

[5] S.V. Hanly and P.A. Whiting, "Information-theoretic capacity of multi-receiver networks," *Telecommun. Syst.*, vol. 1, pp. 1–42, 1993.

[6] A.D. Wyner, "Shannon-theoretic approach to a Gaussian cellular multiple-access channel," *IEEE Transactions on Information Theory*, vol. 40, pp. 1713–1727, Nov. 1994.

[7] O. Somekh, B.M. Zaidel, and S. Shamai (Shitz), "Spectral efficiency of joint multiple cell-site processors for randomly spread DS-CDMA systems," in *Proceedings of the 2004 IEEE International Symposium on Information Theory (ISIT'04)*, (Chicago), p. 278, June 27–July 2, 2004.

[8] O. Somekh, B.M. Zaidel, and S. Shamai (Shitz), "Spectral efficiency of joint multiple cell-site processors for randomly spread DS-CDMA systems." to apear in the IEEE Transactions on Information Theory.

[9] N. Laneman, D. Tse, and G. Wornell, "Cooperative diversity in wireless networks: efficient protocols and outage behavior," *IEEE Transactions on Information Theory*, vol. 50, pp. 3062–3080, Dec. 2004.

[10] E. Aktas, J. Evans, and S. Hanly, "Distributed decoding in a cellular multiple-access channel," in *Proceedings of the 2004 IEEE International Symposium on Information Theory (ISIT'04)*, (Chicago), p. 484, June 27–July 2, 2004.

[11] A. Grant, S. Hanly, J. Evans, and R. Müller, "Distributed decoding for Wyner cellular systems," in *Proceedings of the 2004 Australian Communication Theory Workshop (AusCTW'04)*, (Newcastle, Australia), Feb. 4–6, 2004.

[12] S. Verdú, "Spectral efficiency in the wideband regime," *IEEE Transactions on Information Theory*, vol. 48, pp. 1329–1343, June 2002.

[13] A. Lozano, A. Tulino, and S. Verdú, "High-snr power offset in multi-antenna communications," *IEEE Transactions on Information Theory*, vol. 51, pp. 4134–4151, Dec. 2005.

[14] M. Sharif and B. Hassibi, "On the capacity of MIMO broadcast channel with partial side information," *IEEE Transactions on Information Theory*, vol. 51, pp. 506–522, Feb. 2005.

[15] S. Shamai (Shitz) and A.D. Wyner, "Information-theoretic considerations for symmetric, cellular, multiple-access fading channels – part i," *IEEE Transactions on Information Theory*, vol. 43, pp. 1877–1894, Nov. 1997.

[16] R.M. Gray, "On the asymptotic eigenvalue distribution of Toeplitz matrices," *IEEE Transactions on Information Theory*, vol. IT-18, pp. 725–730, Nov. 1972.

[17] O. Somekh and S. Shamai (Shitz), "Shannon-theoretic approach to a Gaussian cellular multi-access channel with fading," *IEEE Transactions on Information Theory*, vol. 46, pp. 1401–1425, July 2000.

[18] R. Durrett, *Probability: Theory and Examples*. Duxbury Press, Belmont, California, 3rd ed., 1991.

[19] M.G. Krein and A.A. Nudelman, *The Markov Moment Problem and External Problems*, vol. 50, Providence, RI: American Mathematical Society, 1977.

[20] D. Tse, "Optimal power allocation over parallel Gaussian channels," in *Proceeding of the 1997 IEEE International Symposium on Information Theory (ISIT'97)*, (Ulm, Germany), June 1997.

[21] W. Yu and T. Lan, "Minimax duality of Gaussian vector broadcast channels," in *Proceedings of the 2004 IEEE International Symposium on Information Theory (ISIT'2004)*, (Chicago), p. 177, June 27–July 2, 2004.

[22] O. Somekh, B.M. Zaidel, and S. Shamai (Shitz), "Sum-rate characterization of multi-cell processing," in *Proceedings of the Canadian Workshop on Information Theory (CWIT'05)*, (McGill University, Montreal, Quebec, Canada), June 5–8, 2005.

[23] O. Somekh, B.M. Zaidel, and S. Shamai (Shitz), "Sum rate characterization of joint multiple cell-site processing," submitted to the IEEE Transactions on Information Theory, 2005.

[24] S.N. Diggavi and T.M. Cover, "The worst additive noise under a covariance constraint," *IEEE Transactions on Information Theory*, vol. 47, pp. 3072–3081, Nov. 2001.

[25] T. Yoo and A. Goldsmith, "Optimality of zero-forcing beam forming with multiuser diversity," in *Proceedings of the ICC 2005 Wireless Communications Theory (ICC2005)*, (Seoul, Korea), May 16–20, 2005.

[26] O. Somekh, O. Simeone, Y. Bar-Ness, and A.M. Haimovich, "Distributed multi-cell zero-forcing beamforming in cellular downlink channels," in *Proceedings of Globecom 2006*, San Francisco, CA, Nov. 27–Dec. 1, 2006.

[27] O. Simeone, O. Somekh, Y. Bar-Ness, and U. Spagnolini, "Low-SNR analysis of cellular systems with cooperative base stations and mobiles." in *Proc. Asilomar Conference on Signals, Systems and Computers*, Pacific Grove, CA, Oct. 29–31, 2006.

[28] S. Shamai (Shitz) and A.D. Wyner, "Information-theoretic considerations for symmetric, cellular, multiple-access fading channels – parts i and ii," *IEEE Transactions on Information Theory*, vol. 43, pp. 1877–1911, Nov. 1997.

[29] B.M. Zaidel, S. Shamai (Shitz), and S. Verdú, "Multi-cell uplink spectral efficiency of coded DS-CDMA with random signatures," *IEEE Journal on Selected Areas in Communications,* vol. 19, pp. 1556–1569, Aug. 2001. See also: "Spectral efficiency of randomly spread DS-CDMA in a multi-cell model," *Proceedings of the 37th Annual Allerton Conference on Communication, Control and Computing,* (Monticello, IL), pp. 841–850, Sept. 1999.

[30] B.M. Zaidel, S. Shamai (Shitz), and S. Verdú, "Multi-cell uplink spectral efficiency of randomly spread DS-CDMA in Rayleigh fading channels," in *Proceedings of the 6th International Symposium on Communication Techniques and Applications (ISCTA'01),* (Ambleside, U.K.), pp. 499–504, July 15–20, 2001. See also: "Random CDMA in the multiple cell uplink environment: the effect of fading on various receivers," *Proceedings of the 2001 IEEE Information Theory Workshop,* (Cairns, Australia), pp. 42–45, Sept. 2–7, 2001.

[31] H. Dai and V.H. Poor, "Asymptotic spectral efficiency of multi-cell MIMO systems with frequency-flat fading," *IEEE Transactions on Signal Processing,* vol. 51, pp. 2976–2988, Nov. 2003.

[32] O. Shental, A.J. Weiss, N. Shental, and Y. Weiss, "Generalized belief propagation receiver for near-optimal detection of two-dimensional channels with memory," in *Proceedings of the 2004 Information Theory Workshop (ITW'04),* (San Antonio, TX), October 24–29, 2004.

[33] B.L. Ng, J. Evans, S. Hanly, and A. Grant, "Distributed linear multiuser detection in cellular networks," in *Proceedings of the 2004 Australian Communication Theory Workshop (AusCTW'04),* (Newcastle, Australia), Feb. 4–6, 2004.

[34] O. Shental, N. Shental, S. Shamai (Shitz), I. Kanter, A.J. Weiss, and Y. Weiss, "Finite-state input two-dimensional Gaussian channels with memory: estimation and information via graphical models and statistical mechanics." Submitted to the IEEE Transactions on Information Theory, 2006.

[35] B.M. Zaidel, S. Shamai (Shitz), and S. Verdú, "Impact of out-of-cell interference on strongest-users-only CDMA detectors," in *Proceedings of the International Symposium on Spread Spectrum Techniques and Applications (ISSSTA'02),* vol. 1, (Prague, Czech Republic), pp. 258–262, Sept. 2–5, 2002.

[36] B.M. Zaidel, S. Shamai (Shitz), and S. Verdú, "Outage capacities and spectral efficiencies of multiuser receivers in the CDMA cellular environment." Submitted to the IEEE transactions on Wireless Communications, 2006.

[37] S. Shamai (Shitz), B.M. Zaidel, and S. Verdú, "Strongest-users-only detectors for randomly spread CDMA," in *Proceedings of the 2002 IEEE International Symposium on Information Theory (ISIT'02),* (Lausanne, Switzerland), p. 20, June 30 – July 5, 2002.

[38] B.L. Ng, J. Evans, S. Hanly, and A. Grant, "Information capacity of Wyners' cellular network with LMMSE receivers," in *Proceedings of the IEEE International Conference on Communications (ICC'04),* (Paris, France), 2004.

[39] Y. Liang and A. Goldsmith, "Symmetric rate capacity of cellular systems with cooperative base stations," in *Proceedings of Globecom 2006,* San Francisco, CA, Nov. 27–Dec. 1, 2006.

[40] A. Narula, *Information Theoretic Analysis of Multiple-Antenna Transmission Diversity.* Ph.D. thesis, Massachusetts Institute of Technology (MIT), Boston, June 1997.

[41] D. Aktas, M.N. Bacha, J. Evans, and S. Hanly, "Scaling results on the sum capacity of multiuser MIMO systems," in *IEEE Transactions on Information Theory,* vol. 52, no. 7, pp. 3264–3274, 2004.

[42] S. Shamai (Shitz) and B.M. Zaidel, "Enhancing the cellular downlink capacity via co-processing at the transmitting end," in *Proceedings of the IEEE 53rd Vehicular Technology Conference (VTC 2001 Spring),* vol. 3, (Rhodes, Greece), pp. 1745–1749, May 6–9, 2001.

[43] G. Caire and S. Shamai (Shitz), "On the achievable throughput of a multi-antenna Gaussian broadcast channel," *IEEE Transactions on Information Theory,* vol. 49, no. 7, pp. 1691–1706, 2003.

[44] S.A. Jafar and A.J. Goldsmith, "Transmitter optimization for multiple antenna cellular systems," in *Proceedings of the 2002 IEEE International Symposium on Information Theory (ISIT'02)*, (Lausanne, Switzerland), p. 50, June 30–July 5, 2002.

[45] H. Viswanathan, S. Venkatesan, and H. Huang, "Downlink capacity evaluation for cellular networks with known interference cancellation," *IEEE Journal on Selected Areas in Communications*, vol. 21, pp. 802–811, June 2003.

[46] S.A. Jafar, G. Foschini, and A.J. Goldsmith, "Phantomnet: exploring optimal multicellular multiple antenna systems," *EURASIP Journal on Applied Signal Processing, Special Issue on MIMO Communications and Signal Processing*, pp. 591–605, May 2004.

[47] H. Huang and S. Venkatesan, "Asymptotic downlink capacity of coordinated cellular networks," in *Proceedings of Asilomar Conference Signals, Systems, and Computers*, (Pacific Grove, CA), 2004.

[48] H. Dai, A.F. Molisch, and H.V. Poor, "Downlink capacity of interference-limited MIMO systems with joint detection," *IEEE Transactions on Wireless Communications*, vol. 3, pp. 442–453, Mar. 2004.

[49] H. Zhang, H. Dai, and Q. Zhou, "Base station cooperation for multiuser MIMO: joint transmission and BS selection," in *Proceedings of the Conference on Information Sciences and Systems (CISS'05)*, (Baltimore, MD), Mar. 16–18, 2005.

[50] H. Zhang, H. Dai, and Q. Zhou, "Base station cooperation for multiuser MIMO: joint transmission and BS selection," in *Proceedings of the 2004 Conference on Information Sciences and Systems (CISS'04)*, (Princeton University, Princeton, NJ), Mar. 17–19, 2004.

[51] G. Foschini, H.C. Huang, K. Karakayali, R.A. Valenzuela, and S. Venkatesan, "The value of coherent base station coordination," in *Proceedings of the 2005 Conference on Information Sciences and Systems (CISS'05)*, (John Hopkins University, Baltimore, MD), Mar. 16–18, 2005.

[52] A. Ekbal and J.M. Cioffi, "Distributed transmit beamforming in cellular networks," in *Proceedings of the ICC 2005 Wireless Communications Theory (ICC'05)*, (Seoul, Korea), May 16–20, 2005.

[53] B.L. Ng, J.S. Evans, and S.V. Hanly, "Transmit beamforming with cooperative base stations," in *Proceedings of the IEEE International Symposium on Information Theory (ISIT'05)*, (Adelaide, Australia), Sept. 4–9, 2005.

[54] S. Shamai (Shitz), O. Somekh, and B.M. Zaidel, "Multi-cell communications: an information theoretic perspective," in *Proceedings of the Joint Workshop on Communications and Coding (JWCC'04)*, (Donnini (Florence), Italy), Oct. 14–17, 2004.

[55] G.J. Foschini, H.C. Huang, S.J. Mullender, S. Venkatesan, and H. Viswanathan, "Physical-layer design for next-generation cellular wireless systems," *Bell Labs Technical Journal*, vol. 10, no. 2, pp. 157–172, 2005.

3

THEORETICAL LIMITS OF CELLULAR SYSTEMS WITH DISTRIBUTED ANTENNAS[1]

Wan Choi and Jeffrey G. Andrews

Contents

Distributed antenna systems have been shown to possess advantages in terms of power, signal-to-interference-plus-noise ratio (SINR), and capacity owing to macrodiversity and the reduced access distance. Based on these advantages, many cellular service providers or system manufacturers are seriously considering replacing legacy cellular systems with distributed antenna systems or adopting the distributed antenna architecture in the future.

---

[1] This chapter includes parts of the work published in *IEEE Transactions on Wireless Communications*, January 2007. – W. Choi and J. G. Andrews, "Downlink Performance and Capacity of Distributed Antenna Systems in a Multi-cell Environment."

This chapter will highlight advantages of multi-cellular distributed antenna systems and study the achievable capacity/performance of multi-cellular distributed antenna systems. The interference reduction capability of cellular systems with distributed antennas results in capacity gains over conventional cellular systems. Even though the focus of this chapter is on investigating theoretical limits of cellular systems with distributed antennas, this chapter will also discuss possible practical ways to achieve the theoretical limits.

3.1 INTRODUCTION

The distributed antenna architecture was originally proposed in [1] to cover the dead spots in cellular systems with the help of spotted unshielded coaxial cables to simulcast signal. In a distributed antenna system (DAS), antennas are geographically distributed to reduce access distance and connected to a home base station (or central unit) via dedicated wires, fiber optics, or an exclusive radio frequency (RF) link. In addition to covering the dead spots, DASs can reduce the cost of installing systems and simplify maintenance because they can reduce the required number of base stations within a target service area. Blocking probability is also improved owing to the principle of trunking efficiency because resources for signal processing, such as channel cards/elements, are centralized and shared. Various studies on DASs have shown that they possess advantages in terms of power, signal-to-interference-plus-noise ratio (SINR), and capacity owing to macrodiversity and the reduced access distance [2–7].

Based on the fact that geographically distributed antennas construct macroscopic multiple input multiple output (MIMO) channels, Roh and Paulraj [8–10] looked into distributed antenna systems in the context of macroscopic MIMO. Their studies motivated several studies on DAS from the perspective of macroscopic MIMO [11–15]. The approaches within the framework of a macroscopic MIMO system show great potential of the DAS as a future cellular architecture. This chapter highlights advantages of DASs in the multi-cell macroscopic MIMO context and investigates theoretical limits of multi-cellular distributed antenna systems. We first consider a system where each distributed antenna module and mobile station has a single antenna, and then generalize the results in Section 3.5 by employing multiple antennas at each distributed antenna module and mobile unit.

3.2 ARCHITECTURAL ADVANTAGES OF DASs

In this section, we first provide basic tutorials on cellular systems with distributed antennas and then investigate basic architectural advantages of DASs over a conventional cellular system in terms of power efficiency, signal-to-interference-plus-noise ratio (SINR), and outage probability.

3.2.1 Cellular Architecture with Distributed Antennas

An example of multi-cell distributed antenna architecture is given in Figure 3.1 as in [16], where a cell is covered by a small base station and six distributed antenna modules. In contrast, the same area is covered by only a single high-power base station in traditional cellular systems. The radius of a cell (a bold dotted circle) is R and the coverage radius of

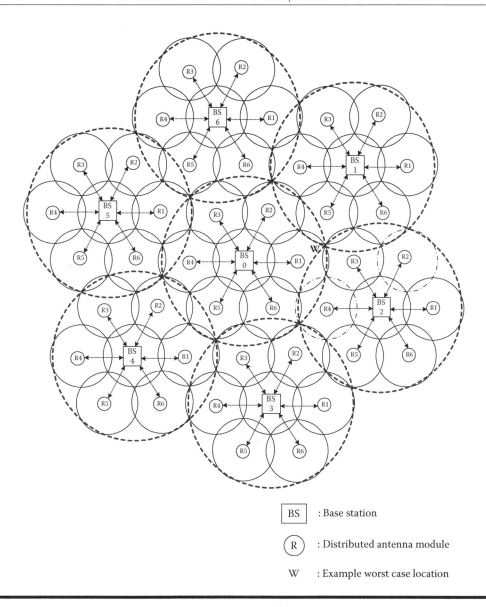

BS	: Base station
(R)	: Distributed antenna module
W	: Example worst case location

Figure 3.1   Cellular Architecture with Distributed Antennas [16].

a distributed antenna module is r. For MIMO systems, a distributed antenna architecture can effectively increase the total number of antennas with low spatial correlations at the cost of feeder lines through geographically distributing antennas instead of centralizing them. The distributed antennas can be jointly processed for data cooperation, although power cooperation across distributed antenna modules is not allowed due to per antenna module power constraints.

3.2.2   Transmission Strategy and Multiple Access Scenario

There are several possible transmission strategies using distributed antennas, but this chapter considers the two most likely transmission strategies as in [16]:

- Blanket transmission scheme — signals are transmitted through all the distributed antennas. Geographically distributed antennas construct a macroscopic multiple antenna system.
- Single transmit selection scheme — only a single distributed antenna is selected for transmission by the criterion of minimizing propagation pathloss.[2] This scheme exploits macroscopic selection diversity and is expected to additionally reduce other-cell interference (OCI) because the number of OCI sources is reduced if other cells adopt the same single transmit selection scheme.

For simplicity, a single user scenario is considered. This scenario, however, corresponds to most practical multi-user systems such as time division multiple access (TDMA), frequency division multiple access (FDMA), orthogonal code division multiple access (CDMA), and orthogonal frequency division multiple access (OFDMA) systems, where only a single user transmitting in any time/frequency/code dimension exits.

3.2.3 Power Efficiency of DASs

Power efficiency is computed by comparing the required power for supporting the coverage area of a single cell, and is calculated based on propagation pathloss. In the distributed antenna architecture given in Figure 3.1, the area covered by a small base station and six distributed antenna modules is $(3\pi + 6\sqrt{3})r^2$. The radius of a circle with the same area becomes $r\sqrt{12\sqrt{3}/\pi - 1}$. For a fair comparison, we regard this circle as the effective coverage of a high-power base station in a traditional cellular system instead of the bold dotted circle in Figure 3.1. Let the required transmit power for each distributed antenna module and the small base station to support the circular area with radius r be P. Hence, the total required power to support the effective coverage area is $7P$ in the distributed antenna structure, whereas for the big base station with the same effective coverage area in the traditional cellular structure it becomes $(12\sqrt{3}/\pi - 1)^{\frac{\alpha}{2}}P$, if the propagation pathloss is assumed to be $L = d^{\alpha}$ where α is the pathloss exponent. Then, the power efficiency of the distributed antenna structure is given by

$$\eta = \frac{(12\sqrt{3}/\pi - 1)^{\frac{\alpha}{2}}P}{7P} = \frac{(12\sqrt{3}/\pi - 1)^{\frac{\alpha}{2}}}{7} \tag{3.1}$$

For the case that the pathloss exponent α is 4, the power efficiency gain η is about 6.54 dB. This result shows that a distributed antenna structure requires much less power to support the same coverage area.

3.2.4 Signal-to-Interference-Plus-Noise Ratio (SINR)

In this section, we show how the distributed antenna structure can reduce OCI and improve SINR in the downlink by increasing the received strength of the desired signal and reducing the power of the received interference. Since the outage-limiting scenario

---

[2] Although smarter selection algorithms can potentially be used, such as maximizing SINR or capacity, we consider this one for simplicity of analysis and because it should minimize the required transmit power (and hence the interference caused to other cells).

in cellular networks is usually a mobile station at the cell boundary, maintaining the SINR of the users near cell boundaries above a given level is particularly important [17].

If we consider blanket transmission in each cell and the same uncoded data is transmitted through distributed antennas, the expected SINR over short term fading for a given location of the target mobile station is given by

$$\mathbb{E}[\gamma] = \frac{\sum_{i=0}^{6} L_i^{(0)} P_i^{(0)}}{\sum_{j=1}^{6} \sum_{i=0}^{6} L_i^{(j)} P_i^{(j)} + \sigma_n^2} \tag{3.2}$$

where $L_i^{(j)}$ and $P_i^{(j)}$ denote propagation pathloss and transmit power from the ith distributed antenna module in the jth cell, where the small base station of each cell and home cell are indexed by $i = 0$ and $j = 0$, respectively.

Figure 3.2 shows the expected SINRs of different transmission schemes according to the normalized distance from the home base station in the direction of the worst-case position W in Figure 3.1. The SINR of each transmission scheme is calculated when the pathloss exponent is 4.0 but lognormal shadowing is not considered. For a fair comparison with the traditional cellular structure, we assume that the transmit power of each distributed antenna module is $0.1P$ and the transmit power of the home base station is $0.4P$, for a total transmit power of P in both the distributed antenna structure and the traditional cell structure. Figure 3.2 demonstrates that DASs with the single transmit selection scheme outperform a conventional cellular system at all the distances owing to their extra OCI reduction capability, whereas DASs with blanket transmissions are worse than conventional cellular systems only near the home base station. However,

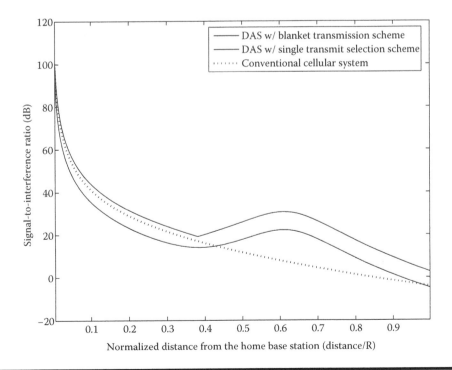

Figure 3.2 Signal-to-Interference-Plus-Noise Ratio (SINR) Versus the Normalized Distance from the Home Base Station.

all the transmission schemes of DASs have substantially higher SINRs than conventional cellular systems beyond the normalized distance 0.5. When a mobile unit moves from the origin to the worst-case position W in Figure 3.1, the signal strength from the distributed antenna module located 0.6R on the line toward W becomes dominant as a mobile unit approaches the distributed antenna module, while the signal strength from the home base station declines. As a mobile unit approaches to the cell edge and becomes far away from the distributed antenna module, the signal strength from the distributed antenna module also decreases, so the curves show convex shapes around the location of the distributed antenna module.

3.2.5 Outage Probability

If we consider lognormal shadowing, the instantaneous SINR at a location for a cellular DAS with blanket transmission is given by

$$\gamma_{inst} = \frac{\sum_{i=0}^{6} L_i^{(0)} \chi_i^{(0)} \left(b_i^{(0)}\right)^2 P_i^{(0)}}{\sum_{j=1}^{6} \sum_{i=0}^{6} L_i^{(j)} \chi_i^{(j)} \left(b_i^{(j)}\right)^2 P_i^{(j)} + \sigma_n^2} \tag{3.3}$$

where $b_i^{(j)}$ and $\chi_i^{(j)}$ denote short term fading and lognormal shadowing from the ith distributed antenna module in the jth cell, respectively. The short-term fading follows an independent and identically distributed i.i.d. complex Gaussian distribution $\sim \mathcal{CN}(0, 1)$. Because the background noise power σ_n^2 is negligible compared to the interference power, the SNIR averaged over the fast fading[3] can be given by

$$\gamma = \frac{\sum_{i=0}^{6} L_i^{(0)} \chi_i^{(0)} P_i^{(0)}}{\sum_{j=1}^{6} \sum_{i=0}^{6} L_i^{(j)} \chi_i^{(j)} P_i^{(j)}} \tag{3.4}$$

The sum of lognormal random variables is well-approximated as a lognormal random variable [19] and hence the numerator in (3.4) is approximated as a lognormal random variable with mean μ_d and variance σ_d^2 as

$$\mu_d = \sum_{i=0}^{6} L_i^{(0)} P_i^{(0)} \exp\left(\frac{\ln 10}{10} \mu_i^{(0)} + \left(\frac{\ln 10}{10}\right)^2 \frac{\left(\sigma_i^{(0)}\right)^2}{2}\right) \tag{3.5}$$

$$\sigma_d^2 = \sum_{i=0}^{6} \left(L_i^{(0)} P_i^{(0)}\right)^2 \exp\left(\frac{2\ln 10}{10} \mu_i^{(0)} + \left(\frac{\ln 10}{10}\right)^2 \left(\sigma_i^{(0)}\right)^2\right) \left\{\exp\left(\left(\frac{\ln 10}{10}\right)^2 \left(\sigma_i^{(0)}\right)^2\right) - 1\right\} \tag{3.6}$$

---

[3] When the fading environment is a superposition of both fast and slow fading, a common performance metric is combined outage and error probability, where the outage occurs when averaged SINR over fast fading falls below some target value [18].

and the denominator is also approximated as a lognormal random variable with mean μ_I and variance σ_I^2 as

$$\mu_I = \sum_{j=1}^{6} \sum_{i=0}^{6} L_i^{(j)} P_i^{(j)} \exp\left(\frac{\ln 10}{10} \mu_i^{(j)} + \left(\frac{\ln 10}{10}\right)^2 \frac{(\sigma_i^{(j)})^2}{2}\right) \tag{3.7}$$

$$\sigma_I^2 = \sum_{j=1}^{6} \sum_{i=0}^{6} (L_i^{(j)} P_i^{(j)})^2 \exp\left(\frac{2\ln 10}{10} \mu_i^{(j)} + \left(\frac{\ln 10}{10}\right)^2 (\sigma_i^{(j)})^2\right) \left\{\exp\left(\left(\frac{\ln 10}{10}\right)^2 (\sigma_i^{(j)})^2\right) - 1\right\}$$

$$\tag{3.8}$$

where $\mu_i^{(j)}$ and $\sigma_i^{(j)}$ are the mean and the standard deviation of $\chi_i^{(j)}$ in dB value, respectively. Because the ratio of lognormal random variables is also a lognormal random variable, the SIR γ becomes a lognormal random variable with mean $\tilde{\mu}_\gamma$ and variance $\tilde{\sigma}_\gamma^2$ in dB by

$$\tilde{\mu}_\gamma = \frac{10\ln(\mu_d)}{\ln 10} - \frac{10}{2\ln 10} \ln\left(\frac{\sigma_d^2}{\mu_d^2} + 1\right) - \frac{10\ln(\mu_I)}{\ln 10} + \frac{10}{2\ln 10} \ln\left(\frac{\sigma_I^2}{\mu_I^2} + 1\right) \tag{3.9}$$

$$\tilde{\sigma}_\gamma^2 = \left(\frac{10}{\ln 10}\right)^2 \ln\left(\frac{\sigma_d^2}{\mu_d^2} + 1\right) + \left(\frac{10}{\ln 10}\right)^2 \ln\left(\frac{\sigma_I^2}{\mu_I^2} + 1\right) \tag{3.10}$$

Then, the outage probability of the mobile station at a given location is given by

$$\Pr[\gamma < \gamma_{req}] = 1 - Q\left(\frac{10\log_{10}(\gamma_{req}) - \tilde{\mu}_\gamma}{\tilde{\sigma}_\gamma}\right) \tag{3.11}$$

where γ_{req} is the required SINR to guarantee a desired quality of service (QoS). Because the analysis of the blanket transmission scheme in the DAS is a general case and includes analysis of other schemes, the outage probabilities of other transmission schemes in DASs and conventional cellular systems can be obtained in the same way.

Figure 3.3 shows the outage probability versus the normalized distance from the home base station in the direction of the worst-case position W when the pathloss exponent is 4.0. We assume power control compensates for the lognormal shadowing in the desired cell, but the standard deviation of the lognormal shadowing for the signals from neighbor cells is 6 dB. The outage threshold is assumed as 10 dB. Figure 3.3 shows that all the transmission schemes in the cellular DAS substantially outperform conventional cellular systems.

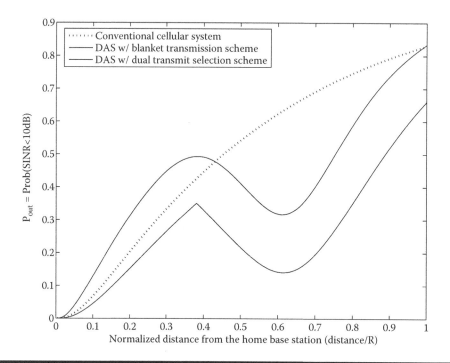

Figure 3.3   Outage Probability Versus the Normalized Distance from the Home Base Station (Outage Threshold = 10 dB).

3.3   ACHIEVABLE TRANSMIT DIVERSITY PERFORMANCE FOR MULTI-CELL DISTRIBUTED ANTENNA SYSTEMS

As previously mentioned, the DAS (particularly the blanket transmission scheme) constructs a macroscopic multiple input single output (MISO) vector channel given by

$$\mathbf{h} = \begin{bmatrix} \sqrt{L_0^{(0)}} b_0^{(0)} & \sqrt{L_1^{(0)}} b_1^{(0)} & \cdots & \sqrt{L_6^{(0)}} b_6^{(0)} \end{bmatrix} \qquad (3.12)$$

When the channel is known to the transmitter, transmit-maximal ratio combining (transmit-MRC) [20] is known to maximize transmit diversity gain in a MISO vector channel. However, if there are per antenna module power constraints, transmit-MRC may not be the best strategy. Therefore, this section investigates the optimal transmit diversity scheme for DASs in terms of symbol error probability performance. The optimal transmit diversity scheme can be thought of as an upper bound to other schemes such as distributed space–time block coding (STBC).

In transmit diversity schemes, the signal is transmitted from each distributed antenna module after being weighted appropriately. Then, the signal at the mobile station during a symbol period is given by

$$y = \mathbf{h}\mathbf{w}x + z \qquad (3.13)$$

where x is the transmitted data with $\mathbb{E}[|x|^2] = 1$, and $\mathbf{w}$ is a 7×1 transmit weight vector with per antenna module power constraints such as $|w_i|^2 \leq P_i^{(0)}$, $\forall i \in \{0, 1, \cdots, 6\}$.

The interference-plus-noise z is assumed to be a complex Gaussian random variable with variance $\sigma_z^2 = \sum_{j=1}^{6} \sum_{i=0}^{6} L_i^{(j)} P_i^{(j)} + \sigma_n^2$ by the central limit theorem (CLT), whereas the interference-plus-noise power of the conventional cellular system is given by $\sigma_c^2 = \sum_{j=1}^{6} L_0^{(j)} P_0^{(j)} + \sigma_w^2$, where σ_n^2 is variance of additive Gaussian noise. Correspondingly, the received SINR is given by

$$\gamma = \frac{\left| b_0^{(0)} \sqrt{L_0^{(0)}} w_0 + b_1^{(0)} \sqrt{L_1^{(0)}} w_1 + \cdots + b_6^{(0)} \sqrt{L_6^{(0)}} w_6 \right|^2}{\sigma_z^2} \tag{3.14}$$

and the weight vector should be chosen to maximize the received SINR under per antenna module power constraints instead of a sum power constraint. By the triangular inequality of $|a + b| \le |a| + |b|$, it holds that

$$\gamma \le \frac{\left(|b_0^{(0)}| \sqrt{L_0^{(0)}} w_0| + |b_1^{(0)}| \sqrt{L_1^{(0)}} w_1| + \cdots + |b_6^{(0)}| \sqrt{L_6^{(0)}} w_6| \right)^2}{\sigma_z^2} \tag{3.15}$$

Therefore, γ is maximized when $w_i = \sqrt{P_i^{(0)}} e^{j\left(\xi - \angle b_i^{(0)} \right)}$, where $\xi \in [0, 2\pi)$ and $\angle b$ denotes the phase of $b \in \mathbb{C}$, $\angle b \in [0, 2\pi)$, and the corresponding SINR is given by

$$\gamma = \frac{\sum_{i=0}^{6} |b_i^{(0)}|^2 L_i^{(0)} P_i^{(0)} + \sum_{i=0}^{6} \sum_{j=0, j\neq i}^{6} |b_i^{(0)}||b_j^{(0)}| \sqrt{L_i^{(0)} L_j^{(0)} P_i^{(0)} P_j^{(0)}}}{\sigma_z^2} \tag{3.16}$$

This result indicates that the optimal transmit diversity strategy in DASs with per antenna module power constraints is phase steering with allowable full power transmission.

For simplicity of analysis, we assume that the transmit power of each distributed antenna module and the home base station in DAS is $P/7$, whereas the transmit power of the base station in the traditional cellular system is P. Then, the symbol error probability is obtained as in [21]

$$P_{e, blanket} \approx \mathbb{E}\left[\overline{N}_e Q \left(\sqrt{\frac{d_{min}^2 \|\mathbf{h}\|_1^2 P}{14\sigma_z^2}} \right) \right] \tag{3.17}$$

where $\overline{N}_e$ and d_{min} are the average number and minimum distance of nearest neighbors of the given signal constellation, and $\| \cdot \|_1$ is the one-norm. When the single transmit selection scheme is used in cellular DAS, the symbol error probability can be obtained as in [22]

$$P_{e, sel} \approx \int_0^\infty \overline{N}_e Q \left(\sqrt{\frac{\gamma d_{min}^2}{2}} \right) f_\gamma(\gamma) d\gamma \tag{3.18}$$

$$= \frac{\overline{N}_e}{2} \left[1 - \sqrt{\frac{d_{min}^2 L_m^{(0)} P}{d_{min}^2 L_m^{(0)} P + 28\sigma_{z1}^2}} \right] \tag{3.19}$$

where $m = \arg\max_{i \in \{0,1,\cdots,6\}} \{L_0^{(j)}, L_1^{(j)}, \cdots, L_6^{(j)}\}$ and $\sigma_{z_1}^2 (= \sum_{j=1}^{6} L_m^{(j)} P_m^{(j)} + \sigma_n^2)$ is variance of interference plus noise for single transmit selection. Similarly, the symbol error probability for the conventional cellular system is given by

$$P_{e,conv} = \frac{\overline{N}_e}{2} \left[1 - \sqrt{\frac{d_{min}^2 L_0^{(0)} P}{d_{min}^2 L_0^{(0)} P + 4\sigma_c^2}} \right] \tag{3.20}$$

where σ_c^2 is the interference-plus-noise power in the conventional cellular structure.

In Figure 3.4(a/b), symbol error probability (SER) of cellular DAS with the optimal transmit diversity scheme is shown versus the normalized distance from the home base station in the direction of the worst-case position W. In this figure, QPSK modulation is used and the effects of lognormal shadowing are not considered. Figure 3.4(a/b) demonstrates that the cellular DASs exploit macroscopic transmit diversity and thus show substantially better symbol error performance than the conventional cellular system beyond the normalized distance 0.4. Because the blanket transmission scheme in DASs achieves maximum transmit diversity gain by employing the phase steering with allowable full power transmission, the blanket transmission scheme outperforms the single selection scheme near the home base station, whereas the single transmit selection scheme shows similar or slightly better performance at the cell boundary because performance is limited by multi-cell interference at the cell boundary.

3.4 ACHIEVABLE CAPACITY OF MULTI-CELL DASs

This section investigates the achievable capacity of cellular systems with distributed antennas from an information-theory standpoint in order to upper bound the theoretically achievable system capacity. The capacity analysis is provided for both when the channel state information is known only at the receiver (CSIR) and when the full channel state information is known to the transmitter (CSIT). The emphasis of this section will be on analyzing the achievable capacity of the distributed antenna systems and showing capacity gains over conventional cellular systems. Eventually, this section helps the readers understand the capacity gains possible from cellular systems with distributed antennas.

3.4.1 Ergodic Capacity with CSIT

If channel state information is available at the transmitter, the achievable ergodic capacity at a given location of the target mobile station is obtained by a proper design of the transmit covariance matrix, such that

$$C_e = \mathbb{E}_\mathbf{h} \left[\max_{S_{ii} \leq P_i^{(0)} \forall i} \log_2 \left(1 + \frac{1}{\sigma_z^2} \mathbf{h} \mathbf{S} \mathbf{h}^H \right) \right] \tag{3.21}$$

where S_{ii} is the ith diagonal element of the transmit covariance matrix $\mathbf{S}$. If power cooperation among distributed modules is allowable, the problem is solved by a standard water filling as in conventional colocated MISO systems even though water filling should be done over a macroscopic MISO vector channel. In this case, the capacity achieving transmission strategy is the transmit-MRC. However, the per antenna module power constraints require a new solution instead of the conventional water-filling algorithm.

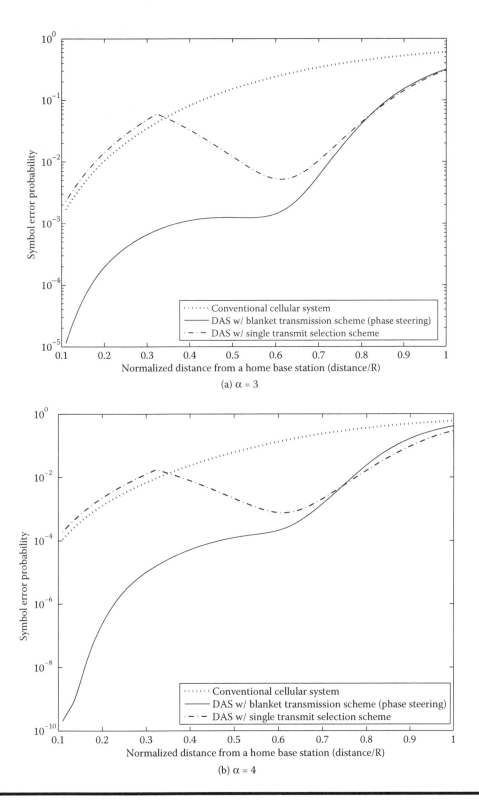

(a) $\alpha = 3$

(b) $\alpha = 4$

Figure 3.4   Uncoded Symbol Error Probability Versus the Normalized Distance from the Home Base Station.

Recent work [16] provided an intuitive solution of (3.21). It showed that per antenna module power constraints affect the off-diagonal components of the transmit covariance matrix $\mathbf{S}$, whereas only diagonal components determine the optimal transmit covariance matrix in a conventional colocated MISO system with a single sum power constraint. Because the log function is a monotonic increasing function with respect to $\mathbf{h}\mathbf{S}\mathbf{h}^H$, the problem (3.21) is reformulated as in [16]

$$\max_{S_{ii} \leq P_i^{(0)}, \forall i} \sum_{i=0}^{6} |h_i^{(0)}|^2 L_i^{(0)} S_{ii} + \sum_{i=0}^{6} \sum_{j=0, j \neq i}^{6} h_i h_j^* \sqrt{L_i^{(0)}} \sqrt{L_j^{(0)}} S_{ij} \qquad (3.22)$$

where h^* denotes the conjugate of h and S_{ij} is the (i, j) entry of the matrix $\mathbf{S}$. This reformulation suggests a two stage approach to find the optimal transmit covariance matrix:

1. The first term of the objective function (3.22) is maximized by setting the diagonal components of $\mathbf{S}_{opt}$ to be $\mathrm{diag}(\mathbf{S}_{opt}) = (P_0^{(0)}, P_1^{(0)}, \cdots, P_6^{(0)})$.
2. Although there are no explicit constraints on off-diagonal components S_{ij}, they should satisfy $S_{ij} \leq \sqrt{P_i^{(0)} P_j^{(0)}}$ for $\forall i, j, i \neq j$ because $\mathbf{S}_{opt}$ is a covariance matrix. Therefore, the second term of the objective function (3.22) is maximized by setting $S_{ij} = \sqrt{P_i^{(0)} P_j^{(0)}} e^{-j\angle h_i h_j^*}\angle$ for $\forall i, j, i \neq j$.

As a result, the achievable ergodic capacity with per antenna module power constraints is given by [16]

$$C_e = \mathbb{E}_{\mathbf{h}} \left[\log_2 \left(1 + \frac{\sum_{i=0}^{6} |h_i^{(0)}|^2 L_i^{(0)} P_i^{(0)} + \sum_{i=0}^{6} \sum_{j=0, j \neq i}^{6} |h_i h_j^*| \sqrt{L_i^{(0)} L_j^{(0)} P_i^{(0)} P_j^{(0)}}}{\sigma_z^2} \right) \right]$$

$$(3.23)$$

We should emphasize here that the SINR term of the achievable ergodic capacity is the same as (3.16), which indicates that the best transmission strategy of DASs *with* per antenna module power constraints is for each antenna module and home base station to transmit allowable full power with phase steering, whereas the best transmission strategy of DASs *without* per antenna module power constraints is transmit-MRC. This result also indicates that DASs with per antenna module power constraints do not require any channel amplitude information at the transmitter to exploit the advantage of CSIT.

Although it is difficult to obtain a closed-form expression for the achievable ergodic capacity, we can obtain an upper bound on the achievable capacity by applying Jensen's inequality:

$$C_e \leq \log_2 \left(1 + \frac{\sum_{i=0}^{6} 4L_i^{(0)} P_i^{(0)} + \sum_{i=0}^{6} \sum_{j=0, j \neq i}^{6} \pi \sqrt{L_i^{(0)} L_j^{(0)} P_i^{(0)} P_j^{(0)}}}{4\sigma_z^2} \right) \qquad (3.24)$$

The ergodic capacities of the single transmit selection scheme and the conventional cellular system can be obtained similarly because derivation for a MISO vector channel is a generalization of a single-input single-output (SISO) channel.

Figure 3.5(a) shows the ergodic capacity when channel state information is known to the transmitter according to the normalized distance from the home base station when the pathloss exponent is 3.5. We assume that the transmit power of each distributed antenna module and the home base station in the DAS is $P/7$, whereas the transmit power of the base station in the traditional cellular system is P. Both blanket transmission and single transmit selection of DASs have substantially higher capacity beyond the normalized distance 0.5 owing to the reduced access distance to the distributed antenna module located around $0.6R$. The single transmit selection has superior other-cell-interference (OCI) reduction capability to the blanket transmission scheme and, hence, outperforms the blanket transmission scheme except in the region of $0.25R <$ normalized distance $< 0.4R$. In a single transmit antenna module selection scheme, only a distributed antenna module or a home base station is selected for transmission in each cell and other distributed antenna modules shut off their transmission. So there is a switch between a home base station and a distributed antenna module located at $0.6R$ in the region of $0.25R <$ normalized distance $< 0.4R$, and the curve has a cusp around $0.3R$. On the other hand, all the distributed antenna modules and home base stations are transmitting their signals in the blanket transmission, so the curve looks dull and is better than the selection scheme in the region of $0.25R <$ normalized distance $< 0.4R$. If we consider uniformly distributed mobile units in a cell, the ergodic capacity averaged over user geometry is plotted versus the pathloss exponent in Figure 3.5(b). DASs show a large capacity gain over conventional cellular systems because three times more users are outside of the radius $1/2R$ within a cell assuming uniform distribution.

3.4.2 Ergodic Capacity Only with CSIR

This subsection investigates the achievable capacity of DAS when the channel state information is known only at the receiver (CSIR). Because CSIT is not available, the transmit covariance matrix $\mathbf{S}$ should be $\mathrm{diag}(P_0^{(0)}, P_1^{(0)}, \cdots, P_6^{(0)})$. Then, the ergodic capacity is given by [16]

$$C_e = \mathbb{E}_{\mathbf{h}}\left[\log_2\left(1 + \frac{1}{\sigma_z^2}\sum_{i=0}^{6}|h_i^{(0)}|^2 L_i^{(0)} P_i^{(0)}\right)\right]$$

$$= \int_{x=0}^{\infty} \log_2(1 + \gamma) f_\gamma(\gamma)d\gamma \tag{3.25}$$

where $\gamma = \frac{1}{\sigma_z^2}\sum_{i=0}^{6}|h_i^{(0)}|^2 L_i^{(0)} P_i^{(0)}$ is a weighted chi-squared distributed random variable with p.d.f. given by

$$f_\gamma(\gamma) = \sum_{i=0}^{6}\frac{\sigma_z^2 \pi_i}{L_i^{(0)} P_i^{(0)}}\exp\left(-\frac{\sigma_z^2 \gamma}{L_i^{(0)} P_i^{(0)}}\right) \quad \text{where } \pi_i = \prod_{k=0, k\neq i}^{6}\frac{L_i^{(0)} P_i^{(0)}}{L_i^{(0)} P_i^{(0)} - L_k^{(0)} P_k^{(0)}} \tag{3.26}$$

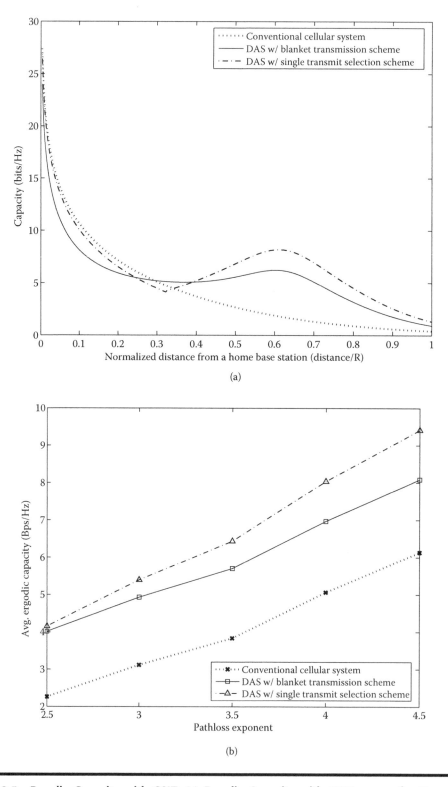

Figure 3.5   Ergodic Capacity with CSIT. (a) Ergodic Capacity with CSIT versus the Normalized Distance from the Home Base Station ($\alpha = 4$); (b) Average Ergodic Capacity with CSIT versus the Pathloss Exponent.

A closed-form expression for (3.25) is provided in [16] as

$$C_e = -\frac{1}{\ln 2} \sum_{i=0}^{6} \pi_i \exp\left(-\frac{\sigma_z^2}{L_i^{(0)} P_i^{(0)}}\right) Ei\left(-\frac{\sigma_z^2}{L_i^{(0)} P_i^{(0)}}\right) \tag{3.27}$$

where $Ei(x)$ is the exponential integral function ($Ei(x) = -\int_{-x}^{\infty} e^{-t}/t \, dt$).

Figure 3.6(a) shows the ergodic capacity when channel state information is known to the transmitter, with the same setup as Figure 3.5(a). When the target mobile stations are uniformly distributed in space, the ergodic capacity averaged over uniform user distribution is plotted versus the pathloss exponent in Figure 3.6(b). These figures overall show similar trends to the CSIT case, although the gap between blanket transmission and single transmit selection is smaller because the absence of CSIT makes blanket transmission less effective.

3.5 MULTI-CELL DISTRIBUTED MIMO SYSTEMS

In this section, we extend the concept of distributed antenna systems to distributed MIMO systems, where each distributed antenna module and mobile unit has multiple antennas. Distributed MIMO systems can effectively increase the degrees of freedom via multiple transmit antennas to the level which cannot be achieved in a conventional MIMO cellular system, where the number of antennas installable at a base station is practically limited due to zoning restriction. Furthermore, data cooperation among distributed antenna modules and a home base station enables distributed MIMO systems to achieve higher capacity than the system not allowing data cooperation, such as a pico-cell system.

As in distributed antenna systems with a single antenna per distributed antenna module, achievable capacities with CSIT and with CSIR are investigated. Although distributed MIMO systems construct macroscopic MIMO channels, per distributed antenna module power constraints make it difficult to apply the previously known MIMO results in the analysis [24]. Per distributed antenna module power constraints in a distributed MIMO system correspond to antenna group power constraints in a conventional colocated MIMO system. This section mainly relies on the results of [24].

3.5.1 System Model

In a distributed MIMO system, the ith distributed antenna module in the jth cell is assumed to have $t_i^{(j)}$ transmit antennas, so the effective number of transmit antennas in a distributed MIMO system becomes $M_t = \sum_{i=0}^{N} t_i^{(0)}$. The number of receive antennas per mobile station is assumed to be M_r. Correspondingly, a distributed MIMO system constructs an $M_r \times M_t$ composite MIMO channel given by

$$\mathcal{H} = \begin{bmatrix} \mathbf{H}_0^{(0)} & \mathbf{H}_1^{(0)} & \cdots & \mathbf{H}_N^{(0)} \end{bmatrix} \begin{bmatrix} \mathbf{L}_0^{(0)} & 0 & \cdots & 0 \\ 0 & \mathbf{L}_1^{(0)} & \vdots & \vdots \\ \vdots & \cdots & \ddots & 0 \\ 0 & \cdots & 0 & \mathbf{L}_N^{(0)} \end{bmatrix}$$

$$\triangleq \mathbf{HL} \tag{3.28}$$

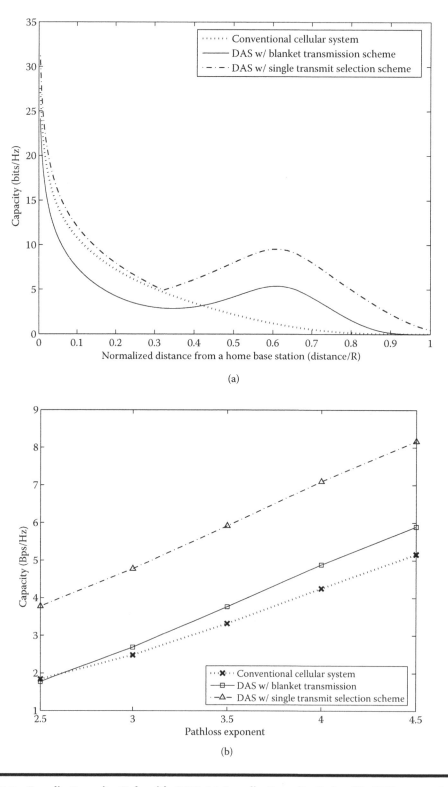

Figure 3.6   Ergodic Capacity Only with CSIR. (a) Ergodic Capacity Only with CSIR versus the Normalized Distance from the Home Base Station ($\alpha - 4$); (b) Average Ergodic Capacity only with CSIR versus the Pathloss Exponent.

where $\mathbf{H}_i^{(j)} \in \mathbb{C}^{M_r \times t_i^{(j)}}$, $i, j \in \{0, 1, \cdots, N\}$, and $\mathbf{L}_i^{(j)} = \sqrt{L_i^{(j)}} \mathbf{I}_{t_i^{(j)}}$ denote the MIMO channel matrix representing short term fading and propagation pathloss from the ith distributed antenna module of the jth adjacent cell, respectively. We assume that each MIMO channel matrix $\mathbf{H}_i^{(j)}$ is static and independent during a symbol period. The entries of $\mathbf{H}_i^{(j)}$ are independent and follow a zero mean complex Gaussian distribution $\sim \mathcal{CN}(0, 1)$.

In a distributed MIMO system, each distributed antenna module has its own power constraint, and power cooperation across the distributed antenna modules is not allowed for practical consideration. A composite transmitted vector $\mathbf{x} = \left[(\mathbf{x}_0^{(0)})^T \, (\mathbf{x}_1^{(0)})^T \right.$ $\left. \cdots (\mathbf{x}_N^{(0)})^T \right]^T \in \mathbb{C}^{M_t}$, where $\mathbf{x}_i^{(0)}$ is a $t_i^{(0)} \times 1$ transmitted vector from the ith distributed antenna module in the home cell, has zero mean with a covariance matrix $\mathbb{E}[\mathbf{x}\mathbf{x}^H] = \mathbf{S}$. The per antenna module power constraints restrict the partial sums of diagonal entries of the transmit covariance matrix $\mathbf{S}$, such that

$$\sum_{\sum_{m=1}^{i-1} t_m^{(0)}+1}^{\sum_{m=1}^{i} t_m^{(0)}} S_{k,k} \leq P_i^{(0)}, \quad \forall i \in \{0, 1, \cdots, N\} \tag{3.29}$$

where $S_{k,k}$ is the (k, k) entry of the matrix $\mathbf{S}$. If we define a $t_i^{(0)} \times M_t$ matrix such that

$$\Xi_i = \left[\mathbf{0}_{t_i^{(0)} \times t_0^{(0)}} \cdots \mathbf{0}_{t_i^{(0)} \times t_{i-1}^{(0)}} \, \mathbf{I}_{t_i^{(0)}} \, \mathbf{0}_{t_i^{(0)} \times t_{i+1}^{(0)}} \cdots \mathbf{0}_{t_i^{(0)} \times t_N^{(0)}} \right] \tag{3.30}$$

each power constraint given as a partial sum of diagonal entries can be simply represented by

$$\mathrm{tr}(\Xi_i \mathbf{S} \Xi_i^H) \leq P_i^{(0)}, \quad \forall i \in \{0, 1, \cdots, N\} \tag{3.31}$$

3.5.2 Ergodic Capacity with CSIT

This subsection derives the achievable capacity when channel state information is available at the transmitter. For time varying fading channels, the optimal transmit covariance matrix $\mathbf{S}$ should be derived for each realization of $\mathbf{H}$, and then the ergodic capacity is obtained by averaging the achieved capacities such that

$$C_e = \mathbb{E}\left[\max_{\mathrm{tr}(\Xi_i \mathbf{S} \Xi_i^H) \leq P_i^{(0)}, \forall i} \log \left| \mathbf{S}_z + \mathbf{H}\mathbf{L}\mathbf{S}\mathbf{L}^H\mathbf{H}^H \right| - \log |\mathbf{S}_z| \right] \tag{3.32}$$

where $\mathbf{S}_z$ is the covariance matrix of interference-plus-noise $\mathbf{z} = \sum_{j \in \mathcal{I}} \sum_{i=0}^{N} \mathbf{H}_i^{(j)} \mathbf{L}_i^{(j)} \mathbf{x}_i^{(j)} + \mathbf{n}$, the set $\mathcal{I}$ contains the indices of interfering base stations, and $\mathbf{n}$ is the additive Gaussian noise vector with a covariance matrix $\mathbb{E}[\mathbf{n}\mathbf{n}^H] = \sigma_n^2 \mathbf{I}_r$. Although distributed MIMO systems construct macroscopic MIMO channels, per distributed antenna module power constraints make it difficult to apply the previously known MIMO results [24]. Even generating numerical results for the systems with per distributed antenna module power constraints[4] has been regarded as quite challenging [25].

To accommodate different power constraints per distributed antenna module, we can apply the rate splitting algorithm [26,27] or the submatrix decomposing algorithms [24]. Both of those algorithms lead to the same result, but the latter algorithm does not rely on the conceptual rate splitting method based on dirty paper coding (DPC) [28]. In this chapter, we skip the details of the algorithms and just provide numerical results. The readers may refer to the references [24,26,27] for the details of the algorithms.

Figure 3.7(a) and Figure 3.7(b) show the ergodic capacity with CSIT of cellular distributed MIMO systems versus the normalized distance from the home base station when the number of receive antennas at a mobile unit is 2 and 14, respectively. In these figures, the following conditions are assumed:

- Each distributed antenna module and a home base station has two antennas, $l_i^{(j)} = 2$.
- The pathloss exponent is $\alpha = 3.5$.
- The transmit power of each distributed antenna module and the home base station in DAS is $P/7$, whereas the transmit power of the base station in the traditional cellular system is P.
- A two-tier cellular structure with universal frequency reuse, where a given cell is surrounded by two continuous tiers of 18 cells.

As shown in previous figures, Figure 3.7(a) and Figure 3.7(b) show a nonmonotonic relationship between capacity and the normalized distance from the base station because the signal from a distributed antenna module becomes dominant as a mobile unit approaches the distributed antenna module located at $0.6R$. When the number of receive antennas is small, the single transmit selection scheme achieves the highest capacity because of the multi-cell interference reduction and macroscopic selection diversity. As the number of receive antennas increases, blanket transmission outperforms single transmit selection owing to increased spatial dimensions of macroscopic MIMO channels. Only if we are able to deploy a large number of receive antennas at the mobile station (which is unlikely) does blanket transmission become more effective than other discussed schemes.

In Figure 3.7(a), the achieved capacity by blanket transmission is slightly lower than that of a conventional cellular system near the home base station due to reduced transmit power. The achieved capacity of blanket transmission has substantially higher capacity beyond the normalized distance 0.5 because 80% of the users will be in $(0.5R, R)$ assuming a uniform distribution, while the other 20% have very high SINR because they are close to the home base station and farther from the interfering cells.

3.5.3 Ergodic Capacity Only with CSIR

When channel state information is available only at the receiver, the transmit covariance matrix **S** should be determined without knowledge of channel state information to maximize (3.32) under the per distributed antenna module power constraints. According

---

[4] Per distributed antenna module power constraints correspond to per antenna group power constraints in a conventional co-located MIMO system.

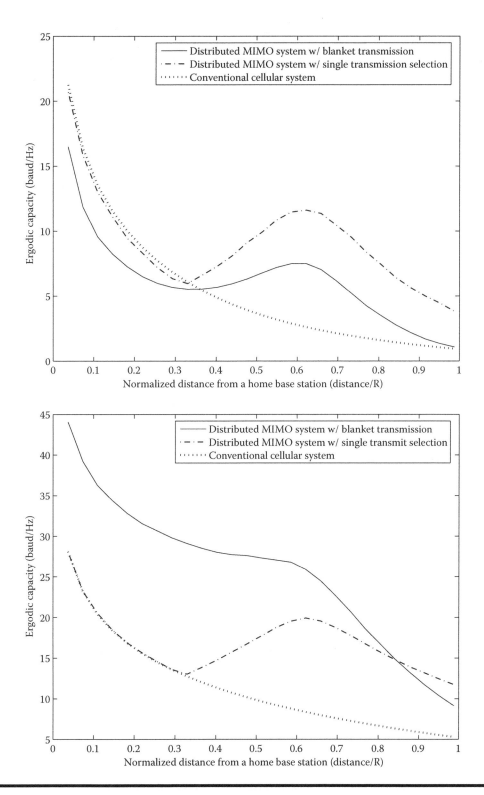

Figure 3.7   Ergodic Capacity with CSIT Versus the Normalized Distance from the Home Base Station.

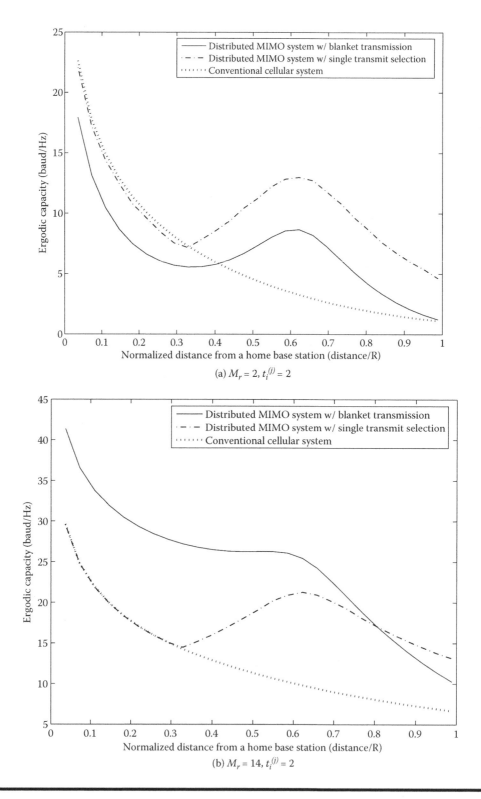

(a) $M_r = 2,\ t_i^{(j)} = 2$

(b) $M_r = 14,\ t_i^{(j)} = 2$

Figure 3.8 Ergodic Capacity Only with CSIR Versus the Normalized Distance from the Home Base Station.

to [24], the optimal transmit covariance matrix is given by

$$
\mathbf{S}^{opt} = \begin{bmatrix} \frac{P_0^{(0)}}{t_0^{(0)}}\mathbf{I}_{t_0^{(0)}} & 0 & \cdots & 0 \\ 0 & \frac{P_1^{(0)}}{t_1^{(0)}}\mathbf{I}_{t_1^{(0)}} & \vdots & \vdots \\ \vdots & \cdots & \ddots & 0 \\ 0 & \cdots & 0 & \frac{P_N^{(0)}}{t_N^{(0)}}\mathbf{I}_{t_N^{(0)}} \end{bmatrix} \tag{3.33}
$$

Then, the ergodic capacity only with CSIR is given by

$$
C_e = \mathbb{E}\left[\log\left|\mathbf{S}_z + \mathbf{H}\mathbf{L}\mathbf{S}^{opt}\mathbf{L}^H\mathbf{H}^H\right| - \log|\mathbf{S}_z|\right] \tag{3.34}
$$

The capacity gains of distributed MIMO systems when only CSIR is available are shown in Figure 3.8(a) and Figure 3.8(b). We consider the same setup as Figure 3.7(a) and Figure 3.7(b). Similar results to those with CSIT are observed in Figure 3.8(a) and Figure 3.8(b), but the gains of blanket transmission are smaller compared to CSIT cases because the increased spatial dimensions cannot be fully exploited when only CSIR is available.

3.6 CONCLUSION

This chapter has highlighted advantages of multi-cellular distributed antenna systems and analyzed achievable capacity/performance of multi-cellular systems with distributed antennas in the context of a macroscopic MIMO system. Analysis and simulations have shown that cellular systems with distributed antennas achieve a nontrivial capacity increase and symbol error probability improvement over conventional cellular systems, especially at the cell boundaries. These notable gains result from effectively increased spatial dimensions and the interference reduction capabilities of distributed multiple antenna architecture, whereas increasing the number of antennas is significantly limited by zoning restrictions in conventional colocated MIMO systems. Considering that one of the greatest challenges facing MIMO in the context of cellular systems is that present MIMO systems do not cope gracefully with high levels of interference, the large capacity gain achieved from deploying distributed antennas could be the key to the success of large scale cellular MIMO systems with high spectral efficiency.

REFERENCES

[1] A.A.M. Saleh, A.J. Rustako, and R.S. Roman, "Distributed antennas for indoor radio communications," *IEEE Trans. Commn.*, vol. 35, pp. 1245–1251, Dec. 1987.

[2] A. Salmasi and K. Gilhousen, "On the system design aspects of code division multiple access (CDAM) applied to digital cellular and personal communication networks," in *Proc., IEEE Veh. Technol. Conf.*, pp. 57–62, St. Louis, MO, 1991.

[3] R. Ohmoto, H. Ohtsuka, and H. Ichikawa, "Fiber-optic microcell radio system with a spectrum delivery scheme," *IEEE J. Sel. Areas Commn.*, vol. 11, no. 7, pp. 1108–1117, Sept. 1993.

[4] H. Yanikomeroglu and E.S. Sousa, "CDMA Sectorized distributed antenna system," in *Proc., IEEE Int. Symposium on Spread Spectrum Techniques and Applications*, pp. 792–797, Sun City, South Africa, Sept. 1998.

[5] M.V. Clark et al., "Distributed versus centralized antenna arrays in broadband wireless networks", in *Proc., IEEE Veh. Technol. Conf.*, pp. 33–37, Rhodes Island, Greece, May 2001.

[6] R.E. Schuh and M. Sommer, "WCDMA coverage and capacity analysis for active and passive distributed antenna systems," in *Proc., IEEE Veh. Technol. Conf.*, pp. 434–438, Birmingham, AL, May 2002.

[7] R. Hasegawa et al., "Downlink performance of a CDMA system with distributed base station," in *Proc., IEEE Veh. Technol. Conf.*, pp. 882–886, Oct. 2003.

[8] W. Rho and A. Paulraj, "MIMO channel capacity for the distributed antenna," in *Proc., IEEE Veh. Technol. Conf.*, pp. 706–709, Vancouver, Canada, Sept. 2002.

[9] W. Roh and A. Paulraj, "Outage performance of the distributed antenna systems in a composite fading channel," in *Proc., IEEE Veh. Technol. Conf.*, pp. 1520–1524, Vancouver, Canada, Sept. 2002.

[10] W. Rho and A. Paulraj, "Performance of the distributed antenna systems in a multi-cell environment," in *Proc., IEEE Veh. Technol. Conf.*, pp. 587–591, Jeiu, Korea, Apr. 2003.

[11] H. Zhuang, L. Dai, L. Xiao, and Y. Yao, "Spectral efficiency of distributed antenna systems with random antenna layout," *Elec. Lett.*, vol. 39, no. 6, pp. 495–496, Mar. 2003.

[12] H. Zhang and H. Dai, "On the capacity of distributed MIMO systems," in *Proc., Conf. Inform. Syst.*, Mar. 2004.

[13] Z. Ni and D. Li, "Effect of fading correlation on capacity of distributed MIMO," in *Proc., IEEE PIMRC*, pp. 1637–1641, Barcelona, Spain, Sept. 2004.

[14] Q. Zhou, H. Zhang, and H. Dai, "Adaptive spatial multiplexing techniques for distributed MIMO systems," in *Proc., Conf. Inform. Sys.*, Mar. 2004.

[15] L. Dai, S. Zhou, and Y. Yao, "Capacity analysis in CDMA distributed antenna systems," *IEEE Trans. Wireless Commn.*, vol. 4, no. 6, pp. 2613–2620, Nov. 2005.

[16] W. Choi and J.G. Andrews, "Downlink performance and capacity of distributed antenna systems in a multi-cell environment," *IEEE Trans. on Wireless Commn.*, Jan. 2007.

[17] A. Ghosh, J.G. Andrews, D.R. Wolter, and R. Chen, "Broadband wireless access with WiMax/802.16: current performance benchmarks and future potential," *IEEE Commun. Mag.*, vol. 43, no. 2, pp. 129–136, Feb. 2005.

[18] A. Goldsmith, *Wireless Communications*, 1st ed., New York: Cambridge University Press, 2005.

[19] S. Schwartz and Y. Yeh, "On the distribution function and moments of power sums with lognormal components," *Bell Syst. Tech. J.*, vol. 61, no. 7, pp. 1441–1462, Sept. 1982.

[20] T. Lo, "Maximal ratio transmission," *IEEE Trans. Commn.*, vol. 47, no. 10, pp. 1458–1461, Oct. 1999.

[21] J. Proakis, *Digital Communications,* New York: McGraw-Hill, 1995.

[22] I. Gradshteyn and I. Ryzhik, *Table of Integrals, Series, and Products*, London: Academic Press, 2003.

[23] T. Cover and J. Thomas, *Elements of Information Theory*, New York: Wiley-Interscience, 1991.

[24] W. Choi, J.G. Andrews, and R.W. Heath, Jr. "Achievable capacity of cellular systems with distributed multiple antennas," in preparation.

[25] A. Goldsmith, S. Jafar, N. Jindal, and S. Vishwanath, "Capacity limits of MIMO channels," *IEEE J. Sel. Areas Commn.*, vol. 21, no. 5, pp. 684–702, June 2003.

[26] S.A. Jafar, G. Foschini, and A.J. Goldsmith, "Phantomnet: exploring optimal multicellular multiple antenna systems," *EURASIP J. Appl. Signal Process., Spec. Issue MIMO Commun. Signal Process.*, pp. 591–605, May 2004.

[27] S. Jafar and A. Goldsmith, "Tramsmitter optimization for multiple antenna cellular systems," in *Proc., IEEE Intl. Symp. Inform. Theory*, Lausanne, Switzerland July 2002, p. 50.

[28] M. Costa, "Writing on dirty paper," *IEEE Trans. Info. Theory*, vol. 29, no. 3, pp. 439-441, May 1983.

4

COOPERATIVE COMMUNICATIONS
IN MOBILE AD HOC NETWORKS:
RETHINKING THE LINK
ABSTRACTION

Anna Scaglione, Dennis L. Goeckel and J. Nicholas Laneman

Contents

This chapter rethinks the link abstraction for wireless networks in the context of co-operative communications, which has recently received interest as an untapped means for improving the performance of relay transmission systems operating over the ever-challenging wireless medium. The common theme of most research in this area is to optimize physical layer performance measures without considering in much detail how cooperation interacts with higher layers and improves network performance measures. Because these issues are important for enabling cooperative communications to prac-tice in real-world networks, especially for the increasingly important class of mobile ad hoc networks (MANETs), the goals of this chapter are to survey basic cooperative com-munications and outline two potential architectures for cooperative MANETs. The first architecture relies on an existing clustered infrastructure: cooperative relays are centrally controlled by cluster heads. In another, without explicit clustering, cooperative links are formed by request of a source node in an ad hoc, decentralized fashion. In either case, cooperative communication considerably improves the network connectivity. Although far from a complete study, these architectures provide modified wireless link abstractions and suggest tradeoffs in complexity at the physical and higher layers.[1]

4.1 INTRODUCTION

Network architecture and the process of abstraction go hand in hand. For most wired networks, the notion of a link has been a useful abstraction directly tied to the physi-cal propagation medium. Indeed, this link abstraction has remained robust even under recent developments such as network coding [6]. For wireless networks, especially the increasingly important class of mobile ad hoc networks (MANETs), the classical notion of a link is more nebulous than in the wired case. Even so, two constraints are often imposed on network architectures to maintain it. These constraints include: (Constraint I) A functional physical layer communication link can originate from only one transmitter; (Constraint II) Concurrent transmissions of multiple transmitters result in interference that, if not sufficiently attenuated by spatial or channel multiplexing, produces a col-lision, i.e., a level of distortion for the useful signal that is irreversible at the ultimate receiver.

At various levels, many current MANET protocols attempt to create, adapt, and man-age a network based on a maze of point-to-point links, all conforming to Constraints I and II. Multi-hop transmission along routes consists of several intermediate links among pairs of nodes, and nodes use buffer space, power, and bandwidth to route their own data as well as data from other sources. Although an architecture based upon the classi-cal link abstraction leads to many advantages that should not be underestimated, it also creates several challenges in terms of collisions in medium access, as well as complexity and overhead in routing. Furthermore, there are a number of issues that arise in wire-less communications that fundamentally challenge the classical link abstraction upon which these architectures are based. Well-known examples include the broadcast nature of the wireless medium, interference from multiple simultaneous transmissions, coupled queues, and so forth. These observations, along with the emerging area of cooperation

---

[1] A shortened version of this work has been published as: "Cooperative communications in mobile ad hoc networks," Scaglione, A.; Goeckel, D.; Laneman, J.N. *IEEE Signal Processing Magazine*, Volume 23, Issue 5, Sept. 2006, pp. 18–20.

communications to be described in the next section, suggest that it is worthwhile to explore a broader solution space in which Constraint I or Constraint II is relaxed.

4.1.1 Cooperative Communications: A Top-Down Motivation

Multi-hop transmission is a special case of a broader class of transmission protocols called *cooperative communications* that have recently received significant attention in various communities. Within prevalent models for cooperation, Constraints I and II correspond to additional constraints on the transmission protocols, imposed for practical or architectural reasons. Much of the work on cooperative communications demonstrates improved performance largely from physical layer perspectives; however, because many of the advantages essentially result from violating either Constraint I or II, there is a great deal of room for design of network architectures that integrate cooperation, especially for MANETs. The goal of this chapter is to help bridge this gap by summarizing key ingredients of cooperative communications and illustrating two approaches for cooperative MANETs.

A cooperative link consists of separate radios encoding and transmitting their messages at the physical layer in coordination; these nodes could be a single source and relay, or they can be a group or relays, or both. As described in Section 4.2, physical layer researchers have championed the use of cooperative diversity in wireless networks, arguing that nodes equipped with a single antenna, through physical layer coding and signal processing, could achieve similar diversity and coding gains to those of colocated multi-antenna systems [75], while leveraging the distributed hardware and battery resources that are already available. Such arguments are based mostly on link quality metrics, such as the average error probability and the outage probability. As indicated by the two network models described briefly in the next section, this point of view should be expanded because cooperative communications is inherently a network solution, and there are issues of protocol layering and cross-layer architecture that must be explored jointly by a broad community of researchers. In addition to offering performance improvements in terms of network metrics such as connectivity, cooperation alleviates certain collision resolution and routing problems because it allows for simpler networks of more complicated links, rather than complicated networks of simple links.

4.1.2 Network Models

As further developed in the sequel, we consider cooperative communications within two wireless network models. We focus the discussion on how the cooperative groups are activated and supported. Although this perspective is insufficient to claim that we specify an entire architecture, it does suggest tradeoffs between centralized and decentralized architectures as well as complexity among the physical, link, medium access, and network layers of the protocol stack.

In Section 4.3, the first network model is a MANET with an existing clustered infrastructure, in which cooperative transmission is centrally activated and controlled by the cluster access points (APs). All terminals communicate through a cluster AP, which handles routing to other clusters. In the classical multi-hop architecture, each cluster is responsible for transmitting the message to a "gateway" node in the next cluster. In our cooperative network architecture, the AP uses multiple gateway nodes between clusters

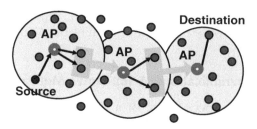

Figure 4.1 Cooperative Gateways.

(Figure 4.1), which propagate the message providing cooperative gains compared to the single gateway solution. Better links translate into better network connectivity compared to multi-hop solutions. Relying on existing techniques to determine the clustering structure, our objective is to describe how the AP can select the cooperative nodes by means of matching algorithms and how this benefits the network connectivity.

In Section 4.4, the second network model is a MANET in which a random source conveys extra control information and link parameters in the message to enable recipients to self-select and form a random cooperative cluster. The nodes recruited in this cluster can rely on the synchronization data available in the source packet. Within their estimation inaccuracies and propagation delays, the nodes can infer their transmission schedule. They can be ignorant of the codes chosen by the other nodes, but the resulting cooperative gains are close to those of a centralized scenario in which codes are explicitly assigned to the nodes. A small cluster of nodes can act as a source and recruit additional nodes to form a larger cluster, and so forth, to create multi-stage cooperation (Figure 4.2). Section 4.4 presents two main ideas concerning this architecture. First, we show that, as in a traditional channel access problem, multi-stage cooperative access can be randomized although not quite in the same way as traditional random access. Second, we demonstrate in multicast applications that multi-stage cooperative access requires up to 50% less power compared to multi-hop solutions.

As we will see, the new ingredient of cooperative communications suggests a re-thinking of the link abstraction and creates many opportunities and challenges from the physical layer to higher layers. In the clustered architectures, more work at the network

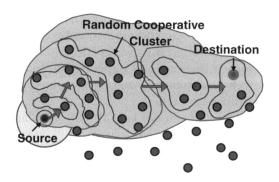

Figure 4.2 Randomized Distributed Cooperation.

layer is necessary in order to support cooperation. In the distributed approach we describe, the brunt of the work lies in the physical layer.

4.2 ELEMENTS OF COOPERATIVE COMMUNICATIONS

Early formulations of general relaying problems appeared in the information theory community [12,78] and were inspired by the concurrent development of the ALOHA system at the University of Hawaii. The classical relay channel model is comprised of three terminals (Figure 4.3): a source that transmits information, a destination that receives information, and a relay that both receives and transmits information in order to enhance communication between the source and destination. More recently, models with multiple relays have been examined [39,63,64]. Cooperation [40,41,65,66] is a generalization of the relay channel to multiple sources with information to transmit that also serve as relays for each other. Combinations of relaying and cooperation are also possible, and are often referred to generically as "cooperative communications." Less well-known is the fact that all of these models fall within the broader class of channels with generalized feedback [11,37,81,82].

Even after 40 years of intense study, the relay channel capacity is not known in general. Although useful bounds on capacity have been obtained for various approaches (see, e.g., the summary in [39]), it is thanks to our increased understanding of the benefits of multi-antenna systems in wireless channels [75] that many have come to realize that multiple relays can emulate the strategies designed for multiple transmit antenna systems and offer significant network performance enhancements in terms of various metrics, including: increased capacity (or larger capacity region); improved reliability in terms of diversity gain, diversity-multiplexing tradeoff, and packet- or symbol-error probabilities. The interest, therefore, has percolated to other communities, and today there are many specific practical solutions to harvest diversity from a network. Early examples are: [3,4,9,14,18,27,28,30,31,36,40–43,51,60,61,65–68] and many more in recent years ([54] provides other useful references). Multipath diversity instead of antenna diversity is exploited in [7,18,60].

In this section, we summarize the main elements of cooperative communication protocols, and we illustrate their performance advantages. Like the large part of models that are studied in the cooperative communication literature, we assume that the cooperative nodes do not know the channel response at the transmitter, but that it can be estimated

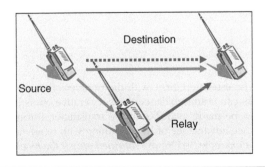

Figure 4.3 Three Nodes Model.

at the receiver; we assume that the estimate is without error.[2] Receiver cooperation in the form of compress and forward schemes [38] is not considered.

4.2.1 Physical Layer Model for Cooperative Radios

We assume that each radio has a baseband equivalent, discrete-time transmit signal $X_i[k]$, with average power constraint $\sum_{k=0}^{K-1} |X_i[k]|^2 \leq K P_i$, where K is the duration of the signal, and the receive signal is $Y_i[k]$, $i = 1, 2, ..., N$. Hardware limitations introduce the so-called "half-duplex" constraint, namely, the impossibility of concurrent radio transmission and reception. Incorporating this constraint, we model the discrete-time received signal at radio i and time sample k as [38]

$$
Y_i[k] = \begin{cases} \sum_{j=1, j\neq i}^{N} H_{ij}[k] \times X_j[k] + W_i[k] & \text{if radio } i \text{ receives at time } k \\ 0 & \text{if radio } i \text{ transmits at time } k \end{cases} \tag{4.1}
$$

where $H_{i,j}[k]$ captures the combined effects of symbol asynchronism, frequency-selective quasi-static multipath fading, shadowing, and pathloss between radios i and j; $W_i[k]$ is a sequence of mutually independent, circularly symmetric, complex Gaussian random variables with common variance N_0 that models the thermal noise and other interference received at radio i. Note that $H_{i,j}[k]$ is assumed to be fixed during the block length. Radio i knows the realized $H_{i,j}[k]$ but not $H_{p,j}[k]$, for $p \neq$ g, and $j = 1, 2, ..., t$. The $H_{i,j}[k]$ are modeled as independent complex-valued random variables for different j, which is reasonable for scenarios in which the radios are separated by a number of carrier wavelengths (in all cases, each transmitter has an independent random phase due to its local oscillator). Nodes that cooperate share a common message, which was transmitted previously by one or more nodes and received by the group of cooperating nodes. General relaying is done by mapping the message embedded in the received vector $\mathbf{y}_i = (Y_i[0], \ldots, Y_i[K-1])^T$ onto a matrix code where each column is the new relay signal. Specifically, we can consider a portion or the entire decoded message as a vector of length M denoted by $\mathbf{s} = (S[0], \ldots, S[M-1])^T$; each one of the T cooperating relay nodes transmits a column $\mathbf{x}_r = (X_r[0], \ldots X_r[K-1])^T$ of a $K \times T$ matrix code $\mathbf{X} = \mathbf{G}_{K \times T}(\mathbf{s})$ (Figure 4.4). Denoting by $\log_2(|S|)$ the number of bits per symbol, $(M/K)\log_2(|S|)$ is the spectral efficiency of the code. The number of columns T is the number of cooperating nodes. Different cooperative schemes correspond to different instantiations of the mapping $\mathbf{s} \rightarrow \mathbf{G}(\mathbf{s})$. Hence, cooperative transmission is equivalent to a multi-input single-output system (MISO) with a per antenna power constraint.

---

[2] The channel parameters to be estimated grow with the number of cooperating nodes, and the effect of channel estimation errors can counterbalance the cooperative gains [24]. For situations in which the channel parameters can be made available at the transmitter, through feedback or by duality, one can earn the performance advantages of beamforming with bandwidth efficiency equal to one. MIMO gains prospected, for example, in [76] *are attained only if the receivers process the data jointly.* When, instead, the receivers process the data independently the spectral efficiency is always ≤ 1 and, when the channel is known only at the receiver, for blocks of finite size > 2, the spectral efficiency is always strictly less than one [75].

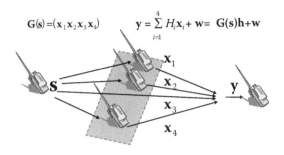

Figure 4.4   Codes for Cooperative Transmission.

4.2.1.1 Modulation and Channel Coding for Cooperation

The simplest setting possible to isolate the benefits of spatial diversity is that of frequency flat fading channels $H_{i,j}[k] = H_{i,j}\delta[k]$. The received data vector $\mathbf{y}_i = (Y_i[0], \ldots, Y_i[K-1])^T$ is:

$$\mathbf{y}_i = \sum_{r=1}^{T} H_{i,r}\mathbf{x}_r + \mathbf{w}_i = \mathbf{X}\mathbf{h}_i + \mathbf{w}_i \tag{4.2}$$

where $\mathbf{h}_i = (H_{i,1}, \ldots, H_{i,T})^T$ is the vector of the relays' fading coefficients. The simplest forms of cooperative diversity are the so-called *amplify and forward* (AF) and *decode and forward* (DF). In the AF strategy, for each transmit symbol S the nodes retransmit a scaled version of the samples received over orthogonal channels. This can be expressed in our general model by the following coding rule with $\mathbf{s} = S$ and $\mathbf{G(s)}$:

$$\mathbf{G}(S) = diag\,(\beta_1 Z_1, \ldots, \beta_T Z_T)$$

$$Z_r = \mathbf{h}_r^H \mathbf{y}_r, \; \beta_r \leq \sqrt{\frac{P_r}{E\left\{\mathbf{h}_r^H \mathbf{y}_r \mathbf{y}_r^H \mathbf{h}_r\right\}}} \tag{4.3}$$

where $\mathbf{y}_r$ in $Z_r = \mathbf{h}_r^H \mathbf{y}_r$ is the received vector containing the symbol S of the message, and the constraints on the scaling coefficients β_r guarantee that the node transmit power is P_r. For the DF strategy, the nodes decode each symbol of the message and transmit the decoded symbol over orthogonal channels. Thus, the code matrix that corresponds to the DF is:

$$\mathbf{G}(S) = diag\left(\sqrt{P_1}\hat{S}_1, \ldots, \sqrt{P_T}\hat{S}_T\right) \tag{4.4}$$

In both cases it is assumed that each relay transmits in an orthogonal channel, so $K = T$ and $M = 1$, resulting in a spectral efficiency equal to $(1/T)\log_2(|S|)$ that decreases with the number of nodes. Greater spectral efficiency can be achieved using space–time codes $\mathbf{X} = \mathbf{G}_{K \times T}(\mathbf{s})$ (e.g., [1]) instead of AF or DF, which tend to attain diversity gains that are similar to those of DF but have both M and K growing in the same order.

4.2.1.2 Channel Synchronization for Cooperative Transmission

In reality the relays will not transmit in perfectly orthogonal channels. In this section we show how timing offset and carrier frequency offset (CFO) can be incorporated in the model and how they can be managed with conventional designs and synchronization algorithms.

First, we shall consider the effect of timing offset. A relative time offset among the nodes produces signal dispersion analogous to that of a frequency selective channel. Given that synchronization algorithms at the physical layer can achieve subsymbol synchronization, network synchronization algorithms should not be used to synchronize cooperative relays. With subsymbol synchronization accuracy attained through training at the physical layer, the timing offset effect will be dominated by the difference in propagation delay (which can be contained through network management of the size of the cooperative clusters) and the clock jitter of the devices (which is negligible if the retransmissions are not procrastinated). In an equivalent discrete time model, the effect can be modeled approximately with a complex-valued, possibly time-varying finite impulse response (FIR) filter of order D. To write a received vector $\mathbf{y}_i = (Y_i[0], \ldots, Y_i[K - D - 1])$ without having to consider possible inter-block interference (IBI), we can set the code matrix duration K so that $D \leq 2K$. Let us denote by $\mathbf{H}_{i,r}$ the channel convolution matrix between the i-th and the r-th terminals:

$$\mathbf{y}_i = \sum_{r=1}^{T} \mathbf{H}_{i,r}\mathbf{x}_r + \mathbf{w}_i \tag{4.5}$$

where $\mathbf{w}$ is the additive white Gaussian noise (AWGN) vector and $\mathbf{H}_{i,r}$ are $(P - D) \times P$ Toeplitz convolution matrices with first column $(H_{i,r}[D], 0, .., 0)^T$ and first row $(H_{i,r}[D], ..., H_{i,r}[0], 0, .., 0)^T$. Channels with $D > 2K$ will incur additional IBI, unless adequate guards are inserted between blocks. If these guards are inserted after the transmissions of several blocks, the matrix $\mathbf{G}(\mathbf{s})$ can represent several subsequent length K blocks, transmitted consecutively between guards or training symbols so that (4.5) is still valid. One can rearrange (4.5) by forming Toeplitz matrices such that $\mathbf{H}_{i,r}\mathbf{x}_r = T(\mathbf{x}_r)\mathbf{h}_{i,r}$, where in this case $\mathbf{h}_{i,r} = (H_{i,r}[0], ..., H_{i,r}[D])^T$, so that:

$$\mathbf{y}_i = \sum_{r=1}^{T} T(\mathbf{x}_r)\mathbf{h}_{i,r} + \mathbf{w}_i = \mathcal{X}\mathbf{h}_i + \mathbf{w}_i; \quad \mathcal{X} \triangleq (T(\mathbf{x}_1), ..., T(\mathbf{x}_T)); \quad \mathbf{h}_i \triangleq (\mathbf{h}_{i,1}^T, ..., \mathbf{h}_{i,T}^T)^T \tag{4.6}$$

Then it can be easily recognized how similar (4.2) and (4.6) are, leading to the following conclusion: the dispersive medium effectively operates as an additional source of diversity that can be exploited by a judicious design of the matrix code $\mathbf{X} \rightarrow \mathcal{X}$.

In fact, the cooperative multipath designs in [7,61,83] achieve diversity by having the cooperative nodes behave as active multipath scatterers and require no prior channel (in this case path delay) assignment. Each node can choose a specific delay, which amounts to selecting a column of the matrix:

$$\mathbf{G}_{ij}(\mathbf{s}) = X[i - j] \tag{4.7}$$

where the sequence $X[k]$ could simply equal the message $S[k]$ or could be encoded to guarantee the extraction of diversity from it. More specifically, (4.7) can be combined with spread spectrum techniques, or orthogonal frequency division multiplexing (OFDM) (see, e.g., [7]) and, correspondingly, $\mathbf{x}$ is constructed in the two following ways:

$$\mathbf{x} = \mathbf{cs}^T, \quad \mathbf{c}: \text{ spreading code;}$$

$$\mathbf{x} = \mathbf{Fs}, \quad \mathbf{F}: \text{ inverse fast Fourier transform (IFFT) matrix + prefix} \qquad (4.8)$$

Cooperative multipath has numerous benefits: (1) it enables receiver architectures that have reduced complexity; (2) it provides simple options for multiplexing sources using different spreading codes or subcarriers; and (3) OFDM can have very high spectral efficiency,[3] and simplifies the problem of designing large space–time code matrices $\mathbf{G(s)}$.

As mentioned earlier, cooperative nodes have distinct oscillators in their radio frequency (RF) front ends which cannot be perfectly tuned. CFO is not a unique problem of cooperative transmission; it is present in all wireless uplink communication channels. The CFO introduces time variations that hinder the modeling done above in one aspect: the $\mathbf{H}_{i,r}$ are actually $(P - D) \times P$ time-varying convolution matrices. The higher the CFO, the shorter the channel coherence time, and the smaller the block size P for which the effective channel is approximately time invariant. In considering issues that may arise in implementing cooperation, one needs to recognize that our ability to design decoding algorithms today is limited to so-called under-spread channels, having a small product of delay spread and channel spread compared to the time-bandwidth product of the modulation. As for a variety of wireless standards used today, neglecting inter-symbol interference (ISI) or CFO issues is a valid zero-order approximation to identify codes and trends; including the effect of an underspread ISI channel is a good first-order approximation of reality.

4.2.2 Performance Benefits

Having described some basic relaying algorithms, we now turn to illustrating their performance benefits at the physical layer. For simplicity, we refer to the perfectly synchronous model, although via (4.6) several observations can be generalized. We first use a simple argument based on large scale attenuation only, to demonstrate that cooperation is more power efficient than routing when multicasting to several destinations. The performance benefits of cooperation are described more often for a link with a single termination. In this case, we consider as performance metrics the outage probability and average error probability. Outage probability allows for analysis of systems independent of a specific code design, because it is an information theoretic framework based on random coding. It gives an asymptotic bound on the rate of outage (packet loss) of a link at a given spectral efficiency, where the limit is taken over the code length. Error probability allows for analysis and design of specific codes of limited coding block lengths at a given spectral efficiency. Both frameworks can account for additional temporal or frequency diversity in the system.

4.2.2.1 Power Efficiency of Cooperative Links

Assume that the goal is to have both R1 and R2 receive a power normalized to be equal to 1, and that the path loss is simply $d_{ij}^{-\alpha}$ (Figure 4.5). The power to be sent

---

[3] OFDM requires the addition of a cyclic prefix, which, based on (4.7), should be of length T, so $K = M + T$ and the spectral efficiency is $\log_2(|S|)M/(M + T)\rho$, where ρ is the encoding rate of the symbols that is necessary to harvest the frequency diversity. For a large M, it tends to be $\log_2(|S|)\rho$.

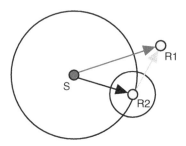

Figure 4.5   Cooperative Wireless Advantage.

with routing is the solution of a *minimum spanning tree* problem and is $P_{ROUTING} = \min(d_{S1}^\alpha + d_{S2}^\alpha, d_{S1}^\alpha + d_{12}^\alpha)$.

In a broadcast medium, as noted by [80], when the farthest node R1 is reached the closer one is in range, and thus one should spend a total power equal to $P_{MULTICAST} = \min(d_{S1}^\alpha, d_{S1}^\alpha + d_{12}^\alpha) \leq P_{ROUTING}$. Further observed in [45,46] and [27] is that node R1 could accumulate the power sent to reach the closer node R2 and only ask for the residual power to be sent from the less attenuated of the sources, with total power expenditure $P_{COOPERATIVE} = \min(d_{S1}^\alpha, d_{S1}^\alpha + (1 - d_{S1}^\alpha/d_{S2}^\alpha)d_{12}^\alpha) \leq P_{MULTICAST} \leq P_{ROUTING}$. This argument for three nodes is sufficient to establish the power efficiency of cooperation over routing. Note that routing is a special case of cooperation and, therefore, cooperating does not restrict the solution space.

4.2.2.2   Outage Probability

We study the outage probability 55 using the simple model in Figure 4.3 and use the indices *s*, *d*, and *r* to denote the source, the destination, and the relay nodes, respectively. Assuming that the channels are quasi-static over the transmission of each message, the channel mutual information becomes a random variable as a function of the fading coefficients, and the outage probability is then the probability that the mutual information random variable falls below the rate chosen *a priori* to encode the message. Focusing on outage probability allows us to easily account for the decreased spectral efficiency required by half-duplex operation in the relays. In the following $\gamma_s \triangleq P_s/N_0$, $\gamma_r \triangleq P_r/N_0$.

4.2.2.2.1   Noncooperative Transmission

To be more precise, and for comparison with the results to follow, let us compute the outage probability of a system without cooperative diversity in the model (4.1) from radio *s* to radio *d*. In this case, the mutual information, in bits per channel use,[4] viewed as a function of the fading coefficient $H_{d,s}$, satisfies [13,76]:

$$I_{NC} = \log\left(1 + |H_{ds}^2|\gamma_s\right) \tag{4.9}$$

The outage probability for rate R, in bits per channel use, is then given by 55:

$$p_{out}^{NC} := Pr[I_{NC} \leq R] = Pr\left[|H_{ds}|^2 \leq (2^R - 1)\gamma_s^{-1}\right] \tag{4.10}$$

---

[4] All logarithms are taken to the base 2 unless otherwise indicated.

Note that if radios s and d transmit and receive, respectively, in only L out of the K channel uses, the mutual information random variable becomes:

$$I_{NC} = (L/K)\log\left(1 + (K/L)|H_{ds}^2|\gamma_s\right) \tag{4.11}$$

Because the number of channel uses is reduced by the factor (L/K), radio s can increase its transmitted power per channel use by the factor (K/L) and remain within its average power constraint for the entire block. (More details can be found in [55].) This observation is useful for studying half-duplex relaying.

4.2.2.2.2 Cooperative Transmission

Outage results for cooperative transmission can be obtained by extending similar results for multiple-input multiple-output (MIMO) systems [76]. The simplest AF algorithm for a single source and relay produces an equivalent one-input two-output, complex Gaussian noise channel with different noise levels in the outputs. As 43 details, the mutual information random variable is

$$I_{AF} = \frac{1}{2}\log\left(1 + 2|H_{ds}|^2\gamma_s + f(2|H_{rs}|^2\gamma_s, 2|H_{dr}|^2\gamma_r)\right) \tag{4.12}$$

as a function of the fading coefficients, where:

$$f(x, y) = \frac{xy}{x + y + 1} \tag{4.13}$$

For the simplest selection DF algorithm with repetition coding 43, the mutual information random variable is

$$I_{RDF} = \begin{cases} \frac{1}{2}\log(1 + 2|H_{ds}|^2\gamma_s) & \text{if } \frac{1}{2}\log(1 + 2|H_{rs}|^2\gamma_s) \le R \\ \frac{1}{2}\log(1 + 2|H_{ds}|^2\gamma_s + |H_{dr}|^2\gamma_r) & \text{if } \frac{1}{2}\log(1 + 2|H_{rs}|^2\gamma_s) > R \end{cases} \tag{4.14}$$

The two cases in (4.14) correspond to the relays not being able to decode and being able to decode, respectively. More sophisticated space–time coding in distributed form can be employed (see e.g., [42]). For comparison, we show their outage probability in Figure 4.6 (labeling them as space–time decode and forward (STC-DF)), but leave a detailed analysis of the error probability of specific codes in Section 4.2.2.3.

The outage probabilities $p_{\text{Out}}^{AF} := \Pr[I_{AF} \le R]$ for (4.12) and $p_{\text{Out}}^{RDF} := \Pr[I_{RDF} \le R]$ for (4.14) can be evaluated numerically (Figure 4.6). The term *diversity order* that we have so far used informally, is in this case defined as the negative slope of a plot of log outage versus the signal-to-noise ratio (SNR) in dB:

$$d^{(out)} \overset{\Delta}{=} \lim_{SNR \to \infty} \frac{-\log p_{\text{out}}(SNR)}{\log SNR}$$

It is the sum of the SNR random variables $|H_{i,j}|^2 P_j/N_0$ in (4.12) and (4.14) that leads to diversity gains when compared to (4.9). In fact, even for such simple relaying algorithms, one can often show that full diversity order 2 can be achieved [43].

Figure 4.6 illustrates example outage performance for noncooperative transmission and cooperative transmission with up to two relays for no cooperation (4.9), AF (4.12), repetition decode-and-forward (RDF) (4.14), and parallel/STC-DF (see, e.g., [42]).

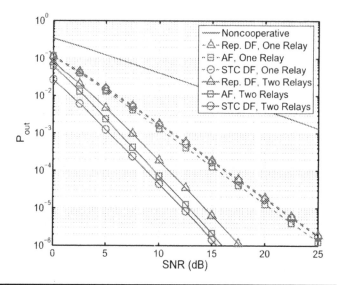

Figure 4.6 Outage Performance of Noncooperative and Cooperative Transmission. Path-Loss Exponent α = 3; i.i.d. Rayleigh Fading; Relays Placed at the Midpoint Between the Source and Destination, Spectral Efficiency $R = 1/2$; Uniform Power Allocation.

We observe from Figure 4.6 that cooperation increases the diversity order, and provides full spatial diversity in the number of cooperating nodes (source plus the relays). Although the two forms of DF have similar performance for the case of one relay for the particular network geometry, path-loss exponent, and spectral efficiency considered, for two relays the advantages of parallel/STC-OF are apparent in Figure 4.6.

4.2.2.3 Probability of Error

As done in [75], cooperative diversity gains can be demonstrated using a probability of error metric. For links affected by Rayleigh flat fading $\mathbf{h} \sim \mathcal{CN}(\mathbf{0}, \Phi_{\mathbf{h}})$, using (4.2) the following Chernoff bound on the pairwise error probability holds:

$$P(\mathbf{s}_k \rightarrow \mathbf{s}_i) \leq \left| \mathbf{I} + SNR/4(\mathbf{G}(\mathbf{s}_k) - \mathbf{G}(\mathbf{s}_i))^* \Phi_{\mathbf{h}}(\mathbf{G}(\mathbf{s}_k - \mathbf{G}(\mathbf{s}_i))) \right|^{-1} \qquad (4.15)$$

Hence, denoting by d the number of nonzero eigenvalues of the combined matrix $(\mathbf{G}(\mathbf{s}_k) - \mathbf{G}(\mathbf{s}_i))^* \Phi_{\mathbf{h}}(\mathbf{G}(\mathbf{s}_k) - \mathbf{G}(\mathbf{s}_i))$, we have $P(\mathbf{s}_k \rightarrow \mathbf{s}_i) = O\left(SNR^{-d}\right)$ for $SNR >> 1$. Using lower and upper bounds on the error probability the corresponding diversity order is:

$$d^{(Pe)} = \lim_{SNR \rightarrow \infty} \frac{-\log P_e(SNR)}{\log SNR} \qquad (4.16)$$

$$\rightarrow d^{(Pe)} = \min_{k,i} \left(rank\left((\mathbf{G}(\mathbf{s}_k) - \mathbf{G}(\mathbf{s}_i))^* \Phi_{\mathbf{h}}(\mathbf{G}(\mathbf{s}_k) - \mathbf{G}(\mathbf{s}_i))\right) \right) \qquad (4.17)$$

Note that the maximum diversity order is equal to the number of cooperating nodes $d \leq T$. Code matrices that maximize the product of the eigenvalues of $(\mathbf{G}(\mathbf{s}_k) - \mathbf{G}(\mathbf{s}_i))^* \Phi_{\mathbf{h}}(\mathbf{G}(\mathbf{s}_k) - \mathbf{G}(\mathbf{s}_i))$ for all possible message pairs are those that are expected to provide the best coding gain, because they maximize (4.16). The generalization to the frequency selective channel case is straightforward using the model in (4.6), because (4.6) and (4.2) look exactly alike, although $\mathcal{X}$ in (4.6) has a constrained structure; combinations of block space–time codes, OFDM, and trellis coding provide excellent practical solutions for this case [4,7,47,51].

4.3 COOPERATIVE LINKS IN EXISTING NETWORK ARCHITECTURES

In a cluster-based MANET, all terminals communicate through a cluster head or AP. In such scenarios, the AP can gather information about the state of the network, e.g., the path losses among terminals, select a cooperative mode based upon some network performance criterion, and feed back its decision on the appropriate control channels. Here cooperative diversity lives across the medium access control, and physical layers; routing is not considered. Each cluster involved in the route is responsible for getting the signal to some destination "gateway" node, serving as the source node for the next cluster.

The dashed links in Figure 4.7 illustrate how the APs communicate information between terminals in different clusters. In our cooperative network model *the gateways between clusters are cooperative links* (indicated by the solid links in the Figure 4.7). In this context, the cost of cooperation compared to using a noncooperative gateway amounts to a loss in spectral efficiency, that depends on the code selection $\mathbf{s} \to \mathbf{G(s)}$, and also to the additional cost of the AP control overhead. The architectural benefits expected are similar to those of a two-tier network that offers more reliable and longer range connections for intercluster communications but uses point-to-point communications within the limits of the cluster.

Note that there are too many tradeoffs in the design of clustering algorithms to fully address here. For instance, clustering schemes can be designed in order to reduce the complexity and overhead of routing [21]; they can be designed to harmonize sleeping

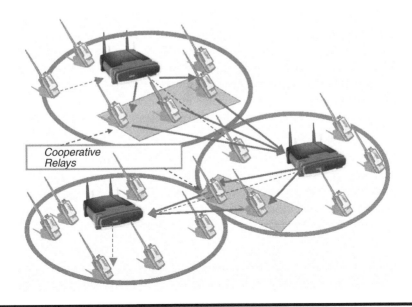

Figure 4.7   Clustering with Direct and Cooperative Transmission.

schedules and reduce power consumption in the network [15] or to facilitate the fusion of measurements in sensor networks [25,26]. Below we describe some approaches that the AP can use to *match terminals* and activate cooperative links given an existing clustered infrastructure.

4.3.1 Centralized Partitioning for Infrastructure Networks

In this section, we consider grouping terminals into cooperating *pairs*. Additional studies of grouping algorithms appear in [32,44]. Choosing pairs of cooperating terminals is an instance of a more general set of problems known as *matching* problems on graphs [58]. In the general matching framework, $\mathcal{G} = (\mathcal{V}, \mathcal{E})$ is a graph, $\mathcal{V}$ a set of vertices, and $\mathcal{E} \subseteq \mathcal{V} \times \mathcal{V}$ a set of edges between vertices. A subset $\mathcal{M}$ of $\mathcal{E}$ is called a *matching* if edges in $\mathcal{M}$ are pairwise disjoint, i.e., no two edges in $\mathcal{M}$ are incident on the same vertex. Note that $|\mathcal{M}| < \lfloor |\mathcal{V}|/2 \rfloor$, where $|\mathcal{M}|$ is again the cardinality of the set $\mathcal{M}$ and $\lfloor x \rfloor$ denotes the usual floor function. When this bound is achieved with equality, the matching is called a *perfect matching*. Since we will be working with *complete* graphs, i.e., there is an edge between each pair of vertices, there will always be a perfect matching for $|\mathcal{V}|$ even. As a result, we will not be concerned with so-called *maximal matching* problems. Instead, we will focus on *weighted matching* problems. Given an edge e in E, the *weight* of the edge is some real number $w(e)$. Given a subset S of $\mathcal{E}$, we denote its sum weight by

$$w(\mathcal{S}) = \sum_{e \in \mathcal{S}} w(e) \qquad (4.18)$$

The *minimal weighted matching*[5] problem is to find a matching of minimal weight [58]. Other matching algorithms with lower complexity are possible. Specifically, we consider:

- **Minimal Weighted Matching:** These algorithms are well-studied and readily available in, e.g., [2,58]. The simplest algorithms have cubic complexity in the number of nodes [58].
- **Greedy Matching:** In this low complexity alternative algorithm we randomly select a free node and match it with its best remaining partner. The process continues until all of the vertices have been matched. The complexity of all such greedy algorithms is quadratic in the number of nodes.
- **Random Matching:** An even simpler alternative is to match nodes randomly. The complexity is linear in the number of nodes.

The rewards for the added complexity of solving the matching problems are the enhanced physical layer performance and the reduction by half of the order of the networking problem. To illustrate the performance benefits, Figure 4.8 shows a set of example results from the various matching algorithms described above and for AF cooperation (cf. Section 4.2.2.2). We note several features of the results in Figure 4.8. First, all the

---

[5] There are several alternatives to the weighted matching approach. For example, we can randomly partition the terminals into two sets and utilize bipartite weighted matching algorithms, with lower complexity (albeit in the same order) than the matching approaches in [58]. Also suitable for decentralized implementation, we can randomly partition the terminals into two sets and use so-called stable marriage algorithms [2].

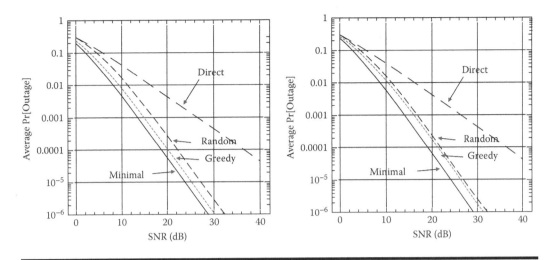

Figure 4.8 Outage Probability vs. SNR for Different Matching Algorithms (Averaged Over 100 Random Trial Networks Uniformly Distributed in a Square of Side 2000 m, with the Base-Station/Access Point Located in the Center). Fading Variances Are Computed Using a d^{-a} Path-Loss Model, with $a = 3$. The Weight Between a Pair of Nodes Is the Average of the Outage Probabilities for One Terminal Using the Other as a Relay, and Vice Versa. The Received SNR for Direct Transmission Averaged Over All the Terminals in the Network Is Normalized to Be the Horizontal Axis.

matching algorithms exhibit full diversity gain of order two with respect to direct transmission. As we would expect, random, greedy, and minimal matching perform increasingly better, but only in terms of SNR gain. Although diversity gain remains constant because we only group terminals into cooperating pairs, the relative SNR gain does improve slightly with increasing network size. This effect is most pronounced in the case of greedy matching, suggesting that optimal matching is crucial to good performance in small networks, offering fewer choices among a small number of terminals.

In general, the SNR gains of the more computationally demanding matching algorithms are most beneficial in low to moderate SNR regimes where the diversity benefits of the diversity gains are smallest. For higher SNR, the diversity gains increase and the rewards for complex matching algorithms diminish.

4.3.2 Connectivity in Clustered Networks with Cooperative Gateways

Whereas capacity measures the aggregate amount of information that can be sent across a wireless network, the connectivity of a network identifies the pairs of nodes between which information can be transferred (i.e., those that can exploit a portion of that capacity). Because it depends upon individual link metrics and channel access is not considered, connectivity is usually simpler to determine than the network capacity. It can be measured in various senses, depending on the criterion that determines if a link is available or not. In this section, we indicate to what degree cooperative transmission in a clustered network can improve connectivity. We assume that all transmissions are affected by a deterministic path loss and random independent fading such that two identical signals transmitted simultaneously from the same distance result in a signal that has twice the average power. A link is available if the receive average SNR is above a fixed threshold ($SNR \geq \tau$). For comparison, the next section highlights results on connectivity that apply to classical point-to-point MANETs.

4.3.2.1 Connectivity in Point-to-Point Networks

Connectivity has been well-studied for ad hoc networks in the limit of an infinite number of nodes placed randomly on a two-dimensional surface. Assuming that the path loss is a monotonic increasing function of the distance d_{ij} between two nodes i and j (such as $d_{ij}^{-\alpha}$), and denoting by $\mathcal{A}$ the circular area centered at a node i, where all nodes $j \in \mathcal{A}$ have $SNR(d_{ij}) \propto d_{ij}^{-\alpha}$ above the threshold set for connectivity $SNR(d_{ij}) > \tau$, any two nodes that are within a distance from each other that is smaller than the radius of $\mathcal{A}$ can be wirelessly connected. The graph obtained by drawing a line between any two nodes of separation less than the radius of $\mathcal{A}$ reveals sets of nodes that can communicate with each other directly or through a path consisting of multiple hops, and such a set is termed a cluster. In such a setting, there have been two separate definitions of what it means for a network to be connected. In the "sparse" network setting, the network is defined as connected if a cluster containing an infinite number of nodes (termed the "infinite cluster") is present in the network. In the "dense" network setting, the network is defined as being connected once all pairs of nodes can communicate with one another. We will refer to the latter definition as the networks being "fully connected." Both definitions are intrinsic properties of the network graph, and in clustered multi-hop networks the edges of this graph are used to communicate.

In the large sparse network scenario, analysis is generally performed for nodes distributed on the infinite two-dimensional plane with some density λ nodes per unit squared. In such a scenario, connectivity is amenable to analysis via "percolation theory" [49]. Clearly, increasing the node density λ must improve the connectivity. Interestingly, there exists a node density λ_0 (termed the "percolation threshold") such that, for $\lambda < \lambda_0$, networks with density λ will almost never exhibit an infinite cluster, whereas for $\lambda > \lambda_0$, networks with density λ will almost surely exhibit an infinite cluster [22,49].

For dense networks, analysis is generally performed for N nodes distributed randomly on a surface of unit area. The seminal work by Gupta and Kumar [19], considering large N on a unit disk, provides a necessary and sufficient condition to guarantee full connectivity of the network: the area of radio coverage of each node should be at least $\mathcal{A}^{(non-coop)} = N^{-1}[\log N + c(N)]$, where $\liminf c(N) = +\infty$. The condition is necessary in the sense that a network with nodes communicating with coverage area $< N^{-1}[\log N + c(N)]$ (where $\limsup c(N) < +\infty$) is proven to be not fully connected. Critical to the proof of the above result is the powerful theorem that, asymptotically, the probability that the network is not fully connected is dominated by the probability of an isolated node.

4.3.2.2 Connectivity in Clustered Networks with Cooperation

In cooperative networks, clustering can help connectivity by essentially increasing the area in which to search for new neighbors, as shown in Figure 4.9.

In fact, if not all nodes are isolated, there will be nodes in a connected cluster that the AP can recruit into finding new neighbors. As a result, it is intuitive that the necessary condition in [19] need not be satisfied for the cooperative network to be fully connected with high probability. To prove it, we assume that the cooperative signals add up in power and that the SNRs at the receiver can be calculated as done in Section 4.2.2. We assume that the link is symmetric, although in practice an AF algorithm should be used

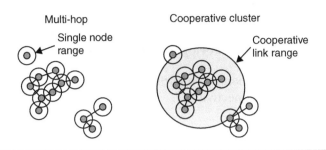

Figure 4.9   Connectivity with Cooperative Radios in a Cluster.

in the reverse link from the faraway node to the cluster. Such a simple model does not account for the fact that cooperation can bring diversity and lower the *SNR** threshold necessary to attain a certain outage or average error rate probability (cf. Section 4.2); however, it is amenable to large scale analysis and provides bounds to the connectivity that can be expected from diversity-achieving schemes.

In the sparse network case, it can be shown through simulation (see Figure 4.10) that the percolation threshold $\lambda_0^{(coop)}$ is significantly reduced from the value of $\lambda_0^{(non-coop)} = 4.5$ of noncooperative networks [69]. In dense networks, it can be shown that the cooperative network can be fully connected with high probability without satisfying the necessity condition for full connectivity in the noncooperative network. The proof construction relies on subdividing the network region into small sections, all of which are likely to have a large cluster of nodes within the area $\mathcal{A}$ with high probability. These clusters can connect not only with all nodes within the section, but also with clusters in neighboring sections, thus fully connecting the network. The required radio coverage

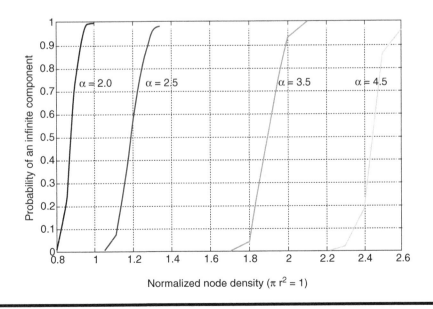

Figure 4.10   Probability of the Existence of an Infinite Cluster vs. Node Density for Collaborative Networks with Pathloss Exponent α.

area of a given node for such connectivity is given by Theorem 6 of [69]:

$$\mathcal{A}^{(coop)} \geq N^{-1} 4\pi (4 \log N)^{\frac{\alpha}{\alpha+2}} (\log \log N + \log 2)^{\frac{2}{\alpha+2}} \tag{4.19}$$

where α is the path-loss exponent. Comparing (4.19) with the result in [19] we have a gain in required power for connectivity of:

$$Cooperative - Gain = \frac{\mathcal{A}^{(non-coop)}}{\mathcal{A}^{(coop)}} = \left(\frac{\log N}{\log \log N} \right)^{\frac{2}{\alpha+2}} \tag{4.20}$$

Furthermore, in contrast to the results of [19] that are restricted to the unit disk, the power in (4.19) is sufficient to fully connect collaborative networks with nodes distributed on a wide variety of unit-area planar shapes; roughly, any network occupying a region whose interior points form a connected set and whose boundary is smooth will be fully connected.

4.4 COOPERATION FOR NEW AD HOC ARCHITECTURES

Section 4.3 illustrates how cooperative communications can serve a MANET, clustered in the traditional sense, as a tool to improve performance. However, cluster heads were required to perform two additional operations: (1) the encoding strategy for the co-operating nodes, and (2) deciding which nodes are involved in a given cooperative transmission. Both issues involve the physical, multiple access, and network layers, and require additional complexity in the network. In this section, we illustrate one way in which these operations can become distributed, leading to a sketch of a new cooperative architecture for MANETs. Section 4.4.1 deals with point (1) and discusses randomized cooperative coding, and Section 4.4.2 deals with point (2) and discusses randomized clustering.

4.4.1 Randomized Cooperative Coding

Let us assume that there are T cooperating nodes. As explained in Section 4.3, in the presence of a central control like the cluster AP, each of the cooperative nodes is as-signed to transmit a column $\mathbf{x}_l \in \mathbf{X}$ of a predetermined code matrix $\mathbf{X} = \mathbf{G}_{K \times T}(\mathbf{s})$. This section shows how the code assignment can be randomized, when the nodes are unaware of how many nodes are going to cooperate and there is not central code as-signment. A randomized coding rule targets a fixed maximum diversity order L, which is independent of the actual number of nodes cooperating. In randomized cooperation [70] each node projects the rows of the code matrix $\mathbf{X} = \mathbf{G}_{K \times L}(\mathbf{s})$ over a random, in-dependently generated, $L \times 1$ vector $\mathbf{r}_r$, $r = 1, \ldots, T$, generating a randomized code $\tilde{\mathbf{x}}_r = \mathbf{X}\mathbf{r}_r = \mathbf{G}_{K \times L}(\mathbf{s})\mathbf{r}_r$. Special cases are the schemes in [60,83], while the same idea using a set of deterministic vectors $\mathbf{r}_r$, $r = 1, \ldots, T$, is proposed in [85]. Like in (4.5), the received vector is the mixture of each of these randomized codes convolved with their respective channel impulse response:

$$\mathbf{y}_i = \sum_{j=1}^{T} \mathbf{H}_{ij} \mathbf{G}(\mathbf{s}) \mathbf{r}_j = \sum_{l=1}^{L} \left(\sum_{j=1}^{T} \mathbf{H}_{ij} r_{jl} \right) \mathbf{x}_l + \mathbf{w}_i = \sum_{l=1}^{L} \tilde{\mathbf{H}}_{ij} r_{jl} \mathbf{x}_l + \mathbf{w}_i \tag{4.21}$$

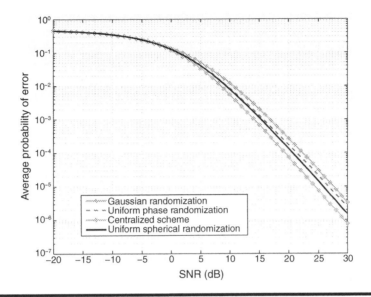

Figure 4.11 **BER of $T = 3$ Cooperative Nodes Using Alamouti Code ($L = 2$) [1] for Different Distributions of R.**

where in the last equation, denoted by $\tilde{\mathbf{H}}_l$, $l = 1, ..., L$ are the equivalent convolution matrices. The received vector is equivalent to that of L cooperative nodes, each transmitting a column $\mathbf{x}_l \in \mathbf{X}$, like in the centralized matching scheme and where each link is characterized by the effective channel response $\tilde{\mathbf{H}}_l$, $l = 1, ..., L$; the latter is the randomized mixture of the true channel responses $\mathbf{H}_j$, $j = 1, ..., T$. The diversity that can be obtained through this scheme depends on the statistics of the resulting equivalent channels $\tilde{\mathbf{H}}_l$, $l = 1, ..., L$ and on the particular selection of the code $\mathbf{G(s)}$ just as it does for the deterministic assignment discussed in Section 4.2. For channels that are frequency flat:

$$\mathbf{y} = \mathbf{G(s)Rh} + \mathbf{w} = \mathbf{G(s)}\tilde{\mathbf{h}} + \mathbf{w} \qquad (4.22)$$

and under the assumption of Rayleigh fading $\mathbf{h} \sim \mathcal{CN}(\mathbf{0}, \Phi_{\mathbf{h}})$ (cf. Section 4.2.2.3).

As shown in [70] (see Figure 4.11), there are several options for the randomization matrix $\mathbf{R}$ to achieve the full diversity L of the code $\mathbf{G(s)}$ when the number of nodes exceeds L even by only one extra node, i.e., if $T = L + 1$. If $T \leq L$ the same random selection rules give a diversity that is $O(T)$. Using the same definition of diversity the main observation in [70] is that for several distributions for $\mathbf{R}$:

$$d = \begin{cases} T \text{ if } L \geq T+1 \\ L \text{ if } L \leq T-1 \end{cases} \qquad (4.23)$$

To assess what potential performance gains can be attained by randomized cooperation in multipath channels with asynchronous cooperative relays, the key step is to rewrite the model in such a way that it can be mapped one to one in a special instance of (4.22). If the channels are all linear time invariant, each equivalent channel matrix $\tilde{\mathbf{H}}_{il}$ has Sylvester structure and therefore the product $\tilde{\mathbf{H}}_{il}\mathbf{x}_l = \mathcal{T}(\mathbf{x}_l)\tilde{\mathbf{h}}_{il}$, where $\mathcal{T}(\mathbf{x}_l)$ has a Toeplitz structure analogous to the one described in (4.7). Now the size depends not only on the equivalent channel order D, but on the design parameter L as well. Hence, with simple manipulations (4.21) can be rewritten as follows:

$$\mathbf{y}_i = \sum_{l=1}^{L} T(\mathbf{x}_l)\tilde{\mathbf{h}}_{il} + \mathbf{w}_i = \mathcal{X}\tilde{\eta}_i + \mathbf{w}_i \qquad (4.24)$$

$$= \mathcal{X}(\mathbf{I}_{D \times D} \otimes \mathbf{R})\eta_i + \mathbf{w}_i \qquad (4.25)$$

where $\mathcal{X} = (T(\mathbf{x}_1), ..., T(\mathbf{x}_L))$; $\eta_i = (\mathbf{h}_{i1}^T, ..., \mathbf{h}_{iT}^T)^T$ and $\tilde{\eta}_i = (\mathbf{I}_{D \times D} \otimes \mathbf{R})\eta_i$. Comparing equation (4.24) with (4.22), we can see that the only difference is that both the equivalent code matrix and random mapping have a very peculiar structure. If, within the constraints for the structure of $\mathcal{X}$, it is possible to find codes that attain maximum diversity without randomization, there are results in [74] that can be extended to work in the ISI model. Hence, the diversity can potentially be as large as the channel order times the number of cooperating users.

Designs that use linear combinations of ST-codes are discussed in [34,85], designs that do not require prior knowledge of the number of nodes in [18,85] and designs that are fully decentralized in [61,70,83]. An interesting observation is that the idea of choosing the delay randomly [83] is equivalent to the random selection scheme, which has several preferable alternatives in terms of providing diversity [74]. Finally, this framework provides a means to tradeoff diversity performance with receiver complexity. The cost in performance is a potential loss of diversity compared to the centralized rule in Section 4.2, with the probability of such a loss decreasing with increasing L [74]. There are benefits and drawbacks in targeting large or small degrees L of diversity. Having large degrees of diversity allows harvesting the greatest gains if the nodes cooperating are $T > L$, but if $T < L$, because the maximum diversity attainable is in the order of T, coding for large L requires an investment in complexity, increased bandwidth, and latency that are strong disincentives toward choosing a large L.

4.4.2 Randomized Clustering in Physical Layer Cooperation

This section overviews a method for forming cooperative groups in a distributed and randomized fashion based upon source requests. In an infrastructureless network, a source can include an appropriate preamble sequence to provide a request for cooperation as well as the sync signal (see Section 4.2). The key differences in layering the transmission functions are: (1) the relay traffic should be stored in a separate buffer, and due to the timing restrictions necessary for coordinating the transmission with other nodes, the control of the relay traffic buffer is primarily at the physical layer and timed through the synchronization preamble in the cooperative request message; (2) the network layer, based on parameters relevant to the traffic flow, enables or inhibits cooperation for relay traffic that identifies a specific source/destination pair; and (3) the link and multiple access sublayers can inhibit or enable cooperation and can mandate a different coding rule for different cooperating radios, but cannot change the schedule.

Figure 4.12 represents a functional block diagram for the physical layer that reflects these points. The two cooperative transmitter and receiver modules, respectively, encode and modulate (transmitter) and demodulate and decode (receiver) the data to the D/A (digital-to-analog) and from the A/D (analog-to-digital) converters (IF (intermediate frequency) or baseband) interfacing the RF front end. The buffers connected to the

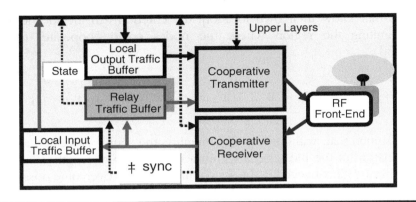

Figure 4.12   Physical Layer of a Cooperative Radio.

receiver contain decoded data. A message that contains a request for cooperation is stored in the relay buffer, whose transmission is synchronized by the preamble sequence received in the message containing the request. The upper layers need to be informed about the state of the relay buffer. The dashed lines going to the upper layers indicate a control that can enable or inhibit the transmitter and receiver modules (and can reject a cooperation request). In general, the half-duplex constraint mandates that the receiver be inactive when the transmitter is busy. But the upper layers can also prevent cooperative transmission for other reasons, exercising network and access control as needed to manage the network.

Let us denote by $\mathcal{S}$ the cooperating group. Because there is no infrastructure, the group cannot rely on an AP to forward its message as in Section 4.3.1. It can, instead, repeat the request for cooperation and recruit a second group and so on. To avoid cycles, the selection rule should exclude from the set $\mathcal{S}[k]$ at the k-th iteration all points that have been in previous sets $(x_j, y_j) \in \mathcal{S}[i]$, $i < k$. Assuming that connectivity is defined in the same sense as in Section 4.3.2, so that the node is connected to receive a message in the k-th iteration if the receive $SNR_k(x_j, y_j) > \tau$, a broadcast architecture could use all such nodes to cooperate, and the set $\mathcal{S}[k]$ (Figure 4.13 (b)) is defined as:

$$\mathcal{S}[k] = \left\{ (x_j, y_j) : (SNR_k(x_j, y_j) > \tau) \cap \Theta \right\}; \Theta = (x_j, y_j) \notin \overset{k}{\underset{i=0}{\cap}} \mathcal{S}[i] \qquad (4.26)$$

where only nodes whose receive $SNR_k(x_j, y_j) > \tau$ retransmit, and the additional condition $\Theta = (x_j, y_j) \notin \cap_{i=0}^{k} \mathcal{S}[i]$ simply verifies that the node has not transmitted before. (Because this condition is always present, we omit it in the following discussion.)

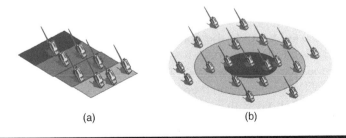

(a)   (b)

Figure 4.13   (a) Cooperation Over a Strip and (b) Cooperative Broadcasting. Multiple Levels of Cooperative Relay Replace the Multi-Hop Strategy Used in Conventional Ad Hoc Networks.

For nodes that know their own location, the criterion can include network layer parameters limiting the region where the nodes could cooperate, for example (Figure 4.13 (a)):

$$\mathcal{S}[k] = \{(x_j, y_j) : (|y_j| < W/2 \cap (0 \le x_j \le D) \cap (SNR(x_j, y_j) > \tau)\} \qquad (4.27)$$

where it is assumed that, without loss of generality, the source is located at point (0,0) and the destination of the message is within a certain strip of length D and width W (parameters that are all conveyed in the source message). To overcome possible failures, ARQ protocols can adaptively expand W, on a step-by-step basis or end to end.

Under the new reference model in Figure 4.12, random access policies can resolve contention between groups that carry uncorrelated data. In other words, the contention and the collision models in Constraint II still stand (see, e.g., [59]), but it applies between relay groups.

Note that going through one level of relay requires a time in the order of the message duration plus some processing time. Hence, the rules that select larger cooperative groups tend to reach the destination faster. This presents a very interesting tradeoff between the spatial and temporal use of network resources by each group.

Perhaps the most dramatic change of this clustering model in a MANET is that individual relays are treated as physical layer resources rather than network layer entities; the network and multiple access layers can deal with each group of cooperative nodes as a single entity, and this actually simplifies their decisions. This new abstraction of "link," how to model it, and how to adjust the medium access control and routing algorithms are questions for the networking community to consider.

4.4.3 Connectivity in Randomized Cooperation

Randomized cooperation can also be shown to improve connectivity, as demonstrated in the asymptotic analysis of the cooperative network first introduced in [71–73]. The limit nearly corresponds to the dense scenario discussed previously, i.e., infinite node density, but here normalizing the relay power per unit area [73]. Specifically, denoting by P_r the relay power and by N the total number of nodes in a region of finite size denoted by A, it is assumed that:

$$\lim_{N \to \infty, P_r \to 0} \frac{P_r N}{A} = \lim_{\rho \to \infty, P_r \to 0} P_r \rho = \bar{P}_r < \infty \qquad (4.28)$$

In [73] the limit is used to study the connectivity under the rule in (4.26) in random small scale fading and deterministic path loss. Let us denote by $l(x - u, y - v)$ the path loss between the nodes in positions (u, v) and (x, y) (for example, $l(x - u, y - v) = ((x - u)^2 + (y - v)^2)^{\alpha/2}$), consider the thermal noise variance to be normalized to unity, and denote the SNR threshold with τ as in Section 4.3.2. In [73] it is proven that if the diversity order L of the code $\mathbf{X} = \mathbf{G}_{K \times L}(\mathbf{S})$ grows proportionally to N, and the path loss is monotonic increasing with the distance, the groups $\mathcal{S}[k]$ tend to occupy fixed areas A_k. In other words, the probability $\pi_k(x, y)$ of a node located at point (x, y) being in a certain set $\mathcal{S}[k]$ can be calculated solving the following recursive equations, where A_k

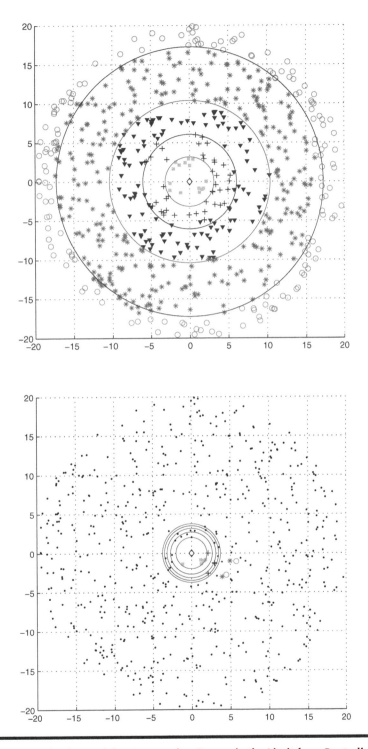

Figure 4.14 Asymptotic Shape of the Cooperative Groups in the Limit for a Centralized Assignment of Orthogonal Cooperation Channels Under Free Space Pathloss ($1/d^2$, d Is the Node Distance). P_s Is the Power of the Source, P_r the Power Density of the Relays and τ Is the SNR Threshold. A Phase Transition Is Observed: in the Figure on the Left the Concentric Annuli Increase in Area While They Shrink in the Right Figure. In the Second Case the Threshold is Too High for Connectivity.

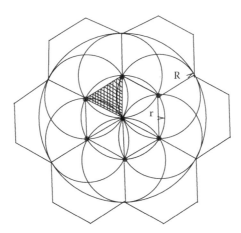

Figure 4.15   Multi-Hop Broadcast.

are the subareas of the network in one-to-one correspondence with the sets $\mathcal{S}[k]$:

$$A_k: \ SNR_k(x, y) = \bar{P}_r \int_{A_{k-1}} \frac{\pi_{k-1}(u, v)}{l(x-u, y-v)} \, du dv > \tau k = 1, 2, \ldots$$

$$\lim_{\rho \to \infty} \pi_k(x, y) = \begin{cases} 1 & \text{if } (x, y) \in A_k \\ 0 & \text{else} \end{cases} \quad k = 2, \ldots \tag{4.29}$$

The equations in (4.29) indicate that, except for the first group, the k-th cooperative group lies with probability one in an area A_k. If the network is a disc with the transmitting node at the center for free space path loss $l(x - u, y - v) = (x - u)^2 + (y - v)^2$, (4.29) can be solved in closed form, A_k are discs, and the analysis highlights a phase transition [73, Theorem 2]: only for $\tau < (\pi \ln 2)\bar{P}_r$ do all nodes receive the message, otherwise the areas A_k progressively shrink. This is illustrated in Figure 4.14.

For a given threshold we can use the result in [73, Theorem 2] to calculate the minimum power density necessary for connectivity $\bar{P}_{r-\min} = \tau(\pi \ln 2)^{-1}$ resulting in a total power expenditure to cover an area A equal to $P^{(coop)} = A\bar{P}_{r-\min} = A\tau(\pi \ln 2)^{-1}$. For the multi-hop scheme, the positions of the nodes that can cover the region with minimum power are shown in Figure 4.15. Ignoring the interference from nearby nodes, each hexagon-center node needs to transmit $P_r \geq \tau r^2$. This implies that $P^{(non-coop)} = N_r P_{r-\min} = N_r \tau r^2 = 2A\tau/\sqrt{3}$. The simple comparison demonstrates that the percentage gain of $P^{(coop)}$ versus $P^{(non-coop)}$ approaches 50%. Other interesting properties of cooperative schemes that we do not have the space to discuss here are their resilience to error propagation documented in [62], and the opportunistic beam-forming effects that arise between multiple levels of relay, also discussed in [73].

4.5   CONCLUSIONS

This chapter illustrates two options for including cooperative designs in MANETs. These examples illustrate how incorporating cooperation at the physical layer offers a number of advantages in flexibility over standard MANETs that go beyond simply providing a more reliable physical layer link. Studies on the diversity of cooperative links are numerous and will continue to develop practical schemes with improved performance.

We argue that since cooperation is essentially a network solution, instead of a point-to-point solution, finding the appropriate link abstraction for cooperative transmission raises a number of important but nontraditional research problems for networking and physical layer researchers to investigate further.

REFERENCES

[1] S. Alamouti, "A simple transmit diversity technique for wireless communications," *IEEE J. Selected Areas Commn.*, vol. 16, no. 8, pp. 1451–1458, Oct. 1998.

[2] R.K. Ahuja, T.L. Magnanti, and J.B. Orlin, *Network Flows: Theory, Algorithms, and Applications.* Englewood Cliffs, NJ: Prentice-Hall, 1993.

[3] P.A. Anghel, G. Leus, and M. Kaveh, "Distributed space-time coding in cooperative networks," in *Proc. of 5th NORDIC Signal Processing Symposium*, Tromso-Trondheim, Norway, p. **, 2002.

[4] P.A. Anghel and M. Kaveh, "Relay assisted uplink communication over frequency-selective channels," *IEEE SPAWC 2003*, Rome, Italy, June 15–18, 2003.

[5] P.A. Anghel and M. Kaveh, "Exact symbol error probability of a cooperative network in a Rayleigh-fading environment," *IEEE Trans. Wireless Commn.*, vol. 3, no. 5, p. 1416–1421, Sept. 2004.

[6] R. Ahlswede, N. Cai, S.Y.R. Li, and R.W. Yeung, "Network Information Flow," *IEEE Trans. Inf. Theory*, IT-46, pp. 1204–1216, 2000.

[7] S. Barbarossa and G. Scutari, "Distributed space-time coding for multi-hop networks," in *Proc. of IEEE International Conference on Communications*, 2004.

[8] , "Distributed space-time coding strategies for wideband multi-hop networks: regenerative vs. non-regenerative relays," in *Proc. of IEEE Int. Conf. on Acoustics, Speech, and Signal Process.* (ICASSP), 2004.

[9] J. Boyer, D.D. Falconer, and H. Yanikomeroglu, "A theoretical characterization of the multi-hop wireless communications channel with diversity," in *Proc. of Global Telecommunications Conference*, vol. 2, 2001, pp. 841–845.

[10] J. Boyer, D.D. Falconer, and H. Yanikomeroglu, "Multi-hop diversity in wireless relaying channels," in *IEEE Trans. Commn.*, vol. 52, pp. 1820–1830, Oct. 2004.

[11] A.B. Carleial, "Multiple-access channels with different generalized feedback signals," *IEEE Trans. Inform. Theory*, vol. 28, no. 6, pp. 841–850, Nov. 1982.

[12] T. Cover and A.E. Gamal, "Capacity theorems for the relay channel," *IEEE Trans. Inform. Theory*, vol. 25, no. 5, pp. 572–584, Sept. 1979.

[13] T.M. Cover and J.A. Thomas, *Elements of Information Theory.* New York: John Wiley & Sons, 1991.

[14] Y. Chang and Y. Hua, "Application of space-time linear block codes to parallel wireless relays in mobile ad hoc networks," in *Signals, Systems and Computers, 2003 The Thirty-Seventh Asilomar Conference*, vol. 1, Pacific Grove, CA, Nov. 2003, pp. 1002–1006.

[15] B. Chen, K. Jamieson, H. Balakrishnan, and R. Morris, "Span: an energy-efficient coordination algorithm for topology maintenance in ad hoc wireless networks," *ACM Wireless Netw. J.*, vol. 8, no. 5, Sept. 2002.

[16] S.H. Chen, U. Mitra, and B. Krishnamachari, "Cooperative communication and routing over fading channels in wireless sensor networks," in *IEEE WirelessCom*, Maui, HI, June 2005.

[17] S. Cui, A.J. Goldsmith, and A. Bahai, "Energy-efficiency of MIMO and cooperative MIMO techniques in sensor networks," *IEEE J. Selected Areas Commn.*, vol. 22, no. 6, Aug. 2004, pp. 1089–1098.

[18] H.E. Gamal and D. Aktas, "Distributed space-time filtering for cooperative wireless networks," in *Proc. of Globecom*, 2003, vol. 4, Dec. 2003, pp. 1826–1830.

[19] P. Gupta and P.R. Kumar, "Critical power for asymptotic connectivity in wireless networks," in *Stochastic Analysis, Control, and Optimization and Applications: A Volume in Honor of W. H. Fleming*, Eds., W.M. McEneany, G. Yin, and Q. Zhang, Birkhauser, Boston, 1998, pp. 547–566.

[20] P. Gupta and P.R. Kumar, "The capacity of wireless networks," *IEEE Trans. Inform. Theory*, vol. IT-46, no. 2, pp. 388–404, Mar. 2000.

[21] B. Das and V. Bharghavan, "Routing in ad hoc networks using minimum connected dominating sets," in *Proc. IEEE Int. Conf. Communications (ICC)*, vol. 1, Montreal, Canada, June 1997, pp. 376–380.

[22] O. Dousse, P. Thiran, and M. Hasler, "Connectivity in ad hoc and hybrid networks," in *Proceedings of IEEE Infocom*, New York, June 2002, pp. 1079–1088.

[23] Z.J. Haas, M.R. Pearlman, and P. Samar, "The Interzone Routing Protocol (IERP) for Ad Hoc Networks," draft-ietf-manet-zoneierp -01. txt, IETF MANET working group, Dec. 2001.

[24] B. Hassibi and B.M. Hochwald, "How much training is needed in multiple-antenna wireless links?," *IEEE Trans. Inform. Theory*, vol. 49, pp. 951–963, Apr. 2003.

[25] W. Heinzelman, A. Chandrakasan, and H. Balakrishnan, "Energy-Efficient Communication Protocols for Wireless Microsensor Networks," in *Proc. of the Hawaii Int. Conf. on System Sciences*, Maui, HI, Jan. 2000, pp. 3005–3014.

[26] W.B. Heinzelman, A.P. Chandrakasan, and H. Balakrishnan, "An application-specific protocol architecture for wireless microsensor networks," *IEEE Trans. Wireless Commn.*, vol. 1, no. 4, pp. 660–670, Mar. 2002.

[27] Y.W. Hong and A. Scaglione, "Energy-efficient broadcasting with cooperative transmission in wireless ad hoc networks," *Proc. Allerton Conference*, Monticello, IL, Oct. 1–3, 2003, to appear in *IEEE Trans. on Wireless Commn.*, 2005.

[28] A. Host-Madsen, "A new achievable rate for cooperative diversity based on generalized writing on dirty paper," in *Proc. of IEEE (ISIT)*, June 2003, p. 317.

[29] A. Hottinen and O. Tirkkonen, "A randomization technique for non-orthogonal space-time block codes," in *Proc. of IEEE Vehicular Technology Conference*, Rhodes Island, Greece, 2001, vol. 12, pp. 1479–1482.

[30] Y. Hua, Y. Mei, and Y. Chang, "Parallel wireless mobile relays with space-time modulations," in *2003 IEEE Workshop on Statistical Signal Processing*, St. Louis, MO, Sept.–Oct. 2003, pp. 375–378.

[31] Y. Hua, Y. Mei, and Y. Chang, "Wireless-antennas making wireless communications perform like wireline communications," in *IEEE AP-S Topical Conference on Wireless Communication Technology*, Honolulu, HI, Oct. 2003.

[32] T. Hunter and A. Nosratinia, "Coded cooperation in multi-user wireless networks," to appear in: *IEEE Trans. on Wireless Communications*.

[33] N. Jindal, U. Mitra, A. Goldsmith, "Capacity of ad hoc networks with node cooperation," *Proc. of Int. Symp. Inf. Theory*, Chicago, IL, June 2004, p. 217.

[34] Y. Jing and B. Hassibi, "Wireless networks, diversity and space-time codes," in *Proc. of IEEE Information Theory Workshop*, San Antonio, TX, 2004.

[35] M. Joa-Ng and I. Tai-Lu "A peer-to-peer zone-based two-level link state routing for mobile ad hoc networks," *IEEE J. Selected Areas Commn.*, vol. 17, no. 8, pp. 1415–1425, August 1999.

[36] A. Khandani, J. Abounadi, E. Modiano, and L. Zhang, "Cooperative routing in wireless," in *Proc. of Allerton Conference on Communications Control and Computing*, Oct. 2003.

[37] R.C. King, "Multiple Access Channels with Generalized Feedback," Ph.D. dissertation, Stanford University, Palo Alto, CA, Mar. 1978.

[38] G. Kramer, "Models and theory for relay channels with receive constraints," in *Proc. Allerton Conf. Communications, Control, and Computing*, Monticello, IL, Oct. 2004. Available: http://cm.belllabs.com/cm/ms/who/gkr/Papers/relayAllerton04.pdf

[39] G. Kramer, M. Gastpar, and P. Gupta. "Cooperative strategies and capacity theorems for relay networks," to appear in *IEEE Trans. on Inf. Theory*, September 2005.

[40] J.N. Laneman, G.W. Wornell, and D.N.C. Tse, "An efficient protocol for realizing cooperative diversity in wireless networks," in *Proc. IEEE Int. Symp. Information Theory* (ISIT), Washington, DC, June 2001. Available online: http://www.nd.edu/ jnl/pubs/isit 2001.pdf

[41] J.N. Laneman, "Cooperative Diversity in Wireless Networks: Algorithms and Architectures," Ph.D. dissertation, Massachusetts Institute of Technology, Cambridge, Aug. 2002. Available online: http://www.nd.edu/jnl/pubs/thesis.pdf

[42] J.N. Laneman and G.W. Wornell, "Distributed space-time coded protocols for exploiting cooperative diversity in wireless networks," in *IEEE Trans. Inf. Theory*, vol. 59, no. 10, Oct. 2003.

[43] J.N. Laneman, D. Tse, and G. Wornell, "Cooperative diversity in wireless networks: efficient protocols and outage behavior," in *IEEE Trans. Inf. Theory*, vol. 50, no. 12, Dec. 2004.

[44] Z. Lin, E. Erkip, and A. Stefanov, "Cooperative regions and partner choice in coded cooperative systems," to appear in *IEEE Trans. on Commn.*

[45] I. Maric and R. Yates, "Efficient Multi-hop Broadcast for Wideband Systems, Book Chapter: Multiantenna Channels: Capacity, Coding and Signal Processing," DIMACS Workshop, Piscataway, NJ, Oct. 2002.

[46] I. Maric and R. Yates, "Efficient multi-hop broadcast for wideband systems," *IEEE JSAC, Spec. Issue on Fundam. Performance Limits of Wireless Sensor Networks*, vol. 22, no. 6, Aug. 2004.

[47] H. Mheidat and M. Uysal, "Equalization techniques for space-time coded cooperative systems," in *IEEE VTC'04-Fall*, Los Angeles, CA, Sept. 2004.

[48] G. Mergen, V. Naware, and L. Tong, "Asymptotic detection performance of type-based multiple access in sensor networks," *Proc. of SPAWC'05*, New York, June 2005.

[49] R. Meester and R. Roy, *Continuum Percolation*, Cambridge: Cambridge University Press, 1996.

[50] J. Mietzner, R. Thobaben, and P.A. Hoeher, "Analysis of the expected error performance of cooperative wireless networks employing distributed space-time codes," in *Proc. of Globecom*, Dallas, TX, Nov.–Dec. 2004.

[51] R.U. Nabar and H. Boelcskei, "Space-time signal design for fading relay channels," *IEEE GLOBECOM*, San Francisco, vol. 4, pp. 1952–1956, Dec. 2003.

[52] R. Nabar, H. Bolcskei, and V. Morgenshtern, "On the robustness of distributed orthogonalization in sense wireless networks," submitted to *IEEE Tran. Inform. Theory*, March 2005.

[53] C.T.K. Ng and A. Goldsmith, "Transmitter cooperation in ad hoc wireless networks: does dirty-payer coding beat relaying," *Proceedings: IEEE ITW*, San Antonio, TX, pp. 297–298, Oct. 2004.

[54] A. Nosratinia, T.E. Hunter, and A. Hedayat, "Cooperative communication in wireless networks," *IEEE Commn. Magazine*, vol. 42, no. 10, pp. 74–80, Oct. 2004.

[55] L.H. Ozarow, S. Shamai (Shitz), and A.D. Wyner, "Information theoretic considerations for cellular mobile radio," *IEEE Trans. Veh. Technol.*, vol. 43, no. 5, pp. 359–378, May 1994.

[56] A. Ribeiro, X. Cai, and G.B. Giannakis, "Symbol error probabilities for general cooperative links," *IEEE Trans. Wireless Commn.*, vol. 4, no. 3, pp. 1264–1273, May 2005.

[57] J.A. Roberts and T.J. Healy, "Packet radio performance over slow Rayleigh fading channels," *IEEE Trans. Commn.*, vol. 28, pp. 279–286, Feb. 1980.

[58] K.H. Rosen, Ed., *Handbook of Discrete and Combinatorial Mathematics*. Boca Raton, FL: CRC Press, 2000.

[59] A. Salhotra and A. Scaglione, "Multiple access in connectionless networks using cooperative transmission," in *Proceedings of Allerton Conference*, Monticello, IL, Oct. 2003.

[60] A. Scaglione and Y.W. Hong, "Opportunistic large arrays" *IEEE International Symposium on Advances in Wireless Communications*, ISWC02, Victoria, BC, Canada, Sept. 23–24, 2002.

[61] A. Scaglione and Y.W. Hong, "Opportunistic large arrays: cooperative transmission in wireless multi-hop ad hoc networks to reach far distances," *IEEE Trans. Signal Process.*, vol. 8, pp. 2082–2092, Aug. 2003.

[62] A. Scaglione, S. Kirti, and B. Sirkeci Mergen, "Error propagation in dense wireless networks with cooperation," in *Proc. of Intl. Conf. on ISPN 2006*, Nashville, Tennessee, April 14–19, pp. 126–133, 2006.

[63] B. Schein and R.G. Gallager, "The gaussian parallel relay network," in *Proc. IEEE Int. Symp. Inf. Theory (ISIT)*, Sorrento, Italy, June 2000, p. 22.

[64] B. Schein, "Distributed Coordination in Network Information Theory," Ph.D. dissertation, Massachusetts Institute of Technology, Cambridge, Aug. 2001.

[65] A. Sendonaris, E. Erkip, and B. Aazhang, "Increasing uplink capacity via user cooperation diversity," in *Proc. IEEE Int. Symp. Inf. Theory (ISIT)*, Cambridge, MA, Aug. 1998.

[66] A. Sendonaris, "Advanced Techniques for Next-Generation Wireless Systems," Ph.D. dissertation, Rice University, Houston, TX, Aug. 1999.

[67] A. Sendonaris, E. Erkip, and B. Aazhang, "User cooperation diversity – part 1: system description," *IEEE Trans. Commn.*, vol. 51, no. 11, Nov. 2003.

[68] A. Sendonaris, E. Erkip, and B. Aazhang, "User cooperation diversity – part 2: implementation aspects and performance analysis," *IEEE Trans. Commn.*, vol. 51, no. 11, Nov. 2003.

[69] S. Song, D. Goeckel, and D. Towsley, "Collaboration improves the connectivity of wireless networks," *Proceedings of IEEE Infocom*, April 2006.

[70] B. Sirkeci-Mergen and A. Scaglione, "Randomized distributed space-time coding for cooperative communication in self-organized networks," in *Proc. of IEEE Workshop Signal Proc. Adv. Wireless Commn.* (SPAWC), June 2005.

[71] B. Sirkeci-Mergen and A. Scaglione, "Message propagation in a cooperative network with asynchronous receptions," *ICASSP 2005*, March 19–23, Philadelphia.

[72] B. Sirkeci-Mergen and A. Scaglione, "A continuum approach to dense wireless networks with cooperation," *IEEE Infocom*, March 13–17, 2005, Miami.

[73] B. Sirkeci-Mergen, A. Scaglione, and G. Mergen, "Asymptotic analysis of multi-stage cooperative broadcast in wireless networks," *Joint Special Issue of the IEEE Trans. on Inf. Theory and IEEE/ACM Trans. on Networking*, June 2006.

[74] B. Sirkeci Mergen and A. Scaglione, "Randomized space-time coding for distributed cooperative communication," *IEEE Trans. Signal Process.*, submitted August 2005; revised March 2006. See also: *Proc. of IEEE International Conference on Communications (ICC)*, June 11–15, 2006, Istanbul, Turkey.

[75] V. Tarokh, H. Jafarkhani, and A. Calderbank, "Space-time block codes from orthogonal designs," *IEEE Trans. Inf. Theory*, vol. 45, no. 5, pp. 1456–1467, July 1999.

[76] I.E. Telatar, "Capacity of multi-antenna Gaussian channels," *European Trans. on Telecomm.*, vol. 10, no. 6, pp. 585–596, Nov.–Dec. 1999.

[77] Y.-C. Tseng, S.-Y. Ni, Y.-S. Chen, and J.-P. Sheu, "The broadcast storm problem in a mobile ad hoc network," *ACM Wireless Networks*, vol. 8, pp. 153–167, Mar. 2002.

[78] E.C. van der Meulen, "Transmission of Information in a T-Terminal Discrete Memoryless Channel," Ph.D. thesis, Dept. of Statistics, University of California, Berkeley, 1968.

[79] E.C. van der Meulen, "Three-terminal communication channels," *Adv. Appl. Prob.*, vol. 3, pp. 120–154, 1971.

[80] J.E. Wieselthier, G.D. Nguyen, and A. Ephremides, "On the construction of energy-efficient broadcast and multicast trees in wireless networks," in *IEEE Proceedings on INFOCOM*, vol. 2, 2000, pp. 585-594.

[81] F.M.J. Willems, "Information Theoretical Results for the Discrete Memoryless Multiple Access Channel," Ph.D. dissertation, Katholieke Universiteit Leuven, Leuven, Belgium, Oct. 1982.

[82]   F.M.J. Willems, E.C. van der Meulen, and J.P.M. Schalkwijk, "An achievable rate region for the multiple access channel with generalized feedback," in *Proc. Allerton Conf. Communications, Control, and Computing*, Monticello, IL, Oct. 1983, pp. 284–292.

[83]   S. Wei, D. Goeckel, and M. Valenti, "Asynchronous cooperative diversity," in *Proc. of 2004 Conference on Information Sciences and Systems*, Princeton, NJ, March 2004.

[84]   S. Yatawatta and A.P. Petropulu, "A multiuser OFDM system with user cooperation," *Proc. of 38th Asilomar Conference*, Pacific Europe, CA, Nov. 2004.

[85]   S. Yiu, R. Schober, and L. Lampe, "Distributed space-time block coding," submitted *IEEE Trans. Commn.*, 2005.

5

DISTRIBUTED ANTENNA SYSTEMS AND LINEAR RELAYING FOR RANK-DEFICIENT MIMO SYSTEMS

Boris Rankov, Jörg Wagner and Armin Wittneben

Contents

We study the impact of multiple amplify-and-forward relays on the achievable rate of wireless multiple-input multiple-output (MIMO) channels. For wireless networks with one source/destination pair (equipped with multiple antennas) and several single antenna amplify-and-forward relays (AF) we determine the compound (over two time slots) channel matrix of the relay-assisted MIMO channel. We show that with two-hop relaying, one can shape the channel matrix to be better conditioned and therefore increase the capacity of rank-deficient MIMO channels. We propose to use AF relays that act as active scatterers and assist the communication between the source and the destination terminal. We consider two cases: First, the relay antennas are connected through a wired backbone (distributed relay array) and the linear signal processing is done in a central unit also connected to this backbone. Second, we look at the case where the relays operate stand-alone, i.e, without a backbone connection. The goal of relaying, in both cases, is to increase the rank of the compound (two time slots) channel matrix and to shape the eigenvalue distribution such that the channel matrix becomes well-conditioned and the capacity of the MIMO channel increases.

5.1 INTRODUCTION

Multiple antennas at a transmitter and a receiver introduce spatial degrees of freedom into a wireless communication system. Space–time signal processing at the receiver or transmitter utilizes these degrees of freedom to boost link capacity or to enhance link reliability of multiple-input multiple-output (MIMO) communication systems. With *spatial multiplexing,* one can increase the data rate without additional cost of bandwidth or power by transmitting data streams simultaneously over spatial subchannels which are available in a rich scattering environment [1]. With *space–time coding* it is possible to mitigate the fading effects by utilizing the spatial diversity of the MIMO channel [2]. It is expected that future wireless broadband communication systems will operate beyond 5 GHz, for example, wireless local area networks (WLANs) at 17 GHz (Hiperlan) or at 24/60 GHz (ISM bands). In higher frequency bands it is possible to accommodate a larger number of antennas in a given volume ("rich array") because the array size scales down with increasing frequency. Further on, the array gain of the system can compensate the path loss which is inversely proportional to the square of the frequency [3].

For zero-mean i.i.d. Gaussian channel coefficients, the ergodic capacity of a MIMO channel with M transmit and N receive antennas scales linearly with min$\{M, N\}$ compared to a corresponding single-input single-output (SISO) channel [4]. However, there is a major obstacle in the practical exploitation of MIMO technology: the capacity gain depends strongly on the propagation environment and diminishes with increasing correlation of the channel coefficients [5]. In higher frequency bands we expect an increase in correlation because the propagation channel tends to become a line-of-sight (LOS) channel and we are confronted with a *rich array–poor scattering* situation [3].

To illustrate the effect of a strong LOS component on the MIMO capacity we consider a Ricean fading MIMO channel with N transmit and N receive antennas, which is given by [6]

$$\mathbf{H} = \sqrt{\frac{K}{K+1}}\overline{\mathbf{H}} + \sqrt{\frac{1}{K+1}}\mathbf{H}_{\mathrm{w}}, \tag{5.1}$$

where the $N \times N$ matrix $\overline{\mathbf{H}}$ is the LOS component and the $N \times N$ matrix $\mathbf{H}_{\mathrm{w}}$ is the fading component of the channel. For simplicity we assume for the LOS component $\overline{\mathbf{H}} = \mathbf{1}_N$, which is the all-one matrix, i.e., we neglect the phase differences between the channel gains.[1] For the fading component we assume a Rayleigh fading matrix, i.e., all entries of the matrix $\mathbf{H}_{\mathrm{w}}$ are circular symmetric i.i.d. Gaussian with zero mean and unit variance. The factor K is the Rice factor and denotes the ratio of the total power in the LOS component to the total power in the fading component. The ergodic capacity of the

---

[1] However, in [7] it is shown that under certain geometrical assumptions about the distance between the transmitter and receiver and the antenna separations the phase differences are able to provide a full rank matrix in MIMO LOS channels. If the transmitter–receiver distance is much larger than the antenna separation at transmitter and receiver, it is a reasonable assumption to neglect the phase differences.

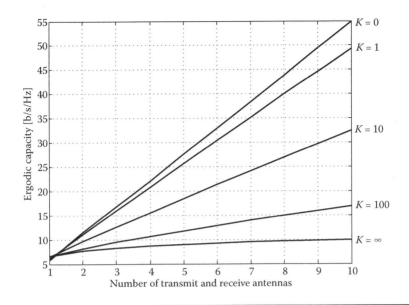

Figure 5.1 Ergodic Capacity of Ricean Fading MIMO Channels Where *K* Is the Ricean Factor. *K* = 0 Corresponds to the Capacity of the Rayleigh Fading MIMO Channel. *K* = ∞ Corresponds to the Capacity of the Rank-1 Line-of-Sight MIMO Channel. Signal-to-Noise Ratio Was Chosen to Be 100, i.e., 20 dB.

Ricean MIMO channel is then given by[2]

$$C_{\text{Rice}} = \mathcal{E} \left\{ \log \det \left(\mathbf{I}_N + \frac{\text{SNR}}{N} \mathbf{H}\mathbf{H}^H \right) \right\}, \tag{5.2}$$

where SNR is the signal-to-noise ratio (SNR) at each receive antenna and $\mathcal{E}\{\cdot\}$ the expectation operator.

Figure 5.1 shows the ergodic capacity as a function of the number of antennas N for different Rice factors K. The highest capacity is achieved for $K = 0$, i.e., Rayleigh fading. For this case we can observe a linear increase with respect to the number of antennas. The smallest capacity is achieved for $K = \infty$, i.e., a nonfading MIMO channel. In this case the channel matrix has rank one and no spatial multiplexing gain is available. The logarithmic increase of the capacity is due to the receive array gain, i.e, the SNR scales linearly with the number of receive antennas N. In Figure 5.2 we plot the ergodic capacity as a function of the Rice factor K for different number of antennas N. Again, we see the decrease of capacity with an increasing LOS component. Other examples of rank-deficient or degenerated MIMO channels are channels with correlated fading [5] or keyhole channels [8].

In this chapter we will show that with two-hop relaying one can increase the rank and therefore the capacity of rank-deficient MIMO channels. We propose to use *amplify-and-forward relays* (AF) that act as active scatterers and assist the communication between a source terminal and a destination terminal: The relays receive the signal from the source in the first time slot and forward an amplified version to the destination in the second time slot. This way of relaying leads to low-complexity relay transceivers and to lower

---

[2] All logarithms are taken to the base 2.

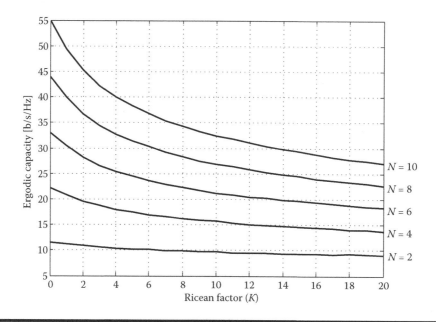

Figure 5.2   Ergodic Capacity vs. Ricean Factor K for Different Numbers of Transmit and Receive Antennas N.

power consumption because there is no signal processing for decoding procedures. The goal of relaying here is to increase the rank of the compound (two time slots) channel matrix and to shape the eigenvalue distribution such that the channel matrix becomes well-conditioned and the capacity of the MIMO channel improves. Note that the use of half-duplex relays[3] induces a factor one half in front of the achievable rate, because two channel uses are necessary to transmit information from source to destination. It is therefore not clear whether the achievable rate of a relay-assisted MIMO channel with full rank is larger than the conventional MIMO channel with low rank. However, in [9], two protocols are proposed that can recover the factor one-half loss due to the half-duplex operation of the relays and, therefore, the rates achievable in the relay-assisted MIMO channel can be substantially higher than in the conventional rank-deficient MIMO channel.

In this chapter, we first look at the case when the relay antennas are connected through a wired backbone (distributed relay array) and the linear signal processing is done in a central unit. We then consider the case where the relays operate stand-alone, i.e., without a backbone connection.

5.2   PROTOCOLS FOR AMPLIFY-AND-FORWARD RELAYS

We consider a MIMO system with N transmit antennas at the source terminal and N receive antennas at the destination terminal. The transmission of a data packet from the source to the destination terminal occupies two time slots of equal length. K single-antenna relay terminals receive signals during the first time slot and simultaneously

---

[3] Relays cannot transmit and receive in the same time and frequency slot.

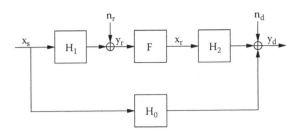

Figure 5.3   System Model for Relay-Assisted MIMO Channel.

forward a linearly modified version of the received signal during the second time slot. We assume that the relays cannot transmit and receive at the same time in the same frequency band (half-duplex mode). The variables of the system, cf. Figure 5.3, are:

- $\mathbf{x}_s$: $N \times 1$ transmit vector at the source with elements i.i.d. $\mathcal{CN}(0, P_s/N)$[4] and P_s the total power of the source terminal.
- $\mathbf{y}_d$: $N \times 1$ receive vector at the destination.
- $\mathbf{x}_r$: $K \times 1$ vector containing all noisy transmit symbols at the K relays with $\mathcal{E}\left\{\mathbf{x}_r^H\mathbf{x}_r\right\} = P_r$ with P_r denoting the total average transmit power.[5]
- $\mathbf{y}_r$: $K \times 1$ vector containing all received symbols at the K relays.
- $\mathbf{H}_0$: $N \times N$ matrix denoting the channel between source and destination (direct channel).
- $\mathbf{H}_1$: $K \times N$ matrix denoting the channel between source and relays (first hop).
- $\mathbf{H}_2$: $N \times K$ matrix denoting the channel between relays and destination (second hop).
- $\mathbf{n}_d$: $N \times 1$ additive white Gaussian noise (AWGN) vector at the destination with elements i.i.d. $\mathcal{CN}(0, \sigma_d^2)$.
- $\mathbf{n}_r$: $K \times 1$ vector containing AWGN of all relays with elements i.i.d. $\mathcal{CN}(0, \sigma_r^2)$.
- $\mathbf{F}$: $K \times K$ forwarding matrix of the AF relays.

5.2.1   Signal and Channel Model

In time slot k, the relay terminals receive the $K \times 1$ vector

$$\mathbf{y}_r[k] = \mathbf{H}_1[k]\mathbf{x}_s[k] + \mathbf{n}_r[k] \tag{5.3}$$

and the destination terminal receives the $N \times 1$ vector

$$\mathbf{y}_d[k] = \mathbf{H}_0[k]\mathbf{x}_s[k] + \mathbf{n}_d[k]. \tag{5.4}$$

---

[4] $\mathcal{CN}(m, \sigma^2)$ denotes the distribution of a complex Gaussian random variable $Z = X + jY$, where $X \sim \mathcal{N}\left(m_X, \frac{\sigma^2}{2}\right)$ and $Y \sim \mathcal{N}\left(m_Y, \frac{\sigma^2}{2}\right)$ are independent and $m = m_X + j \cdot m_Y$.
[5] Averaging is done over source symbols and relay noise.

In time slot $k+1$, the destination terminal receives

$$\mathbf{y}_d[k+1] = \mathbf{H}_0[k+1]\mathbf{x}_s[k+1] + \mathbf{H}_2[k+1]\mathbf{x}_r[k+1] + \mathbf{n}_d[k+1]$$

$$= \mathbf{H}_0[k+1]\mathbf{x}_s[k+1] + \mathbf{H}_2[k+1]\mathbf{F}[k]\mathbf{y}_r[k] + \mathbf{n}_d[k+1]$$

$$= \mathbf{H}_0[k+1]\mathbf{x}_s[k+1] + \mathbf{H}_2[k+1]\mathbf{F}[k]\mathbf{H}_1[k]\mathbf{x}_1[k]$$

$$+ \mathbf{H}_2[k+1]\mathbf{F}[k]\mathbf{n}_r[k] + \mathbf{n}_d[k+1]. \tag{5.5}$$

The forwarding matrix $\mathbf{F} \in \mathcal{C}^{K \times K}$ maps[6] the relay receive vector $\mathbf{y}_r$ to the relay transmit vector $\mathbf{x}_r = \mathbf{F}\mathbf{y}_r$. We will distinguish two different cases:

1. The relays cannot cooperate, i.e, they do not exchange channel or received signal information. This model is appropriate if we assume K *ad hoc relays* that operate stand-alone and assist the MIMO transmission between source and destination. The forwarding matrix $\mathbf{F}$ has a diagonal structure in this case.
2. The relays can fully and perfectly cooperate, i.e., each relay reports its received signal and its local channel information (first or second hop) to a central unit where joint signal processing of all received signals can be established. This model is appropriate when the MIMO transmission between source and destination is assisted by a *distributed relay antenna system*. In this case the forwarding matrix $\mathbf{F}$ may have arbitrary structure.

From now on, we assume that the channel matrices $\mathbf{H}_0$, $\mathbf{H}_1$, and $\mathbf{H}_2$ remain constant over at least two time slots and that $\mathbf{n}_r[k]$ and $\mathbf{n}_d[k]$ are i.i.d. in time and space. Therefore, in the sequel we skip the time slot index k for the channels. Besides the assumption that we transmit over frequency-flat channels, we do not impose a specific model on the channel matrices in this section. We stack the receive vectors $\mathbf{y}_d[k]$ and $\mathbf{y}_d[k+1]$ into one vector and obtain the following description of the two-hop MIMO relay channel:

$$\begin{pmatrix} \mathbf{y}_d[k] \\ \mathbf{y}_d[k+1] \end{pmatrix} = \begin{bmatrix} \mathbf{H}_0 & \mathbf{0} \\ \mathbf{H}_2\mathbf{F}\mathbf{H}_1 & \mathbf{H}_0 \end{bmatrix} \begin{pmatrix} \mathbf{x}_s[k] \\ \mathbf{x}_s[k+1] \end{pmatrix} + \begin{bmatrix} \mathbf{1} & \mathbf{0} & \mathbf{0} \\ \mathbf{0} & \mathbf{1} & \mathbf{H}_2\mathbf{F} \end{bmatrix} \begin{pmatrix} \mathbf{n}_d[k] \\ \mathbf{n}_d[k+1] \\ \mathbf{n}_r[k] \end{pmatrix}, \tag{5.6}$$

where $\mathbf{0}$ and $\mathbf{1}$ denote all-zero and all-one matrices, respectively. Different transmission protocols can be specified:

Protocol P1: The source transmits $\mathbf{x}_s[k]$ in the first time slot to the destination and the relays. In the second time slot, both the relays and the source simultaneously transmit the signals $\mathbf{x}_r[k+1]$ and $\mathbf{x}_s[k+1]$, respectively, to the destination. The signal structure is given in (5.6) and corresponds to the most general transmission protocol.

Protocol P2: The source transmits $\mathbf{x}_s[k]$ in the first time slot to the destination and the relays. In the second time slot the relays forward the signal $\mathbf{x}_r[k+1]$ to the destination,

---

[6] $\mathcal{C}$ denotes the set of complex numbers.

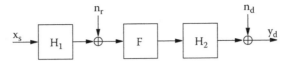

Figure 5.4 System Model for Protocol P3.

whereas the source does not transmit. The signal structure simplifies to

$$
\begin{pmatrix} \mathbf{y}_d[k] \\ \mathbf{y}_d[k+1] \end{pmatrix} = \begin{bmatrix} \mathbf{H}_0 \\ \mathbf{H}_2\mathbf{F}\mathbf{H}_1 \end{bmatrix} \mathbf{x}_s[k] + \begin{bmatrix} 1 & 0 & 0 \\ 0 & 1 & \mathbf{H}_2\mathbf{F} \end{bmatrix} \begin{pmatrix} \mathbf{n}_d[k] \\ \mathbf{n}_d[k+1] \\ \mathbf{n}_r[k] \end{pmatrix}.
\tag{5.7}
$$

Protocol P3: The source transmits $\mathbf{x}_s[k]$ in the first time slot only to the relays (the direct source–destination link is blocked, for example, due to shadowing). In the second time slot the relays forward the signal $\mathbf{x}_r[k+1]$ to the destination, whereas the source does not transmit. The signal structure then follows as

$$
\mathbf{y}_d[k] = \mathbf{0}
\tag{5.8}
$$

$$
\mathbf{y}_d[k+1] = \mathbf{H}_2\mathbf{F}\mathbf{H}_1\mathbf{x}_s[k] + \mathbf{H}_2\mathbf{F}\mathbf{n}_r[k] + \mathbf{n}_d[k+1].
\tag{5.9}
$$

Another protocol proposed in [10] assumes that the source transmits in every time slot, but the destination does not observe the source signal in the first time slot (for example, because the destination is busy due to other tasks). In [10,11], it was shown that protocol P1 performs best in terms of achievable rate (see also Section 5.2.5). However, we will assume that the direct channel $\mathbf{H}_0$ is rank deficient and does not provide enough spatial degrees of freedom to provide spatial multiplexing gain. Including the observation from the direct channel by combining the signals from the first and the second time slot (for example, by maximum ratio combining) leads to an SNR gain. Because we are interested in how AF relays can increase the spatial multiplexing capability we focus on protocol P3, illustrated in Figure 5.4.

Let

$$
\mathbf{H}_1 = \mathbf{U}_1\boldsymbol{\Sigma}_1\mathbf{V}_1^H
\tag{5.10}
$$

$$
\mathbf{H}_2 = \mathbf{U}_2\boldsymbol{\Sigma}_2\mathbf{V}_2^H
\tag{5.11}
$$

be the singular value decompositions (SVDs) of the first-hop and second-hop channel matrices, respectively. The receive vector $\mathbf{y}_d$ is then rewritten as[7]

$$
\mathbf{y}_d = \mathbf{U}_2\boldsymbol{\Sigma}_2\mathbf{V}_2^H\mathbf{F}\mathbf{U}_1\boldsymbol{\Sigma}_1\mathbf{V}_1^H\mathbf{x}_s + \mathbf{U}_2\boldsymbol{\Sigma}_2\mathbf{V}_2^H\mathbf{F}\mathbf{n}_r + \mathbf{n}_d.
\tag{5.12}
$$

For $\widetilde{\mathbf{x}}_s = \mathbf{V}_1^H\mathbf{x}_s$ we have $\mathcal{E}\left\{\widetilde{\mathbf{x}}_s\widetilde{\mathbf{x}}_s^H\right\} = \mathcal{E}\left\{\mathbf{x}_s\mathbf{x}_s^H\right\} = P_s/N\mathbf{I}_N$, i.e., multiplication of $\mathbf{x}_s$ by the unitary matrix $\mathbf{V}_1^H$ does not change the covariance matrix of $\mathbf{x}_s$. It follows for the

---

[7] We drop the time index k completely.

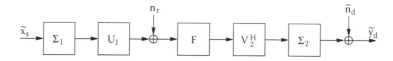

Figure 5.5 **Equivalent System Model for Protocol P3.**

mutual information $I\left(\widetilde{\mathbf{x}}_s; \mathbf{y}_d\right) = I(\mathbf{x}_s; \mathbf{y}_d)$. For the same reasons we may multiply (5.9) at the destination terminal with the unitary matrix $\mathbf{U}_2^H$ to obtain an equivalent system model (see Figure 5.4 and Figure 5.5):

$$\widetilde{\mathbf{y}}_d = \Sigma_2 \mathbf{V}_2^H \mathbf{F} \mathbf{U}_1 \Sigma_1 \widetilde{\mathbf{x}}_s + \Sigma_2 \mathbf{V}_2^H \mathbf{F} \mathbf{n}_r + \widetilde{\mathbf{n}}_d, \qquad (5.13)$$

where $\widetilde{\mathbf{y}}_d = \mathbf{U}_2^H \mathbf{y}_d$ and $\widetilde{\mathbf{n}}_d = \mathbf{U}_2^H \mathbf{n}_d$. Note that $I(\widetilde{\mathbf{x}}_s; \widetilde{\mathbf{y}}_d) = I\left(\widetilde{\mathbf{x}}_s; \mathbf{y}_d\right) = I(\mathbf{x}_s; \mathbf{y}_d)$. In the following we will drop the tilde notation for the equivalent signals $\widetilde{\mathbf{x}}_s$, $\widetilde{\mathbf{y}}_d$, and $\widetilde{\mathbf{n}}_d$, and use $\mathbf{x}_s$, $\mathbf{y}_d$, and $\mathbf{n}_d$ instead. The equivalent system model (5.13) is useful when we assume complete channel state information at the relays for the design of the forwarding matrix $\mathbf{F}$.

5.2.2 Achievable Rates

We assume that the elements of $\mathbf{H}_1$ and $\mathbf{H}_2$ are frequency flat and time varying. We use a block-fading channel model [12] where a channel gain remains constant during a certain time interval (coherence interval) and changes independently from interval to interval. We refer to *slow fading* when the coherence interval is too long for a codeword to span several coherence intervals. In this case, the ϵ-outage capacity [13] is an appropriate performance measure. We speak of *fast fading* when the codeword length spans several coherence intervals of the channel. That might be due to less stringent delay constraints or to a fast varying channel. In this situation the ergodic capacity [13] is the appropriate performance measure. In this chapter we focus on ergodic capacity, because spatial multiplexing gains are usually expressed in terms of ergodic capacity.

The mutual information in bits/s/Hz for perfectly known $\mathbf{H}_1$ and $\mathbf{H}_2$ at the destination terminal is given by

$$I(\mathbf{x}_s; \mathbf{y}_d) = \frac{1}{2} \log \det \left(\mathbf{I}_N + \frac{P_s}{N} \mathbf{R}^{-1} \mathbf{H} \mathbf{H}^H \right) \qquad (5.14)$$

with the equivalent $N \times N$ MIMO channel (over two hops)

$$\mathbf{H} = \Sigma_2 \mathbf{V}_2^H \mathbf{F} \mathbf{U}_1 \Sigma_1 \qquad (5.15)$$

and the noise covariance matrix

$$\mathbf{R} = \mathcal{E} \left\{ \left(\Sigma_2 \mathbf{V}_2^H \mathbf{F} \mathbf{n}_r + \mathbf{n}_d \right) \left(\Sigma_2 \mathbf{V}_2^H \mathbf{F} \mathbf{n}_r + \mathbf{n}_d \right)^H \right\}$$

$$= \sigma_r^2 \Sigma_2 \mathbf{V}_2^H \mathbf{F} \mathbf{F}^H \mathbf{V}_2 \Sigma_2^H + \sigma_d^2 \mathbf{I}_N. \qquad (5.16)$$

The factor $1/2$ in front of the log in (5.14) is due to the half-duplex operation of the relays. This loss in spectral efficiency can be recovered by the protocols proposed in [9].

Reliable communication from source to destination terminal is possible whenever

$$\frac{1}{2} \log \det \left(\mathbf{I}_N + \frac{P_s}{N} \mathbf{R}^{-1} \mathbf{H} \mathbf{H}^H \right) > R, \tag{5.17}$$

where R is the target rate in bits/s/Hz. When the equivalent MIMO channel does not satisfy condition (5.17) the system is in outage. The probability of an outage event is given by:

$$p_{\text{out}}(R) = \mathbb{P}\left[I(\mathbf{x}_s; \mathbf{y}_d) < R \right] = \mathbb{P}\left[\frac{1}{2} \log \det \left(\mathbf{I}_N + \frac{P_s}{N} \mathbf{R}^{-1} \mathbf{H} \mathbf{H}^H \right) < R \right]. \tag{5.18}$$

The ϵ-outage capacity C_ϵ is obtained by solving $p_{\text{out}}(R = C_\epsilon) = \epsilon$.

By coding over a large number of coherence intervals of the channel, the ergodic capacity serves as a long-term rate of reliable communication:

$$C_{\text{erg}} = \mathcal{E}\{I(\mathbf{x}_s; \mathbf{y}_d)\} = \mathcal{E}\left\{ \frac{1}{2} \log_2 \det \left(\mathbf{I}_N + \frac{P_s}{N} \mathbf{R}^{-1} \mathbf{H} \mathbf{H}^H \right) \right\}. \tag{5.19}$$

Depending on the degree of channel state information available at the relay terminals, different possibilities for the optimum choice of the forwarding matrix exist. Different solutions arise depending on whether the relays know both channels $\mathbf{H}_1$ and $\mathbf{H}_2$ or only $\mathbf{H}_1$.

In the following, we first consider the case of a *distributed relay array* (DRA) where the relay terminals are connected through a wired backbone and the signals can be jointly processed at a central processor attached to the backbone. Then, we look at the case of *ad hoc relays* (AR) where the relay terminals do not cooperate with each other, i.e., each relay terminal only processes its own receive signal and forwards it to the destination. In this situation the forwarding matrix $\mathbf{F}$ is diagonal.

5.2.3 System with Distributed Relay Array (DRA)

If the relays are connected through a wired backbone the forwarding matrix can have arbitrary structure, because the receive signal at each relay is known to all relays as well as all channel coefficients of the network. We assume that transmissions over the wired backbone are distortion free, i.e., the receive signal (including noise) at each relay is perfectly known at a central processor. We first look at the case where the relay has perfect knowledge of each channel realization of $\mathbf{H}_1$ and $\mathbf{H}_2$. Then we consider the case where the relays have only knowledge about $\mathbf{H}_1$.

Case 1: We assume that the destination terminal and the relay terminals have perfect knowledge of $\mathbf{H}_1$ and $\mathbf{H}_2$ and the source terminal has only knowledge about the distributions of $\mathbf{H}_1$ and $\mathbf{H}_2$. Furthermore, the signals received at the relay terminals may be processed jointly at a central processor connected to the backbone of the distributed relay array. Without loss of generality, we may expand the forwarding matrix $\mathbf{F}$ as [14]

$$\mathbf{F} = \mathbf{V}_2 \mathbf{F}' \mathbf{U}_1^H, \tag{5.20}$$

where $\mathbf{V}_2$ and $\mathbf{U}_1^H$ are taken from (5.11) and (5.10). $\mathbf{F}' \in \mathbb{R}^{K \times K}$ is called the *inner forwarding matrix*. The receive signal at the destination terminal is then given as

$$\mathbf{y}_d = \Sigma_2 \mathbf{F}' \Sigma_1 \mathbf{x}_s + \Sigma_2 \mathbf{F}' \tilde{\mathbf{n}}_r + \mathbf{n}_d, \tag{5.21}$$

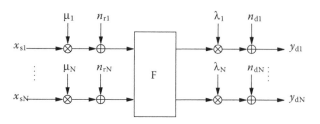

Figure 5.6 $K = N$.

where $\tilde{\mathbf{n}}_r = \mathbf{U}_1^H \mathbf{n}_r$. Note that (5.20) can be interpreted as distributed receive beamforming (with $\mathbf{U}_1^H$) in the first time slot and distributed transmit beamforming (with $\mathbf{V}_2$) in the second time slot. The equivalent $N \times N$ MIMO channel (over two hops) follows as

$$\mathbf{H}(\mathbf{F}') = \Sigma_2 \mathbf{F}' \Sigma_1 \tag{5.22}$$

and the noise covariance matrix at the destination terminal as

$$\mathbf{R}(\mathbf{F}') = \sigma_r^2 \Sigma_2 \mathbf{F}'^2 \Sigma_2^H + \sigma_d^2 \mathbf{I}_N. \tag{5.23}$$

The equivalent system model is given in Figure 5.6 for the case $K = N$, where μ_i is the ith singular value of the first-hop channel $\mathbf{H}_1$ and λ_i the ith singular value of the second-hop channel $\mathbf{H}_2$. The mutual information between $\mathbf{x}_s$ and $\mathbf{y}_d$ is given by

$$I(\mathbf{x}_s; \mathbf{y}_d) = \frac{1}{2} \log \det \left(\mathbf{I}_N + \frac{\text{SNR}}{N} \left(\Sigma_2 \mathbf{F}'^2 \Sigma_2 + \mathbf{I}_N \right)^{-1} \Sigma_2 \mathbf{F}' \Sigma_1^2 \mathbf{F}'^T \Sigma_2 \right), \tag{5.24}$$

where $\text{SNR} = P/\sigma^2$, $P = P_s = P_r$ and $\sigma^2 = \sigma_d^2 = \sigma_r^2$ for simplicity. When we choose the inner forwarding matrix to be diagonal, it has been shown in [15] that the optimal diagonal elements are given by

$$|f_i'|^2 = \frac{1}{\frac{P}{N}\mu_i^2 + \sigma^2} \cdot p_i \tag{5.25}$$

for $i = 1, 2, \ldots, N$ with

$$p_i = \left[0, \sqrt{\frac{P\mu_i^2}{\eta N \lambda_i^2} + \left(\frac{P\mu_i^2}{2N\lambda_i^2} \right)^2} - \frac{P\mu_i^2}{2N\lambda_i^2} - \frac{\sigma^2}{\lambda_i^2} \right]^+, \tag{5.26}$$

where $[0, x]^+ = \max(0, x)$ and η is the water-filling level, which is chosen such that the total relay power constraint is met.

Case 2: In the following, we consider the case that the central unit knows the first-hop channel $\mathbf{H}_1$ perfectly, but has no channel state information (CSI) about the second-hop channel $\mathbf{H}_2$. Furthermore, we assume that the second-hop channel has no preferred direction, i.e., the elements of the channel matrix are zero mean and i.i.d. For this scenario, the optimal forwarding matrix

$$\mathbf{F}^{\text{opt}} = \arg\max_{\mathbf{F}} \mathcal{E}_{\mathbf{H}_2} \left\{ \log \det \left(\mathbf{I}_N + \frac{\text{SNR}}{N} \left(\mathbf{I}_N + \mathbf{H}_2 \mathbf{F} \mathbf{F}^H \mathbf{H}_2^H \right)^{-1} \mathbf{H}_2 \mathbf{F} \mathbf{H}_1 \mathbf{H}_1^H \mathbf{F}^H \mathbf{H}_2^H \right) \right\} \tag{5.27}$$

subject to the relay power constraint[8]

$$\mathcal{E}\left\{\text{Tr}\left(\mathbf{x}_r \mathbf{x}_r^{\text{H}}\right)\right\} = \text{Tr}\left(\frac{\text{SNR}}{N}\mathbf{F}\mathbf{H}_1\mathbf{H}_1^{\text{H}}\mathbf{F}^{\text{H}} + \mathbf{F}\mathbf{F}^{\text{H}}\right) = \text{SNR} \tag{5.28}$$

is not known in the general case.[9] However, some insight into the structure of the optimal matrix can be obtained by considering the asymptotic regimes $\text{SNR} \to 0$, $\text{SNR} \to \infty$, and $N \to \infty$. First, we investigate the low SNR case. Here, it can be shown that choosing the forwarding matrix as a matched filter with respect to the channel of the first hop, i.e.,

$$\mathbf{F} = \sqrt{\frac{\text{SNR}}{\frac{\text{SNR}}{N} \cdot \text{Tr}\left(\mathbf{H}_1^{\text{H}}\mathbf{H}_1\mathbf{H}_1^{\text{H}}\mathbf{H}_1 + \mathbf{H}_1^{\text{H}}\mathbf{H}_1\right)}} \cdot \mathbf{H}_1^{\text{H}}, \tag{5.29}$$

is asymptotically ($\text{SNR} \to 0$) optimal. In order to verify this, we use an approximation (which becomes tight in the limit) of the objective function in (5.27) as follows:

$$\mathcal{E}_{\mathbf{H}_2}\left\{I(\mathbf{x}_s; \mathbf{y}_d)\right\} \approx \mathcal{E}_{\mathbf{H}_2}\left\{\frac{1}{2}\log\det\left(\mathbf{I}_N + \frac{\text{SNR}}{N}\cdot\mathbf{H}_2\mathbf{F}\mathbf{H}_1\mathbf{H}_1^{\text{H}}\mathbf{F}^{\text{H}}\mathbf{H}_2^{\text{H}}\right)\right\}$$

$$\approx \frac{1}{2}\log(e)\cdot\frac{\text{SNR}}{N}\cdot\mathcal{E}_{\mathbf{H}_2}\left\{\text{Tr}\left(\mathbf{H}_2\mathbf{F}\mathbf{H}_1\mathbf{H}_1^{\text{H}}\mathbf{F}^{\text{H}}\mathbf{H}_2^{\text{H}}\right)\right\}$$

$$= \frac{1}{2}\log(e)\cdot\frac{\text{SNR}}{N}\cdot\text{Tr}\left(\mathbf{H}_1\mathbf{F}\mathcal{E}\left\{\mathbf{H}_2\mathbf{H}_2^{\text{H}}\right\}\mathbf{F}^{\text{H}}\mathbf{H}_1^{\text{H}}\right)$$

$$= \frac{1}{2}\log(e)\cdot\text{SNR}\cdot\text{Tr}\left(\mathbf{F}\mathbf{H}_1\mathbf{H}_1^{\text{H}}\mathbf{F}^{\text{H}}\right). \tag{5.30}$$

The first approximation follows from $\lim_{\text{SNR}\to 0}(\mathbf{I}_N + \mathbf{H}_2\mathbf{F}\mathbf{F}^{\text{H}}\mathbf{H}_2^{\text{H}}) = \mathbf{I}_N$; the second one is the well-known MIMO low SNR approximation of mutual information. We may swap expectation and trace due to linearity, and the rearrangement of $\mathbf{H}_1$ and $\mathbf{H}_2$ follows as matrices commute under multiplication inside the trace. Thus, it remains to solve the following problem, whose solution in (5.29) follows immediately from the fact that matched filtering maximizes SNR:

$$\mathbf{F}_{\text{lowSNR}}^{\text{opt}} = \arg\max_{\mathbf{F}}\text{Tr}(\mathbf{F}\mathbf{H}_1\mathbf{H}_1^{\text{H}}\mathbf{F}^{\text{H}}), \tag{5.31}$$

where $\mathbf{F}$ has to fulfill the relay power constraint (5.28). This solution is also intuitive as spatial multiplexing gain vanishes in the low SNR regime, and thus SNR maximization becomes crucial for rate maximization.

In the high SNR regime, again, we use an appropriate approximation which allows for the identification of the important design criteria for the forwarding matrix:

$$\mathcal{E}_{\mathbf{H}_2}\left\{I(\mathbf{x}_s; \mathbf{y}_{ds})\right\} \approx \frac{1}{2}\left(N\log\frac{\text{SNR}}{N} + \log\det(\mathbf{F}\mathbf{H}_1\mathbf{H}_1^{\text{H}}\mathbf{F}^{\text{H}})\right.$$

$$\left. + \mathcal{E}_{\mathbf{H}_2}\left\{\log\det(\mathbf{H}_2\mathbf{H}_2^{\text{H}})\right\} - \mathcal{E}_{\mathbf{H}_2}\left\{\log\det(\mathbf{I}_N + \mathbf{H}_2\mathbf{F}\mathbf{F}^{\text{H}}\mathbf{H}_2^{\text{H}})\right\}\right). \tag{5.32}$$

---

[8] We choose $\sigma^2 = 1$ and, thus, $P = \text{SNR}$.

[9] $\text{Tr}(\cdot)$ denotes the trace operator.

The optimization problem thus reduces to

$$\mathbf{F}_{\text{highSNR}}^{\text{opt}} = \arg\max_{\mathbf{F}} \log \det(\mathbf{FH}_1\mathbf{H}_1^{\text{H}}\mathbf{F}^{\text{H}}) - \mathcal{E}_{\mathbf{H}_2}\left\{ \log \det\left(\mathbf{I}_N + \mathbf{H}_2\mathbf{FF}^{\text{H}}\mathbf{H}_2^{\text{H}}\right)\right\}, \quad (5.33)$$

such that

$$\text{Tr}\left(\frac{\text{SNR}}{N}\mathbf{H}_1\mathbf{FF}^{\text{H}}\mathbf{H}_1^{\text{H}}\right) = \text{SNR}. \quad (5.34)$$

Although we are not aware of an analytic solution, both terms can be identified to reflect two conflicting requirements. By applying the arithmetic mean–geometric mean inequality, the first term is found to be maximized by a zero-forcing (ZF) matrix $\mathbf{F} \propto (\mathbf{H}_1^{\text{H}}\mathbf{H}_1)^{-1}\mathbf{H}_1^{\text{H}}$. Consequently, it accentuates the need for isotropic signal radiation into the second-hop channel. The second term, on the other hand, is nothing else but the ergodic mutual information $\mathcal{E}\{I(\mathbf{n}_{\text{r}}; \mathbf{y}_{\text{d}})\}$ evaluated at a low SNR which is mainly kept small by avoiding noise enhancement at the relay. In conclusion, the optimal forwarding matrix thus has to balance isotropic radiation with noise enhancement. A filter well-known to accomplish a closely related tradeoff, namely, the one between multistream interference and noise enhancement, is realized by the minimum mean-squared error (MMSE) matrix. Such a forwarding matrix demonstrates better performance than the ZF matrix in our computer experiments.

5.2.4 System with Ad Hoc Relays

Here we assume that the relays are not connected through a wired backbone and, therefore, are not aware of any received signals other than their own. Thus, the forwarding matrix is constrained to be diagonal. We will assume that $N \gg 1$ in this section, such that the law of large numbers becomes effective and each relay receives roughly the same signal power. The optimal forwarding matrix — due to symmetry — then is $\mathbf{F} \propto \mathbf{I}_K$.

We are interested in the performance of this particularly simple choice of $\mathbf{F}$ compared to the forwarding strategies introduced in the previous section. While it is clearly inferior to the matched filter in the low SNR regime as it does not collect the signal power efficiently, the result in the medium and high SNR region looks different. Denoting the rate under $\mathbf{F} \propto \mathbf{I}_K$ by $R_{\mathbf{I}}$ and the rate under an MMSE forwarding matrix by R_{MMSE}, we have the following:

Given a certain SNR there is always an N_0, such that for $N > N_0$ we have $R_{\mathbf{I}} > R_{\text{MMSE}}$. To gain insight into the reason behind this, we let $N = K$ grow large, such that

$$\frac{1}{N} \cdot \mathbf{H}_i\mathbf{H}_i^{\text{H}} \longrightarrow \mathbf{I}_N \text{ as } N \to \infty, \text{ for } i = 1, 2 \quad (5.35)$$

again by the law of large numbers. Thus, choosing $\mathbf{F} = \alpha\mathbf{I}_K$ (the same derivations hold if $\mathbf{F}$ is unitary) yields

$$\mathbf{H} = \frac{1}{N}\mathbf{H}_2\mathbf{FH}_1\mathbf{H}_1^{\text{H}}\mathbf{F}^{\text{H}}\mathbf{H}_2^{\text{H}} \longrightarrow \mathbf{H}_2\mathbf{FF}^{\text{H}}\mathbf{H}_2^{\text{H}} = \alpha\mathbf{H}_2\mathbf{H}_2^{\text{H}} \quad (5.36)$$

and

$$\mathbf{R} = \frac{1}{N}\mathbf{H}_2\mathbf{FF}^{\text{H}}\mathbf{H}_2^{\text{H}} + \mathbf{I}_N = \frac{\alpha}{N}\mathbf{H}_2\mathbf{H}_2^{\text{H}} + \mathbf{I}_N \longrightarrow (1 + \alpha)\mathbf{I}_N. \quad (5.37)$$

Consequently, this choice of **F** yields an (almost) isotropic radiation into the second-hop channel while — in contrast to the MMSE matrix — avoiding both noise enhancement and coloring. Thus, it will demonstrate better performance for sufficiently large N, if $\mathbf{H}_2$ has no preferred direction. Our computer simulations even show that for moderate SNR we have $N_0 = 1$, although convergence in (5.35) is rather slow. Thus, the loss in performance compared to the distributed relay array with knowledge of the first-hop channel at the relay can only be expected to be small for sufficiently large N under moderate and high SNR. This result is also appealing for the linear distributed array system scenario due to its simplicity. In Section 5.3 we will look at the ad hoc relay case in more detail for a specific channel model and show how ad hoc AF relays can help to provide spatial multiplexing gain in poor scattering environments.

5.2.5 Numerical Examples

In this section we give some numerical examples for Gaussian i.i.d. fading channels. Figure 5.7 compares the ergodic capacities of protocols P1–P3 for a $4 \times 4 \times 4$ system, i.e., four antennas at source and destination and four relays. We can see that protocol P1 performs best in terms of ergodic capacity. We normalized the transmit powers such that each node consumes an average transmit energy of PT over two time slots each of length T. SNR is defined as P/σ^2, where σ^2 is the noise variance at each antenna (relay or destination). Figure 5.8 shows the cumulative distribution function (CDFs) of the mutual information of protocols P1–P3. We observe that protocols P2 and P3 have slightly higher diversity orders ("steeper" CDF) than protocol P1. The reason is that in protocol P3 every second transmit vector interferes with the transmit vector of the relays. Because both vectors do not contain the same data, different degrees of freedom have to be used for separation of the signals rather than combining them to achieve higher diversity gains.

We also confirm some of the results obtained in the previous section through numerical examples. In Figure 5.9 and Figure 5.10, we plot the ergodic capacity of the relay

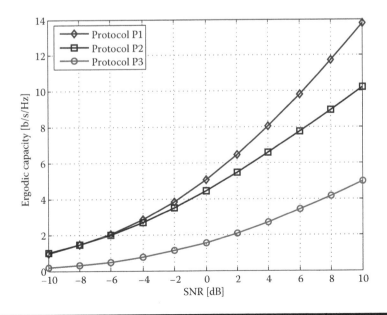

Figure 5.7   Ergodic Capacities for Protocols P1–P3 for $N = K = 4$ and Ad Hoc Relays.

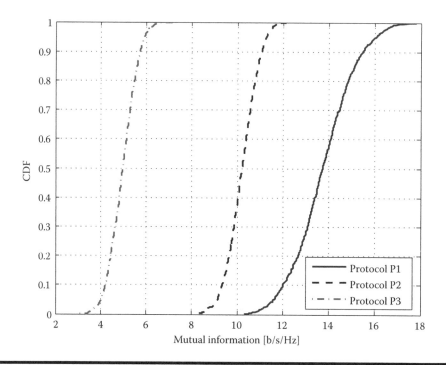

Figure 5.8   CDFs of Mutual Information of Protocols P1–P3 for $N = K = 4$ and Ad Hoc Relays.

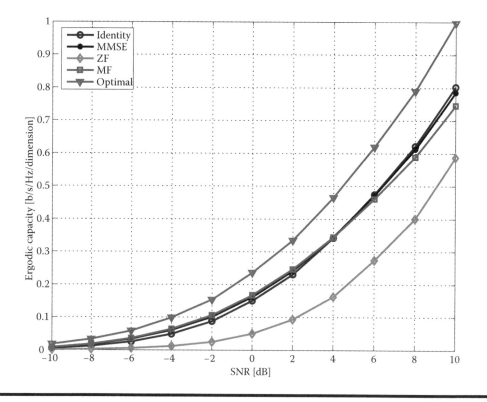

Figure 5.9   Ergodic Capacities for Low SNR and Different Choices of F and $N = K = 4$.

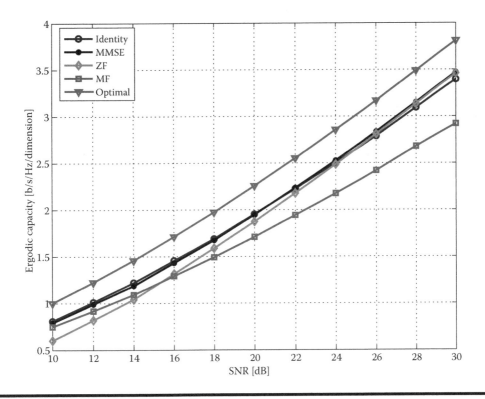

Figure 5.10   Ergodic Capacities for High SNR and Different Choices of F and $N = K = 4$.

channel under different forwarding matrices. While $\mathbf{F} \propto \mathbf{I}$ can be realized in both linear distributed array and ad hoc relay systems, all other strategies are applicable to the first case only. As a reference, the capacity in the case that the relay array knows both $\mathbf{H}_1$ and $\mathbf{H}_2$ and chooses the forwarding matrix in an optimal way (5.25) is plotted.

Under low SNR, we see that the (asymptotically optimal) matched filter (MF) matrix performs best among the matrices realizable without channel knowledge about $\mathbf{H}_2$ and approaches the performance of the MMSE matrix from above. $\mathbf{F} \propto \mathbf{I}$ performs poorly as it does not collect the signal power efficiently, the ZF matrix even worse as it enhances the noise. Under high SNR the MMSE matrix confirms its ability to balance noise enhancement and isotropic radiation into the second hop and shows the best performance as expected. The ZF matrix ensures perfect isotropic radiation, however, the noise enhancement — though not crucial in this case — makes it inferior to the MMSE matrix. We can also see that the identity matrix becomes the best choice among the considered candidates under medium SNR. The intersection point of its curve with the curve of the MMSE matrix in Figure 5.9 will be shifted even further to the right, if N is increased. This leads to the conclusion that whenever the second-hop channel is not known at the relay array, the joint processing of the received signals at the relay antennas does not lead to significant performance gains for sufficiently high SNR. Using a diagonal forwarding matrix is sufficient, which means that the relay antennas do not have to be connected to a central unit. An intuitive argument strengthening the derivations in the large array limit in the previous section is that unitary matrices do not diminish the rank of the compound channel matrix while simultaneously keeping the noise level constant.

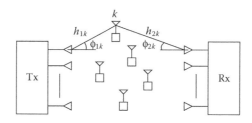

Figure 5.11   Relay-Assisted MIMO Communication System.

5.3   ERGODIC PERFORMANCE OF A RELAY-ASSISTED RANK-DEFICIENT MIMO CHANNEL

In this section, we assume that the relays are not connected to a central unit (ad hoc relays) and we assume a specific channel model for the first-hop and second-hop matrix channels. We assume that the channel gain from one source antenna to one relay antenna[10] is characterized by an angle of departure (AoD) and a path gain that is a complex Gaussian random variable. The channel gain from the relay antenna to one destination antenna is characterized by an angle of arrival (AoA) and again a path gain that is a complex Gaussian random variable. We assume that the relays do not move during the time of interest and that the path gains of the first- and the second-hop channels are independent. Further, we look at the case where the relay operates stand-alone, i.e., they are not connected to a central processor where joint signal processing can be done. Hence, the relay can only operate in an AF mode because single-antenna decode-and-forward relays cannot decode the MIMO encoded source signal.

5.3.1   System Model

We consider uniform linear antenna arrays at the source, destination, and single-antenna relays. For the antenna array depicted in Figure 5.11, we denote the AoD vector with respect to relay k as

$$\Phi_{1k} = \left[1, e^{j\,2\pi d \sin \phi_{1k}/\lambda}, \ldots, e^{j\,2\pi(N-1)d \sin \phi_{1k}/\lambda} \right]^{\mathrm{T}}, \qquad (5.38)$$

where d is the antenna spacing at source and destination, λ the operational wavelength, and $\phi_{1k} \in [-\pi/2, \pi/2]$ the horizontal angle characterizing the paths to relay k. Equation (5.38) follows from a narrowband signal and planar wavefront assumption [16]. The array response for a plane wave arriving at angle ϕ_{2k} follows accordingly as

$$\Phi_{2k} = \left[1, e^{-j\,2\pi d \sin \phi_{2k}/\lambda}, \ldots, e^{-j\,2\pi(N-1)d \sin \phi_{2k}/\lambda} \right]^{\mathrm{T}}, \qquad (5.39)$$

where $\phi_{2k} \in [-\pi/2, \pi/2]$ are the horizontal angles characterizing the paths from relay k, respectively (Figure 5.11). The signal received in the first time slot by relay k is

$$y_{\mathrm{r},k} = h_{1k}\Phi_{1k}^{\mathrm{T}}\mathbf{x}_{\mathrm{s}} + n_{\mathrm{r}k}, \qquad (5.40)$$

---

[10] Each relay is equipped with only one antenna.

where b_{1k} is the first-hop channel coefficient of relay k, which usually accounts for path loss, shadowing, and small-scale fading. We assume that the fading coefficient is approximately the same for all transmit antennas (spacing of antenna elements at both sides sufficiently small), i.e.,

$$b_{1k} = b_{1k}^{(1)} \approx b_{1k}^{(2)} \approx \ldots \approx b_{1k}^{(N)}, \tag{5.41}$$

where $b_{1k}^{(j)}$ denotes the channel coefficient between transmit antenna j and relay antenna k. Note that we consider protocol P3 and, therefore, the destination does not receive any signal from the source in the first time slot. In the second time slot the destination receives

$$\mathbf{y}_d = \sum_{k=1}^{K} b_{2k} f_k b_{1k} \Phi_{2k} \Phi_{1k}^T \mathbf{x}_s + \sum_{k=1}^{K} b_{2k} f_k \Phi_{2k} n_{rk} + \mathbf{n}_d, \tag{5.42}$$

where b_{2k} is the second-hop channel coefficient of relay k (again approximately the same for all receive antennas), f_k is the gain factor of relay k and $\mathbf{y}_d = [y_{d,1}, \ldots, y_{d,N}]^T$ is the receive vector at the destination in time slot 2. In matrix notation, we write

$$\mathbf{y}_d = \Phi_2 \Gamma_{12} \Phi_1^T \mathbf{x}_s + \Phi_2 \Gamma_2 \mathbf{n}_r + \mathbf{n}_d, \tag{5.43}$$

where Γ_{12} and Γ_2 are diagonal with $[\Gamma_{12}]_{k,k} = b_{2k} f_k b_{1k}$ and $[\Gamma_2]_{k,k} = b_{2k} f_k$. The columns of Φ_1 are the steering vectors (5.38) and the columns of Φ_2 are the array response vectors (5.39) and $\mathbf{n}_3 = [n_{31}, \ldots, n_{rK}]$.

5.3.2 Achievable Rates

The mutual information of (5.43) is given by

$$I\left(\mathbf{x}_s; \mathbf{y}_d | \Phi_2 \Gamma_{12} \Phi_1^T, \mathbf{R}\right) = b\left(\mathbf{y}_d | \Phi_2 \Gamma_{12} \Phi_1^T, \mathbf{R}\right) - b\left(\mathbf{n} | \Phi_2 \Gamma_{12} \Phi_1^T, \mathbf{R}\right), \tag{5.44}$$

where $b(\cdot)$ denotes the differential entropy of a random vector, $\mathbf{n} = \Phi_2 \Gamma_2 \mathbf{n}_r + \mathbf{n}_d$ and $\mathbf{R}$ is the covariance matrix of the effective noise at the destination

$$\mathbf{R} = \Phi_2 \Gamma_{12} \Gamma_{12}^H \Phi_2^H \sigma_r^2 + \sigma_d^2 \mathbf{I}_N. \tag{5.45}$$

The mutual information is maximized when $b\left(\mathbf{y}_d | \Phi_2 \Gamma_{12} \Phi_1^T, \mathbf{R}\right)$ is maximized, i.e., the receive vector $\mathbf{y}_d$ has to be Gaussian for a given two-hop channel matrix $\Phi_2 \Gamma_{12} \Phi_1^T$ and its differential entropy is [17]

$$b\left(\mathbf{y}_d | \Phi_2 \Gamma_{12} \Phi_1^T, \mathbf{R}\right) = \log_2 \det\left(\pi e \left(\Phi_2 \Gamma_{12} \Phi_1^T \mathbf{R}_s \Phi_1^* \Gamma_{12}^H \Phi_2^H\right) + \mathbf{R}\right), \tag{5.46}$$

where $\mathbf{R}_s = E\left\{\mathbf{x}_s \mathbf{x}_s^H\right\}$. The mutual information measured in bits per channel use follows then as

$$I\left(\mathbf{x}_s; \mathbf{y}_d | \Phi_2 \Gamma_{12} \Phi_1^T, \mathbf{R}\right) = \frac{1}{2} \sum_{k=1}^{r} \log_2\left(1 + \frac{P_s}{N} \lambda_k \left(\mathbf{R}^{-1} \Phi_2 \Gamma_{12} \Phi_1^T \Phi_1^* \Gamma_{12}^H \Phi_2^H\right)\right), \tag{5.47}$$

where

$$r = \mathrm{rk}\left(\mathbf{R}^{-1} \Phi_2 \Gamma_{12} \Phi_1^T \Phi_1^* \Gamma_{12}^H \Phi_2^H\right) = \min\{N, K\}. \tag{5.48}$$

We used $\mathbf{R}_s = \frac{P}{N}\mathbf{I}_N$ because we assume no channel knowledge at source and relays, hence, optimal power allocation is not possible, i.e., in every second time slot the source distributes the power P equally among the antennas. The factor $1/2$ is due to the use of two time slots.

To evaluate the ergodic capacity performance of this scheme we determine the eigenvalues of the channel covariance matrix

$$\frac{1}{N}\mathbf{R}^{-1}\mathbf{H}\mathbf{H}^{\mathrm{H}} = \left(\Phi_2\Gamma_2\Gamma_2^{\mathrm{H}}\Phi_2^{\mathrm{H}}\sigma_r^2 + \sigma_d^2\mathbf{I}_N\right)^{-1}\Phi_2\Gamma_{12}\Phi_1^{\mathrm{T}}\Phi_1^*\Gamma_{12}^{\mathrm{H}}\Phi_2^{\mathrm{H}} \qquad (5.49)$$

when the number of antennas N goes to infinity (large-array limit), the antenna separation d remains constant, and $\mathbf{H} = \Phi_2\Gamma_{12}\Phi_1^{\mathrm{T}}$. The eigenvalues are asymptotically accurate as $N \to \infty$ and serve as an approximation in the nonasymptotic regime. In the large-array limit we obtain for the eigenvalues:

Theorem. For $N \to \infty$ the eigenvalues of the channel covariance matrix are given by

$$\lambda_k\left(\frac{1}{N}\left(\Phi_2\Gamma_2\Gamma_2^{\mathrm{H}}\Phi_2^{\mathrm{H}}\sigma_r^2 + \sigma_d^2\mathbf{I}_N\right)^{-1}\Phi_2\Gamma_{12}\Phi_1^{\mathrm{T}}\Phi_1^*\Gamma_{12}^{\mathrm{H}}\Phi_2^{\mathrm{H}}\right) \xrightarrow{N\to\infty} \frac{|b_{1k}|^2}{\sigma_r^2} \qquad (5.50)$$

for $k = 1, \ldots, K$ and $b_{2k}, f_k \neq 0 \ \forall k$.

Proof of the Theorem. Can be found in [18].

Corollary. For finite N the eigenvalues of $\frac{1}{N}\mathbf{R}^{-1}\mathbf{H}\mathbf{H}^{\mathrm{H}}$ are approximated by

$$\lambda_k\left(\frac{1}{N}\mathbf{R}^{-1}\mathbf{H}\mathbf{H}^{\mathrm{H}}\right) \approx \frac{N|b_{1k}g_k b_{2k}|^2}{\sigma_d^2 + N|g_k b_{2k}|^2\sigma_r^2} \qquad (5.51)$$

for $k = 1, \ldots, K$.

The theorem and corollary stated above have an interesting physical interpretation: From [16], we know that for a uniform linear array with weighting vector $\mathbf{w} = \frac{P}{N}\Phi_k$ at the source, the 3 dB beamwidth (half-power points) is $\Delta_{\mathrm{3dB}} = 0.891\frac{\lambda}{Nd}$ and the Rayleigh resolution limit (null-to-null beamwidth) is $\Delta_{00} = 2\frac{\lambda}{Nd}$, i.e., the beams from source to relays become narrower with an increasing number of transmit antennas and the spatial overlaps between the beams disappear. The same holds for the receive side, if $\mathbf{w} = \Theta_k$ (see Figure 5.12). In our case the "beams" (eigenvectors of the channel correlation matrix (5.49)) become spatially orthogonal in the large-array limit and the kth eigenvalue of the compound channel matrix depends only on the channel parameters of relay k. Actually, the kth eigenvalue depends only on the relay's first hop channel and the relay noise variance: due to array gain at the destination, the SNR at the destination is determined by the relay noise only.

Asymptotic Ergodic Capacity. We assume here a static relay topology, i.e., the relays do not change their positions during the time of interest. The channel coefficients in the relay-uplink and relay-downlink are random variables that are constant during one block of transmission. With random coding over a large number of independent blocks one can achieve the ergodic capacity of the system.

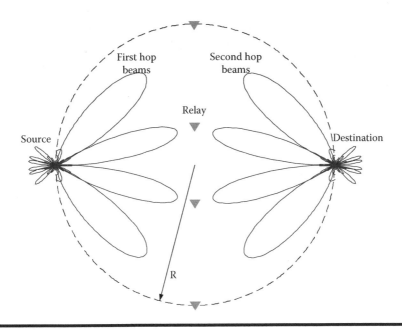

Figure 5.12 Source and Destination with Multiple Antennas, Relays Are Single Antenna Nodes. At 17 GHz We Have $R = 500\lambda \approx 8.8$m.

Combining (5.47), (5.49), and (5.50), we obtain for the asymptotic mutual information:

$$I\left(\mathbf{x}_{\mathrm{s}}; \mathbf{y}_{\mathrm{d}} | \Phi_2 \Gamma_{12} \Phi_1^{\mathrm{T}}, \mathbf{R}\right) = \frac{1}{2} \sum_{k=1}^{K} \log_2\left(1 + \mathrm{SNR} |b_{1k}|^2\right). \tag{5.52}$$

To determine the asymptotic ergodic capacity, we assume for the first-hop channel coefficients a model that includes path loss and small-scale fading:

$$b_{1k} = \frac{1}{(1 + r_{1k})^{\alpha/2}} x_{1k}. \tag{5.53}$$

Here, $1 + r_{1k}$ is the normalized distance between source and relay k, and $x_{1k} \sim \mathcal{CN}(m, \sigma_{\mathrm{h}}^2)$.

Rayleigh Fading. We assume an i.i.d. x_{1k} with zero mean $m = 0$, i.e., $|x_{1k}|^2$ has an exponential probability density function and the asymptotic ergodic capacity is obtained by taking the expectation over the channel statistics

$$C^\infty = \frac{1}{2} \sum_{k=1}^{K} \int_0^\infty \log_2\left(1 + \frac{\mathrm{SNR}|x_{1k}|^2}{(1 + r_{1k})^\alpha}\right) \frac{1}{2\sigma_{\mathrm{h}}^2} e^{-\frac{|x_{1k}|^2}{2\sigma_{\mathrm{h}}^2}} \, \mathrm{d}|x_{1k}|^2$$

$$= \frac{\ln 2}{2} \sum_{k=1}^{K} e^{\frac{1}{\mathrm{SNR}_k 2\sigma_{\mathrm{h}}^2}} \mathrm{Ei}\left(-\frac{1}{\mathrm{SNR}_k 2\sigma_{\mathrm{h}}^2}\right) \tag{5.54}$$

with $\mathrm{SNR}_k = \mathrm{SNR}/(1 + r_{1k})^\alpha$ and where $\mathrm{Ei}(x)$ is the exponential integral defined as $\mathrm{Ei}(x) = -\int_{-x}^\infty \frac{e^{-t}}{t} \mathrm{d}t$.

Rice Fading. We assume an i.i.d. x_{1k} with nonzero mean m per real dimension, i.e., $z = |x_{1k}|^2$ has a noncentral chi-square distribution with two degrees of freedom and an upper bound on the asymptotic ergodic capacity is then

$$C_{T3}^{\infty} = \frac{1}{2} \sum_{k=1}^{K} \int_0^{\infty} \log_2 \left(1 + \text{SNR}_k z \right) \frac{1}{2\sigma_h^2} e^{-\frac{2m^2 + z}{2\sigma_h^2}} J_0 \left(\sqrt{2z} \frac{m}{\sigma_h^2} \right) dz$$

$$\leq \frac{1}{2} \sum_{k=1}^{K} \int_0^{\infty} \log_2 \left(\text{SNR}_k z \right) \frac{1}{2\sigma_h^2} e^{-\frac{2m^2 + z}{2\sigma_h^2}} J_0 \left(\sqrt{2z} \frac{m}{\sigma_h^2} \right) dz$$

$$= \frac{1}{2 \ln 2} \sum_{k=1}^{K} \left(\ln \left(2 \text{SNR}_k m^2 \right) - \text{Ei} \left(2m^2 \right) \right), \tag{5.55}$$

where we used in the inequality a high SNR approximation and in the last equality a result from [19] about the expected log of a noncentral chi-square random variable. $J_n(x)$ denotes the nth-order modified Bessel function of the first kind.

5.3.3 Numerical Results

In this section, we present numerical examples to demonstrate the accuracy of the asymptotic and approximate eigenvalues given in (5.50) and (5.51) and the ergodic capacities given in (5.54) and (5.55).

Simulation Setup. The setup of the relay network is depicted in Figure 5.12. As previously mentioned, we will consider only deterministic relay positions. For random relay topologies, i.e., the relays change their position during the data transmission according to a predefined probability distribution, the capacity results have to be averaged over the relay topology. Here the relays are placed so that the angle difference between the two relays, with respect to source and destination, is constant. By increasing the angle difference (decreasing the relay density), we obtain less spatial cross talk between the "beams" (eigenvectors of channel correlation matrix) for a finite number of antennas and, with that, the results in (5.50) and (5.51) become more accurate. Note that in the large-array limit the beams become infinitely narrow and no spatial overlap between the beams occurs (channel correlation matrix becomes diagonal).

The first-hop channel coefficients are chosen according to (5.53), where in the Rayleigh case we choose $m = 0$ and $\sigma^2 = 1/2$. For Rice channel simulations we choose $m = \sqrt{K_R/2}$, where K_R denotes the Ricean K-factor. The same holds for the second-hop channel coefficients. Under the assumption that the relays can measure the receive power, the gain coefficients in the AF relays are chosen according to $f_k = \sqrt{P_{rk} / \left(|b_{1k}|^2 P_s + \sigma_r^2 \right)}$, where P_{rk} denotes the maximum transmit power of relay k. This is in general a suboptimal power allocation and other strategies can achieve a better performance [11].

Channel Normalization. To obtain defined average SNR values at the destination, we normalize the channel matrix for the simulation such that the average channel gain is equal to the array gain:

$$\widetilde{\mathbf{H}} = \frac{\mathbf{R}^{-1/2}\mathbf{H}}{\sqrt{\mathrm{E}\left\{\|\mathbf{R}^{-1/2}\mathbf{H}\|_{\mathrm{F}}^2\right\}}} N. \tag{5.56}$$

The total average received power is then

$$\mathcal{E}\left\{\|\widetilde{\mathbf{H}}\|_{\mathrm{F}}^2\right\}\frac{P_{\mathrm{s}}}{N} = NP_{\mathrm{s}}, \tag{5.57}$$

i.e., it equals the total transmitted power times the receive array gain N. This implies that the eigenvalues also have to be normalized according to

$$\widetilde{\lambda}_k = \frac{\lambda_k\left(\mathbf{R}^{-1}\mathbf{H}\mathbf{H}^{\mathrm{H}}\right)}{\mathcal{E}\left\{\sum_{k=1}^K \lambda_k\left(\mathbf{R}^{-1}\mathbf{H}\mathbf{H}^{\mathrm{H}}\right)\right\}} N^2, \tag{5.58}$$

where we use for $\lambda_k\left(\mathbf{R}^{-1}\mathbf{H}\mathbf{H}^{\mathrm{H}}\right)$ either (5.50) or (5.51). Further, we choose an operational frequency of 17 GHz, an antenna separation of $d = \lambda/2$, and an average destination SNR of 20 dB (averaged over the small-scale fading).

Figure 5.13 shows the CDFs of the nonzero eigenvalues of a relay-assisted MIMO system with $K = 4$ relays and $N = 20$ transmit/receive antennas as an example. We see that the CDFs based on the asymptotic eigenvalues are quite close to the corresponding empirical distributions.

In Figure 5.14 we plot capacity versus number of antennas assuming Rayleigh fading (in relay-uplink and relay-downlink) for the different number of relays. We compare the results obtained via the asymptotic (5.50) and the approximated (5.51) eigenvalues with the empirical capacity curve. Capacity scales linearly with the number of relays when

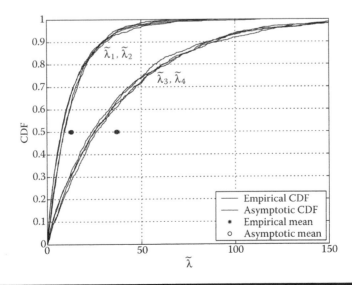

Figure 5.13 Eigenvalue CDFs of the Relay-Assisted MIMO Rayleigh Channel for $K = 4$ Relays and $N = 20$ Antennas.

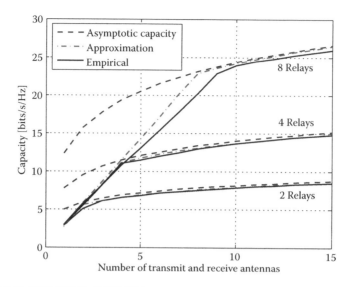

Figure 5.14 Capacity of the Relay-Assisted MIMO Rayleigh Channel vs. Number of Antennas N for K = 2, 4, 8 Relays. Asymptotic Capacity Refers to 5.54, the Approximation Refers to 5.51 and Empirical Means Monte-Carlo Simulation of the Relay Assisted MIMO Channel 5.43.

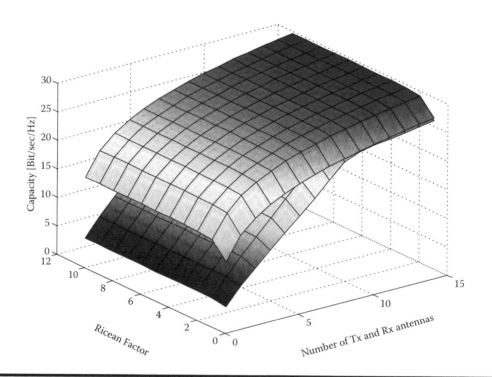

Figure 5.15 Capacity of the Relay-Assisted MIMO Rice Channel vs. Number of Antennas N and Rice Factor K_R for 8 Relays. The Upper Surface Corresponds to 5.55 and the Lower Surface is an Empirical Monte-Carlo Simulation.

$K \geq N$, and logarithmic when $K \leq N$ (array gain). Note that the approximative result is very accurate even when the number of antennas is small.

Figure 5.15 shows the results for Rice fading. A key observation here is that the capacity is independent of the Ricean factor (note that the effect of an increased receive power due to the LOS component is removed due to our normalization (5.56)). The rank of the channel matrix and the eigenvalue distribution are determined by the number of relays and their locations. The relays play a role of *active channel shapers* and make the performance of the system insensitive to the small-scale fading statistics.

5.4 CONCLUSIONS AND OUTLOOK

We showed that with two-hop relaying one can shape the channel matrix to be better conditioned and therefore increase the capacity of rank-deficient MIMO channels. We proposed to use amplify-and-forward relays that act as active scatterers and assist the communication between a source terminal and a destination terminal. We considered two cases: First, the relay antennas were connected through a wired backbone (distributed relay array) and the linear signal processing was done in a central unit also connected to this backbone. Second, we looked at the case where the relays operated stand-alone, i.e, without a backbone connection. The goal of relaying in both cases was to increase the rank of the compound (two time slots) channel matrix and to shape the eigenvalue distribution such that the channel matrix became well-conditioned and the capacity of the MIMO channel improved.

Further work includes the theoretical analysis of the ergodic and outage capacity of the two-hop MIMO relay channel for different channel models like, for example, Rayleigh fading (no line-of-sight) and Rice fading (line-of-sight). Asymptotic results, with respect to the number of antennas, can be obtained by the use of random matrix theory or replica analysis. Often the asymptotic results then serve as good approximations for systems with a finite number of antennas. A further issue of interest is the choice of the gains at the relays, such that the condition number of the channel matrix is maximized. When the relays operate stand-alone it might be interesting to extend the protocols to compress-and-forward relays, where the relays quantize and compress their received signals and use a channel code to transmit their quantized signals to the destination.

REFERENCES

[1] G.J. Foschini, "Layered space-time architecture for wireless communication in a fading environment when using multi-element antennas," *Bell Labs Tech. J.*, vol. 1, pp. 41–59, Autumn 1996.

[2] V. Tarokh, N. Seshadri, and A.R. Calderbank, "Space-time codes for high data rate wireless communication: performance criterion and code construction," *IEEE Trans. Inform. Theory*, vol. 44, pp. 744–765, Mar. 1998.

[3] A. Wittneben and B. Rankov, "MIMO signaling for low rank channels," in *Proc. of 2004 URSI, Inter. Symp. on Electromagn. Theory*, (Pisa, Italy), pp. 72–74, May 2004.

[4] I.E. Telatar, "Capacity of multi-antenna Gaussian channels," *European Trans. on Telecommn.*, vol. 10, pp. 585–595, Nov./Dec. 1999.

[5] C. Chuah, D. Tse, J. Kahn, and R. Valenzuela, "Capacity scaling in MIMO wireless systems under correlated fading," *IEEE Trans. Inform. Theory*, vol. 48, pp. 637–650, Mar. 2002.

[6] A. Paulraj, R. Nabar, and D. Gore, *Introduction to Space-Time Wireless Communications*, Cambridge, UK: Cambridge University Press, 2003.

[7] D. Gesbert, H. Boelcskei, D.A. Gore, and A.J. Paulraj, "Outdoor MIMO wireless channels: models and performance prediction," *IEEE Trans. Commn.*, vol. 50, pp. 1926–1934, Dec. 2002.

[8] D. Chizhik, G.J. Foschini, M.J. Gans, and R. Valenzuela, "Keyholes, correlations, and capacities of multielement transmit and receive antennas," *IEEE Trans. Commn.*, vol. 1, pp. 361–368, Apr. 2002.

[9] B. Rankov and A. Wittneben, "Spectral efficient protocols for half-duplex fading relay channels," *IEEE J. Select. Areas Commn.*, Feb. 2007.

[10] R.U. Nabar, H. Bölcskei, and F. Kneubühler, "Fading relay channels: performance limits and space-time signal design," *IEEE J. Select. Areas Commn.*, vol. 22, pp. 1099–1109, Aug. 2004.

[11] A. Wittneben and B. Rankov, "Impact of cooperative relays on the capacity of rank-deficient MIMO channels," in *Proc. Mobile and Wireless Communications Summit (IST)*, (Aveiro, Portugal), pp. 421–425, June 2003.

[12] L. Ozarow, S. Shamai (Shitz), and A.D. Wyner, "Information theoretic considerations for cellular mobile radio," *IEEE Trans. Veh. Technol.*, vol. 43, pp. 359–378, May 1994.

[13] E. Biglieri, J. Proakis, and S. Shamai (Shitz), "Fading channels: information-theoretic and communications aspects," *IEEE Trans. Inform. Theory*, vol. 44, pp. 2619–2692, Oct. 1998.

[14] A. Wittneben and B. Rankov, "Distributed antenna systems and linear relaying for gigabit MIMO wireless," in *Proc. IEEE Veh. Tech. Conf.*, (Los Angeles), pp. 3624–3630, Sep. 2004.

[15] O. Munoz, J. Vidal, and A. Augustin, "Non-regenerative MIMO relaying with channel state information," in *Proc. IEEE ICASSP*, (Philadelphia), pp. 361–364, Mar. 2005.

[16] H.L. Van Trees, *Optimum Array Processing: Part IV of Detection, Estimation and Modulation Theory*, New York: John Wiley & Sons, 2002.

[17] T.M. Cover and J.A. Thomas, *Elements of Information Theory*, New York: John Wiley & Sons, 1991.

[18] B. Rankov and A. Wittneben, "On the capacity of relay-assisted wireless MIMO channels," in *Proc. IEEE SPAWC*, (Lisbon, Portugal), July 2004.

[19] A. Lapidoth and S.M. Moser, "Capacity bounds via duality with applications to multiple-antenna systems on flat fading channels," *IEEE Trans. Inform. Theory*, vol. 49, pp. 2426–2467, Oct. 2003.

PART II

MAC AND PROTOCOLS

6

DISTRIBUTED SIGNAL PROCESSING
IN WIRELESS SENSOR NETWORKS

Sudharman K. Jayaweera and Ramanarayanan Viswanathan

Contents

While distributed sensor signal processing goes back a long way, a recent development is the implementation of wireless links for sensor communication. Typically, distributed nodes in a wireless sensor network (WSN) are battery powered and the network has access to only a finite portion of the spectrum. This leads to both power and bandwidth constrained wireless communication between sensing nodes and the fusion nodes. Securing reliable communication over a wireless channel is a challenging task due to physical properties of the wireless medium. Hence, channel-induced errors need to be taken into account in order to achieve optimal data fusion/detection in distributed wireless sensor systems. In recent years, many authors have investigated various aspects of distributed processing in resource-constrained wireless sensor networks. In this chapter, we provide a tutorial exposition of those recent contributions.[1]

6.1 THE PROBLEM OF DISTRIBUTED DETECTION AND DATA FUSION

Gathering and processing of information through a large number of networked sensors has potential applications in a number of areas, including environmental monitoring (e.g., traffic, habitat, security), industrial sensing (e.g., nuclear power plants), infrastructure integrity monitoring (e.g., health monitoring of bridges, power grid), homeland security (e.g., remote surveillance of ports and airports), military applications (e.g., target tracking) [1–7], and monitoring people for detecting any suspicious activity [3]. Availability of microsensors with miniature batteries, processors with built-in computation, and wireless connectivity capabilities has made such a paradigm a reality. Sensor nodes (because a sensor has computation and communication capabilities apart from sensing, it is termed a *sensor node*) can be deployed almost anywhere: on the ground and in the air, inside buildings, on vehicles, and under water. In some applications, they can even be worn by humans. Realizing the full potential of sensor networks, however, presents a number of challenges, including the limitations posed by finite battery life, limited processing capability due to power constraints, and limitations posed by unreliable wireless link quality.

Typically, each individual distributed node in a wireless sensor network (WSN) can sense in multiple modalities, but has limited communication and computation capabilities. There are two issues related to reliable information gathering: (1) efficient methods for exchanging information between nodes and (2) collaborative processing of useful information about the environment being monitored. A successful design of a sensor network involves addressing layers of design issues: computational capability of a sensor node, network architecture, and routing of information between nodes [1,2,4,8]. All these issues must be resolved so that reliable information is gathered in an efficient and affordable manner while extending the whole network lifetime.

In this chapter, we restrict our attention to the problem of distributed detection and data fusion in wireless sensor networks [9,10]. This problem primarily touches on the computational aspects of a sensor node, the exchange of information between nodes (link layer issues, as termed in communications terminology), and the routing architecture. Information processing in any application can be broadly classified into two categories, namely, detection and estimation [11–13]. In a detection problem, one is interested in knowing whether a particular phenomenon of interest (POI), say the presence of a biological spill or the presence of a particular object or an individual in a specified location, is present. The answer to such a query is binary in nature, yes or no. Myriad sensors gather and process information about the POI, before passing them on to a fusion center where a final answer to the query is arrived at [9,10,14–17]. As in any situation with uncertain and incomplete information, the final answer arrived at could be different from the true situation of the POI. An acceptably reliable operation is achieved by guaranteeing that the number of incorrect decisions made over a period of time remains below an acceptable number. In an estimation problem, on the other

---

[1] Parts of this chapter have appeared in *IEEE Communication Letters*, vol. 9, pp. 769–771, Sept. 2005, and IEEE International Conference on Acoustics, Speech and Signal Processing (ICASSP'06), Toulouse, France, May 2006. S. K. Jayaweera was supported in part by a Kansas National Science Foundation (NSF) EPSCOR First Award grant KUCR # NSF32241.

hand, one would like to know the characteristics of a POI, e.g., a precise location of a target or the strength of a biological spill, etc. Moreover, one may want to know the characteristics of a POI as a function of time or space, e.g., the spatial distribution of a biological spill over a region of interest or the movement of a target in a region as a function of time (e.g., target tracking) [2, 18–20]. Reliable performance is assured by specifying the estimated value to be within a fraction of the true value. In a wireless sensor network, the communication between two nodes is typically unreliable due to channel fading/shadowing, transmission bandwidth limitations, and transmitter and receiver processing power constraints. The quality of sensed data, the quality of processed data at a node, and the quality of information passed between nodes all play important roles in the overall performance of a sensor network. A number of papers have addressed the interplay between these issues within the context of distributed detection [14–16, 21–26] and estimation [18, 19, 27–32].

This chapter is organized as follows: In Section 6.2, we briefly outline various architectures that have been considered for data fusion in sensor networks. This is followed by a discussion of recent results in distributed detection and data fusion in wireless sensor networks in Section 6.3. Note that the emphasis of Section 6.3 is on recent work that has addressed strictly resource-constrained large wireless sensor networks. Thus, results on large system analysis and performance in fading channels will be the primary focus of our discussion. However, we will also briefly outline recent advances in distributed estimation in wireless sensor networks. Next, in Section 6.4 we detail recently established performance results for distributed detection and data fusion systems with analog-relay amplifier local processing. In particular we consider sensor system optimization techniques based on large system performance measures. Section 6.6 provides a summary of the chapter.

6.2 FUSION ARCHITECTURES

Consider the generic wireless sensor network architecture shown in Figure 6.1. The sensors monitor the environment to provide inference regarding a POI. Here $Z_1, Z_2, \cdots, Z_n$ represent the observations at sensor nodes $1, 2, \cdots, n$, respectively. Unless otherwise stated, it is assumed that each sensor observation is statistically independent of others, conditioned on the true state of nature of the environment (some studies have addressed correlation among sensor data and these will be discussed in the sequel). The locally processed (e.g., quantized) data sent from these nodes are represented as $U_1, U_2, \cdots, U_n$, where $U_k = \delta_k(Y_k)$ with $\delta_k(.)$ being the local processing (decision) rule at node k. For example, if node k quantizes its observation Y_k to D_k number of levels, $U_k \in [1, 2, \cdots, D_k]$. Using a particular modulation and coding scheme, the node k transmits its data U_k to a *central* node, called the cluster head or the access point (AP) (see Figure 6.2).

Depending on the application under consideration, different fusion architectures are possible in these wireless sensor networks. Most of them stem from the architectures that were originally considered in decentralized detection problems and from the architectures present in mobile ad hoc networks [17]. Broadly speaking, there are three types of fusion architectures in decentralized detection/estimation, namely, the parallel, the serial, and the tree structure [14]. In the parallel configuration, all sensors pass their locally processed data (i.e., quantized, compressed, or amplified data) to a central site called the fusion center, where a final inference regarding the POI is made. In the serial configuration, sensors communicate in a tandem fashion, with sensor S_1 sending

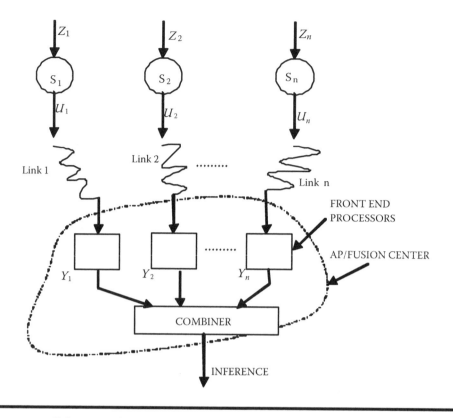

Figure 6.1   Parallel Data Fusion Model in Sensor Network.

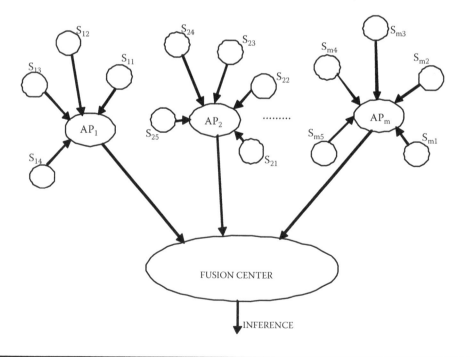

Figure 6.2   Sensor Network Topology.

its quantized (locally processed) data to sensor S_2, which in turn sends a quantized data, which is derived based on its own data and the data from S_1, to the next sensor S_3 in the tandem chain. Data progression along the chain continues until the last sensor is reached, where the final inference is made [33–35]. In general, for decentralized detection, the performance of the serial configuration is inferior to that of the parallel configuration [14]. The tree structure is similar to the architecture used in ad hoc networks, where data from neighboring nodes is transmitted to a *central* node (or the AP) as in Figure 6.2. Several such AP neighborhoods might be monitoring the environment with regard to a POI. These APs then transmit their quantized data to the next level of cluster head nodes in the hierarchy. If more than one such second level cluster head node exists, then these will in turn transmit their data to a third level cluster head node. Data progression continues until a final cluster node, called the fusion center, is reached.

In wireless sensor networks, a typical architecture is similar to the tree structure described above. In a number of applications, the architecture will have only two levels, with the first level APs (i.e., distributed sensing nodes) sending their data to the fusion center for the final assessment. Of course, because data is wirelessly transmitted from a node to an AP, this connection could be established through intermediate nodes, which is then called a multi-hop transmission. Another possibility is for two or more nodes to transmit data cooperatively to an AP, which is called cooperative relaying [36]. Cooperative relaying seems to provide a definitive advantage in performance, especially when the direct link between a node and an AP experiences severe signal degradation. Pertinent questions in many of these wireless sensor networks include how to designate APs among a large number of sensor nodes distributed in a given geographical area, how to identify neighborhood nodes that form the cluster around an AP, and how to route information to an AP or from APs to the fusion center. Optimal solutions to these questions become more difficult when one considers energy constraints on nodes as well as any possible mobility of nodes with time. A flurry of research has been done to answer these questions in recent years, but a comprehensive consideration of these results is beyond the scope of this chapter. In the next two sections, we consider primarily the quality of the data as received at an AP from a node and the quality of the final inference arrived at the fusion center. A bulk of the discussion deals with the detection problem, although some results from distributed estimation research are also provided briefly at the end of Section 6.3.3.

6.3 RECENT ADVANCES IN DISTRIBUTED DETECTION/ESTIMATION IN RESOURCE-CONSTRAINED WIRELESS SENSOR NETWORKS

Up until recently, most work on decentralized detection/estimation assumed that the senor data was transmitted to the fusion center error free [14,15]. In a WSN, however, this assumption of perfect transmission fails as data is sent over a typically unreliable channel [4,37]. To account for such channel-induced error, some recent studies have addressed the performance of decentralized detection and fusion schemes in resource-constrained WSNs. Due to the assumption of a large number of sensors in a WSN, a number of these studies has specifically dealt with the asymptotic (infinite number of sensors) performance issues [16,21,24,38–43].

To be specific, let us consider the architecture shown in Figure 6.1. The sensors monitor the environment to determine the presence or the absence of a POI based on local observations $Z_1, Z_2, \cdots, Z_n$. We may assume that the local processing at distributed nodes is a form of quantization such that the data sent from distributed nodes to the fusion center is $U_1, U_2, \cdots, U_n$, where $U_k = \delta_k(Z_k) \in [1, 2, \cdots, D_k]$. In general, this local processing at distributed nodes could either be continuous or discrete mappings, in the sense that U_k's could either be analog or discrete (quantized) valued. The node k sends its quantized data U_k to the AP using a particular modulation and coding scheme.

Until recently, it has been usual to assume the existence of parallel and noninterfering communication links between each sensor node and the AP. However, in the context of wireless sensor networks it is more realistic to assume that there could be bandwidth or power constraints on these links. Partial or complete interference among transmitted signals from different sensors could also be an interesting topic to consider, especially because of savings in bandwidth or power. A receiver at the AP processes the signal received from node k and outputs Y_k. Due to channel degradation such as fading and additive white Gaussian noise, Y_k can be considered as a corrupted version of U_k. Using all of the data, $\{Y_k\}_{k=1}^{n}$, the AP then makes a final decision regarding the presence or the absence of a POI.

6.3.1 Detection in Large Systems

Assuming a total capacity constraint $\sum_{k=1}^{n} \log_2(D_k) \le R$ on the communication channel, the allocation of optimal numbers of bits to each sensor was considered in [21]. For the problem of detecting deterministic signals in additive Gaussian noise, it was shown in [21] that having a set of identical binary sensors is asymptotically optimal, as the number of observations per sensor goes to infinity. Thus, the gain offered by having more sensors exceeds the benefits of getting detailed information from each sensor. It must be mentioned that this result was obtained with the restriction that each $\log_2(D_k)$ was an integer, i.e., no fractional bit was allowed, and that each $\log_2(D_k)$ was rounded off to the nearest higher integer. Previously large decentralized system performance has been considered in [16], with the emphasis on optimal processing at the sensors and at the fusion center. However, in [16], the links between the sensors and the fusion center were assumed to be error free.

With a joint power constraint on the channels between sensors and the AP, and assuming additive white Gaussian noise (AWGN), it was shown, using large deviation theory, that having identical sensor nodes, i.e., each node having the same transmission scheme, is asymptotically (as the total transmit power is allowed to go to infinity or equivalently, as the number of sensors, each with finite power, is allowed to go to infinity) optimal [38]. With reference to Figure 6.1 terminology, in this work, $Y_k = U_k + W_k$, where W_k represents the AWGN. For any reasonable coding (local processing) scheme ($\delta_k, k = 1, 2, \cdots, n$) and a fusion rule, which is a function of ($Y_k, for k = 1, 2, \cdots, n$), the Bayes error probability goes to zero as the total energy is allowed to go to infinity. Having established the optimality of identical sensor nodes, an appropriate measure of efficiency is the normalized Chernoff information [38]:

$$S = -\frac{1}{f(\delta)} \min_{\lambda \in [0, 1]} \left(\log \mathbb{E}_0 \left\{ e^{\lambda \log(\mathcal{L}_k)} \right\} \right), \qquad (6.1)$$

where $\mathcal{L}_k$ is the likelihood ratio of Y_k, $\mathbb{E}_0\{.\}$ denotes the expectation with respect to induced variable Y_k, under the no POI hypothesis H_0 and (null hypothesis), and $f(\delta)$ denotes and the expected power spent at a node. In [38], the measure S was computed for two transmission schemes, namely, binary sensor nodes and analog sensor nodes. A binary sensor node employs a binary mapping and sends U_k, where $U_k = m$ if Z_k

exceeds a threshold and $U_k = -m$ if Z_k falls below the threshold (in [38] the threshold was taken as 0 for the case of a specific observation model for Z_k). In the case of analog sensor nodes, as will be discussed in detail in the next section, $U_k = gZ_k$, where g is the amplification gain of the sensor node [36,39,44]. A plot of normalized Chernoff information S against the observation SNR (i.e., the signal-to-noise ratio of the observation Z_k at the sensor) showed that there exists a threshold SNR below which the analog sensor nodes perform better than their binary counterparts. Indeed, the authors observed that for some detection applications, wireless sensor nodes with continuous transmission mappings may outperform sensor nodes with finite-valued transmission mappings. They also pointed out that for the Neyman–Pearson criterion of signal detection, the normalized relative entropy measure (Kullback–Leibler distance) plays a role analogous to that of the Chernoff measure for Bayes error criterion.

The problem of decentralized detection in a sensor network subjected to a total average power constraint and all nodes sharing a common bandwidth was considered in [24]. The bandwidth constraint was taken into account by assuming nonorthogonal communication between sensors and the data fusion center via direct-sequence codedivision multiple access (DS-CDMA) spreading. In the case of large sensor systems and random spreading, the asymptotic decentralized detection performance was derived assuming independent and identically distributed sensor observations via random matrix theory. The results showed that, even under both power and bandwidth constraints, it is better to combine many not-so-good local decisions rather than relying on one (or a few) very-good local decisions. Using large deviation analysis similar to that in [38], the question of allocating two bits per sensor versus one bit per sensor was addressed in [45]. A general conclusion is that a higher SNR at a sensor would dictate a larger number of bits per sensor for achieving higher Chernoff information at the fusion center.

The impact of specific binary modulation schemes on the overall performance of fusion rules was examined in [41] and [42], by modeling each link between a sensor and the fusion center as an independent and identically distributed slow Rayleigh-fading AWGN channel. While [41] addressed only the performance of counting rules (fusion center counts the number of decisions received in favor of the presence of a POI and compares it to a threshold) at the fusion center, performances of other combiners were addressed in [42]. For three standard modulation techniques (binary phase shift keying (BPSK), on/off keying (OOK), and frequency shift keying (FSK)) this study considered (1) the impact of the sensor-fusion center link on the quality of the decision received at the fusion center and (2) the minimum required sensor decision quality, given the availability of a minimum sensor-to-fusion center link SNR, in order that the asymptotic (large number of sensors) error in the counting rule classification goes to zero. With a proper choice of threshold for noncoherent OOK detection, it was shown that an asymptotic performance comparable to that of FSK, while achieving some energy savings, is possible. Asymptotic error exponents of the probability of false alarm and the probability of miss at the fusion center were derived in [42] for the following cases: BPSK modulation and (1) maximal ratio combining (MRC), (2) equal gain combining (EGC), and (3) decision fusion (DF) and BFSK modulation and (1) square law combining and (2) decision fusion. In the case of BPSK, the EGC performs the best for low and moderate SNR, with DF achieving the next best performance. The DF scheme performs the best for large SNR values, whereas the MRC performs the best for very low SNR values. Similar relative performance results were obtained earlier for the case of a finite number of sensors (see discussion below and [37]). In the case of BFSK, square law combining was shown to outperform DF, except for large SNR values.

6.3.2 Detection Performance in Fading Channels

An analysis of the performance of different fusion rules in the presence of Rayleigh fading has been carried out in [37]. In this work, the sensor nodes are assumed to transmit their binary decisions ($U_k \in \pm 1$) over parallel noninterfering, but slow Raleigh fading channels to the fusion center. The BPSK modulation with coherent detection was assumed. Assuming that the fusion center has the knowledge of channel state information (CSI), the authors derived the optimal likelihood ratio test (LRT) at the fusion center. For large SNR, the LRT was shown to approach the Chair–Varshney rule [10], which is based on individual decisions made from the matched filter outputs of each link. Note that the Chair–Varshney rule requires the knowledge of sensor quality information, i.e., individual (local) sensor probability of false alarm and the probability of detection. For identical sensors, i.e., all having the same decision quality, they also showed that the LRT approaches the MRC as the average SNR of the fading link approaches zero.

They also pointed out another interesting result: In traditional combining where there is one source and many diversity paths, the MRC maximizes the SNR among all linear combiners; MRC does not exhibit any such optimality in distributed sensor networks, where a consensus of decisions of all the sensors does not occur with probability one. Interestingly, except for very small SNR, both EGC and the Chair–Varshney rule outperform MRC. The EGC also performs better than the Chair–Varshney rule, except for large SNR values. Similar to this analysis, an exercise in finding different statistics for the fusion of censored decisions was carried out in [46]. Although the authors termed this as censoring, it is essentially an OOK modulation for transmitting the U_ks. Certainly, OOK allows for noncoherent detection, thereby eliminating any need for phase tracking of an individual link. Assuming complete CSI and sensor information quality, the authors derived the LRT, which is a function of the energy detector outputs of the individual links. For very small SNR and independent and identically distributed (iid) sensors, the LRT becomes a weighted energy detector, with weights being proportional to the individual channel gains. Censoring strategy with resource constraints (expected cost arising from transmission and measurement at each sensor) was considered in [47]. It was found that the randomization over the choice of measurement and when to transmit achieves the best performance (in Bayesian, Neyman–Pearson, and Ali–Silvey sense).

The effect of link quality on the performance of a counting rule has been investigated in [41,48]. Assuming iid sensor observations and a specific quality of sensor decision, [48] showed how the fusion false alarm probability could change several orders of magnitude as the link error rate changes from low to high. For a counting rule at the fusion center and a slow fading channel, [48] established the correctness of using an average bit error rate for a link, averaged with respect to the fading distribution. The effect of correlation on the performance of a wireless sensor detection system subjected to a total transmission power constraint was studied in [44]. Assuming that the sensors are placed as a linear array, such that the correlation coefficient between any two sensors is exponentially decreasing with distance separation, they studied the fusion performance with analog-relay amplifier local processing at the sensors. It is found that the optimal number of sensor nodes in the system increases as the correlation coefficient decreases. In general, systems with many low-power nodes appear to perform better in the case of a deterministic signal detection, regardless of the specific correlation coefficient. In contrast, the effect of correlation on the detection of a stochastic (random) signal in a total power constrained sensor network was investigated in [39]. An important observation was that the average fusion probability of error does not improve monotonically with the number of sensors, unlike in the case of deterministic signal detection reported earlier in [44,47]. In particular, [39] showed that there is an optimal number of sensors that minimizes the probability of fusion error, which depends on both the local observation SNR as well as channel SNR (we will discuss these results in detail in Section 6.4 below).

6.3.3 Estimation with Distributed Sensors

In this section, we briefly discuss some important results in distributed estimation in sensor networks. In estimation problems, the nodes periodically transmit their processed information to the fusion center where the estimation of a POI takes place (see Figure 6.1). Early work in this area dealt with target tracking based on distributed data [49]. Essentially, the problem boiled down to aggregating different Kalman filter estimates that were obtained at several sensors. Performance analysis of parameter estimation with distributed sensors was carried out in [50]. A specific design of a decentralized estimation system was considered in [50]. The local processors at the sensor nodes were taken to be quantizers and the aim was to minimize a certain distortion function. Necessary conditions for the optimum systems based on Bayes distortion measure and Fisher information were derived. The numerical results in [50] also compared the resulting quantizers obtained by different distortion criteria. Another early work considered the estimation of an unknown constant, using distributed estimators [51]. For estimating a constant parameter in Laplace noise density at each sensor, it was assumed that the sample medians of a set of iid observations at each sensor were obtained. These local median estimates were then combined in some fashion at the fusion center. The obtained results revealed that the mean of local medians exhibits a slightly smaller mean-squared error (MSE) than the median of local medians. It is noteworthy that in all these early papers, the links between sensors and the fusion center were assumed to be error free.

Some recent work has considered distributed estimation of a constant parameter within the context of bandwidth constrained sensor networks [52,53]. In this setup, each sensor observes a corrupted version of the parameter and quantizes its data. The analysis in [53] applies only to the case of the observation noise at a sensor being over a bounded interval. Bandwidth constraint is indirectly met by allowing approximately 1/2 of the sensors to send one bit quantized data, 1/4 of the sensors to send two bits quantized data, and so on. In [53] both the quantization rule at a sensor (all sensors employ identical quantizers) and the fusion rule were completely distribution free, thereby making the scheme highly suitable for ad hoc networks. Moreover, the author showed that the MSE of the proposed distributed quantizer is almost within a factor of four of the Cramer–Rao lower bound of the centralized counterpart. Subsequently, [54] has shown that the MSE of any universal decentralized estimator is lower bounded by 1/16-th of the MSE of the scheme in [53]. For the case of distributed estimation of a constant parameter in Gaussian noise, results in [52] showed that a class of maximum likelihood (ML) estimators requires sending just one bit from each sensor, when the dynamic range of the parameter is small or comparable to the noise standard deviation. Moreover, such a scheme yields a fused estimator whose variance is close to that of the sample mean estimator based on all (unquantized) sensor samples. When the dynamic range is comparable or larger than the noise standard deviation, there exists an optimum quantization scheme that achieves the best possible variance for a given bandwidth constraint.

Another interesting result in distributed estimation is the use of dithering to reduce MSE [55]. In [55], the authors showed that the addition of independent random noise to sensor observations before quantization helps to reduce the MSE of the estimate at the fusion center. In the past, such dithering was shown to be beneficial in the quantization of speech signals.

The theory and methodology of estimating inhomogeneous, two-dimensional fields using wireless sensor networks have been addressed in [56]. The sensors make noisy measurements of the field, and the goal is to obtain an accurate estimate of the field at some desired location (typically remote from the sensor network). Key questions are the accuracy attainable in estimation and the energy consumption for communication. This paper also presented a practical strategy for estimation and communication. So far, all analysis in estimation has assumed that the links between sensors and the fusion center are error free. The impact of link errors on the overall estimation accuracy needs to be investigated.

6.4 RECENT RESULTS ON ANALOG DATA FUSION IN WIRELESS SENSOR NETWORKS

Having provided a summary of recent advances in distributed detection/estimation under resource constraints for wireless sensor networks, in this section we will consider some of those results in detail. Specifically, we consider large wireless sensor networks with so-called analog-relay amplifier local processing schemes. We will look at large system analysis based fusion performance and sensor system optimization results for resource-constrained wireless sensor networks.

Let us consider a binary hypothesis testing problem in an n-node wireless sensor network connected to a data fusion center via distributed parallel architecture [14,17]. Denote by H_0 and H_1 the null and alternative hypotheses, respectively, having corresponding prior probabilities $P(H_0) = \pi_0$ and $P(H_1) = \pi_1 = 1 - \pi_0$. Note that, unless otherwise stated, we will assume equal priors so that $\pi_0 = \pi_1 = 1/2$. To be specific, the observed POI is a Gaussian signal denoted by $X_k \sim \mathcal{N}(m, \sigma_x^2)$ corrupted by AWGN. The k-th local sensor observation Z_k, for $k = 1, \cdots, n$, can be written as

$$H_0 : \quad Z_k = V_k$$

$$H_1 : \quad Z_k = X_k + V_k \tag{6.2}$$

where observation noise $V_k \sim \mathcal{N}(0, \sigma_v^2)$ is zero-mean Gaussian with the collection of noise samples $\mathbf{V} = [V_1, V_2, \cdots, V_n]^T \sim \mathcal{N}(\mathbf{0}, \Sigma_v)$. Each sensor locally processes its observations to generate a local decision $U_k = \delta_k(Z_k)$ which is sent to the fusion center. Denote by $\mathbf{Y}(U_1(Z_1), U_2(Z_2), \cdots, U_n(Z_n))$ the received signal at the fusion center. The fusion center makes a final decision U_0 based on the decision rule $U_0 = \delta_0(\mathbf{Y})$. The problem at hand is to choose $\delta_0(\mathbf{Y}), \delta_1(Z_1), \delta_2(Z_2), \cdots, \delta_n(Z_n)$ to optimize a given performance metric (e.g., Bayesian or Neyman–Pearson criterion). If local observations are independent, conditioned on the true hypothesis, then all local decision rules simplify to a set of likelihood ratio (LR) based tests but with possibly coupled thresholds [9]. While this assumption is commonly found in most work on distributed detection, in the context of dense wireless sensor networks it may not be justified. Once the conditional independence assumption is dropped, the optimality of simple threshold tests may be lost and the analysis could get unwieldy.

In this section, we exclusively consider a simple but important continuous local mapping called the amplify-and-relay processing, according to which the local observations are amplified before retransmission to the fusion center [44]:

$$U_k = g_k Z_k, \quad \text{for} \quad k = 1, \cdots n, \tag{6.3}$$

where $g_k > 0$ is the analog-relay amplifier gain at the k-th node. Interestingly, as mentioned in the previous section, it has been shown in [38] that below a certain threshold SNR value, such continuous local mappings may outperform binary (or discrete valued) local mappings for certain detection problems. This makes analog-relay local processing a good candidate for emerging low-power, wireless sensor networks.

In modeling dense distributed wireless sensor networks it is more appropriate to consider nonorthogonal sensor-to-fusion center communication over noisy channels. Thus, let the k-th sensor node be assigned a signaling waveform (code) $\mathbf{s}_k$ normalized such that $\mathbf{s}_k^T \mathbf{s}_k = 1$, for $k = 1, \cdots, n$. We assume the number of degrees of freedom (DoF) in the signaling waveform to be N (for example, the number of chips per symbol in DS-CDMA signaling) so that $\mathbf{s}_k$ is a length N vector. The message U_k of the k-th sensor is transmitted to the fusion center over a noisy, bandlimited wireless channel by modulating onto the signaling waveform $\mathbf{s}_k$. Hence, k-th sensor's transmitted signal is given by $\mathbf{s}_k U_k = g_k \mathbf{s}_k Z_k$. Throughout this section we assume an AWGN channel with double-sided spectral density σ_w^2 and ignore the effects of fading for simplicity. Assuming synchronized sensor transmissions, the signal received at the fusion center is a superposition of signals transmitted from all the nodes $\sum_{k=1}^{n} g_k \mathbf{s}_k Z_k$ corrupted by additive noise. The output of a bank of matched filters at the fusion center (each matched to a signaling waveform $\mathbf{s}_k$ of a particular node) can be written in vector notation as

$$\mathbf{Y} = \mathbf{R}\mathbf{U} + \mathbf{W}, \tag{6.4}$$

$$= \mathbf{R}\mathbf{A}\mathbf{Z} + \mathbf{W}, \tag{6.5}$$

where we have defined $\mathbf{A} = diag(g_1, g_2, \cdots, g_n)$, $\mathbf{U} = [U_1, \cdots, U_n]^T$, $\mathbf{Z} = [Z_1, \cdots, Z_n]^T$, and $\mathbf{R}$ is the $n \times n$ symmetric and normalized received signal correlation matrix in which the (k, k')-th element is given by $\mathbf{s}_k^T \mathbf{s}_{k'}$. If we define the $N \times n$ matrix $\mathbf{S}$ such that its k-th column is the waveform $\mathbf{s}_k$, then it is easily shown that

$$\mathbf{R} = \mathbf{S}^T \mathbf{S}. \tag{6.6}$$

Note that, in the special case of orthogonal sensor-to-fusion center communication, the received signal model (6.4) simplifies such that $\mathbf{R} = \mathbf{I}$. In (6.4), $\mathbf{w} \sim \mathcal{N}(\mathbf{0}, \sigma_w^2 \mathbf{R})$ is the n-dimensional filtered noise vector.

A sensible way to model a system in which the most important objective is to extend the whole network lifetime is to impose a total average power constraint P on the whole sensor system. According to this model, as the number of nodes in the system increases, the power available for each node correspondingly decreases. This allows trading off individual node power against the number of nodes in the network and vice versa. For example, in certain applications the cost of a node may be dominated by the cost of batteries. In such situations it may be necessary to determine whether to deploy a few nodes with high power or a large number of nodes with low power. Also, when the sensor system is powered by a distributed power source with a certain power density per unit area the total available power may be constant, justifying application of a global power model. With this model, g_k, for $k = 1, \cdots n$, depends on the total average power constraint P. For simplicity, throughout this discussion we assume $g_k = g$ for all k. With these assumptions the average radiated power of node k is given by $\mathbb{E}\{|U_k|^2\} = g^2 \mathbb{E}\{|Z_k|^2\} = g^2(\frac{m^2 + \sigma_x^2}{2} + \sigma_v^2)$, where σ_x^2 and σ_v^2 are the variances of the signal of interest and the observation noise, respectively, and we have assumed that $\pi_0 = \frac{1}{2}$. Hence, the local amplifier gain g is given by

$$g^2 = \frac{P}{n\left(\sigma_v^2 + \frac{m^2 + \sigma_x^2}{2}\right)}. \tag{6.7}$$

Observe that, as more nodes are introduced the gain at each node correspondingly decreases. With equal amplifier gains at the nodes (6.5) simplifies to

$$\mathbf{Y} = g\,\mathbf{RZ} + \mathbf{W}. \tag{6.8}$$

The detection problem at the fusion center is then given by the following binary hypothesis testing problem:

$$H_0: \quad \mathbf{Y} \sim \mathcal{N}(\mathbf{0}, \Sigma_0)$$

$$H_1: \quad \mathbf{Y} \sim \mathcal{N}(\mathbf{m}, \Sigma_1), \tag{6.9}$$

where $\Sigma_0 = g^2 \mathbf{R}\Sigma_v \mathbf{R} + \sigma_w^2 \mathbf{I}$, $\Sigma_1 = g^2 \mathbf{R}(\Sigma_x + \Sigma_v)\mathbf{R} + \sigma_w^2 \mathbf{I} = g^2 \mathbf{R}\Sigma_x \mathbf{R} + \Sigma_0$, $\mathbf{m} = g\,\mathbf{R}\mathbb{E}\{\mathbf{X}\} = gm\mathbf{R1}$ and $\mathbf{1}$ is the vector of all ones. Because the quantity $\frac{P}{\sigma_w^2}$ is a measure of how good the channel is, let us define the channel quality SNR as $\gamma_c \triangleq \frac{P}{\sigma_w^2}$.

6.4.1 Distributed Detection of a Deterministic Signal in a Total Power and Bandwidth Constrained System

In this section, we consider the detection of a deterministic signal in uncorrelated observation noise so that $\mathbf{X} = m\mathbf{1}$, for known $m > 0$, and $\Sigma_v = \sigma_v^2 \mathbf{I}$. Then $\Sigma_0 = \Sigma_1 = \Sigma$, where

$$\Sigma = g^2 \sigma_v^2 \mathbf{R}^2 + \sigma_w^2 \mathbf{R}. \tag{6.10}$$

Accordingly, the amplifier gain g in (6.7) is simplified as

$$g = \sqrt{\frac{P}{n\left(\frac{m^2}{2} + \sigma_v^2\right)}}. \tag{6.11}$$

The quality of local observations is then characterized by the ratio $\frac{m^2}{\sigma_v^2}$. Hence, let us define the local observation quality SNR as $\gamma_0 \triangleq \frac{m^2}{\sigma_v^2}$.

The fusion center design problem is then a standard Gaussian hypothesis testing problem with the only additional caveat being that the gain g depends on the total power constraint P as in (6.11). It is well-known that the the optimal fusion rule is a LR threshold test of the form of

$$\delta_0(\mathbf{y}) = \begin{cases} 1 \\ 0 \end{cases} \quad \text{if } T(\mathbf{y}) \begin{array}{c} \geq \\ < \end{array} \tau', \tag{6.12}$$

where we have defined the decision variable T as $T(\mathbf{y}) = \mathbf{m}^T \Sigma^{-1} \mathbf{y} = gm\mathbf{1}^T \mathbf{R} (g^2 \sigma_v^2 \mathbf{R}^2 + \sigma_w^2 \mathbf{R})^{-1} \mathbf{y}$ and τ' is the threshold that depends on the specific optimality criteria. It can be shown that the false-alarm P_f and miss P_m probabilities of the detector (6.12) are given by

$$P_f = Q\left(\frac{\tau'}{gm\sqrt{\mathbf{1}^T \mathbf{R} \Sigma^{-1} \mathbf{R} \mathbf{1}}}\right), \tag{6.13}$$

and

$$P_m = Q\left(\frac{g^2 m^2 \mathbf{1}^T \mathbf{R} \Sigma^{-1} \mathbf{R} \mathbf{1} - \tau'}{gm\sqrt{\mathbf{1}^T \mathbf{R} \Sigma^{-1} \mathbf{R} \mathbf{1}}}\right). \tag{6.14}$$

In the case of Neyman–Pearson optimality at the fusion center, τ' is chosen to minimize P_m subject to an upper bound on P_f. On the other hand, under Bayesian minimum probability of error optimality one would choose τ' to minimize $P_e = \pi_0 P_f + \pi_1 P_m$. In the following we explicitly consider Bayesian optimality with equal prior probabilities (i.e., $\pi_0 = \pi_1 = \frac{1}{2}$), in which case the threshold simplifies to

$$\tau' = \frac{1}{2} g^2 m^2 \mathbf{1}^T \mathbf{R} \Sigma^{-1} \mathbf{R} \mathbf{1}. \tag{6.15}$$

The resulting minimum fusion probability of error is given by

$$P_e = Q\left(\frac{gm}{2}\sqrt{\mathbf{1}^T \mathbf{R} \Sigma^{-1} \mathbf{R} \mathbf{1}}\right). \tag{6.16}$$

The above analysis characterizes the fusion performance for a deterministic signal in a resource-constrained, noisy, bandlimited wireless sensor network. Of course, to say anything beyond this point we need to specify the particular signaling scheme used to share the total available bandwidth because the performance depends on the particular waveforms (or codes) assigned to each sensor node as seen from (6.16). This hinders drawing general conclusions regarding the fusion system. However, such conclusions can be reached for large systems through asymptotic (in large n) analysis, as we show next.

Let us assume that the signaling codes are chosen randomly, so that each element of $\mathbf{s}_k$ takes either $\frac{1}{\sqrt{N}}$ or $-\frac{1}{\sqrt{N}}$ with equal probability, and that, as assumed above, sensor observations are independent so that $\Sigma_v = \sigma_v^2 \mathbf{I}$. Consider a large sensor system in which both n and N are large, such that $\lim_{N \to \infty} \frac{n}{N} = \alpha$. Using the definitions of Σ, $\mathbf{R}$, $\mathbf{S}$, and $\mathbf{1}$, we can show that

$$g^2 \mathbf{1}^T \mathbf{R} \Sigma^{-1} \mathbf{R} \mathbf{1} = g^2 \mathbf{1}^T \mathbf{S}^T \mathbf{C}^{-1} \mathbf{S} \mathbf{1}$$

$$= g^2 \left(\sum_{k=1}^{n} \mathbf{s}_k^T \mathbf{C}^{-1} \mathbf{s}_k + \sum_{k=1}^{n} \sum_{\substack{k'=1 \\ k' \neq k}}^{n} \mathbf{s}_k^T \mathbf{C}^{-1} \mathbf{s}_{k'}\right), \tag{6.17}$$

where we have defined $\mathbf{C} = g^2\sigma_v^2\mathbf{S}\mathbf{S}^T + \sigma_w^2\mathbf{I}$. Let $\mathcal{I}$ denote a set of sensor indices (i.e., $\mathcal{I} \subset \{1, 2, \cdots, n\}$), $\mathbf{S}_{\mathcal{A}}$ denote the matrix $\mathbf{S}$ with columns specified by the indices in set $\mathcal{A}$ deleted, $\Lambda_k = g^2\sigma_v^2\mathbf{I}_k$ and $\mathbf{Q}_{\mathcal{A}} = \left(\mathbf{S}_{\mathcal{A}}\Lambda_{n-|\mathcal{A}|}\mathbf{S}_{\mathcal{A}} + \sigma_w^2\mathbf{I}_n\right)$, where $\mathbf{I}_k$ and $|\mathcal{A}|$ denote the $k \times k$ identity matrix and the cardinality of set $\mathcal{A}$, respectively. Then, using the matrix inversion lemma[2] we can show that, for $k = 1, \cdots, n$,

$$\mathbf{s}_k^T\mathbf{C}^{-1}\mathbf{s}_k = \mathbf{s}_k^T\mathbf{Q}_{\{k\}}^{-1}\mathbf{s}_k/\left(1 + g^2\sigma_v^2\mathbf{s}_k^T\mathbf{Q}_{\{k\}}^{-1}\mathbf{s}_k\right). \tag{6.18}$$

The key to large system asymptotic analysis in this situation is the theory of large random matrices [57]. In particular, under the assumed conditions for signaling codes, the empirical distribution of eigenvalues of the large random matrix $\mathbf{R}$ converges almost surely to a deterministic distribution characterized by the parameter α [58–60]. Applying Theorem 7 of [60], which essentially relies on the above result, and using (6.11), we can show that [24],

$$\mathbf{s}_k^T\mathbf{Q}_{\{k\}}^{-1}\mathbf{s}_k \overset{\text{a.s.}}{\longrightarrow} \frac{\beta_0}{\sigma_w^2}, \tag{6.19}$$

where $\beta_0 = \frac{\sqrt{(\gamma+\sigma_w^2)^2\alpha^2+2\gamma(\sigma_w^2-\gamma)\alpha+\gamma^2}-(\gamma+\sigma_w^2)\alpha+\gamma}{2\gamma}$ and $\gamma = \frac{P}{N(1+\frac{\gamma_0}{2})}$. Substituting (6.19) in (6.18) we have, for $k = 1, \cdots, n$,

$$\mathbf{s}_k^T\mathbf{C}^{-1}\mathbf{s}_k \overset{\text{a.s.}}{\longrightarrow} \left(\frac{\sigma_w^2}{\beta_0} + g^2\sigma_v^2\right)^{-1}. \tag{6.20}$$

Similarly, repeated application of the matrix inversion lemma twice yields, for $k \neq k'$

$$\mathbf{s}_k^T\mathbf{C}^{-1}\mathbf{s}_{k'} = \frac{\mathbf{s}_k^T\mathbf{Q}_{\{k,k'\}}^{-1}\mathbf{s}_{k'}}{\left(1 + g^2\sigma_v^2\mathbf{s}_k^T\mathbf{Q}_{\{k\}}^{-1}\mathbf{s}_k\right)\left(1 + g^2\sigma_v^2\mathbf{s}_{k'}^T\mathbf{Q}_{\{k,k'\}}^{-1}\mathbf{s}_{k'}\right)} \overset{\text{a.s.}}{\longrightarrow} 0 \tag{6.21}$$

where we have again used Theorem 7 of [60] in the last step to obtain (6.21). Substituting (6.20) and (6.21) in (6.17) gives

$$g^2\mathbf{1}^T\mathbf{R}\Sigma^{-1}\mathbf{R}\mathbf{1} \overset{\text{a.s.}}{\longrightarrow} \left(\frac{\sigma_v^2}{n} + \frac{\sigma_v^2\left(1+\frac{\gamma_0}{2}\right)}{\gamma_c\beta_0}\right)^{-1}. \tag{6.22}$$

This asymptotic convergence result can be used to characterize the large sensor system Bayesian fusion error probability when $\lim_{N\to\infty}\frac{n}{N} = \alpha$. Substituting (6.22) in (6.16) gives

$$P_e(\alpha) \overset{\text{a.s.}}{\longrightarrow} Q\left(\frac{m}{2\sqrt{\frac{\sigma_v^2}{n} + \frac{\sigma_v^2(1+\frac{\gamma_0}{2})}{\gamma_c\beta_0}}}\right). \tag{6.23}$$

---

[2] If $\mathbf{A}$, $\mathbf{C}$ and $\left(\mathbf{C}^{-1} + \mathbf{D}\mathbf{A}^{-1}\mathbf{B}\right)$ are all nonsingular square matrices, then $(\mathbf{A} + \mathbf{B}\mathbf{C}\mathbf{D})^{-1} = \mathbf{A}^{-1} - \mathbf{A}^{-1}\mathbf{B}\left(\mathbf{C}^{-1} + \mathbf{D}\mathbf{A}^{-1}\mathbf{B}\right)^{-1}\mathbf{D}\mathbf{A}^{-1}$.

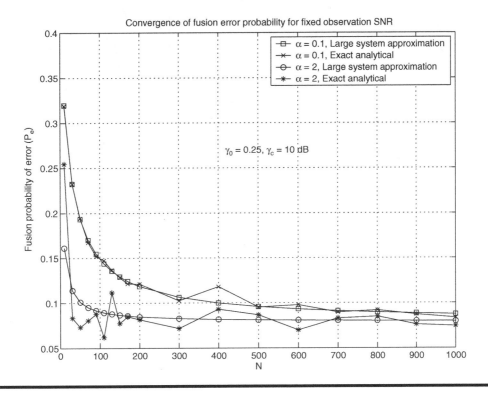

Figure 6.3 Large Sensor System Fusion Performance in a Noisy, Bandlimited Channel Subjected to a Total Power Constraint.

Figure 6.3 shows the convergence of the random waveform based decentralized detection performance as predicted by (6.23). Note that the exact analytical result in Figure 6.3 was obtained from (6.16) by using a random choice of the code matrix **S** where the large system approximation results are from (6.23). As can be seen from Figure 6.3, (6.23) provides a very good approximation to the fusion performance for large code lengths N, and thus for large-sensor systems (because $n = N\alpha$). More importantly, we can observe from Figure 6.3 that for each fixed N, increasing α improves the decentralized detection performance. Because this is equivalent to increasing the number of sensors n allowed in the system for a fixed bandwidth, we conclude that it is better to allow as many sensors to send their local decisions to the fusion center.

In fact, for large α, one can show that $\beta_0 \overset{a.s.}{\longrightarrow} 1$, and as a result in this case, the error probability in (6.23) goes to (see Figure 6.4).

$$P_e(\alpha) \longrightarrow Q\left(\frac{1}{2}\sqrt{\frac{\gamma_c}{\frac{1}{2} + \frac{1}{\gamma_0}}}\right). \tag{6.24}$$

On the other hand, if one were to allocate all available power P and the total bandwidth to just one sensor node, the fusion center performance will be given by

$$P_{e,1} = Q\left(\sqrt{\frac{\gamma_c}{\frac{\gamma_c}{\gamma_0} + \frac{1}{\gamma_0} + \frac{1}{2}}}\right). \tag{6.25}$$

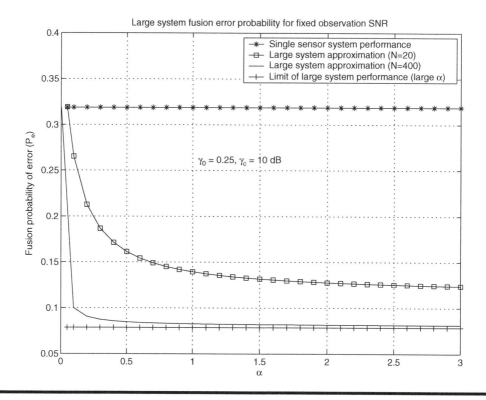

Figure 6.4 **Limit of Large Sensor System Approximation to the Fusion Performance in a Noisy, Bandlimited Channel Subjected to a Total Power Constraint when** $\alpha \longrightarrow \infty$.

Comparison of (6.24) and (6.25) shows that allowing more sensor nodes in the network is better even if the channel is both noisy and bandlimited. This comparison is shown in Figure 6.4, in which the limit of large system performance and the single sensor system performance refer to, respectively, (6.24) and (6.25). The large system approximations for finite N values shown in Figure 6.4 were obtained from (6.23). First, observe from Figure 6.4 that as N increases the fusion center performance improves. Secondly, note that as $N \longrightarrow \infty$, the performance for large α indeed goes to (6.24). Third, Figure 6.4 confirms that combining more local decisions is better than allocating all available power and bandwidth to one sensor. Moreover, the performance improves monotonically with increasing α (for a fixed N) showing that it is better to combine as many local decisions as possible at the fusion center. We should divide the available power among all nodes and allow them to share the available bandwidth, even if they interfere with each other due to nonorthogonality.

6.4.2 Distributed Detection of a Random Gaussian Signal in a Total Power and Bandwidth Constrained System

In the previous section, we could derive the exact fusion error probability in closed form for any finite n due to the assumed simplicity of the model. However, instead of being a deterministic signal, if the POI to be detected happened to be a random signal (e.g., a Gaussian signal), the analysis quickly becomes much more difficult. The present convenience is quickly lost when we consider more involved signaling and channel models. In those situations, asymptotic performance analysis (for large n) becomes a necessity in order to establish any meaningful characterization of performance. Recently, several works have employed large deviations theories and error exponent analysis to achieve this goal. In this section, we illustrate some of these ideas in fusion performance analysis, assuming the POI to be detected in (6.2) to be a zero-mean Gaussian signal such that $X_k \sim \mathcal{N}(0, \sigma_x^2)$, for $k = 1, \cdots, n$. Again assuming both signal of interest and the additive noise V_k are independent at each node, the set of observation noise samples and the set of desired signal samples are then distributed as $\mathbf{V} \sim \mathcal{N}(0, \sigma_v^2 \mathbf{I})$ and $\mathbf{X} \sim \mathcal{N}(0, \sigma_x^2 \mathbf{I})$, respectively. Accordingly, we redefine the local observation quality SNR at each node as $\gamma_0 \stackrel{\Delta}{=} \frac{\sigma_x^2}{\sigma_v^2}$.

As before, assume that the local decisions sent to the fusion center are generated via analog-relay amplifier processing, so that $U_k = g_k Z_k$, for $k = 1, \cdots n$, where $g_k > 0$ is the gain at the k-th node that depends on the global system power constraint P on the whole sensor system. For simplicity, suppose also that $g_k = g$ for all k, and that the k-th node transmits its local decision U_k to the fusion center after modulating it with a normalized signaling waveform $\mathbf{s}_k$. A sufficient statistic for the fusion center processing is again given by $\mathbf{Y}$ in (6.8). The matrix $\mathbf{R} = \mathbf{S}^T \mathbf{S}$ in which the (k, k')-th element is given by $\mathbf{s}_k^T \mathbf{s}_{k'}$ reflects the possible nonorthogonality of signaling due to a finite total bandwidth constraint. Under the global average power constraint P on the system, the local amplifier gain is now given by

$$g^2 = \frac{P}{n\left(\sigma_v^2 + \frac{\sigma_x^2}{2}\right)}. \tag{6.26}$$

With these definitions, the new fusion problem is reduced to the following binary hypothesis testing problem

$$H_0: \quad \mathbf{y} \sim p_0(\mathbf{y}) = \mathcal{N}(\mathbf{0}, \Sigma)$$

$$H_1: \quad \mathbf{y} \sim p_1(\mathbf{y}) = \mathcal{N}\left(\mathbf{0}, g^2 \sigma_x^2 \mathbf{R}^2 + \Sigma\right),$$

where $p_j(\mathbf{y})$ is the density of $\mathbf{Y}$ under the hypothesis H_j, for $j = 0, 1$, and Σ is as defined in (6.10). The optimal (e.g., Bayesian, minimax, or Neyman–Pearson) fusion rules should then be based on the LR $\mathcal{L}(\mathbf{y}) = \frac{p_1(\mathbf{y})}{p_0(\mathbf{y})}$ that can be written as

$$\mathcal{L}(\mathbf{y}) = \left(\frac{|\Sigma|}{|g^2 \sigma_x^2 \mathbf{R}^2 + \Sigma|}\right)^{\frac{1}{2}} \exp\left(\frac{1}{2}\mathbf{y}^T\left(\Sigma^{-1} - (g^2 \sigma_x^2 \mathbf{R}^2 + \Sigma)^{-1}\right)\mathbf{y}\right). \tag{6.27}$$

Let us define the spectral decomposition of $\mathbf{R}$ to be $\mathbf{R} = \sum_{k=1}^{n} \lambda_k \xi_k \xi_k^T$. Under the assumption that the signaling waveforms (equivalently, codes) of the sensors are all linearly independent of each other, the set of orthonormal eigenvectors ξ_ks forms a complete basis for $\mathbb{R}^n$ and λ_k's are the corresponding eigenvalues. In that case, we have that $\mathbf{R}^2 = \sum_{k=1}^{n} \lambda_k^2 \xi_k \xi_k^T$ and $\Sigma = \sum_{k=1}^{n} \left(g^2 \sigma_v^2 \lambda_k + \sigma_w^2\right) \lambda_k \xi_k \xi_k^T$. Using these in (6.27) leads to

$$\mathcal{L}(\mathbf{y}) = \exp\left(\frac{1}{2} \sum_{k=1}^{n} \frac{g^2 \sigma_x^2}{\left(g^2 \sigma_v^2 \lambda_k + \sigma_w^2\right) \left(g^2 \left(\sigma_x^2 + \sigma_v^2\right) \lambda_k + \sigma_w^2\right)} |\xi_k^T \mathbf{y}|^2 \right)$$

$$\times \prod_{k=1}^{n} \left(\frac{g^2 \sigma_v^2 \lambda_k + \sigma_w^2}{g^2 (\sigma_x^2 + \sigma_v^2) \lambda_k + \sigma_w^2} \right)^{\frac{1}{2}}. \tag{6.28}$$

For $k = 1, \cdots, n$, let us define a new set of random variables $\bar{Y}_1, \cdots, \bar{Y}_n$ by projecting the observation vector $\mathbf{Y}$ onto each of the eigenvectors ξ_k followed by scaling:

$$\bar{Y}_k = \sqrt{\frac{g^2 \sigma_x^2}{\left(g^2 \sigma_v^2 \lambda_k + \sigma_w^2\right) \left(g^2 \left(\sigma_x^2 + \sigma_v^2\right) \lambda_k + \sigma_w^2\right)}} \xi_k^T \mathbf{r}. \tag{6.29}$$

Due to the orthonormality of ξ_k's, it is easy to show that $\bar{Y}_k$'s are a set of zero-mean independent Gaussian random variables under both hypotheses, that is equivalent to the original statistic $\mathbf{y}$. However, Y_k's are not identically distributed under either hypothesis. In fact, if the variance of the k-th sample $\bar{Y}_k$ under H_j is $\sigma_{j,k}^2$, for $j = 0, 1,$ and $k = 1, \cdots, n,$ then it can be shown that

$$\sigma_{j,k}^2 = \begin{cases} \frac{g^2 \sigma_x^2 \lambda_k}{g^2 (\sigma_x^2 + \sigma_v^2) \lambda_k + \sigma_w^2} & \text{if } j = 0 \\[3mm] \frac{g^2 \sigma_x^2 \lambda_k}{g^2 \sigma_v^2 \lambda_k + \sigma_w^2} & \text{if } j = 1 \end{cases}. \tag{6.30}$$

Substitution of (6.29) in (6.28) allows us to write the fusion center LR as

$$\mathcal{L}(\mathbf{y}) = \exp\left(\frac{1}{2} \sum_{k=1}^{n} |\bar{y}_k|^2 \right) \prod_{k=1}^{n} \left(\frac{g^2 \sigma_v^2 \lambda_k + \sigma_w^2}{g^2 (\sigma_x^2 + \sigma_v^2) \lambda_k + \sigma_w^2} \right)^{\frac{1}{2}}. \tag{6.31}$$

The optimal fusion decision rule is then given by

$$\delta_{opt}(\mathbf{y}) = \begin{cases} 1 \\ 0 \end{cases} \quad \text{if } T(\mathbf{y}) \underset{<}{\overset{\geq}{}} \tau', \tag{6.32}$$

where

$$\tau' = 2\log\tau + \sum_{k=1}^{n} \log\left(\frac{g^2 (\sigma_x^2 + \sigma_v^2) \lambda_k + \sigma_w^2}{g^2 \sigma_v^2 \lambda_k + \sigma_w^2} \right), \tag{6.33}$$

and the decision variable T is the quadratic form $T(\mathbf{y}) = \sum_{k=1}^{n} |\bar{y}_k|^2$. We again restrict our discussion to the minimum probability of error in Bayes detection with equal priors so that $\tau = 1$.

Only in certain special circumstances can one evaluate the exact probability of error P_e of the optimal quadratic detector (6.32) in closed form. A common approach in other situations is to obtain good error bounds or error exponents. While they may not be exact, in most situations error exponents (and the bounds based on them) can be helpful in characterizing the performance of a detection procedure. The most commonly used bound for Bayesian detection is the Chernoff upper bound to the probability of error which (assuming equal priors) can be written as $P_e \leq \frac{1}{2}e^{\mu_C}$, where the Chernoff error exponent is defined as [11]

$$\mu_C = \min_{s \in [0,1]} \log \mathbb{E} \left\{ \mathcal{L}^s(\mathbf{r}) | H_0 \right\}. \tag{6.34}$$

Although somewhat looser than the Chernoff bound, an easier-to-evaluate related bound is the Bhattacharyya upper bound. Specifically, analogous to the Chernoff error exponent, we define the Bhattacharyya error exponent as

$$\mu_B = \log \mathbb{E} \left\{ \mathcal{L}^{\frac{1}{2}}(\mathbf{r}) | H_0 \right\}, \tag{6.35}$$

so that the Bhattacharyya upper bound to the probability of error is given by $P_e \leq \frac{1}{2}e^{\mu_B}$.

The special case in which performance can be characterized in closed form is the orthogonal signaling: i.e., $\mathbf{R} = \mathbf{I}$. In this case, it is easy to show that the $\bar{Y}_k$'s are a collection of independent Gaussian random variables such that:

$$H_0 : \quad \bar{Y}_k \sim \mathcal{N}\left(0, \sigma_0^2\right)$$

$$H_1 : \quad \bar{Y}_k \sim \mathcal{N}\left(0, \sigma_1^2\right),$$

where

$$\sigma_j^2 = \begin{cases} \dfrac{g^2 \sigma_x^2}{\sigma_w^2 + g^2(\sigma_x^2 + \sigma_v^2)} & \text{if } j = 0 \\[2ex] \dfrac{g^2 \sigma_x^2}{\sigma_w^2 + g^2 \sigma_v^2} & \text{if } j = 1 \end{cases}$$

$$= \begin{cases} \dfrac{\gamma_0}{1 + \gamma_0 + \frac{n}{\gamma_c}\left(1 + \frac{\gamma_0}{2}\right)} & \text{if } j = 0 \\[2ex] \dfrac{\gamma_0}{1 + \frac{n}{\gamma_c}\left(1 + \frac{\gamma_0}{2}\right)} & \text{if } j = 1 \end{cases}. \tag{6.36}$$

Hence, the decision variable T is a Gamma random variable of the form $T \sim G\left(\frac{n}{2}, \frac{1}{2\sigma_j^2}\right)$ under the hypotheses H_j. The false alarm and the miss probabilities of the detector (6.32) can then be computed as

$$P_f = 1 - \frac{\Gamma\left(\frac{n}{2}; \frac{\tau'}{2\sigma_0^2}\right)}{\Gamma\left(\frac{n}{2}\right)}, \tag{6.37}$$

and

$$P_m = \frac{\Gamma\left(\frac{n}{2}; \frac{\tau'}{2\sigma_1^2}\right)}{\Gamma\left(\frac{n}{2}\right)}, \tag{6.38}$$

where $\Gamma(a) = \int_0^\infty e^{-y} y^{a-1} dy$ is the Gamma function and $\Gamma(a, t) = \int_0^t e^{-y} y^{a-1} dy$ is the incomplete Gamma function. In addition, the threshold τ' in (6.33) simplifies to

$$\tau' = 2\log\tau + n\log\left(\frac{g^2\sigma_v^2(1+\gamma_0)+\sigma_w^2}{g^2\sigma_v^2+\sigma_w^2}\right)$$

$$= n\log\left(1 + \frac{\gamma_c\gamma_0}{n\left(1+\frac{\gamma_0}{2}\right)+\gamma_c}\right), \tag{6.39}$$

where in the last step we have used the fact that $\tau = 1$. Substitution of definitions for γ_c and γ_0 gives the minimum error probability achieved by the optimal Bayesian fusion rule for a random Gaussian signal to be

$$P_e = \frac{1}{2}\left[1 + \frac{\Gamma\left(\frac{n}{2}; \frac{\tau'}{2\sigma_1^2}\right) - \Gamma\left(\frac{n}{2}; \frac{\tau'}{2\sigma_0^2}\right)}{\Gamma\left(\frac{n}{2}\right)}\right], \tag{6.40}$$

where σ_0^2 and σ_1^2 are given by (6.36).

Note from (6.39) that for minimum probability of error criterion with equal priors

$$\lim_{n\to\infty} \tau' = \frac{\gamma_c}{\frac{1}{2}+\frac{1}{\gamma_0}}. \tag{6.41}$$

Taking the limit in (6.40) with the aid of (6.41), it can be shown that

$$\lim_{n\to\infty} P_e = 0.5. \tag{6.42}$$

Moreover, as a function of the observation SNR, the minimum fusion error probability exhibits the following asymptotic property.

$$\lim_{\gamma_0\to\infty} P_e = \frac{1}{2}\left(1 - \frac{\Gamma\left(\frac{n}{2}; t_0\right) - \Gamma\left(\frac{n}{2}; t_1\right)}{\Gamma\left(\frac{n}{2}\right)}\right), \tag{6.43}$$

where $t_0 = t_1 + \frac{n}{2}\log\left(1 + \frac{\gamma_c}{n/2}\right)$ and $t_1 = \frac{n^2}{4\gamma_c}\log\left(1 + \frac{\gamma_c}{n/2}\right)$.

Investigating the fusion error behavior given by (6.40) shows that the final fusion performance is not monotonic in the number of nodes n. Additionally, (6.42) shows that in contrast to the deterministic signal fusion considered earlier, dividing the available total power infinitesimally among many sensors is bound to degrade the performance. In fact, as can be seen from Figure 6.5, there is an optimal number of sensor nodes for each γ_0 and γ_c combination beyond which the performance monotonically degrades. Figure 6.6 shows the convergence of fusion probability of error to the asymptotic bound (6.43) for large γ_0 values.

The exact fusion error probability in (6.40), however, is too complicated for determining this optimal number of nodes, $n = n_0$, that leads to the lowest possible error probability. To that purpose, we resort to the error exponents. It can be shown that the Chernoff and Bhattacharyya error exponents for this situation are given by [43],

$$\mu_C = \frac{n}{2}\left[\log\frac{1+\sigma_1^2}{1+(1-s_0)\sigma_1^2} - s_0\log\left(1+\sigma_1^2\right)\right] \tag{6.44}$$

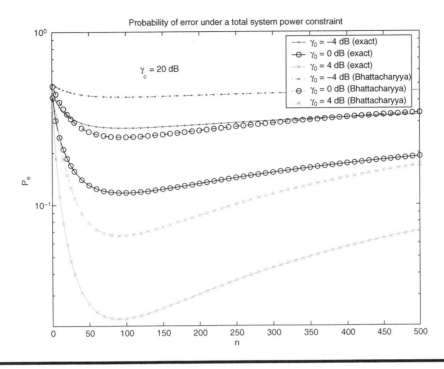

Figure 6.5   Exact and the Bhattacharyya Upper-bound to the Minimum Achievable Fusion Probability of Error in Distributed Detection of a Random Signal with Orthogonal Sensor-to-Fusion Center Communication under a Global Power Constraint. $\gamma_c = 20$ dB.

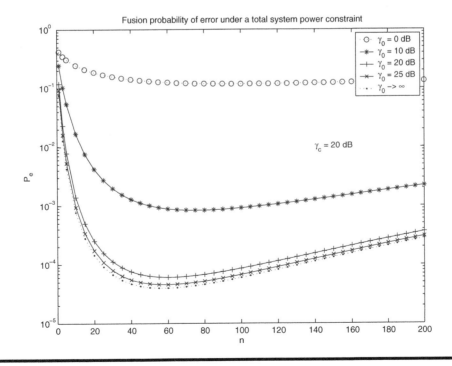

Figure 6.6   Large γ_0 Limiting Behavior of the Minimum Achievable Fusion Probability of Error in Distributed Detection of a Random Signal with Orthogonal Sensor-to-Fusion Center Communication under a Global Power Constraint. $\gamma_c = 20$ dB.

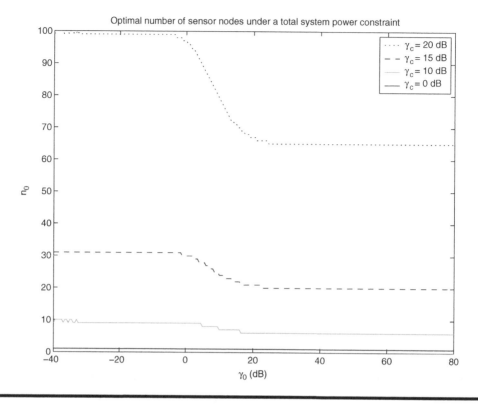

Figure 6.7 **Optimal Number of Sensor Nodes as a Function of the Observation SNR for a Given Channel SNR. The System Is Under a Global System Power Constraint with Orthogonal Sensor-to-Fusion Center Communication.**

and

$$\mu_B = \frac{n}{2}\left[\frac{1}{2}\log\left(1 + \sigma_1^2\right) - \log\left(1 + \frac{\sigma_1^2}{2}\right)\right], \tag{6.45}$$

where $s_0 = 1 + \frac{1}{\sigma_1^2} - \frac{1}{\log(1+\sigma_1^2)}$ in (6.44). Interestingly, using the fact that $\sigma_1^2 \ll 1$ for $n \gg 1$, one can show that $\lim_{n\to\infty}\mu_B = 0$. This indicates that the Bhattacharyya upper bound to the error probability goes to 0.5, suggesting that fusion error may also degrade in large systems. Figure 6.5 includes the behavior of μ_B as a function of n for a fixed γ_c. Clearly there is an optimal value of n for which the μ_B-based bound is also minimized. Although the bound could be somewhat loose, the optimal $n = n_0$ for the Bhattacharyya bound seems to be almost the same as that for the exact error probability. This motivates the use of the Bhattacharyya exponent as the basis for optimizing the sensor system size due to its relative simplicity.

Using standard optimization techniques, it was shown in [43] that in orthogonal signaling ($\rho = 0$) under a global power constraint P, the optimal number of nodes n_0 that results in the minimum Bhattacharyya upper bound to the fusion error probability is given by

$$n_0 = \gamma_c\left(\frac{1}{2x_0} - \frac{1}{\gamma_0}\right)\left(\frac{1}{2} + \frac{1}{\gamma_0}\right)^{-1}, \tag{6.46}$$

where $x = x_0$ is the unique positive solution to the equation $f_{\gamma_0}(x) = 0$ with $f_{\gamma_0}(x) = \log \frac{\sqrt{1+2x}}{1+x} + (1 - 2x/\gamma_0) \frac{x^2}{(1+x)(1+2x)}$. This optimal number of sensors can also be approximated as follows (where $\tilde{x}_0 \approx 1.535$):

$$
n_0 \approx \begin{cases} \gamma_c/\tilde{x}_0 & \text{if } \gamma_0 \gg 1 \\ \gamma_c & \text{if } \gamma_0 \ll 1 \end{cases}. \tag{6.47}
$$

Figure 6.7 shows the optimal number of nodes $n = n_0$ for distributed detection of a stochastic signal under a global power constraint obtained via the exact solution to the zero of f_{γ_0}. Figure 6.7 also shows that the asymptotic solutions given in (6.47) provide a very good approximation except for a small range of values for the observation SNR γ_0. In the case of nonorthogonal communication with an equicorrelated signaling model, [43] generalized the above approach to obtain the final fusion performance as well as the optimal number of sensors to use.

6.5 FUTURE DIRECTIONS

In channel-aware decision fusion, the decentralized detection system needs to be adapted to the conditions of the sensor-to-fusion center communication channel. Recently, [61] pointed out how the design of quantizers at distributed nodes could be optimized depending on the channel state information. However, such an optimization procedure is in general complex, and is restricted to relatively small-size networks. Thus alternative channel-aware decision fusion techniques are to be developed in the future. In particular, instantaneous CSI-based low-complexity, adaptive local decision rules at the distributed nodes as well as adaptive fusion rules are to be investigated.

In large networks, communications between sensors and fusion centers need to be coordinated with a multiple-access channel (MAC) algorithm. With bandwidth constraints, a typical MAC protocol might provide nonorthogonal links thereby causing multiple-access interference. While this issue was addressed for a DS-CDMA-based sensor system in [24], as we discussed in Section 6.4, this remains a topic for further research.

The interplay between sensing, signal processing, and communications in wireless sensor networks was discussed in a recent special issue [62]. Some of the topics presented there have a direct bearing on the research issues presented here. Other future research directions include decentralized estimation/detection with correlated sensor data, when correlation models are derived from realistic physical measurements and a study of reliability achievable through codes, such as low density parity check (LDPC) codes, which may be employed in sensor communication links.

6.6 CHAPTER SUMMARY

In this chapter we reviewed the recent advances in distributed signal processing in resource-constrained wireless sensor networks. The particular attention was on distributed detection and decision fusion in large sensor systems. However, we also briefly outlined recent results on distributed estimation.

We first described the basic problem of distributed detection and fusion in the specific context of wireless sensor networks. The recent work on this topic differs from early work in the sense that, in wireless sensor networks, the communication errors between the distributed nodes and the decision fusion center are nonnegligible. This is due to the unreliable nature of wireless channels (due to fading, shadowing, and interference) as well as limited resources (limited battery power and finite channel bandwidth) in wireless sensor networks. We discussed some important recent work that has specifically taken into account such channel errors in distributed detection and fusion system design and performance analysis. We were particularly interested in outlining large system analysis based results that provide useful insight into the fusion system performance.

In the final section of this chapter, we distinctly considered a specific wireless sensor system in which local processing is assumed to be analog-relay amplifier processing. The fusion center performance was investigated for a distributed binary hypothesis testing problem assuming that the sensor network is both power as well as bandwidth limited. We showed one of the interesting conclusions regarding deterministic versus random signal detection in this context, i.e., while in the case of a deterministic signal it is better to divide the available power and bandwidth among as many nodes as possible, in the case of a random signal there is an optimal number of nodes that provides the best fusion performance. A large system analysis was employed to characterize this fusion performance and obtain the optimal number of nodes to be used.

Finally, we have outlined several open issues and future research directions in Section 6.5.

REFERENCES

[1] S. Kumar, F. Zhao, and D. Sheperd, "Collaborative signal and information processing in micro sensor networks," *IEEE Sig. Process. Mag.*, vol. 19, pp. 13–14, Mar. 2002.

[2] D. Li, K.D. Wong, Y.H. Hu, and A.M. Sayeed, "Detection, classification and tracking of targets," *IEEE Sig. Process. Mag.*, vol. 19, pp. 17–29, Mar. 2002.

[3] J. Kunagi and S. Cheny, "Sensors and Sensibility," *IEEE Spectrum*, pp. 22–28, July 2004.

[4] A.J. Goldsmith and S.B. Wicker, "Design challenges for energy-constrained ad hoc wireless networks," *IEEE Wireless Commn.*, vol. 9, pp. 8–27, Aug. 2002.

[5] R. Min, M. Bhardwaj, S.H. Cho, N. Ickes, E. Shih, A. Sinha, A. Wang, and A. Chandrakasan, "Energy-centric enabling technologies for wireless sensor networks," *IEEE Wireless Commn.*, vol. 9, pp. 28–39, Aug. 2002.

[6] A. Mainwaring, J. Polastre, R. Szewczyk, D. Culler, and J. Anderson, "Wireless sensor networks for habitat monitoring," in *1st ACM Int. Workshop on Wireless Sensor Networks and Applications*, Atlanta, GA, Sept. 2002, pp. 88–97.

[7] P.G. Flikkema and B.W. West, "Wireless sensor networks: from the laboratory to the field," in *National Conf. for Digital Government Research*, Los Angeles, May 2002.

[8] L. Doherty, B.A. Warnake, B. Baser, and K.S.J. Pister, "Energy and performance consideration for SmartDust," *Int. J. Parallel and Distributed Sensor Networks*, vol. 4, no. 3, pp. 121–133, 2001.

[9] R.R. Tenney and N.R. Sandell Jr., "Detection with distributed sensors," *IEEE Trans. Aerosp. Electron. Sys.*, vol. AES-17, no. 4, pp. 501–510, July 1981.

[10] Z. Chair and P.K. Varshney, "Optimal data fusion in multiple sensor detection systems," *IEEE Trans. Aerosp. Electron. Syst.*, vol. AES-22, pp. 98–101, Jan. 1986.

[11] H.V. Poor, *An Introduction to Signal Detection and Estimation.* New York: Springer-Verlag, 1994.

[12] S.M. Kay, *Fundamentals of Statistical Signal Processing: Estimation Theory.* Upper Saddle River, NJ: Prentice Hall, 1993, vol. I.

[13] S.M. Kay, *Fundamentals of Statistical Signal Processing: Detection Theory*. Upper Saddle River, NJ: Prentice Hall, 1998, vol. II.

[14] R. Viswanathan and P.K. Varshney, "Distributed detection with multiple sensors: part I - fundamentals," *Proc. IEEE*, vol. 85, no. 1, pp. 54–63, Jan. 1997.

[15] R.S. Blum, S.A. Kassam, and H.V. Poor, "Distributed detection with multiple sensors: part II – advanced topics," *Proc. IEEE*, vol. 85, no. 1, pp. 64–79, Jan. 1997.

[16] J.N. Tsistsiklis, "Decentralized detection by a large number of sensors," *Math. Control. Signals Syst.*, vol. 1, pp. 167–182, 1988.

[17] P.K. Varshney, *Distributed Detection and Data Fusion*. New York: Springer-Verlag, 1996.

[18] S.S. Pradhan, J. Kusuma, and K. Ramchandran, "Distributed compression in a dense microsensor network," *IEEE Sig. Process. Mag*, vol. 19, pp. 51–60, Mar. 2002.

[19] Z. Xiong, A.D. Liveris, and S. Cheng, "Distributed source coding for sensor networks," *IEEE Sig. Process. Mag*, vol. 21, pp. 80–94, Sep. 2004.

[20] M. Longo, T.D. Lookabaugh, and R.M. Gray, "Quantization for decentralized hypothesis testing under communication constraints," *IEEE Trans. Inform. Theory*, vol. 36, no. 2, pp. 241–255, Mar. 1990.

[21] J. Chamberland and V. Veeravalli, "Decentralized detection in sensor networks," *IEEE Trans. Sig. Process.*, vol. 51, pp. 407–416, Feb. 2003.

[22] Y. Sung, L. Tong, and A. Swami, "Asymptotically locally optimal detector for large scale sensor networks under the Poisson regime," *IEEE Trans. Signal Process.*, vol. 53, no. 6, pp. 2005 –2017, June 2004.

[23] T.M. Duman and M. Salehi, "Decentralized detection over multiple-access channels," *IEEE Trans. Aerosp. Electron. Syst.*, vol. 34, no. 2, pp. 469–476, Apr. 1998.

[24] S.K. Jayaweera, "Large system decentralized detection performance under communication constraints," *IEEE Commn. Letters*, vol. 9, pp. 769–771, Sept. 2005.

[25] C.K. Sestok, M.R. Said, and A.V. Oppenheim, "Randomized data selection in detection with applications to distributed signal processing," *Proc. IEEE*, vol. 85, pp. 1184–1198, Aug. 2003.

[26] J. Xiao and Z. Luo, "Universal decentralized detection in bandwidth constrained sensor network," *IEEE Trans. Sig. Process.*, vol. 53, pp. 2617–2624, Aug. 2005.

[27] J. Chou, D. Petrovic, and K. Ramachandran, "A distributed and adaptive signal processing approach to reducing energy consumption in sensor networks," in *Proc. 21st Annual Joint Conf. IEEE Computer and Commun. Soc., IEEE INFOCOM 2003*, vol. 1, San Francisco, Mar. 2003.

[28] S.C. Draper and G.W. Wornell, "Side information aware coding strategies for sensor networks," *IEEE J. Selected Areas Commn.*, vol. 22, pp. 966–976, Sept. 2004.

[29] A.D. Murugan, P.K. Gopala, and H.E. Gamal, "Correlated sources over wireless channels: cooperative source-channel coding," *IEEE J. Selected Areas Commn.*, vol. 22, pp. 988–998, Aug. 2004.

[30] A. Stefanov and E. Erkip, "Cooperative coding for sensor networks," *IEEE Trans. Commn.*, vol. 52, pp. 1470–1476, Sept. 2004.

[31] V. Aravinthan, S.K. Jayaweera, and K. Altarazi, "Distributed estimation in a power constrained sensor network," in *IEEE 63rd Vehicular Technology Conf. (VTC'06 Spring)*, Melbourne, Australia, May 2006.

[32] J. Xiao, S. Cui, Z. Luo, and A.J. Goldsmith, "Power scheduling of universal decentralized estimation in sensor networks," *IEEE. Trans. Sig. Process.*, vol. 54, No. 2, pp. 413–422, Feb. 2006.

[33] H.R. Hashemi and I.B. Rhodes, "Decentralized sequential detection," *IEEE Trans. Inform. Theory*, vol. 35, no. 3, pp. 509–520, May 1989.

[34] V.V. Veeravalli, T. Basar, and H.V. Poor, "Decentralized sequential detection with a fusion center performing the sequential test," *IEEE Trans. Inform. Theory*, vol. 39, no. 2, 1993.

[35] R. Viswanathan, S.C.A. Thomopoulos, and R. Tumuluri, "Optimal serial distributed decision fusion," *IEEE Trans. Aerosp. Elect. Syst.*, vol. 24, pp. 366–376, July 1988.

[36] J.N. Laneman, D.N.C. Tse, and G.W. Wornell, "Cooperative diversity in wireless networks: efficient protocols and outage behavior," *IEEE Trans. Inform. Theory*, pp. 3062–3080, 2004.

[37] B. Chen, R. Jiang, T. Kasetkasem, and P.K. Varshney, "Channel aware decision fusion in wireless sensor networks," *IEEE Trans. Sig. Process.*, vol. 52, pp. 3454–3458, Dec. 2004.

[38] J. Chamberland and V.V. Veeravalli, "Asymptotic results for decentralized detection in power constrained wireless sensor networks," *IEEE J. Select. Areas Commn.*, vol. 22, no. 6, pp. 1007–1015, Aug. 2004.

[39] S.K. Jayaweera, "Decentralized detection of stochastic signals in power-constrained sensor networks," in *IEEE Workshop on Sig. Process. Adv. Wireless Commn. (SPAWC)*, New York, June 2005.

[40] K. Altarazi, S.K. Jayaweera, and V. Aravinthan, "Performance of decentralized detection in a resource-constrained sensor network," in *39th Annual Asilomar Conf. on Sig., Syst. and Comput.*, Pacific Grove, CA, Nov. 2005.

[41] V.R. Kanchumarthy and R. Viswanathan, "Performance of decentralized detection in large sensor networks: impact of different binary modulation schemes and fading in sensor-to-fusion center link," in *Proc. of 43rd Allerton Conf. Commn. Control Comput.*, Monticello, IL, Sept. 2005, pp. 916–925.

[42] V.R. Kanchumarthy and R. Viswanathan, "Further impacts on the quality of wireless sensor links on decentralized detection performance," in *Proc. of CISS'06*, Princeton, NJ, 2006, pp. 44–49.

[43] S.K. Jayaweera, "Sensor system optimization for Bayesian fusion of distributed stochastic signals under resource constraints," in *IEEE International Conf. on Acoustics, Speech and Signal Processing (ICASSP'06)*, Toulouse, France, May 2006.

[44] J. Chamberland and V.V. Veeravalli, "Decentralized detection in wireless sensor systems with dependent observations," in *Proc. 2nd Intl. Conf. Computing, Commn. Contrl. Technologies*, Austin, TX, Aug. 2004.

[45] S.A. Aldosari and J.M.F. Moura, "Detection in decentralized sensor networks," in *Proc. of IEEE ICASSP*, Montreal, Canada, May 2004.

[46] R. Jiang and B. Chen, "Fusion of censored decisions in wireless sensor networks," *IEEE Trans. Wireless Commn.*, pp. 2668–2673, Nov. 2005.

[47] S. Appadwedula, V.V. Veeravalli, and D.L. Jones, "Energy-efficient detection in sensor networks," *IEEE J. Select. Areas Commn.*, vol. 23, pp. 693–702, Apr. 2005.

[48] M. Madishetty, V. Kanchumarthy, C.H. Gowda, and R. Viswanathan, "Distributed detection with channel errors," in *IEEE 37th Southeast Symp. Systems Theory*, Tuskegee University, Tuskegee, AL, Mar. 2005, pp. 302–306.

[49] K.C. Chang, C.Y. Ching, and Y. Bar-Shalom, "Joint probability data association in distributed sensor networks," in *Proc. American Control Conf.*, 1985, pp. 817–822.

[50] W.M. Lam and A.R. Reibman, "Design of quantizers for decentralized estimation systems," *IEEE Trans. Inform. Theory*, vol. 41, no. 11, pp. 1602–1605, Nov. 1993.

[51] M. Su and R. Viswanathan, "Distributed estimation of location parameter of two example densities," in *Proc. 20th Annual Allerton Conf. Commn. Control Comp.*, Allerton House, Monticello, IL, Oct. 1996, pp. 1009–1018.

[52] A. Ribeiro and G.B. Giannakis, "Bandwidth-constrained distributed estimation for wireless sensor networks – part I: Gaussian case," *IEEE Trans. Sig. Process.*, vol. 54, pp. 1131–1143, Mar. 2006.

[53] Z.Q. Luo, "An isotropic universal decentralized estimation in a bandwidth constrained ad hoc sensor network," *IEEE J. Select. Areas Commn.*, vol. 23, pp. 735–744, Apr. 2005.

[54] J.J. Xiao, Z.Q. Luo, and G.B. Giannakis, "Performance bounds for the rate-constrained universal decentralized estimation in sensor networks," in *IEEE 6th Workshop on Signal Processing Advances in Wireless Communications*, New York, June 2005, pp. 126–130.

[55] H.C. Papadopoulos, G.W. Wornell, and A.V. Oppenheim, "Sequential signal encoding from noisy measurements using quantizers with dynamic bias control," *IEEE Trans. Inform. Theory*, vol. 47, no. 3, pp. 978–1002, 2001.

[56] R. Nowak, U. Mitra, and R. Willett, "Estimating inhomogeneous fields using wireless sensor networks," *IEEE J. Select. Areas Commn.*, vol. 22, no. 6, pp. 999–1006, Aug. 2004.

[57] A.M. Tulino and S. Verdu, *Random Matrix Theory and Wireless Communications*. Hanover, MA: now publishers inc., 2004.

[58] J.W. Silverstein and Z.D. Bai, "On the empirical distribution eigenvalues of a class of large dimensional random matrices," *J. Mult. Anal.*, vol. 54, pp. 175–192, 1995.

[59] J.W. Silverstein, "Strong convergence of the empirical distribution eigenvalues of large dimensional random matrices," *J. Mult. Anal.*, vol. 55, pp. 331–339, 1995.

[60] J. Evans and D.N.C. Tse, "Large system performance of linear multiuser receivers in multipath fading channels," *IEEE Trans. Inform. Theory*, vol. 46, pp. 2059–2078, Sept. 2000.

[61] B. Chen, L. Tong, and P.K. Varshney, "Channel-aware distributed detection in wireless sensor networks," *IEEE Sig. Process. Mag.*, vol. 23, no. 4, pp. 16–26, July 2006.

[62] Z.Q. Luo, M. Gastpar, J. Liu, and A. Swami, "Distributed signal processing in sensor networks," Guest editorial, special issue, *IEEE Sig. Process. Mag.*, vol. 14, pp. 14–25, July 2006.

7

OPTIMAL RESOURCE ALLOCATION OF DAS

Lin Dai

Contents

The distributed antenna system (DAS) has emerged as a promising candidate for the future beyond 3G or 4G mobile communications thanks to its open architecture and flexible resource management. The distributed characteristic of the antennas provides a more efficient utilization of space resources; however, it also raises a crucial challenge for the advanced resource allocation. In this chapter, the optimal resource allocation for DAS networks is investigated. We start with an overview of the current adaptive techniques for wireless systems. The resource allocation strategies in distributed channels are then proposed, and the performance comparison with equal allocation helps us understand why an adaptive resource allocation is indispensable to the DAS. It is further extended to the multi-user scenario and the optimal resource allocation among multiple users is discussed. The chapter concludes by presenting some open research issues in the realization of resource allocation for DAS networks.

7.1 RESOURCE ALLOCATION FOR WIRELESS SYSTEMS

Future wireless networks are expected to support a wide variety of communication services, such as voice, video, and multimedia. However, the wireless environment provides unique challenges to reliable communication: the time-varying nature of the channel and scarcity of the radio resources such as power and bandwidth. Therefore, it is of great interest to investigate how to efficiently allocate the limited radio resources to meet diverse quality-of-service (QoS) requirements of users and maximize the utilization of available bandwidth based on the channel states of users. In this section, we will present an optimal resource allocation framework for the wireless system, and based on it, some adaptive techniques will be introduced.

7.1.1 Optimal Framework of Resource Allocation

There are tremendous ways to perform resource allocation. For instance, adaptive power and rate allocation can provide significant performance gain in fading channels [1]. In code division multiple access (CDMA) systems, the radio resources are usually allocated to the users by regulating their transmit power and spreading gains [65]. In general, if we consider a wireless network with $\mathcal{K}$ users, each with a utility function $U_k(.)$ and a constraint $S_k(.) \leq q_k$, the problem of optimal resource allocation can be formulated as

$$\text{Maximize} \quad \sum_{k=1}^{\mathcal{K}} U_k(P_k, R_k, L_k, \Phi_k)$$

$$\text{Subject to} \quad S_k(P_k, R_k, L_k, \Phi_k) \leq q_k, \qquad k = 1, \ldots, \mathcal{K}, \tag{7.1}$$

where the utility of user k $U_k(.)$ is a function of the allocated resources including: the transmit power P_k, the modulation and coding level R_k, the packet length L_k and Φ_k. Φ_k represents the available resources in specific systems, which can be the number of spreading codes assigned to user k in CDMA systems; the number of subcarriers in orthogonal frequency division multiplexing (OFDM) systems; the number of antennas in multiple-input multiple-output (MIMO) systems; or the number of time slots in time division multiple access (TDMA) systems. q_k is the required QoS for user k, i.e., the required rate, error probability, and delay constraints, etc.

From (7.1) it can be seen that the optimal resource allocation is essentially an optimization problem of maximizing the network utility $\sum_{k=1}^{\mathcal{K}} U_k(.)$ over the available resources, subject to the constraints of all users. Obviously, in wireless systems where users may have diverse QoS requirements and distinct channel statistics, optimal resource allocation can bring huge performance gains.

Optimal resource allocation for a specific user can be performed in time, space, or frequency domain. For example, the transmit power and rate can be adjusted among different fading states (time domain), different antennas (space domain), or different subcarriers (frequency domain) to maximize the throughput or minimize the error rate. Optimal resource allocation among multiple users, however, will be much more complex because the feasible solution field may not always exist. Maximizing the network utility $\sum_{k=1}^{K} U_k(.)$ does not guarantee that the utility of each user is maximized; on the contrary, sometimes it is achieved by sacrificing some users' performance. Therefore, a tradeoff between efficiency and fairness usually has to be addressed in multi-user resource allocation.

7.1.2 Adaptive Techniques

We have shown that in wireless systems, resources should be adaptively allocated according to users' QoS requirements and channel statistics. In this section, we will further introduce some adaptive techniques and show how to perform resource allocation in specific systems.

7.1.2.1 Adaptive Power and Rate Allocation

Adaptive power and rate allocation was proposed a long time ago as an effective means to overcome the detrimental effect of time-varying channels [1, 3–5]. Later, information-theoretic work showed that to maximize the ergodic capacity of a single-user fading channel with channel state information (CSI) at both the transmitter and receiver, the optimal power and rate allocation is a water-filling procedure over the fading states [6]. The ergodic capacity region of a fading multiple-access channel (MAC) and the corresponding optimal power and rate allocation, which is a multi-user version of the single-user water-filling procedure, were obtained in [7] using the polymatroidal structure of the region. [8] further derived the optimal power allocation for maximizing the delay-limited capacity. It was shown that with the proposed channel inversion strategy, there is zero outage probability and the end-to-end delay is independent of the channel variation. The price is that huge power has to be consumed to invert the channel when it is in an unfavorable state.

Another line of work focused on practical schemes, which typically assume a finite number of power levels and modulation and coding schemes [9–15]. For example, adaptive modulation and coding (AMC) has been studied extensively [10–14] and adopted at the physical layer in several standards, e.g., 3GPP, 3GPP2, IEEE 802.11a, IEEE 802.15.3, and IEEE 802.16 [16–18]. Recent work includes the cross-layer optimization combining AMC at the physical layer and automatic request protocol (ARQ) or finite-length queue at the link layer [19,20], and the joint optimization of rate and packet length in cooperative ad hoc networks [21].

7.1.2.2 Adaptive Resource Allocation for MIMO Systems

MIMO systems have recently attracted tremendous interest due to their ability in providing great capacity improvements [22,23]. Different from the traditional power and rate allocation in fading channels, which is performed in time domain, the resources in MIMO systems are usually allocated among the antennas or in space domain.

It has been shown that the optimal power allocation among the multiple antennas is the water-filling strategy [22]. However, to perform this optimal allocation requires full

CSI at the transmitter. Later work focused on transmit beamforming and precoding with limited feedback [24–27], where the transmitter uses a small number of feedback bits to adjust the power and phases of the transmit signals. To further reduce the amount of feedback and complexity, per-antenna rate and power control was proposed [28–32]. By adapting the rate and power for each antenna separately, the performance (error probability [42] or throughput [39–41]) can be improved greatly at a slight cost of complexity.

Antenna selection was proposed to reduce the number of radio frequency (RF) chains and the receiver complexity. Various criteria for receive antenna selection or transmit antenna selection have been presented, aiming at minimizing the error probability [33–40] or maximizing the capacity bounds [32,33]. It was shown that only a small performance loss is suffered when the transmitter/receiver selects a good subset of the available antennas based on the instantaneous CSI. However, recently it is found that in the correlated scenario, proper transmit antenna selection cannot just be used to decrease the number of RF chains, but as an effective means to bring the performance gain [34]. When the channel links present spatial correlation (due to the lack of spacing between antennas or the existence of small angular spread), the degrees of freedom of the channel are usually less than the transmit antennas. Therefore, by the use of transmit antenna selection, the resources are allocated only to the "good" subchannels so that a capacity gain can be achieved.

Most of the above work focused on the peer-to-peer link in the single-user scenario. Resource allocation in a multi-user MIMO scenario is still quite an open issue. [44,45] both considered a multi-user MIMO system and focused on multi-user precoding and turbo space–time multi-user detection, respectively. More recent work includes a cross-layer resource allocation in downlink multi-user MIMO systems [46].

7.1.2.3 Adaptive Resource Allocation for OFDM Systems

OFDM was proposed to combat the intersymbol interference (ISI) problem [47]. Later it was found that adaptive rate allocation can be perfectly performed in OFDM systems, where subcarriers with higher channel gains carry more bits while the ones in deep fade carry few or even zero bits [48,49]. Similar to the per-antenna rate and power allocation, here the rate of each subcarrier is adjusted according to the CSI following the water-filling principle. The optimal power allocation has also been studied [50,51]. Significant performance gain can be achieved through the power and rate adaptation.

In MIMO systems, it is not straightforward to extend the per-antenna rate and power allocation to the multi-user scenario as there is no bijective mapping between the transmit antenna set and the subchannel set. Some complicated interference cancellation techniques have to be developed. In multi-user OFDM systems, however, thanks to the orthogonality among the subcarriers, each subcarrier can be allocated to a user with the best channel condition. Here multi-user diversity gain is further achieved based on the low probability that all the users' signals on the same subcarrier are in deep fading [52–55]. Some recent work includes the adaptive resource allocation for MIMO-OFDM systems [56] and cross-layer optimization for multi-user OFDM systems [57].

7.1.2.4 Adaptive Resource Allocation for CDMA Systems

The available resources in CDMA systems include transmit power and spreading codes. Joint power allocation and base station assignment problems were first analyzed in [58,59]. In these works, the objective is the minimization of the total transmit power subject to the QoS requirements of the sources, without considering the allocation of the spreading gains. Another line of work focuses on multiple classes of service, where users are allocated different class-dependent spreading gains to maximize the throughput [60,61]. The joint optimal allocation of power and spreading gains was considered in [2, 62–66], in which an optimization problem is usually formulated to optimize the total transmit power or the sum of the transmission rate (or say, the network throughput) under the constraint on the maximum transmission power of each user or the minimum spreading gain (or both).

Recently, utility-based power control has received significant attention, where a game theoretic approach is applied to the power control problem for CDMA data networks. Here, the optimization objective is neither the total transmit power nor the sum rate. Instead, a utility function is proposed which quantifies the level of satisfaction a user gets from using the system resources [67], and the resources are allocated to optimize the network utility. Its attractiveness comes partially from the distributed nature: each user can maximize its own utility in a distributed fashion. See [67,68] and references therein for more details.

7.1.2.5 Channel-Aware Scheduling

Efficient resource allocation for multiple users is always an interesting but challenging issue. As we have introduced, in OFDM (or CDMA) wireless systems, subcarriers (or spreading codes) are assigned to users according to their QoS requirements and channel conditions. Another option, however, is to allocate all the system resources to different users, in different time slots. This leads to the so-called *scheduling* problem. In the earliest work on scheduling for wireless systems, the time-varying nature of wireless channels was not taken into full consideration [69,70]. The channel is usually simplified as an "ON-OFF" model and the focus is on the queue statistics. Knopp and Humblet [71] first proposed to always schedule the user with the best channel and showed that significant throughput gain can be brought by multi-user diversity — "when there are many users who fade independently; at any one time there is a high probability that one of the users will have a strong channel" [72]. Obviously, the more users scheduled, the higher throughput that can be obtained.

There have been numerous works on how to exploit this multi-user diversity gain [73–77]. However, to directly implement the idea of multi-user diversity will result in unfairness if users' fading statistics are not identical: The user with a statistically stronger channel has a higher opportunity in acquiring the system resources. From a system aspect, efficiency and fairness are both crucial issues in resource allocation and should be carefully addressed. Several definitions of fairness have been proposed, such as max-min fairness [78] and proportional fairness [79,80]. A scheduler combined with multi-user diversity and proportional fairness has been proposed in [81], which is also the baseline scheduler for the downlink of IS-856. Asymptotic analysis of scheduling can be found in [82].

So far we have presented the optimal resource allocation framework and introduced some representative adaptive techniques. In the next section, we will focus on distributed antenna systems (DASs) and illustrate how to efficiently allocate resources in distributed channels.

7.2 RESOURCE ALLOCATION IN DISTRIBUTED CHANNELS

In DASs, many remote antenna ports are distributed over a large area and connected to a central processor by fiber, coax cable, or microwave link [83]. Basically, resource allocation of DASs is also performed among the antennas, similar to that of the MIMO systems. However, due to some special characteristics of distributed channels, resource allocation is indispensable to a DAS and is not just for performance enhancement. It will be shown that the performance severely deteriorates without proper resource allocation. In this section, we consider the downlink resource allocation in the single-user scenario, i.e., how to assign the transmission phases, rate, and power of different distributed antennas to a specific user. Multi-user resource allocation will be discussed in Section 7.3 from a system perspective.

7.2.1 System and Channel Model

Consider a DAS with $\mathcal{M}$ remote antennas which are randomly distributed around $\mathcal{K}$ users each equipped with n colocated antennas. Here the cells are divided not geographically, but according to the user demands, which are called "virtual cell" [83]. As shown in Figure 7.1, the remote antennas serving for user k form the k-th virtual cell. When user k moves, the remote antennas in the k-th virtual cell will be dynamically modified to adapt to the changes of user k. The central processor continuously tracks the channel between user k and each remote antenna and selects the best m remote antennas to form the virtual cell of user k.

Particularly, user k receives signals from the m remote antennas of its virtual cell. Assume a flat fading and quasi-static channel model and perfect symbol synchronization

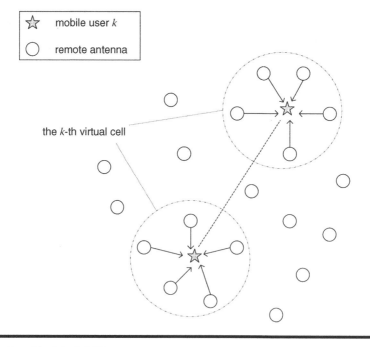

Figure 7.1   System Model of DAS.

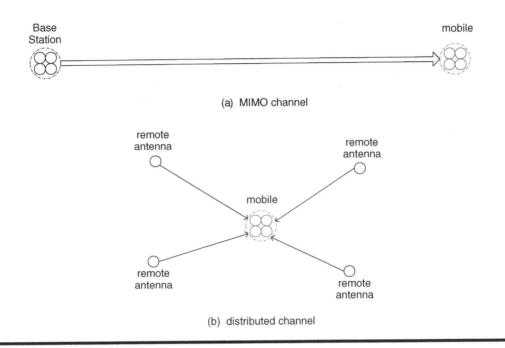

(a) MIMO channel

(b) distributed channel

Figure 7.2 MIMO Channel Vs. Distributed Channel.

at the receiver. The discrete model of the received complex signal vector can be written as

$$\mathbf{y} = \mathbf{H}\mathbf{x} + \mathbf{z} = \mathbf{R}_r^{1/2}\mathbf{H_w}\mathbf{F}\mathbf{x} + \mathbf{z}, \tag{7.2}$$

where $\mathbf{z}$ is the noise vector with i.i.d. $\mathcal{CN}(0, \sigma^2)$ entries. The small-scale fading is denoted by an $n \times m$ matrix $\mathbf{H_w}$ with i.i.d. $\mathcal{CN}(0, 1)$ entries. $\mathbf{R_r}$ denotes the $n \times n$ antenna correlation matrix at the receiver (where n antennas are colocated) [84]. $\mathbf{F}$ is an $m \times m$ diagonal matrix.

Equation (7.2) is reminiscent of a MIMO channel model [85]. The only difference is that the transmit correlation matrix $\mathbf{R_t}$ is replaced by a diagonal matrix $\mathbf{F}$. Illustrated in Figure 7.2 is a comparison between a MIMO channel and a distributed channel. It can be seen that in a MIMO system, the antennas are colocated at both the base station and the user. Therefore, the transmit signals experience similar large-scale fading and resource allocation here is usually adopted to overcome the spatial correlation, which leads to insufficient degrees of freedom of the channels. However, in distributed channels, the signals transmitted from different remote antennas suffer from distinct degrees of large-scale fading, which is denoted by this $m \times m$ diagonal matrix $\mathbf{F}$. In particular, $\mathbf{F} = diag(f_1, \ldots, f_m) = diag(\sqrt{\varsigma_1 d_1^{-\alpha}}, \ldots, \sqrt{\varsigma_m d_m^{-\alpha}})$, where d_i is the access distance between the user and the i-th remote antenna. α is the path loss exponent and ς_i represents the effect of log-normal shadowing, $i = 1, \ldots, m$. In the following discussions, we normalize $trace(\mathbf{FF}^*)$ to be m and let $\eta = \| f_1 \|^2 : \| f_2 \|^2 : \cdots : \| f_m \|^2$.[1] We will show that due to the existence of $\mathbf{F}$, equally allocating resources among the transmit antennas, which is often adopted in MIMO systems, will lead to severe performance degradation. Adaptive resource allocation is highly desired in a DAS.

---

[1] Throughout the chapter, "∗" represents the conjugate and transpose operator.

7.2.2 Water-Filling and Equal Power Allocation

To understand why resource allocation is requisite in a DAS, we start with a simple power allocation problem in a multiple-antenna channel. For simplicity, let us assume that the receive antennas at the user are uncorrelated, i.e., $\mathbf{R_r} = \mathbf{I_n}$. The mutual information is then given by

$$I(\mathbf{Q}) = \log_2 \det\left(\mathbf{I_n} + \frac{1}{\sigma^2}\mathbf{H_w}\mathbf{FQF^*}\mathbf{H_w^*}\right), \tag{7.3}$$

where $\mathbf{Q} = E(\mathbf{xx^*})$ is the transmit covariance matrix.

Without the CSI at the transmitter, equal power allocation, i.e., $\mathbf{Q} = \frac{P_t}{m}\mathbf{I_m}$, would be optimal [22, 23]. In this case, we have

$$C = \log_2 \det\left(\mathbf{I_n} + \frac{\rho}{m}\mathbf{H_w}\mathbf{FF^*}\mathbf{H_w^*}\right) = \log_2 \det\left(\mathbf{I_m} + \frac{\rho}{m}\mathbf{F^*}\mathbf{H_w^*}\mathbf{H_w}\mathbf{F}\right), \tag{7.4}$$

where $\rho = P_t/\sigma^2$ is the average receive single-to-noise ratio (SNR). Since $\mathbf{H_w^*}\mathbf{H_w}$ is Hermitian, it can be diagonalized as $\mathbf{H_w^*}\mathbf{H_w} = \mathbf{U_h^*}\Lambda_\mathbf{h}\mathbf{U_h}$, with a unitary matrix $\mathbf{U_h}$ and a nonnegative diagonal matrix $\Lambda_\mathbf{h}$. Let $\mathbf{X} = \Lambda_\mathbf{h}^{1/2}\mathbf{U_h}\mathbf{F} = \mathbf{V_x^*}\Lambda_\mathbf{x}^{1/2}\mathbf{U_x}$, we see that

$$C_{eq} = \sum_{i=1}^{r}\log_2\left(1 + \frac{\rho}{m}\lambda_i\right), \tag{7.5}$$

where λ_i is the i-th eigenvalue of $\mathbf{X^*X}$ and $r = min(m, n)$.

On the other hand, if CSI is available at the transmitter, [22] has shown that the water-filling policy would maximize the capacity, which requires

$$\tilde{\mathbf{Q}} = \mathbf{U_x}\mathbf{Q}\mathbf{U_x^*} = diag(\mu - \sigma^2\lambda_i^{-1})^+, \tag{7.6}$$

where μ is chosen to satisfy $\sum_{i=1}^{m}\tilde{Q}_{ii} = P_t$, and the capacity is then given by

$$C_{wf} = \sum_{i=1}^{r}(\log_2(\mu\lambda_i/\sigma^2))^+. \tag{7.7}$$

When $\mathbf{X^*X}$ is of full rank and well-conditioned, the water-filling strategy allocates nearly an equal amount of power to all the dimensions, and the capacity is approximated by

$$C_{wf} = \sum_{i=1}^{r}\log_2\left(1 + \frac{\rho}{r}\lambda_i\right). \tag{7.8}$$

Comparing (7.8) and (7.5), we can see that the water-filling strategy can achieve a power gain of a factor of m/r over the equal power allocation. This implies that when there are more transmit antennas than receive antennas, CSI at the transmitter is highly desired so that the transmit energy can be effectively allocated to only r degrees of freedom instead of being spread out equally across all m directions. Figure 7.3 presents the 10% outage capacity results of the water-filling strategy and the equal power allocation in i.i.d. MIMO channels (i.e., $\mathbf{F} = \mathbf{I_m}$). Only a slight capacity gain can be observed with the water-filling strategy when both the number of transmit antennas and receive antennas

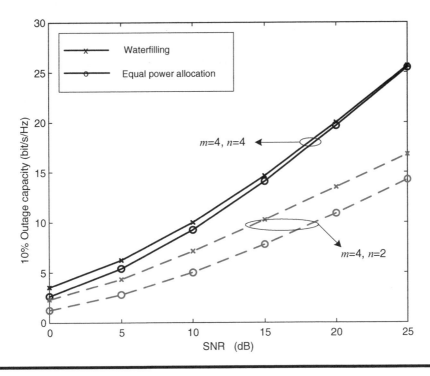

Figure 7.3 **10% Outage Capacity Comparison of Waterfilling Strategy and Equal Power Allocation in MIMO Channels.**

is equal to four. However, the capacity gap is significantly enlarged if there are not as many receive antennas as transmit antennas.

In fact, if we notice that $\mathbf{X}^*\mathbf{X}$ is usually ill-conditioned in distributed channels, we will find that the water-filling strategy can bring even more substantial capacity gains. Again, assume $m = n = 4$ and consider two types of distributed channels with $\eta_1 = 500{:}100{:}20{:}1$ and $\eta_2 = 1000{:}100{:}10{:}1$. Obviously, in the first case two subchannels are significantly better than the others and $\mathbf{X}^*\mathbf{X}$ is rank deficient. The second case is even more asymmetric, and $\mathbf{X}^*\mathbf{X}$ is severely ill-conditioned. Figure 7.4 shows the capacity gains of the water-filling strategy over the equal power allocation, i.e., $(C_{wf} - C_{eq})/C_{eq} \times 100\%$, in both cases. For comparison, the results in MIMO channels are also provided. In distributed channels, the water-filling strategy performs much better than the equal power allocation. In a low SNR regime, over 50% capacity gains can be achieved by the water-filling strategy. This gain, however, will diminish when the SNR is high enough.

Based on the above discussions, we can conclude that resource allocation is highly desired in a DAS. In MIMO systems, all the subchannels suffer from nearly the same large-scale fading; hence, the equal power allocation can provide comparable performance, especially when $m = n.$[2] In DASs, however, due to the large differences among subchannels, equal allocation incurs a significant capacity loss. CSI is highly desired at the transmitter to perform the adaptive resource allocation.[3]

---

[2] Note that here, no space correlation is assumed.
[3] Throughout this chapter, perfect CSI is always assumed available at the receiver.

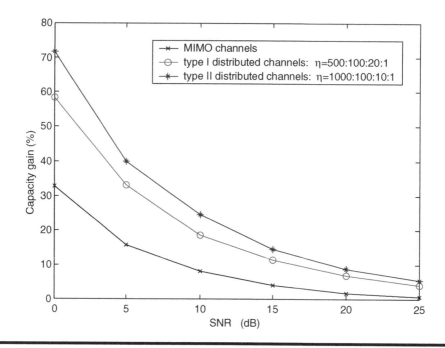

Figure 7.4 Capacity Gains of Waterfilling Strategy Over Equal Power Allocation in Distributed Channels and MIMO Channels. $m = n = 4$.

7.2.3 Full CSI at the Transmitter

So far we have shown that CSI at the transmitter plays an important role in distributed channels. In a DAS, the signals transmitted from different remote antennas suffer from distinct degrees of large-scale fading. Therefore, CSI is required at the transmitter to ensure that the resources are allocated only to those "good" subchannels. In this subsection, we will focus on the case with full CSI at the transmitter.[4]

7.2.3.1 Transmit Precoding

As we know, with full CSI, i.e., $\mathbf{H}$, at the transmitter, the water-filling strategy can achieve the optimal capacity. Therefore, a natural way is to design a precoding matrix based on this water-filling principle.

As shown in Figure 7.5, at the transmitter, the information is split into m parallel data streams and encoded separately. After being modulated, those streams are multiplied by a linear transformation matrix $\mathbf{L} \in \mathbf{C}^{m \times m}$ and then transmitted through m remote antennas. Based on the water-filling principle, $\mathbf{L}$ is given by [86]

$$\mathbf{L} = \mathbf{V}\mathbf{D}^{1/2}\mathbf{W} \tag{7.9}$$

where the columns of $\mathbf{V}$ are eigenvectors of $\mathbf{H}^*\mathbf{H}$. $\mathbf{D} = diag\,(m(\mu - \lambda_i^{-1})/\rho)^+$ and $\mathbf{W}$ is a unitary matrix. It can be easily proved that (7.9) satisfies the constraint condition

---

[4] Although the subsection is entitled "full CSI at the transmitter," the resource allocation can be performed at the receiver. The receiver then feeds back the allocation results, instead of the exact CSI information, to the transmitter.

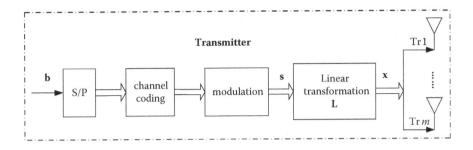

Figure 7.5   Transmitter Structure with Precoding.

$tr(\mathbf{LQL}^*) = tr(\mathbf{Q})$ and achieves the optimal capacity given by (7.7). Although the capacity is maximized as long as $\mathbf{W}$ is a unitary matrix, $\mathbf{W}$ should be carefully selected for particular coding and modulation beacuse the error performance depends on $\mathbf{W}$. We search the optimal $\mathbf{W}$ to minimize the pairwise error probability, namely, to maximize the minimum distance between received vectors.

Figure 7.6 presents the frame error rate (FER) comparison of this precoding scheme and the equal power allocation in MIMO channels. Quadratic phase shift keying (QPSK) modulation is assumed at the transmitter with $m = 2$ remote antennas. Maximum likelihood detection (MLD) is adopted at the receiver. From Figure 7.6 it can be seen that a significant performance gain is brought by precoding at the transmitter. For instance, a 3 dB gain can be observed at the FER of 0.1 with $n = 2$ receive antennas, and this gain increases to 10 dB when there are fewer antennas at the receiver, say, $n = 1$.

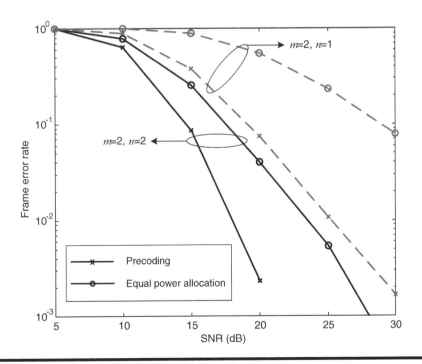

Figure 7.6   FER Comparison of the Precoding Scheme and Equal Power Allocation in MIMO Channels with QPSK Modulation and MLD.

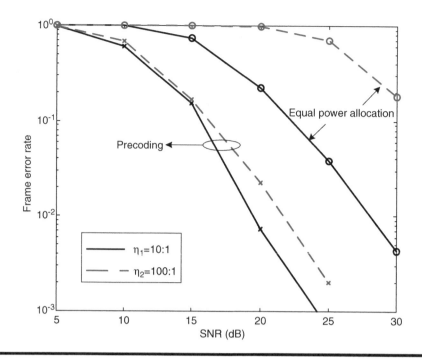

Figure 7.7   FER Comparison of the Precoding Scheme and Equal Power Allocation in Distributed Channels with QPSK Modulation and MLD. $m = n = 2$.

The performance gap is even larger in distributed channels. As shown in Figure 7.7, when $\eta_1 = 10{:}1$, an 8 dB gain can be achieved by the precoding scheme over the equal allocation one. As the variance of subchannels increases, the performance of the equal power allocation deteriorates rapidly. When $\eta_2 = 100{:}1$, the equal power allocation cannot work at all. In contrast, only slight performance degradation is observed with the precoding scheme.

Despite the superior performance, the precoding scheme requires either full CSI or an updated linear precoding matrix $\mathbf{L}$ to be fed back to the transmitter, both of which will incur a large amount of feedback. In addition, the transmission phase of each remote antenna is adjusted based on the feedback information, which makes this scheme highly sensitive to the feedback errors.

7.2.3.2   Per-Antenna Rate and Power Adaptation

With full CSI at the transmitter, the water-filling based precoding scheme has been able to achieve huge performance gains, especially in distributed channels. However, this pre-coding scheme requires a large amount of feedback and is quite sensitive to the feedback errors, which restricts its application in the practical scenarios. In this subsection, we will introduce a more robust resource allocation strategy, where the transmission rate and power are adjusted in a *per-antenna* manner.

As shown in Figure 7.8, the coding, modulation, and average transmit power of each remote antenna are adjusted based on the feedback information. Here we define a *mode* as a combination of specific coding and modulation. Let M_i denote the mode of the i-th antenna and the corresponding spectral efficiency is denoted by $R(M_i)$. Given the total

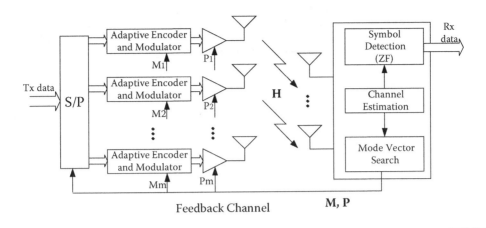

Figure 7.8 Transmitter and Receiver Structures with Per-Antenna Rate and Power Allocation.

required spectral efficiency R_t, we define the mode vector as $\mathbf{M} = [M_1, \ldots, M_m]$ such that $R_t = \sum_{i=1}^{m} R(M_i)$. Likewise, with the total transmit power P_t, we define the power allocation vector as $\mathbf{P} = [P_1, \ldots, P_m]$ such that $P_t = \sum_{i=1}^{m} P_i$, where P_i denotes the average power radiated by the i-th transmit antenna. Although full CSI is required to optimize the mode vector $\mathbf{M}$ and the power allocation vector $\mathbf{P}$, this can be performed at the receiver and only the optimized results on $\mathbf{M}$ and $\mathbf{P}$ are fed back to the transmitter.

There are numerous ways to optimize the mode vector $\mathbf{M}$ and the power allocation vector $\mathbf{P}$, based on different objectives. Here we take the example of minimizing bit error rate (BER) to illustrate how to design the optimization criterion. In particular, define the active antenna set as $\mathcal{A} = \{i \,|\, R(M_i) > 0, \forall i\}$. Denote the BER of the i-th antenna after detection as BER_i. Zero-forcing (ZF) is assumed at the receiver and denote the nulling vector of the i-th substream as $\mathbf{w}_i^{\mathcal{A}} (i \in \mathcal{A})$. The total transmit power can then be expressed as

$$P_t = \sigma^2 \sum_{i \in \mathcal{A}} \xi(M_i, BER_i) \|\mathbf{w}_i^{\mathcal{A}}\|^2 R(M_i), \tag{7.10}$$

where $\xi(M_i, BER_i)$ represents the required E_b/N_0 in additive white Gaussian noise (AWGN) for the target BER_i, with the mode M_i. It can be approximated by $\xi(M_i, BER_i) \approx K(M_i) \cdot F(BER_i)$, where $K(M_i)$ is the coefficient in terms of mode and $F(BER_i)$ is a monotonously decreasing function of BER_i [42]. To optimize the BER performance, we should minimize the maximum BER_i because the overall BER performance is mainly dictated by the worst one. Therefore, the optimal mode vector $\tilde{\mathcal{M}}$ and antenna set $\tilde{\mathcal{A}}$ can be finally obtained as

$$\tilde{\mathcal{A}}, \tilde{\mathcal{M}} = \arg\min_{\mathcal{A}, \mathbf{M}} \sum_{i \in \mathcal{A}} \|\mathbf{w}_i^{\mathcal{A}}\|^2 K(M_i) R(M_i) \tag{7.11}$$

and the corresponding power allocation vector $\tilde{\mathbf{P}}$ satisfies

$$\tilde{P}_i = \begin{cases} P_t \dfrac{\|\mathbf{w}_i^{\mathcal{A}}\|^2 K(M_i) R(M_i)}{\sum_{k \in \mathcal{A}} \|\mathbf{w}_k^{\mathcal{A}}\|^2 K(M_k) R(M_k)} & i \in \mathcal{A} \\ 0 & i \notin \mathcal{A}. \end{cases} \tag{7.12}$$

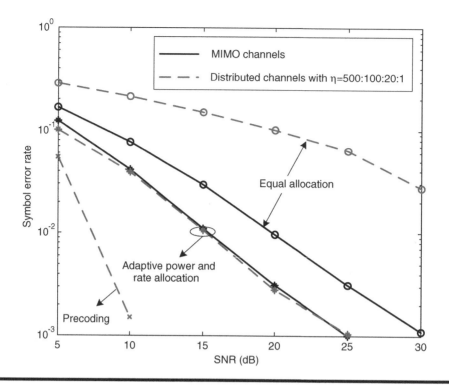

Figure 7.9   SER Comparison of Per-Antenna Adaptive Rate and Power Allocation, Equal Allocation and Precoding. $m = n = 4$.

The transmission rate and power of each antenna can be determined by (7.11 to 7.12). Because the number of available modes is usually small and the power quantization requires only limited bits, the amount of feedback information can be sharply reduced compared to the precoding case. Figure 7.9 shows the symbol error rate (SER) performance of this per-antenna adaptive rate and power allocation strategy. Here four antennas are assumed at both the transmitter and the receiver, i.e., $m = n = 4$. Three modes are considered: uncoded binary phase shift keying (BPSK), uncoded QPSK, and uncoded 16 quadratic amplitude modulation (QAM). The total spectral efficiency R_t is constrained to be 4 b/s/Hz. For comparison, the SER curves of the precoding scheme and the equal allocation are also provided.

From Figure 7.9 it can be seen that in MIMO channels, a 5 dB gain can be achieved by the adaptive power and rate allocation over the equal one at the SER of 10^{-3}. When the variance of subchannels increases significantly, i.e., in distributed channels with $\eta = 500{:}100{:}20{:}1$, the performance of the equal allocation deteriorates sharply while the adaptive one still works well. A closer observation shows that in this case the adaptive power and rate allocation scheme always chooses the best antenna using 16 QAM or the best two antennas using QPSK, because the degrees of freedom of the channel never exceed two. It can be also seen that the performance of the precoding scheme is significantly better than the adaptive one, which is partially attributed to the optimal transmission phase and partially to the optimal MLD receiver. Nevertheless, considering that the precoding scheme requires a large amount of feedback and is sensitive to the feedback error, the adaptive power and rate allocation scheme is still highly attractive in a DAS.

7.2.4 Long-Term Channel Statistics at the Transmitter

So far we have discussed the resource allocation schemes based on full CSI, which require the adaptation results to be updated for each channel instance. In this section, we will further show that in distributed channels, resources can be adaptively allocated based only on the *long-term channel statistics*, i.e., $\mathbf{F}$ and $\mathbf{R_r}$ in (7.2), instead of the instantaneous CSI, with only negligible performance loss.

7.2.4.1 Antenna Selection

As explained in the system model, in the downlink of DASs, the central processor usually selects the "best" m antennas for user i's data transmission. A natural question is how to optimally choose those m antennas. There have been numerous papers on antenna selection algorithms (see Section 7.1.2.2 for an overview). However, most of them are based on the instantaneous CSI at the transmitter. In the following we will introduce an optimal antenna selection criterion for capacity maximization assuming that only the long-term channel statistics, i.e., $\mathbf{F}$ and $\mathbf{R_r}$ in (7.2), are available.

Define the selected transmit antenna subset and selected receive antenna subset as $\mathbf{\Lambda_t}$ and $\mathbf{\Lambda_r}$, respectively, which are both unordered sets with m and n selected antennas. Let $\mathbf{R_{\Lambda_r}}$ denote the cross-correlation matrix of those n selected antennas, and $\mathbf{F_{\Lambda_t}}$ denote the large-scale fading of the m selected antennas. These matrices can be obtained by eliminating the columns and rows of the nondesired antennas from $\mathbf{R_r}$ and $\mathbf{F}$, respectively. Assume that m and n are selected to ensure that $\mathbf{R_{\Lambda_r}}$ and $\mathbf{F_{\Lambda_t}}$ are both of full rank. Now let $\mathbf{\tilde{H}}$ represent the $n \times m$ channel gain matrix between m selected transmit and n selected receive antennas. Then,

$$\mathbf{\tilde{y}} = \mathbf{\tilde{H}\tilde{x}} + \mathbf{\tilde{z}} = \mathbf{R}_{\Lambda_r}^{1/2}\mathbf{\tilde{H}_w}\mathbf{F_{\Lambda_t}}\mathbf{\tilde{x}} + \mathbf{\tilde{z}}. \tag{7.13}$$

Assume equal power allocation among those selected transmit antennas. By applying eigenvalue decomposition to $\mathbf{R_{\Lambda_r}}$, we can obtain

$$C = \log_2\det\left[\mathbf{I_n} + \frac{\rho}{m}\mathbf{\tilde{H}\tilde{H}}^*\right] = \log_2\det\left[\mathbf{I_n} + \frac{\rho}{m}\mathbf{\hat{H}_w}\mathbf{Q_t}\mathbf{\hat{H}_w^*}\mathbf{Q_r}\right], \tag{7.14}$$

where $\mathbf{Q_r}$ is a diagonal matrix whose diagonal entries are the eigenvalues of $\mathbf{R_{\Lambda_r}}$, and $\mathbf{Q_t} = \mathbf{F_{\Lambda_t}}\mathbf{F_{\Lambda_t}^*}$. $\mathbf{\hat{H}_w} = \mathbf{U_r^*}\mathbf{\tilde{H}_w}$, where $\mathbf{U_r}$ is a unitary matrix whose columns are the eigenvectors of $\mathbf{R_{\Lambda_r}}$. Clearly, $\mathbf{\hat{H}_w^*}\mathbf{\hat{H}_w}$ has the same eigenvalues as $\mathbf{\tilde{H}_w^*}\mathbf{\tilde{H}_w}$.

When $m = n$, from (7.14) we have

$$C \approx m\log_2\left(\frac{\rho}{m}\right) + \log_2\det\left[\mathbf{\hat{H}_w}\mathbf{\hat{H}_w^*}\right] + \log_2\det[\mathbf{Q_t}] + \log_2\det[\mathbf{Q_r}] \tag{7.15}$$

at high values of ρ. From (7.15) to maximize the capacity, we should maximize the determinants of $\mathbf{Q_t}$ and $\mathbf{R_{\Lambda_r}}$. In other words, the optimal transmit (or receive) antenna set $\mathbf{\Lambda_t}$(or $\mathbf{\Lambda_r}$), in terms of capacity maximization, should be selected to maximize the determinant of the corresponding matrix $\mathbf{Q_t}$(or $\mathbf{R_{\Lambda_r}}$). When $n \neq m$, however, it is difficult to obtain a closed form of the exact capacity expression. In [34], lower and upper bounds were developed which converge to the same limit. Both bounds can be maximized according to the following antenna selection criterion.

Antenna Selection Criterion: *The optimal selected transmit antenna subset Λ_r^* and receive antenna subset Λ_t^* that maximize the capacity are given by*

$$\Lambda_r^* = \underset{\Lambda_r}{\text{argmax}}\ \det(\mathbf{R}_{\Lambda_r}), \quad and \quad \Lambda_t^* = \underset{\Lambda_t}{\text{argmax}}\ \det(\mathbf{F}_{\Lambda_t}\mathbf{F}_{\Lambda_t}^*). \tag{7.16}$$

The above criterion is based on the assumption that both $\mathbf{F}_{\Lambda_t}\mathbf{F}_{\Lambda_t}^*$ and $\mathbf{R}_{\Lambda_r}$ are of full rank. For the cases when $\mathbf{F}_{\Lambda_t}\mathbf{F}_{\Lambda_t}^*$ and $\mathbf{R}_{\Lambda_r}$ are singular, the criterion is also applicable if we substitute $\det(\mathbf{R}_{\Lambda_r})$ and $\det(\mathbf{F}_{\Lambda_t}\mathbf{F}_{\Lambda_t}^*)$ in (7.16) by $\prod_{i=1}^{rank(\mathbf{R}_{\Lambda_r})} q_r^{(i)}$ and $\prod_{i=1}^{rank(\mathbf{F}_{\Lambda_t}\mathbf{F}_{\Lambda_t}^*)} q_t^{(i)}$, respectively.

Then we can describe a selection process according to the above criterion, namely, long-term selection algorithm (LtSA). This algorithm consists of creating all possible antenna sets Λ_t (or Λ_r) with m (or n) out of $\mathcal{M}$ transmit (or $\mathcal{N}$ receive) antennas. The corresponding $\det(\mathbf{F}_{\Lambda_t}\mathbf{F}_{\Lambda_t}^*)$ (or $\det(\mathbf{R}_{\Lambda_r})$) are computed and the one with the best measure, as described in the criterion, is selected.

The capacity cumulative density function (cdf) curves of the LtSA in distributed channels are provided in Figure 7.10 and Figure 7.11 for $\mathcal{M} = n = 6$ with an SNR of 10 dB. Here, we only consider antenna selection at the transmitter side with m ranging from 2 to 6. For comparison, the capacity cdf results of the instantaneous selected algorithm (ISA), which is based on the exact CSI, are also presented. From Figure 7.10, the LtSA incurs only a negligible capacity loss compared to ISA. When $m = 4$ or 5, the gap between the capacity of the ISA and LtSA is so slight that the two curves overlap. With a smaller m, say, $m = 2$, a 10% outage capacity of the LtSA is only 0.5 b/s/Hz less than that of the ISA. Figure 7.11 presents the capacity comparison in a more asymmetric distributed channel, i.e., $\eta = 1000{:}500{:}200{:}100{:}50{:}1$. In this case, the LtSA can always achieve almost

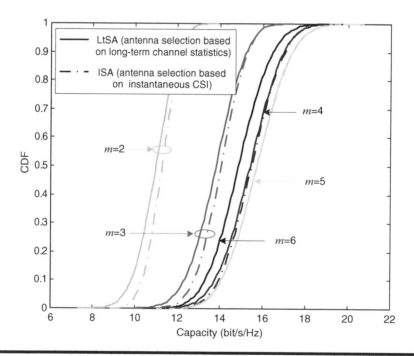

Figure 7.10 Capacity cdf Curves of LtSA and ISA in Distributed Channels with $\eta = 50{:}40{:}30{:}20{:}10{:}1$. $\mathcal{M} = n = 6$. **SNR = 10 dB.**

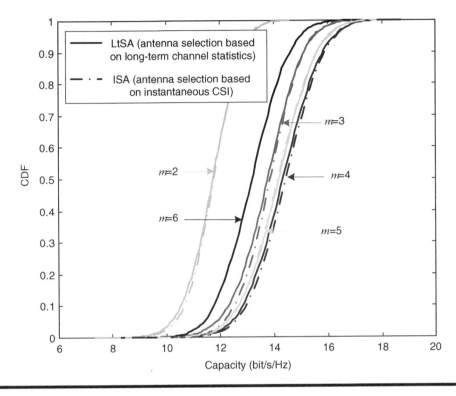

Figure 7.11 Capacity cdf Curves of LtSA and ISA in Distributed Channels with $\eta = 1000{:}500{:}200{:}$ **$100{:}50{:}1$. $\mathcal{M} = n = 6$. SNR $= 10$ dB.**

the same capacity as the ISA. As a result, we conclude that in distributed channels, antenna selection can be performed based on the long-term channel statistics instead of the instantaneous CSI with a very slight capacity loss, but significant complexity reduction.

From Figure 7.10 and Figure 7.11 it can be also seen that in distributed channels, transmit antenna selection can bring significant capacity gains. In Figure 7.10, the highest capacity is achieved when $m = 5$ instead of 6. In distributed channels, there are usually insufficient degrees of freedom of the channel. By allocating the transmit power to only the "good" subchannels, the optimal transmit antenna selection actually performs like a water-filling strategy. In Figure 7.11, the number of degrees of freedom of the channel further decreases to around 4. Therefore, choosing the best 4 transmit antennas can achieve the highest capacity.

7.2.4.2 Adaptive Power Adaptation

In Section 7.2.4.1, we have shown how to choose m antennas for downlink transmission based on the long-term channel statistics. In this section, we will further present an adaptive power allocation scheme which also requires only the information of $\mathbf{F}$ instead of the full CSI.

In particular, recall that for a channel model given by (7.2), the mutual information can be written as

$$I(\mathbf{Q}) = \log_2 \det\left(\mathbf{I_n} + \frac{1}{\sigma^2}\mathbf{H_w}\mathbf{F}\mathbf{Q}\mathbf{F}^*\mathbf{H_w^*}\right) \qquad (7.17)$$

by assuming that the receive antennas at the user are uncorrelated, i.e., $\mathbf{R_r} = \mathbf{I_n}$, and $\mathbf{Q} = E(\mathbf{xx^*})$ is the transmit covariance matrix. Let $\boldsymbol{\Omega} = (1/\sigma^2)\mathbf{QF^*H_w^*H_wF}$. We have

$$E\{I(\mathbf{Q})\} \leq E\left\{\log_2 \prod_{i=1}^m (1 + \Omega_{ii})\right\} \leq \prod_{i=1}^m \log_2 E\{1 + \Omega_{ii}\} = \prod_{i=1}^m \log_2\left(1 + \frac{nP_i}{\sigma^2}\|f_i\|^2\right). \quad (7.18)$$

The suboptimal power allocation scheme that maximizes the upper bound in (7.18) can be solved using the water-filling principle [87], i.e.,

$$P_i = \left(\mu - \frac{\sigma^2}{n\|f_i\|^2}\right)^+, \quad (7.19)$$

$i = 1, \ldots, m$, where μ is chosen to satisfy $\sum_{i=1}^m P_i = P_t$.

Clearly the power allocation given by (7.19) requires only the long-term channel statistics, i.e., $\|f_i\|^2$, $i = 1, \ldots, m$. Figure 7.12 provides the capacity cdf results of the adaptive power allocation in distributed channels with $\eta = 500{:}100{:}20{:}1$ and SNR = 10dB. Despite a slight capacity loss, say, 0.3 b/s/Hz at 10% outage, compared to the optimal water filling strategy, significant capacity gains can be achieved over the equal power allocation. This performance degradation becomes negligible in a more asymmetric channel. As shown in Figure 7.13, the adaptive power allocation achieves almost the same capacity as the optimal waterfilling strategy in distributed channels with $\eta = 1000{:}100{:}10{:}1$.

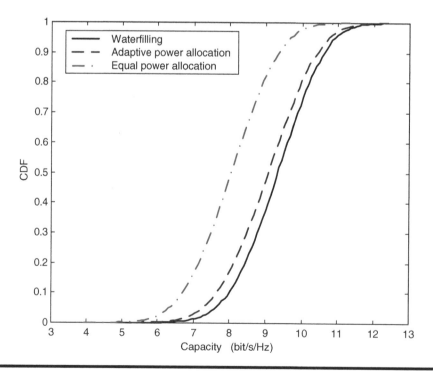

Figure 7.12 Capacity cdf Curves of Waterfilling, Adaptive Power Allocation, and Equal Power Allocation in Distributed Channels with $\eta = 500{:}100{:}20{:}1$ and SNR $= 10$ dB. $m = n = 4$.

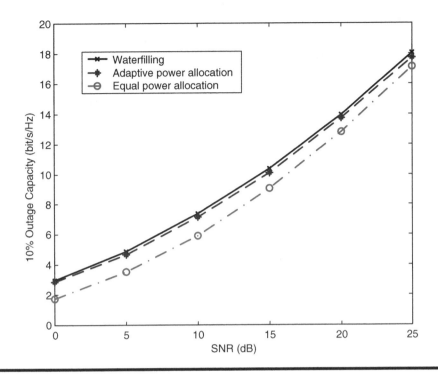

Figure 7.13   10% Outage Capacity Comparison of Waterfilling, Adaptive Power Allocation, and Equal Power Allocation in Distributed Channels with $\eta = 1000{:}100{:}10{:}1$ and SNR = 10 dB. $m = n = 4$.

7.2.5   Summary

In this section, we discussed the resource allocation strategies in distributed channels. The information-theoretic results demonstrated that resource allocation is indispensable to DASs due to the large differences among subchannels. Transmit precoding and the per-antenna rate and power allocation schemes were introduced, which adaptively allocate resources among the remote antennas according to the full CSI. Superior performance has been shown in distributed channels, where equal allocation suffers from severe performance degradation. We further showed that in a DAS, resource allocation can be performed based on only the long-term channel statistics with a negligible capacity loss compared to the ones with full CSI.

Performance evaluation of the above schemes in more practical scenarios, i.e., with the effect of Doppler spread and frequency selectivity, still needs further investigation. Additionally, a two-dimensional resource allocation would be interesting if OFDM is adopted. The cross-layer joint optimization with some link layer techniques such as ARQ is also an attractive issue.

7.3   MULTI-USER RESOURCE ALLOCATION IN DASs

In Section 7.2, we introduced the adaptive resource allocation strategies in the context of a single-user scenario. In this section, we will turn to a network of multiple users and study the optimal multi-user resource allocation strategy for DASs.

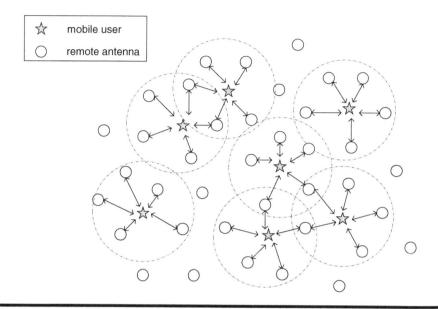

Figure 7.14   A Multi-User DAS Network. The Circle Represents the Virtual Cell for Each User.

In a multi-user DAS network, each user has its own virtual cell and different remote antennas may transmit to (or receive from) different user sets. As shown in Figure 7.14, instead of a one point to multipoint (in downlink) or multipoint to one point (in uplink) channel, here both downlink and uplink resource allocation should be performed based on a multipoint-to-multipoint channel, which is much more complex than the traditional resource allocation in cellular systems. It can be also seen that in a DAS, each remote antenna connects to only a small set of users (instead of the whole user set in cellular systems). Likewise, each user only receives signals from its own virtual cell. This can actually bring significant performance gains, as we will show later. However, it also requires a more complicated resource allocation strategy.

In particular, in a multi-user DAS network, the first issue to be addressed is interference management. The signals transmitted to and from different users need to be distinguished using code division, frequency division, time division, or space division so as to avoid strong interference. Then based on some specific multiplexing/multiple-access scheme, the system resources, such as power, antennas, codes, etc., can be adaptively allocated among multiple users according to their channel states and QoS requirements. In Section 7.3.1, we will focus on a CDMA-based DAS system, i.e., transmit signals to and from users are assigned with different spreading codes. The optimal power allocation strategy will be introduced for the downlink transmission. In Section 7.3.2 opportunistic transmission will be applied to a DAS and the fairness issue will be also addressed.

7.3.1   CDMA-Based Resource Allocation

Let us consider a downlink CDMA-based DAS where the transmit signal to each user is a pseudo-noise (PN) code modulated bit stream with a spreading factor (or processing gain) of ϕ. Assume that each user has only one antenna and maximum ratio combining (MRC) is adopted. The power of the pilot channel, P, is equal to the total allocated power of each user. As explained previously, in DASs each signal from remote antennas to a user propagates through a distinct path and arrives at the user with independent fading. Therefore, some form of resource allocation is required. In the following, we focus on the optimal power allocation.

Assume that the power allocated to user k from antenna $l_{k,i}$ is $\varpi_{k,l_{k,i}}^2 \cdot P$, where $l_{k,i}$ and $\varpi_{k,l_{k,i}}$ represent the i-th antenna in user k's virtual cell and its corresponding weight, respectively. Clearly, we have $\forall k, \sum_{i=0}^{m-1} \varpi_{k,l_{k,i}}^2 = 1$. Then, the received signal of user 0 is given by

$$
x(t) = \sum_{i=0}^{M-1} \sum_{k=0}^{K_i-1} \psi_k \sqrt{P} \varpi_{k,i} \gamma_{0,i} b_k \left(\lfloor \frac{t - \tau_{0,i}}{T} \rfloor \right) c_k(t - \tau_{0,i})
$$

$$
+ \sum_{i=0}^{M-1} \sqrt{P} \gamma_{0,i} b_i' \left(\lfloor \frac{t - \tau_{0,i}}{T} \rfloor \right) c_i'(t - \tau_{0,i}) + n(t), \tag{7.20}
$$

where the first and second items represent the data and pilot signals received by user 0, respectively. In particular, ψ_k is the voice activity variable with an activity factor of v. $b_k(\cdot)$ denotes the transmitted bit of user k in duration T and $c_k(\cdot)$ is the spreading code used by user k. $b_i'(\cdot)$ and $c_i'(\cdot)$ represent the bit and the spreading code used by the pilot of antenna i. K_i is the number of users that communicate with antenna i. $\gamma_{0,i}$ represents the channel gain between user 0 and antenna i, which includes the effect of both the large-scale fading and the small-scale fading. $\tau_{0,i}$ is the propagation delay from antenna i to user 0, $i = 0, \ldots, M-1$. By regarding the signals from different antennas in user 0's virtual cell to user 0 as multiple paths of the desired signal, we can separate the paths with a RAKE receiver. The E_b/I_0 at the receiver can then be derived as [83]

$$
\frac{E_b}{I_0} = \sum_{j=0}^{m-1} \left(\frac{E_b}{I_0} \right)_j = \frac{\phi \cdot \sum_{j=0}^{m-1} \varpi_{0,j}^2 \|\gamma_{0,j}\|^2}{(vK/M + 1) \sum_{i=0}^{M-1} \|\gamma_{0,i}\|^2}. \tag{7.21}
$$

It can be easily proved that the optimal weight vector to maximize (7.21) is given by

$$
\varpi_{0,i} = \begin{cases} 1, & i = \text{argmax} \|\gamma_{0,j}\|^2 \\ 0, & \text{otherwise} \end{cases} \tag{7.22}
$$

$i = 0, \ldots, m-1$. Obviously this is the well-known selective transmission scheme, i.e., the transmit power is allocated to the antenna with the best channel. Figure 7.15 shows the curves of outage probability versus the number of users per antenna. Here we consider a three-tier hexagonal model, i.e., $M = 37$. Both the effect of path loss and shadow fading are included with the path loss exponent $\alpha = 4$ and the standard variance of the log-normal shadowing variable $\sigma_s = 8$ dB. Additionally, the voice activity factor is $v = 0.375$ and the spreading factor $\phi = 127$. Assume adequate performance (i.e., BER $\leq 10^{-3}$) is achieved with $E_b/I_0 = 7$ dB. Figure 7.15 shows that the downlink capacity[5] decreases rapidly as m increases. This is because the received signal power at the user is

---

[5] Here the "downlink capacity" is defined as the number of users that can be supported by the system at a certain outage probability. For example, from Figure 7.15 it can be seen that when $m = 1$, at an outage probability of 10^{-3}, 18 users can be supported. This number drops to 10 when m increases to 4.

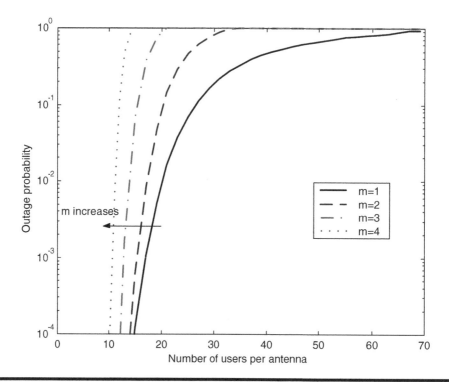

Figure 7.15 Outage Probability Vs. the Number of Users Per Antenna in a CDMA-Based DAS. Selective Transmission ($m = 1$) Performs the Best.

the sum of the power received from each involved antenna. By assuming that the total power allocated to each user is a constant, which implies that the total interference is fixed, it is clear that distributing the transmit power among several antennas will cause a decrease of the received SIR. Therefore, selective power allocation performs the best.

The above conclusion is drawn based on the assumption that only the optimal transmit power allocation is performed. If the phases of the transmit signals can be also jointly adjusted, the downlink capacity can be dramatically improved with an increase of m.

In particular, assume that the desired signals from the antennas in user 0's virtual cell are jointly adjusted so that they arrive at user 0 in phase and simultaneously. The received E_b/I_0 of user 0 can then be derived as

$$\frac{E_b}{I_0} \approx \frac{\phi\left(\sum_{j=0}^{m-1} \varpi_{0,j}\|\gamma_{0,j}\|\right)^2}{(\nu\mathcal{K}/\mathcal{M}+1)\sum_{i=0}^{\mathcal{M}-1}\|\gamma_{0,i}\|^2}. \tag{7.23}$$

It can be proved that when $\varpi_{0,i} = \dfrac{\|\gamma_{0,i}\|}{\sqrt{\sum_{j=0}^{m-1}\|\gamma_{0,j}\|^2}}$, $i=0, \ldots, m-1$, the received E_b/I_0 is

maximized and given by

$$E_b/I_0 = \frac{\phi}{(\nu\mathcal{K}/\mathcal{M}+1)} \cdot \frac{\sum_{i=0}^{m-1}\|\gamma_{0,i}\|^2}{\sum_{i=0}^{\mathcal{M}-1}\|\gamma_{0,i}\|^2}. \tag{7.24}$$

Equation (7.24) shows that E_b/I_0 will increase as m increases. Here the power weight of each antenna is proportional to the channel gain. Therefore, it is also called

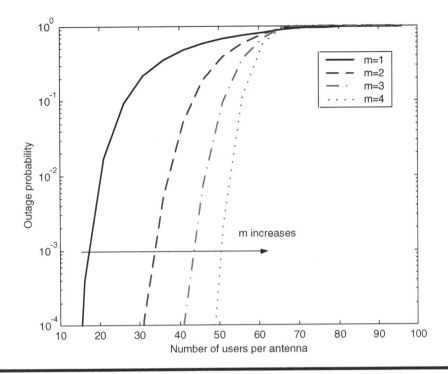

Figure 7.16 **Outage Probability Vs. the Number of Users Per Antenna in a CDMA-Based DAS with Maximum Ratio Transmission.**

maximum ratio transmission. As shown in Figure 7.16, a substantial capacity gain can be achieved with the increase of *m.* Nevertheless, the maximum ratio transmission requires the transmit power and phases to be jointly adjusted, according to the instantaneous CSI of users, which will incur a huge amount of feedback information and is quite sensitive to the feedback errors. This greatly restricts its application in fast fading channels.

So far we have studied the optimal power allocation strategy in a CDMA-based DAS network. It is shown that if the transmit phases are not jointly adjusted, selective transmission, i.e., to put all the transmit power on the best remote antenna, is optimal. Otherwise, maximum ratio transmission achieves the highest capacity where the transmit power of each antenna is proportional to its channel gain. Actually, the selective transmission strategy follows the water-filling principle. With one antenna at the mobile user, only one degree of freedom of the channel is provided, no matter how many remote antennas are included. Therefore, the water-filling strategy in this case suggests that the transmission power should always be allocated to the antenna with the best channel. On the other hand, the maximum ratio transmission is reminiscent of beamforming, although there are no real beams toward users.

7.3.2 Opportunistic Transmission

In Section 7.3.1, we assume that each user is assigned equal transmit power P and a spreading code with the same spreading factor ϕ. In this way the system resources are equally allocated to users and the optimal power allocation is performed among multiple

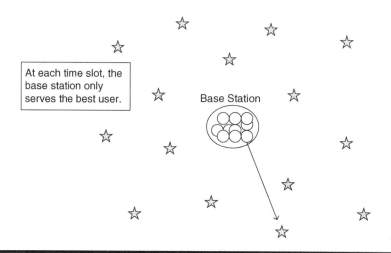

Figure 7.17   Opportunistic Transmission in Cellular Systems.

antennas of each user's virtual cell. In this subsection, we will further address how to efficiently and fairly allocate resources among multiple users.

Opportunistic transmission has been proposed in [71], where in each time slot the system resources are allocated only to the user with the highest instantaneous channel gain. As illustrated in Figure 7.17, the base station tracks the channel variations of all users and schedules transmissions to the best one. Because users are expected to experience independent fading, opportunistic transmission can adaptively exploit the time-varying channel conditions of users and achieve the *multi-user diversity* gain; the network throughput will increase with the number of users [72].

Despite the substantial throughput gain brought by multi-user diversity, opportunistic transmission may not work well when multiple antennas are employed at base stations [72,88]. As we know, multi-user diversity gain has its root in the independent fluctuation of channels of different users, which to some extent exploits the channel fading. However, the conventional multi-antenna transmission techniques aimed at maximizing the diversity gain, i.e., space–time coding, beamforming, etc., are designed to counteract the adverse effect of fading. Therefore, by decreasing the channel fluctuations of different users, opportunistic transmission with multiple antennas may lead to an even lower throughput than the one in the single-antenna scenario.[6]

In the above work, only the channel fluctuations introduced by small-scale fading are taken into account. In cellular systems, power control is usually adopted to counteract the large-scale fading, such as path loss and log-normal shadowing. Otherwise, the users close to the base station will always occupy the system resources and severely impair the performance of the users far away from the base station.[7] In a DAS, however, the large-scale fading can be exploited to *amplify* the fluctuations.

---

[6] To address this issue, [75] proposed to induce large channel fluctuations by using multiple antennas, which is called *opportunistic beamforming using dumb antennas*. In this case, the phases and power allocated to transmit antennas randomly vary and at any time the transmission is scheduled to the user which is currently closest to the beam. In this way the rate of channel fluctuations is artificially increased.

[7] Viswanath et al. [75] proposed a proportional fair opportunistic scheduler to avoid such cases, where data is transmitted to a user when it hits its own "peak." We will discuss it in detail later.

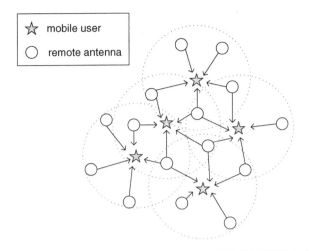

Figure 7.18   Each User Receives Independent Fading Signals from the *m* Remote Antennas of Its Virtual Cell.

We take the example of downlink transmissions. As shown in Figure 7.18, each user receives independent fading signals from the *m* remote antennas of its virtual cell. Assume only one antenna is employed at the mobile user. From (7.2) we know that for user k, the m-dimension channel vector is given by $\mathbf{h}_k = [f_{k,1}\gamma_{k,1}, \ldots, f_{k,m}\gamma_{k,m}]$, where $f_{k,i}$ and $\gamma_{k,i}$ represent the large-scale fading and small-scale Rayleigh fading of the channel from user k to the i-th remote antenna, respectively. In cellular systems, antennas are colocated at the base station and the large-scale fading has been counteracted. Therefore, we have $\|f_{k,i}\| = 1$, for any $k = 1, \ldots, \mathcal{K}$, and $i = 1, \ldots, m$.

Let $\varphi_k = \mathbf{h}_k \mathbf{h}_k^*$. The variance of the sum channel gain is then given by

$$\mathrm{var}(\varphi_k) = \mathrm{var}(\|\gamma_{k,i}\|^2) \sum_{i=1}^{m} \|f_{k,i}\|^4. \tag{7.25}$$

Equation (7.25) shows that when $\|f_{k,1}\| = \cdots = \|f_{k,m}\| = 1$, φ_k has the minimum variance. By introducing different levels of large-scale fading among the different paths, the channel fluctuation will be boosted in distributed channels. Figure 7.19 presents the fluctuations of the sum channel gain φ_k in distributed channels with $\eta = 500:100:20:1$ and multiple-input single-output (MISO) channels (multiple antennas at the base station and one antenna at the mobile user) with $m = 4$. Obviously a much larger channel fluctuation is observed in distributed channels.

It should be noticed that the channel fluctuation is *amplified* instead of *sped up* in distributed channels, because the large-scale fading does not determine the time-varying rate of the channel. Therefore, in DASs, it is still possible that some users with good channels always occupy the system resources while others have no chances to transmit at all (for example, in a slow fading environment). To meet the fairness constraints, a proportional fair opportunistic scheduler has been proposed in [75] where the user with the largest fraction of current channel data rate to its average throughput is scheduled in each time slot, and the average throughput is updated using the following low-pass

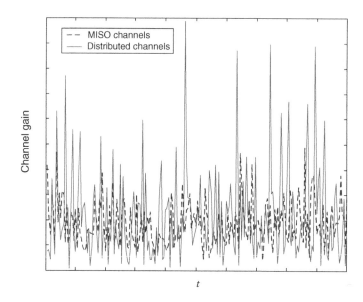

Figure 7.19   Channel Fluctuations in MISO Channels and Distributed Channels. $m = 4$ and $\eta = 500{:}100{:}20{:}1$.

filter:

$$
T_k[i + 1] = \begin{cases} \left(1 - \frac{1}{t_c}\right) T_k[i] + \frac{1}{t_c} R_k[i] & k = k^* \\ \left(1 - \frac{1}{t_c}\right) T_k[i] & k \neq k^* \end{cases}, \tag{7.26}
$$

where $R_k[i]$ is the current channel data rate of user k in time slot i. Clearly, if the scheduling time scale t_c is much larger than the correlation time scale of the channel, each user's throughput converges to the same quantity. Therefore, this scheduling algorithm can guarantee fairness in the long term.

In a DAS, the fairness performance can be further improved by scheduling multiple users simultaneously. An important characteristic of a DAS is that each user connects to only a subset of the remote antennas, i.e., the ones in its virtual cell, instead of all the antennas in the system. Therefore, the whole network usually can be decomposed into several disjoint subnetworks. As shown in Figure 7.20, assume the active user set includes users 1, 2, 4, 5, 7, 8, and 10. Obviously they can be divided into 3 subsets: {1, 2, 10}, {4, 5} and {7, 8}. The users in the same subset share part of the antennas while there are no common antennas shared by different user subsets.[8] In this case, different user subsets can be scheduled at the same time, thanks to a natural frequency reuse pattern. In cellular systems, multiple users can also be scheduled simultaneously [90]; however, either multiple spreading codes or subcarriers are required to differentiate those users, which leads to a lower spectral efficiency.

---

[8] Given an arbitrary active user set, the network can be decomposed into x disjoint subnetworks, $1 \leq x \leq \check{k}$, where $\check{k}$ is the number of active users. Obviously x depends on the network topology. A *network decomposition* methodology has been proposed in [89]. A similar idea can be applied to the distributed antenna case.

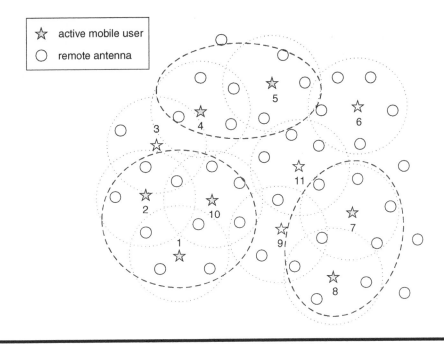

Figure 7.20 In DAS, the Active User Set Usually Can Be Decomposed into Several Disjoint Subsets. Multiple Users Can Be Scheduled at the Same Time.

7.3.3 Summary

In this section, we considered the resource allocation in a multi-user DAS network. We first studied the optimal power allocation strategy for each user in a CDMA-based DAS, and then focused on efficient and fair resource allocation among multiple users. We took the example of opportunistic transmission and showed that a DAS has great potential to fully exploit the multi-user diversity gain, and at the same time to achieve a good balance between efficiency and fairness.

In a DAS, each remote antenna connects to only a small set of users and each user only receives signals from its own virtual cell. This can bring huge performance gains, i.e., better frequency reuse, larger channel fluctuation, better interference management, and less transmit power; however, it also raises a great challenge: how to perform the resource allocation with a reasonable level of complexity, for example, in a distributed way. In addition, to perform a fair and efficient resource allocation among users, not only the channel state information but the users' QoS requirements need to be taken into consideration. A comprehensive cross-layer model for DASs would be helpful for jointly optimizing the resource allocation.

7.4 CONCLUSION

This chapter studied the optimal resource allocation strategies for DAS networks. In contrast to MIMO systems, where resource allocation is usually conducted as performance enhancement, in DASs, resources must be allocated adaptively to the channel states due to the large differences among subchannels. Equal allocation will lead to severe performance degradation. Fortunately, only long-term channel statistics are required to perform the adaptive resource allocation. The resource allocation strategies in a multi-user scenario were also checked, and a DAS is able to fully exploit the multi-user diversity gain and achieve a good tradeoff between efficiency and fairness.

There are still quite a lot of open issues in this field. For example, to perform the proposed resource allocation strategies in practical scenarios, the effect of Doppler spread and frequency selectivity of the channels as well as synchronization and feedback errors needs to be taken into consideration. Furthermore, distributed algorithms have to be developed to realize the multi-user resource allocation in a large scale network, while at the same time central control is also required to balance the efficiency and fairness. Finally, the cross-layer optimization with link layer techniques such as ARQ or application layer requirements would be highly desirable.

REFERENCES

[1] A.J. Goldsmith and S.G. Chua, "Variable-rate variable-power MQAM for fading channels," *IEEE Trans. Commn.*, vol. 45, pp. 1218–1230, Oct. 1997.

[2] S.J. Oh, D. Zhang, and K.M. Wasserman, "Optimal resource allocation in multiservice CDMA networks," *IEEE Trans. Wireless Commn.*, vol. 2, pp. 811–821, July 2003.

[3] J.K. Cavers, "Variable-rate transmission for Rayleigh fading channels," *IEEE Trans. Commn.*, vol. 20, pp. 15–22, Feb. 1972.

[4] B. Vucetic, "An adaptive coding scheme for time-varying channels," *IEEE Trans. Commn.*, vol. 39, pp. 653–663, May 1991.

[5] W.T. Webb and R. Steele, "Variable rate QAM for mobile radio," *IEEE Trans. Commn.*, vol. 43, pp. 2223–2230, July 1995.

[6] A.J. Goldsmith and P.P. Varaiya, "Capacity of fading channels with channel side information," *IEEE Trans. Inform. Theory*, vol. 43, pp. 1986–1992, Nov. 1997.

[7] D.N. Tse and S.V. Hanly, "Multiple-access fading channels – part I: polymatroidal structure, optimal resource allocation and throughput capacities," *IEEE Trans. Inform. Theory*, vol. 44, pp. 2796–2815, Nov. 1998.

[8] S.V. Hanly and D.N. Tse, "Multiple-access fading channels – part II: delay-limited capacities," *IEEE Trans. Inform. Theory*, vol. 44, pp. 2816–2831, Nov. 1998.

[9] C. Kose and D.L. Goeckel, "On power adaptation in adaptive signaling systems," *IEEE Trans. Commn.*, vol. 48, pp. 1769–1773, Nov. 2000.

[10] A.J. Goldsmith and S. Chua, "Adaptive coded modulation for fading channels," *IEEE Trans. Commn.*, vol. 46, pp. 595–602, May 1998.

[11] T. Ue, S. Sampei, N. Morinaga, and K. Hamaguchi, "Symbol rate and modulation level-controlled adaptive modulation/TDMA/TDD system for high-bit-rate wireless data transmission," *IEEE Trans. Veh. Technol.*, vol. 47, pp. 1134–1147, Nov. 1998.

[12] M.S. Alouini and A.J. Goldsmith, "Adaptive modulation over Nakagami fading channels," *Kluwer J. Wireless Commn.*, vol. 13, pp. 119–143, May 2000.

[13] K.J. Hole, H. Holm, and G.E. Oien, "Adaptive multidimensional coded modulation over flat fading channels," *IEEE J. Select. Areas Commn.*, vol. 18, pp. 1153–1158, July 2000.

[14] M.B. Pursley and J.M. Shea, "Adaptive nonuniform phase-shift-key modulation for multimedia traffic in wireless networks," *IEEE J. Select. Areas Commn.*, vol. 18, pp. 1394–1407, Aug. 2000.

[15] S.T. Chung and A. Goldsmith, "Degrees of freedom in adaptive modulation: a unified view," *IEEE Trans. Commn.*, vol. 49, pp. 1561–1571, Sept. 2001.

[16] *Physical Layer Aspects of UTRA High Speed Downlink Packet Access (Release 4)*, 3GPP TR 25.848 V4.0.0, 2001.

[17] *Physical Layer Standard for CDMA 2000 Spread Spectrum Systems*, 3GPP2 C.S0002-0 Ver. 1.0, 1999.

[18] *IEEE Standard 802.16 Working Group, IEEE Standard for Local and Metropolitan Area Networks Part 16: Air Interface for Fixed Broadband Wireless Access Systems,* 2002.

[19] Q. Liu, S. Zhou, and G.B. Giannakis, "Cross-layer combining of adaptive modulation and coding with truncated ARQ over wireless links," *IEEE Trans. Wireless Commn.,* vol. 3, no. 5, pp. 1746–1755, Sept. 2004.

[20] Q. Liu, S. Zhou, and G.B. Giannakis, "Queuing with adaptive modulation and coding over wireless links: cross-layer analysis and design," *IEEE Trans. Wireless Commn.,* vol. 4, pp. 1142–1153, May 2005.

[21] L. Dai and K.B. Letaief, "Throughput maximization of ad hoc wireless networks using adaptive cooperative diversity and truncated ARQ," *IEEE Trans. Commn.,* in revision. (http://www.ee.ust.hk/~eedailin/publications.htm)

[22] E. Telatar, "Capacity of multi-antenna Gaussian channels," *AT&T Bell Labs Internal Tech. Memo,* June 1995.

[23] G.J. Foschini and M.J. Gans, "On limits of wireless communications in a fading environment when using multiple antennas," *Wireless Pers. Commn.,* vol. 6, pp. 311–335, Mar. 1998.

[24] D.J. Love, R.W. Heath Jr., and T. Strohmer, "Grassmannian beamforming for multiple-input multiple-output wireless systems," *IEEE Trans. Inf. Theory,* vol. 49, pp. 2735–2747, Oct. 2003.

[25] K.K. Mukkavilli, A. Sabharwal, E. Erkip, and B. Aazhing, "On beamforming with finite rate feedback in multiple antenna systems," *IEEE Trans. Inf. Theory,* vol. 49, pp. 2562–2579, Oct. 2003.

[26] S. Zhou, W. Wang, and G.B. Giannakis, "Quantifying the power loss when transmit beamforming relies on finite-rate feedback," *IEEE Trans. Wireless Commn.,* vol. 4, pp. 1948–1957, July 2005.

[27] D.J. Love and R.W. Heath Jr., "Limited feedback unitary precoding for spatial multiplexing systems," *IEEE Trans. Inf. Theory,* vol. 51, pp. 2967–2976, Aug. 2005.

[28] S.T. Chung, A. Lozano, and H.C. Huang, "Approaching eigenmode BLAST channel capacity using V-BLAST with rate and power feedback," in *Proc. IEEE VTC'01-Fall,* vol. 2, pp. 915–919, Oct. 2001.

[29] S.T. Chung, A. Lozano, and H.C. Huang, "Low complexity algorithm for rate quantization in extended V-BLAST," in *Proc. IEEE VTC'01-Fall,* Atlantic City, NJ, vol. 2, pp. 910–914, Oct. 2001.

[30] S. Catreux, P.F. Driessen, and L. J. Greestein, "Data throughputs using multiple-input multiple-output (MIMO) techniques in a noise-limited cellular environment," *IEEE Trans. Wireless Commn.,* vol. 1, pp. 226–235, Apr. 2002.

[31] H. Zhuang, L. Dai, S. Zhou, and Y. Yao, "Low complexity per-antenna rate and power control approach for closed-loop V-BLAST," *IEEE Trans. Commn.,* vol. 51, pp. 1783–1787, Nov. 2003.

[32] Z. Zhou, B. Vucetic, M. Dohler, and Y. Li, "MIMO systems with adaptive modulation," *IEEE Trans. Veh. Technol.,* vol. 54, pp. 1828–1842, Sept. 2005.

[33] A.F. Molisch, M.Z. Win, and J.H. Winters, "Reduced-complexing multiple transmit/receive antenna systems," *IEEE Trans. on Signal Processing,* vol. 51, no. 11, pp. 2729–2738, Nov. 2003.

[34] R.W. Heath Jr. and A. Paulraj, "Antenna selection for spatial multiplexing systems based on minimum error rate," in *Proc. ICC'01,* Helsinki: Finland, vol. 7, pp. 2276–2280, June 2001.

[35] X.N. Zeng and A. Ghrayeb, "Performance bounds for space-time block codes with receive antenna selection," *IEEE Trans. Inf. Theory,* vol. 50, no. 9, pp. 2130–2137, Sept. 2004.

[36] A. Ghrayeb and T.M. Duman, "Performance analysis of MIMO systems with antenna selection over quasi-static fading channels," *IEEE Trans. Veh. Technol.,* vol. 52, no. 2, pp. 281–288, Mar. 2003.

[37] I. Bahceci, T.M. Duman, and Y. Altunbasak, "Antenna selection for multiple-antenna transmission systems: performance analysis and code construction," *IEEE Trans. Inf. Theory*, vol. 49, no. 10, pp. 2669–2681, Oct. 2003.

[38] I. Berenguer and X. Wang, "MIMO antenna selection with lattice-reduction-aided linear receivers," *IEEE Trans. Veh. Technol.*, vol. 53, no. 5, pp. 1289–1302, Sept. 2004.

[39] D. Gore, R. Heath, and A. Paulraj, "Statistical antenna selection for spatial multiplexing systems," in *Proc. ICC'02*, New York, pp. 450–454, May 2002.

[40] D.A. Gore and A.J. Paulraj, "MIMO antenna subset selection with space-time coding," *IEEE Trans. Signal Processing*, vol. 50, no. 10, pp. 2580–2588, Oct. 2002.

[41] A. Molisch, M. Win, and J. Winters, "Capacity of MIMO systems with antenna selection," in *Proc. ICC'01*, Helsinki, Finland, pp. 570–574, 2001.

[42] A. Gorokhov, D.A. Gore, and A.J. Paulraj, "Receive antenna selection for MIMO flat fading channels: theory and algorithms," *IEEE Trans. Inf. Theory*, vol. 49, no. 10, pp. 2687–2696, Oct. 2003.

[43] L. Dai, S. Sfar, and K.B. Letaief, "Optimal antenna selection based on capacity maximization for MIMO systems in correlated channels," *IEEE Trans. Commn.*, vol. 54, no. 3, pp. 563–573, Mar. 2006.

[44] K.K Wong, R.D. Murch, and K.B. Letaief, "A joint-channel diagonalization for multiuser MIMO antenna systems," *IEEE Trans. Wireless Commn.*, vol. 2, pp. 773–786, July 2003.

[45] H. Dai, A.F. Molisch, and H.V. Poor, "Downlink capacity of interference-limited MIMO systems with joint detection," *IEEE Trans. Wireless Commn.*, vol. 3, pp. 442–453, Mar. 2004.

[46] C. Wang and R.D. Murch, "Adaptive cross-layer resource allocation for downlink multi-user MIMO wireless system," in *Proc. IEEE VTC'05-Spring*, Stockholm, Sweden, vol. 3, pp. 1628–1632, June 2005.

[47] L.J. Cimini, Jr., "Analysis and simulation of a digital mobile channel using orthogonal frequency division multiplexing," *IEEE Trans. Commn.*, vol. 33, pp. 665–675, July 1995.

[48] A. Czylwik, "Adaptive OFDM for wideband radio channels," in *Proc. IEEE Globecom'96*, London, pp. 713–718, Nov. 1996.

[49] T. Keller and L. Hanzo, "Adaptive modulation techniques for duplex OFDM transmission," *IEEE Trans. Veh. Technol.*, vol. 49, pp. 1893–1906, Sept. 2000.

[50] B.S. Krongold, K. Ramchandran, and D.L. Jones, "Computationally efficient optimal power allocation algorithm for multicarrier communication systems," in *Proc. IEEE ICC'98*, Atlanta, GA, pp. 1018–1022, May 1998.

[51] T.J. Willink and P.H. Wittke, "Optimization and performance evaluation of multicarrier transmission," *IEEE Trans. Inform. Theory*, vol. 43, pp. 426–440, Mar. 1997.

[52] C.Y. Wong, R.S. Cheng, K.B. Letaief, and R.D. Murch, "Multi-user OFDM with adaptive subcarrier, bit, and power allocation," *IEEE J. Sel. Areas Commn.*, vol. 17, pp. 1747–1758, Oct. 1999.

[53] W. Rhee and J.M. Cioffi, "Increase in capacity of multiuser OFDM system using dynamic suchannel allocation," in *Proc. IEEE VTC'00-Spring*, Tokyo, pp. 1085–1089, May 2000.

[54] J. Jang and K.B. Lee, "Transmit power adaptation for multiuser OFDM systems," *IEEE J. Sel. Areas Commn.*, vol. 21, pp. 171–178, Feb. 2003.

[55] Z. Shen, J.G. Andrews, and B.L. Evans, "Adaptive resource allocation in multiuser OFDM systems with proportional rate constraints," *IEEE Trans. Wireless Commn.*, vol. 4, pp. 2726–2737, Nov. 2005.

[56] Y.J. Zhang and K.B. Letaief, "An efficient resource-allocation scheme for spatial multiuser access in MIMO/OFDM systems," *IEEE Trans. Commn.*, vol. 53, pp. 107–116, Jan. 2005.

[57] G. Song and Y. Li, "Cross-layer optimization for OFDM wireless netoworks – part I and II," *IEEE Trans. Wireless Commn.*, vol. 4, pp. 614–634, Mar. 2005.

[58] S.V. Hanly, "An algorithm for combined cell-site selection and power control to maximize cellular spread spectrum capacity," *IEEE J. Select. Areas Commn.*, vol. 13, pp. 1332–1340, Sept. 1995.

[59] R. Yates and C.Y. Huang, "Integrated power control and base station assignment," *IEEE Trans. Veh. Technol.*, vol. 44, pp. 638–644, Aug. 1995.

[60] I.C. Lin and K.K. Sabnani, "Variable spreading gain CDMA with adaptive control for true packet switching wireless network," in *Proc. IEEE ICC'95*, Seattle, WA, pp. 1060–1064, May 1995.

[61] S.J. Oh and K.M. Wasserman, "Dynamic spreading gain control in multi-service CDMA networks," *IEEE J. Select. Areas Commn.*, vol. 17, pp. 918–927, May 1999.

[62] S. Ramakrishna and J.M. Holtzman, "A scheme for throughput maximization in a dual-class CDMA system," *IEEE J. Select. Areas Commn.*, vol. 16, pp. 830–844, Aug. 1998.

[63] F. Berggren, S.L. Kim, R. Jantti, and J. Zander, "Joint power control and intracell scheduling of DS-CDMA nonreal time data," *IEEE J. Sel. Areas Commn.*, vol. 19, pp. 1860–1870, Oct. 2001.

[64] S.J. Oh, T.L. Olsen, and K.M. Wasserman, "Distributed power control and spreading gain allocation in CDMA data networks," in *Proc. IEEE INFOCOM'00*, vol. 2, pp. 379–385, 2000.

[65] A.J. Goldsmith and S.B. Wicker, "Design challenges for energy-constrained ad hoc wireless networks," *IEEE Wireless Commn.*, pp. 8–27, Aug. 2002.

[66] J.W. Lee, R.R. Mazumdar, and N.B. Shroff, "Joint resource allocation and base-station assignment for the downlink in CDMA networks," *IEEE/ACM Trans. Networking*, vol. 14, pp. 1–14, Feb. 2006.

[67] C.U. Saraydar, N.B. Mandayam, and D.J. Goodman, "Efficient power control via pricing in wireless data networks," *IEEE Trans. Commn.*, vol. 50, pp. 291–303, Feb. 2002.

[68] C. Li, X. Wang, and D. Reynold, "Utility-based joint power and rate allocation for downlink CDMA with blind multiuser detection," *IEEE Trans. Wireless Commn.*, vol. 4, pp. 1163–1174, May 2005.

[69] L. Tassiulas and A. Ephremides, "Stability properties of constrained queuing systems and scheduling policies for maximum throughput in multi-hop radio networks," *IEEE Trans. Autom. Contro.*, vol. 37, pp. 1936–1948, Dec. 1992.

[70] L. Tassiulas and A. Ephremides, "Dynamic server allocation to parallel queues with randomly varying connectivity," *IEEE Trans. Inf. Thoery*, vol. 39, pp. 466–478, Mar. 1993.

[71] R. Knopp and P.A. Humblet, "Information capacity and power control in single-cell multiuser communications," in *Proc. IEEE ICC'95*, Seattle, WA, pp. 331–335, June 1995.

[72] D. Tse and P. Viswanath, *Fundamentals of Wireless Communication.* Cambridge University Press, May 2005.

[73] X. Liu, E.K.P. Chong, and N.B. Shroff, "Opportunistic transmission scheduling with resource-sharing constraints in wireless networks," *IEEE J. Select. Areas Commn.*, vol. 19, pp. 2053–2064, Oct. 2001.

[74] S. Borst and P. Whiting, "Dynamic rate control algorithm for HDR throughput optimization," in *Proc. IEEE Infocom'01*, Anchorage, AK, pp. 976–985, 2001.

[75] P. Viswanath, D.N.C. Tse, and R. Laroia, "Opportunistic beamforming using dumb antennas," *IEEE Trans. Inf. Theory*, vol. 48, pp. 1277–1294, June 2002.

[76] D. Wu and R. Negi, "Utilizing multiuser diversity for efficient support of quality of service over a fading channel," *IEEE Trans. Veh. Technol.*, vol. 54, pp. 1198–1206, May 2005.

[77] C.J. Chen and L.C. Wang, "A unified capacity analysis for wireless systems with joint multiuser scheduling and antenna diversity in Nakagami fading channels," *IEEE Trans. Commn.*, vol. 54, pp. 469–478, Mar. 2006.

[78] D. Bertselas and R. Gallager, *Data Networks.* Englewood Cliffs, NJ: Prentice-Hall, 1987.

[79] F. Kelly, "Charging and rate control for elastic traffic," *Eur. Trans. Telecommun.*, vol. 8, pp. 33–37, 1997.

[80] F. Kelly, A. Maulloo, and D. Tan, "Rate control for communication networks: shadow prices, proportional fairness and stability," *J. Oper. Res. Soc.*, vol. 49, pp. 237–252, 1998.

[81] A. Jalali, R. Padovani, and R. Pankaj, "Data throughput of CDMA-HDR: a high efficiency, high data rate personal wireless system," in *Proc. IEEE VTC'00-Spring*, Tokyo, pp. 1854–1858, May 2000.

[82] F. Berggren and R. Jantti, "Asymptotically fair transmission scheduling over fading channels," *IEEE Trans. Wireless Commn.*, vol. 3, pp. 326–336, Jan. 2004.

[83] L. Dai, S. Zhou, and Y. Yao, "Capacity analysis in CDMA distributed antenna systems," *IEEE Trans. Wireless Commn.*, vol. 4, no. 6, pp. 2613–2620, Nov. 2005.

[84] D.S. Shiu, G.J. Foschini, M.J. Gans, and J.M. Kahn, "Fading correlation and its effect on the capacity of multi-element antenna systems," *IEEE Trans. Commn.*, vol. 48, no. 3, pp. 502–513, 2000.

[85] D. Gesbert, H. Bolcskei, D.A. Gore, and A.J. Paulraj, "MIMO wireless channels: capacity and performance prediction," in *Proc. IEEE Globecom'00*, San Francisco, pp. 1083–1088, 2000.

[86] L. Dai, S. Zhou, H. Zhuang, and Y. Yao, "A novel closed-loop MIMO architecture based on water-filling," *Electronics Letters*, vol. 38, no. 25, pp. 1718–1720, Dec. 2002.

[87] H. Zhuang, L. Dai, L. Xiao, and Y. Yao, "Spectral efficiency of distributed antenna system with random antenna layout," *Electronics Letters*, vol. 39, no. 6, pp. 495–496, Mar. 2003.

[88] R. Gozali, R.M. Buehrer and B.D. Woerner, "The impact of multiuser diversity on space-time block coding," *IEEE Commn. Letters*, vol. 7, pp. 213–215, May 2003.

[89] W. Chen, L. Dai, K.B. Letaief, and Z. Cao, "A unified cross-layer framework for resource allocation in cooperative networks," *IEEE Trans. Wireless Commn.*, in revision. (http://www.ee.ust.hk/ ~eedailin/publications.htm)

[90] C. Li and X. Wang, "Adaptive multiuser opportunistic fair transmission scheduling in power-controlled CDMA systems," in *Proc. IEEE ICASSP'04*, Montreal, Canada, vol. 4, pp. 553–556, 2004.

8

COOPERATIVE CONTENTION-BASED MAC PROTOCOLS AND SMART ANTENNAS IN MOBILE AD HOC NETWORKS

John A. Stine

Contents

Wireless mobile ad hoc networks (MANETs) are the proposed solution for networking where infrastructure is not available and most communications are among a mobile set of nodes. MANETs are the ultimate distributed and cooperative communications systems. The free roaming radios that make up MANETs must discover each other, collaboratively organize themselves into a network, and then adapt to the continuous network redesign that is caused by their mobility, all the while sharing the same RF spectrum. Conceptually, MANETs are the most flexible form of networking. Unfortunately, their complexity results in limited performance and so they tend to be applied in situations where other networking approaches cannot be made to work. The more obvious applications of MANETs are tactical military networks and emergency networks used in disaster relief. Additional applications may be manifest if their performance can be improved. In this chapter we explore the requirements for using directional and smart antennas with contention access protocols to improve MANET performance.

8.1   INTRODUCTION

Directional and smart antennas can be used to enhance the performance of MANETs. They can increase the capacity of the networks, increase the range of communications, reduce the susceptibility of network nodes to detection, interception, and jamming, conserve energy, and resolve collisions. Properties of antennas that are exploited to yield these benefits include: antenna directivity, increased gain, and a host of capabilities enabled with arrayed antennas and signal processing techniques including beamforming, null steering, diversity, spatial processing, and multiple-input multiple-output (MIMO).

Most successful deployments of these types of antenna technologies have been in communications systems very much different than MANETs, where the antennas are on stationary or mobile nodes that only communicate with stationary access points. In these cases, there are either fixed pointing solutions or a centralized node controlling access and pointing. MANETs differ in that neither of these conditions exists and pointing solutions must be derived and adapted dynamically in response to node mobility and changing traffic patterns. MANETs can only be formed using protocols designed for distributed implementation and cooperative operation, and using directional and smart antenna technologies in MANETs demands an even higher level of cooperation. The feature of the problem that makes this so hard is that the pointing solutions are hard to resolve in a distributed sense and then are very temporary.

In this chapter we look at the role that medium access control (MAC) protocols play. MAC protocols vary from totally scheduled protocols, better known as Time Division Multiple Access (TDMA), to contention protocols. The perceived ability to control which nodes gain access simultaneously has made TDMA schemes appear the most appropriate for directional and smart antenna use. Unfortunately, TDMA protocols are not well suited for ad hoc networking environments because both connectivity and traffic are dynamic. *A priori* assignments of time slots to transmitter–receiver (TR) pairs may go unused in ad hoc networks for the typical reason that there is no traffic between the nodes and an additional reason that the TR nodes are out of range of each other. And certainly, antenna pointing solutions conceived at the beginning of use will quickly become irrelevant as nodes move around. As an alternative for ad hoc networks, many developers have chosen to combine the features of TDMA and contention protocols where a periodically executed contention mechanism is used to reserve TDMA slots. This approach tries to balance the control of TDMA and the adaptability of contention to create solutions better suited for ad hoc networking environments. Regardless, these schemes still try to schedule slots in anticipation of use. Further, to make schedules that account for antenna pointing requires some level of coordination among the nodes. The overhead this requires may counter the benefit of using directional and smart antennas, unless the solutions persist and there is a sufficient amount of traffic to exploit the solution so that a higher throughput justifies the expense. If the necessary duration of the solution is long, we are back to the disadvantages of using TDMA schemes, unused slots when there is no traffic and TR nodes moving out of range of each other. Thus, the objective of using contention protocols is to avoid these inefficiencies. As we will show, the distinction between using contention protocols as opposed to TDMA with directional antennas is that rather than exchanging information to derive transmission schedules with pointing solutions, pointing solutions are derived by how the nodes interact during contentions.

In this chapter we first review the salient features of directional and smart antenna technology that must be considered to create effective MAC protocols. Then we review the mechanisms used in contention access and assess whether they create the conditions for successful smart antenna employment. We then focus on one approach to designing contention MAC protocols that is particularly well-suited for exploiting smart antennas. We describe two variants of the protocol, one for managing fixed beam antennas and the second for purely adaptive antennas. Next we describe how to model smart antennas in simulations when using these protocols. Finally, we identify the necessary research that remains.

8.2 CONDITIONS FOR SMART ANTENNA USE

Smart antennas are antenna systems that have the intelligence to discern where they should point and then mechanisms to point them [1]. The intelligence that points the smart antennas may be controlled by protocols or be the result of algorithms embedded in the electronics. Although protocol based solutions are very popular, ideally this function is pushed down as low in the protocol stack as possible, and into the electronics of the physical layer so that information exchange between nodes and cross layer communications can be avoided. Further, electronic adaptation is much more effective than the simple management of pointing that protocol solutions address. However, at this level in the stack, the MAC protocols must create the conditions for the algorithms to be successful. To motivate our list of conditions, we describe the critical inputs to the adaptation algorithms, identify some of the imperfections of adaptation results, and provide a timing model for antenna adaptation. We conclude with the conditions.

8.2.1 Inputs

Adaptation is of two types, directional and environmental. In directional adaptation, the receiving antenna employs a signal processing technique to determine the direction of arrival (DOA) and then uses this information to point back toward the source of the transmission. Popular algorithms for DOA estimation are minimum variance distortionless response (MVDR) [2], multiple signal classification (MUSIC) [3], and estimation of signal parameters via rotational invariance techniques (ESPRIT) [4]. For these algorithms to work, the adapting node must receive an adequate signal from the distant end to which it can adapt. Environmental adaptation techniques seek to optimize the signal to interference and noise ratio (SINR). They have two benefits over simple beam steering: they can null out interfering sources and they can exploit multipath to spatially isolate transmitters. Multiple algorithms have been proposed for this purpose and a nice summary of a few is provided in [5]. In these algorithms, the receiving node must sample both the intended signal and the interfering signals. We note that there is an important distinction between these two. In DOA adaptation, once the adaptation is complete it remains relevant for the duration of the transmission regardless of new interferers, but this is not the case with environmental adaptation. This point is clarified in the adaptation timing model.

8.2.2 Adaptation Results

Antenna pointing solutions are imperfect. Figure 8.1 illustrates a hypothetical power pattern (in two dimensions for simplicity) and the parameters we use to define the quality of a directional antenna. The directivity of the main beam may be defined in terms of the half power beamwidth (HPBW) or beamwidth between first nulls (BWFN). The selectivity of the directional antenna is typically quantified as the ratio of the energy in the mainbeam to the energy radiated in all other directions. For simplicity, we will use the ratio of the gain of the largest sidelobe (MSLL) to the mainbeam maximum gain as our measure of selectivity. This latter measure is more easily appreciated by examining the radiation patterns of the antennas. In Figure 8.2, we illustrate the array factor for different directions of steering a circular array. In these illustrations we use the tuple (N, d, α) to specify the antenna design where N is the number of elements, d is the spacing between the elements in units of λ, the wavelength, and α is the angle that the

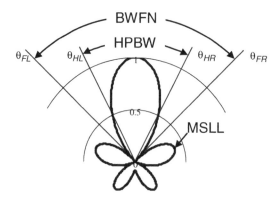

Figure 8.1 Typical Power Pattern Polar Plot of a Directional Antenna.

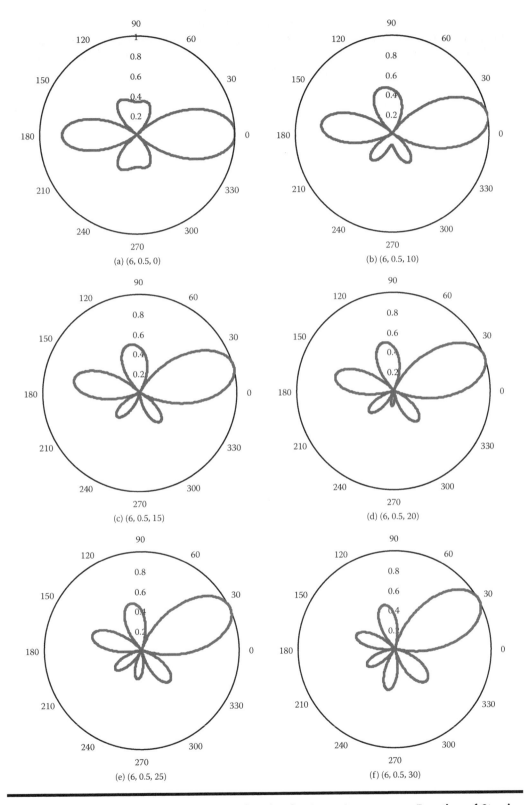

Figure 8.2   Variations of the Array Factor of a Circular Array Antenna as a Function of Steering Direction.

mainbeam is being steered. Both the amplitude and the direction of the sidelobes vary dramatically. Protocols that attempt to track pointing direction usually ignore sidelobe effects. Ideally, MAC protocols use the transmissions of pointed antennas to resolve the set of nodes that can transmit simultaneously so that these sidelobes are not an issue.

Antenna adaptations are also imperfect. In Figure 8.3 we show several adaptations that result from using the Max-SINR algorithm (a.k.a. the ideal beamformer) together with a circular array [6]. We use the tuple $(N, d, \alpha, \gamma_0, \gamma_1, \ldots)$ to define the conditions where N, d, and α are as defined before, and the γ_i are the directions toward the interfering transmitters. These array factors are ideal adaptations to illustrate the issues. We see that adaptations tend to prefer minimizing or nulling interference over amplifying the targeted signal. In many cases the gain in the direction of the distant end is less than the

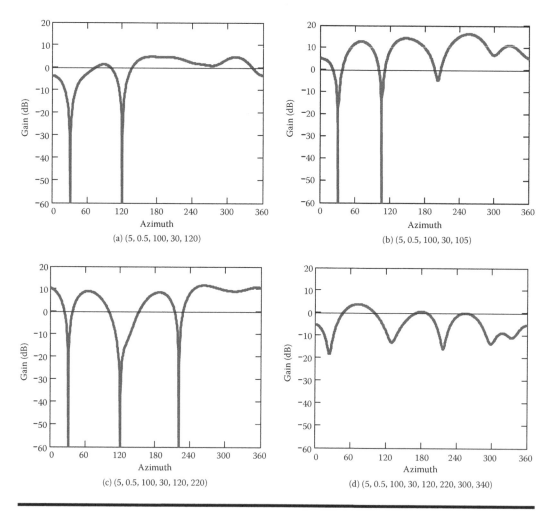

Figure 8.3 Antenna Adaptations Using the Ideal Beamformer With a Circular Array. The Graphs of these Patterns Have Been Normalized so That the Gain Toward the Distant End Is Set at 0 dB. (a) Is a Typical Result. (b) Shows that When an Interferer Comes Close in Direction to the Target that the Adaptation Is Not as Effective, Manifested Here by the Placement of Both Directions on the Same Side of a Lobe and Having a Large Gain in Other Directions. (c) Demonstrates that Additional Directions Also Increase the Relative out of Beam Gain. (d) Demonstrates the Effect of Having as Many Interferers as Elements, Nulls Are Not as Pronounced.

Figure 8.4 Packet Frame for Antenna Adaptation.

maximum gain, and in fact large regions have gains greater than this direction. Thus, if during an ongoing exchange that has followed an adaptation, a new transmission occurs, not only is it not considered in the adaptation but it may also be amplified, resulting in more severe interference than would occur if there had been no adaptation. Clearly, MAC protocols must prevent new interference from occurring. Second, we see that these adaptations are limited in the number of interferers they can null, always being less than the number of elements in the array.

8.2.3 The Antenna Adaptation Model

It is instructive to have a basic model for antenna adaptation to understand MAC protocol requirements. We use a modeling approach first proposed in [7]. Figure 8.4 illustrates the timeline of a received packet. A training sequence at the front end of a packet is used by the antenna to adapt. Adaptation occurs in two parts. First the antenna captures the desired signal which occurs at the beginning of the sequence, and then the antenna optimizes reception of that signal in the adaptation period that follows. Several criteria must be met for adaptation to be successful. There are five parameters:

1. t_s: The time after the first arriving bit of the training sequence that it takes for a transceiver to capture a desired signal.
2. SIR_c: The minimum signal interference ratio (SIR) required to capture a signal.
3. t_{sm}: The minimum time required by a smart antenna to adapt to an interfering signal.
4. t_f: The end of the training sequence used for adaptation.
5. SIR_a: The minimum SIR required to adapt to a signal for either enhancing or nulling.

Let t_a be the time an interfering signal arrives at the receiver. The antenna can determine the DOA if $SIR > SIR_c$ when $t_a < t_s$, $SIR > SIR_a$ when $t_s \leq t_a \leq t_f$, and when $t_a > t_f$. An adaptive antenna can reject interfering signals if $SIR > SIR_c$ when $t_a < t_s$, $SIR > SIR_a$ when $t_s \leq t_a \leq t_f$, and $t_a \leq t_f - t_{sm}$. The significant difference between a DOA adaptation (i.e., beam steering) and environmental adaptation (i.e., null steering) is that the DOA adaptation only needs to adapt to the signal of interest while environmental adaptation must also adapt to the interfering signals. Successful environmental adaptation also depends on whether the quantity of interfering signals is within the adaptation threshold of the antenna.

8.2.4 Conditions for Smart Antenna Use

A successful implementation of smart antennas in a MANET requires that up to six conditions be met. The first two are required for simple pointing schemes. They are deceptively simple.

1. **Know where to point:** The protocol must enable contenders to determine where destinations are.
2. **Know where not to point:** The protocol must enable contenders to avoid destinations that are busy and to avoid transmitting in directions that can interfere with ongoing exchanges. This is very important because sensing ranges become asymmetric with directional antenna use.

The next four conditions are necessary for adaptive strategies to be effective. They are less obvious.

3. **Acquire the condition:** The protocol must enable the adaptive antennas to acquire the conditions for determining the weighting of the antenna elements. Changing conditions mean weights will have a short lifetime and so the MAC should enable weight determination in close proximity to the time they are used.
4. **Prevent congestion:** Antenna arrays have limited degrees of freedom to cancel out interfering nodes.
5. **Prevent coincident transmission:** Adaptive antennas have an angular resolution and cannot differentiate transmitters in the same or near same direction.
6. **Preserve the condition:** The protocol must keep the weighting relevant for the duration of its use. Protocols may not be able to prevent movement of nodes and of objects in the environment, but they should prevent new interference.

In the next section we will evaluate the effectiveness of using popular MAC schemes with smart antennas by whether they can create these conditions.

8.3 CONTENTION MEDIUM ACCESS CONTROL

8.3.1 Mechanisms

A significant difference between wireless and wireline networks is that wireless sources cannot do collision detection. Once a wireless source starts a transmission it commits itself and can only determine if there was a collision through the absence of an expected acknowledgment. Thus, wireless MAC mechanisms seek either to avoid collisions or to resolve them. Three fundamental mechanisms are used: carrier sensing, packet sensing, and random access. In carrier sensing, nodes listen to the channel before transmitting and will not transmit if they perceive it to be busy. In packet sensing, nodes use information that is in packets they have received to assess whether the channel is busy and for how long. It is normally used in addition to carrier sensing to avoid hidden terminal effects. In random access, nodes will either wait for some random amount of time before trying to transmit or will attempt to transmit in a particular time slot with some probability. All wireless contention MAC protocols for ad hoc networks use one or a combination of these to arbitrate access. The important point is that directional and smart antennas do not affect the random access mechanisms, but do affect carrier and packet sensing mechanisms. This will be made clear in the next two sections, where we first review some popular contention MAC protocols for ad hoc networks, and then describe the issues associated with the carrier sensing and packet sensing mechanisms.

A second significant difference between wireless and wireline networks is that the wireless channel is much less reliable than the wireline channel. Wireline MAC protocols rarely use feedback mechanisms to confirm that a packet was received. Reliable delivery is managed by the transport protocols and, in fact, failed deliveries are assumed to be the result of congestion, and the subsequent dropping of packets at routers rather than bit errors that occur in transmission. However, in wireless MAC protocols, the occurrence of bit errors is very common and so the trend is to use MAC packet acknowledgments (ACK) to provide feedback to source nodes that their transmissions were successful. Rules apply to how a receiver should acknowledge, and if an ACK is not received by the source according to these rules, the source assumes the packet transmission was unsuccessful. The significance of using ACK packets in a MAC protocol is that all exchanges require transmissions from both ends, thus extending the effects of a packet exchange to all nodes in range of both ends of that exchange.

The third significant difference between wireline and wireless networks is that in a wireline network all nodes that are affected by or can affect a contention are connected to each other while in wireless ad hoc networks this is not true. This has several ramifications. First, there is ambiguity about what a missing ACK means. It could mean that there were too many bit errors and so a retransmission is appropriate or it could mean the receiver is no longer in range of the source and the appropriate action is to drop or reroute the packet. Second, carrier sensing does not protect receivers because it is possible for nodes to be out of range of the source but in range of the receiver. And third, there is ambiguity as to whether the receiver for which a packet is intended is already busy receiving a packet from another source.

These differences between wireless and wireline environments result in a number of failure modes that uniquely occur in MAC protocols used in ad hoc networking environments. We review these modes after we provide a brief overview of popular MAC protocols and a more detailed view of sensing issues.

8.3.2 Contention MAC Protocols

The most popular MAC protocols for use in ad hoc networks are variants of Aloha and carrier sensing multiple access (CSMA).

In Aloha, sources attempt to transmit packets as they arrive. The assumption is that the media has sufficient capacity and the arrivals are sufficiently random that, on average, most attempts succeed. However, if arrivals overlap, they collide. Source nodes that become aware of collisions will randomly schedule retransmissions. This process results in an optimum capacity of about 18% of the channel capacity, and if the load were to exceed this amount then the capacity would decrease because of congestion collapse. A second variant of Aloha is slotted Aloha. In slotted Aloha the channel is time slotted for packet transmissions. Nodes transmit packets in the time slot that follows their arrival. The benefit of slotting is that it forces collisions to completely overlap. This difference doubles the potential capacity of the protocol with the disadvantages of requiring synchronization in the network and constraining all packets to the size of the slots. Slotted Aloha will also suffer congestion collapse if the load is larger than this optimum capacity.

In CSMA, nodes first listen to the channel and transmit if they perceive an idle channel. In its pure form, CSMA suffers two significant problems in wireless ad hoc networking environments. As described above, carrier sensing alone cannot protect receivers from collisions and, worse, attempting accesses immediately after a channel becomes idle is a strategy for more collisions. Thus, CSMA is enhanced for use in ad hoc networks. The first enhancement is to either employ a collision avoidance scheme or collision resolution scheme to prevent collisions at the point the channel becomes idle. In collision avoidance schemes, nodes wait some random amount of time after the channel becomes free and then attempt to gain access. In the first proposal, called nonpersistent CSMA, nodes would simply reschedule the arrival of a packet to some future time according to a retransmission delay distribution [8]. A more commonly used proposal is to require all contenders to wait for some random amount of silent time before attempting to send a packet. This is the technique that is used in the IEEE 802.11 distributed foundation wireless MAC (DFWMAC) protocol [9]. In collision resolution schemes, contenders participate in a tournament to resolve who can gain access. This is the approach applied in the high performance radio local area network (HIPERLAN) I standard [10]. This approach is not very effective in distributed environments because there is no guarantee that the nodes can interact with each other to play together in the tournament. The second enhancement to CSMA protocols is to extend the protection around the receiver of an exchange. Here, too, there are two techniques. In the first, a busy tone is transmitted out of band by receivers so that all contenders around receivers can again use carrier sensing to avoid causing interference [11]. In the second, a set of short handshake packets is used to coordinate the payload packet exchange. The first of the handshake packets requests to send a packet to a receiver and the return packet from the receiver announces if, and sometimes when, that packet can be transmitted. Because nodes in range of the receiver should hear and understand the return packet, they can learn if the exchange and information in the packet can give them an understanding of how long to wait before attempting a contention themselves. This technique was first proposed in [12] as part of an out-of-band control channel exchange and then later proposed as part of the single channel protocols multiple access collision avoidance (MACA) [13], MACA wireless (MACAW) [14], floor acquisition multiple access (FAMA) [15], and ultimately the IEEE 802.11 DFWMAC. Finally, because wireless channels are unreliable, an acknowledgment is added to the schemes at the conclusion of a successful packet reception. This is part of both the MACAW and the 802.11 DFWMAC protocols. ACKs also serve to indicate that packet exchanges first announced in a receiver response have finished and so neighboring contenders can stop deferring their access attempts.

The 802.11 DFWMAC is currently the most ubiquitous and commercially successful contention MAC protocol for wireless networks, and is at the top end of the evolution of wireless CSMA protocols, hence we provide a more detailed discussion. The 802.11 DFWMAC has two access coordination functions, one where access is coordinated in a completely distributed manner and a second where an access point coordinates access attempts. This first function, called the distributed coordination function (DCF), is the default and what we describe. The DCF is a CSMA protocol that uses all the techniques described above: collision avoidance, request-to-send (RTS) and clear-to-send (CTS) handshaking, and acknowledgments.

The sequence of events of a typical packet exchange begins with the source transmitting a short RTS packet. This packet identifies the station to which it wants to exchange data and announces how long it will be until the transmission is completed. If a collision occurs during this transmission, it will only involve the short RTS packet rather than the larger data packet. If the intended destination hears this packet, it responds with a CTS packet. This CTS packet also includes the duration of the expected exchange. Because all stations within range of these two short packets hear the duration of the exchange, they will, at least virtually, sense the network as busy. The third packet that is transmitted contains the payload of the exchange and is referred to as the MAC protocol data unit (MPDU). Finally, if a destination successfully receives a packet, it responds with an ACK packet addressed to the source. A source will consider an unacknowledged MPDU as a failed transmission.

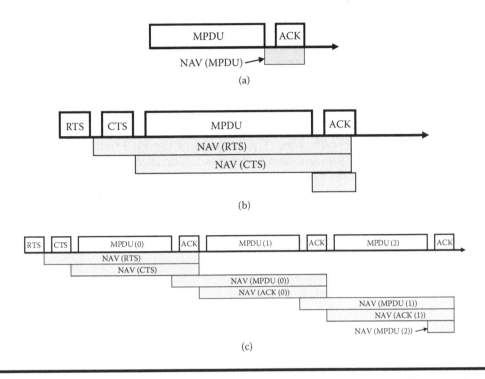

Figure 8.5 Packet Exchange Methods of the 802.11 DFWMAC. (a) Packet Exchange without an RTS/CTS Handshake, (b) Packet Exchange with an RTS/CTS Handshake, (c) Exchange of a Fragmented Packet.

The DFWMAC offers some flexibility in how packets are sent. Figure 8.5 illustrates three methods. The method used depends on the size of the packet. Two thresholds are used to control the method, the RTS threshold and the fragmentation threshold. If the packet is larger than the RTS threshold then it is sent with an RTS–CTS handshake, otherwise the RTS–CTS handshake is not used. If the packet is larger than the fragmentation threshold, then it is fragmented with the first fragment preceded by an RTS–CTS handshake. Figure 8.5 also illustrates the duration of the network allocation vector (NAV) that is included with each packet for the virtual sensing function described above. This is the projected time the channel will be busy. Note that in the fragmented packet, each fragment projects the duration of only the next fragment.

Figure 8.5 illustrates there is a time period between each of the transmissions. This is inevitable due to signal propagation time, the processing time, and transceiver transition time. The 802.11 MAC protocol defines several interframe times and a slot time to account for the cumulative effects of these activity times and to give priority to packets that are part of an ongoing MPDU exchange over packets of subsequent contentions. There are three interframe spaces and one slot time used with the DCF.

Slot time: The slot time is sized to account for the time it takes for the radio to indicate a clear channel, the time to transition from receive to the transmit state, the time for a signal to propagate the maximum range of a transceiver (300 meters is the design range), and the time it takes the MAC protocol to process the observation that the channel is cleared.

Short interframe space (SIFS) time: The slot time is sized to account for the time it takes from the arrival of a signal at the antenna until the modem perceives it is receiving data, then the time it takes the MAC to process a frame and prepare a response, and finally the time to transition from the receive to the transmit state.

DCF interframe space (DIFS) time: SIFS time plus two slot times.

Extended interframe space (EIFS) time: The duration of the EIFS time consists of a SIFS time, a DIFS time, the time for a modem to process the header of a packet, and the time it takes to transmit an ACK PDU at the modem's lowest data rate.

These interframe space times have specific purposes. The SIFS time is used between all packets of a frame. A station that is responding to a RTS will wait a SIFS time from the end of its receipt until it starts to transmit a CTS. Similarly, the SIFS time is used between the CTS and MPDU and between the MPDU and ACK. The DIFS time is the minimum time that stations must wait after they hear a transmission or the end of the virtual sensing time, whichever is later, before they can transmit. Finally, the EIFS time is the time a station must wait after it detects it has received a packet incorrectly. It ensures that another station can acknowledge the packet without interference from this station.

The slot time is the time increment of the collision avoidance backoff mechanism. A station that receives a packet to transmit will randomly select a random number of time slots to back off before attempting to access the channel. This selection is made from a bounded range of time slots called a contention window. So after waiting a DIFS time after the previous transmission, a contending station then waits this number of backoff slots before transmitting its packet. If a station contends and is unsuccessful, it will choose a new backoff time and contend again. The fact that a contention is unsuccessful is an indication that the channel is likely to be busy, and so to avoid congestion, the station randomly selects a backoff time from a larger contention window. This process of selecting a backoff time from a larger contention window is called binary exponential backoff and is defined by the equation:

$$Backoff = \lceil \min \left(\left(2^{i+3} - 1 \right), 255 \right) \times ran \# \rceil \times (a \; Slot \; Time) \qquad (8.1)$$

where *i* is the number of failed access attempts for this packet and *ran #* is a uniformly distributed random number between 0 and 1. Using this equation, the backoff time will be between 1 and 7 slot times the first contention attempt, 1 and 15 the second contention attempt, and then bounded in order by 31, 63, 127, and finally 255 slot times for the subsequent access attempts.

Figure 8.6 illustrates an example of the DCF operation. We illustrate several functions: a successful packet exchange, a failed packet exchange, a collision, and a fragmented packet exchange. We also illustrate the operation of the backoff mechanism. This example scenario starts off with three stations having packets to send, Stations 1, 2, and 4. These stations have backoff times of 5, 2, and 5 slot times and start counting off their backoff slots at the beginning of the contention. Station 2 counts down first and after two backoff slots transmits an RTS packet. Note that all the other contenders stop counting down. Station 2's destination responds with a CTS packet and so the contention is considered to be successful. Station 2 sends an MPDU but receives no ACK in response. The exchange is considered a failure and Station 2 chooses a new backoff time. Note

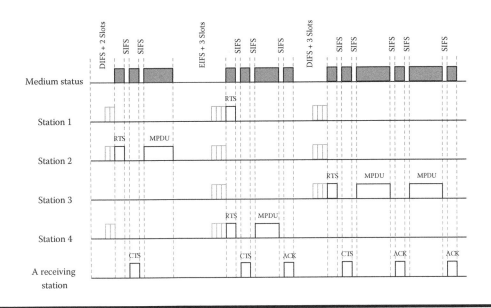

Figure 8.6   Example Cooperative Interaction of Contending Nodes Using the DFWMAC DCF.

that the contention was successful so the new backoff time is selected from the smallest contention window. In this example Station 2 chooses a new backoff time of 7 slots. After the MPDU, all stations wait until the end of the NAV of the MPDU before sensing for the DIFS time that precedes the backoff. If a station sensed, but did not correctly receive the MPDU, it would wait an EIFS before counting down a backoff. Both techniques result in the same start for count down of the backoff slots. Note that during this period a packet arrived at Station 3, which chooses a backoff of 6 slots. All of the contending stations count down their backoff and both Stations 1 and 4 finish counting off and sending an RTS at the same time. This is an example of a collision, but note that Station 4's destination responds with a CTS. This may be possible depending on the geometry of the stations, where this destination is in range of Station 4 but not Station 1. Because Station 1 has a collision, it schedules a new backoff from the next larger contention window. Because Station 4's MPDU is transmitted prior to the required DIFS after the expected ACK, Station 1 does not get the opportunity to backoff right away. Station 4 successfully exchanges its MPDU and then the process repeats itself. The example concludes with Station 3 sending a fragmented packet successfully and Stations 1 and 2 still in the backoff mode.

In this example, we demonstrated the collision avoidance and showed that the backoff is counted down continuously based on when a station perceives the channel is idle. We have shown how each of the interframe spaces gives priority to packets involved in a MPDU exchange. Finally, we have illustrated a collision and the resulting increase in the contention window.

8.3.2.1   Sensing and Interference

As described, the 802.11 MAC uses carrier sensing and packet sensing. Problems can occur with this protocol because of the different ranges of the carrier sensing, packet sensing, and interference. There are two regions about a transmitter, a packet sensing

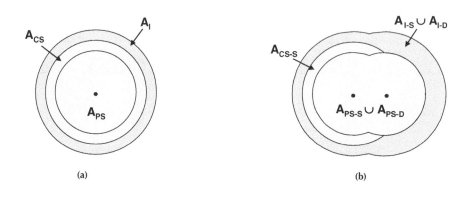

Figure 8.7   Protocol Sensing and Interference Regions.

region (A_{PS}) in which nodes can receive packets and a carrier sensing region (A_{CS}) in which nodes can detect a transmission. Similarly, there are three regions about receivers, the same packet and carrier sensing regions defining where transmissions can originate and be sensed by the receiver, and a third region, an interference region (A_I), in which transmitters may cause unwanted interference to the receiver. (We assume in this explanation that all transceivers in the network use the same transmission power.) Figure 8.7(a) illustrates an example of these regions emanating omnidirectionally from a node. Here, the carrier sensing region subsumes the packet sensing region. The size of the regions and the relative range to the boundaries are affected by the transmission power, receiver thresholds, and in the case of packet sensing, the presence of interference. We also illustrate the interference region subsuming both sensing regions; however, its true boundary is affected by the strength of the received signal with the relation that the stronger the signal, the closer the interfering transmitter must be to be disruptive. Unlike what is drawn, environmental effects cause these boundaries to be irregular and sometimes disconnected.

The 802.11 MAC uses carrier and virtual sensing to arbitrate contention. Carrier sensing is used to prevent new transmissions by nodes that sense the carrier. Virtual sensing is the use of information in packet headers (e.g., the NAV of the 802.11 MAC) to direct neighboring nodes to defer from contention for some period (i.e., the duration of the upcoming packet exchange). Its range is equivalent to the packet sensing range. Figure 8.7(b) illustrates the overlap of the regions that are exploited by the 802.11 MAC when using the RTS–CTS–MPDU–ACK exchange method and when all transmissions use the same power. When functioning properly, the virtual sensing mechanism of the RTS–CTS exchange suppresses new exchanges from originating or terminating in the packet sensing regions ($A_{PS\text{-}S} \cup A_{PS\text{-}D}$).[1] The PDU exchange would only be interfered with by transmitters in ($A_{I\text{-}D} - A_{I\text{-}D} \cap (A_{CS\text{-}S} \cup A_{PS\text{-}D})$). The ACK exchange could be interfered with by transmitters in ($A_{I\text{-}S} - A_{I\text{-}S} \cap (A_{CS\text{-}S} \cup A_{PS\text{-}D})$). We see that neither carrier sensing nor packet sensing can guarantee there are no collisions.

---

[1] The hyphenated extension in the subscript identifies whether the area is associated with the source, -S, or the destination, -D.

8.3.3 Failure Modes of MANET Contention MAC Protocols

Failure modes in access protocols are artifacts of the techniques that are used to resolve contention. Below we describe the most common.

- *Collisions* occur when a receiver cannot receive because of interference. There are two types of collisions, primary and secondary. Primary collisions occur when a node is expected to participate in more than one packet exchange at the same time. Secondary collisions occur when an exchange is interfered with by a distant exchange.

- *Hidden terminals* are an artifact of asynchronous carrier sensing mechanisms. A hidden terminal occurs when a contender in range of a destination of an ongoing exchange is not suppressed by either carrier sensing or virtual sensing. As a result, its contention interferes with the ongoing exchange.

- *Exposed terminals* are an artifact of asynchronous carrier and virtual sensing mechanisms. An exposed terminal is a contender that cannot gain access to a channel because it always senses another exchange in disjoint parts of the network.

- *Misdirected contention* is an artifact of asynchronous access schemes and occurs when a contender attempts to send a packet to an ineligible receiver, a receiver suppressed by virtual sensing (a.k.a. muteness), or a carrier sensing another exchange.

- *Deafness* is a failure of asynchronous virtual sensing. It occurs when interference prevents the packet sensing that is necessary for virtual sensing. As a result, a contender may not defer from contending as would have been appropriate and causes a collision.

- *Blocking* can occur in contention access protocol where destinations are not involved in the contention resolution mechanism. Contenders that are out of range of each other but in range of each other's destinations repeatedly gain access and interfere with each other. Blocking is most likely to occur in synchronous access schemes that do not randomize access attempts.

- *Retry count outs* can occur in any contention access protocol. It is the dropping of packets when a source perceives multiple consecutive successful contentions but failed exchanges. It results because it is not possible for a contending node to differentiate the cause of an exchange failure. Failures that result from the destination being out of range, a condition that warrants dropping the packet, have no different signature than those that fail because of any of the failure mechanisms listed above.

- *Congestion collapse* can occur in any contention protocol where collisions and other failures occur more frequently as the traffic or density of nodes increases. It is especially harmful if collisions occur during long data exchanges. The throughput decreases because the channel is occupied with transmissions that produce no goodput.

8.3.4 Using Smart Antennas with the Popular Contention MAC Protocols

8.3.4.1 CSMA

The dilemma in using directional antennas with CSMA style protocols is that they disable the carrier sensing and virtual sensing mechanisms upon which these access schemes

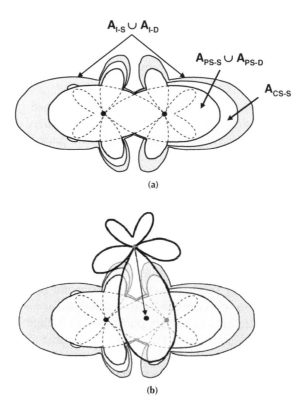

Figure 8.8   Protocol Sensing and Interference Regions When Using Directional Antennas.

are based. Figure 8.8(a) illustrates a representative view of the carrier sensing, packet sensing, and interference regions when directional antennas are used. Carrier sensing and packet sensing do not suppress contentions in a symmetric way and so there is an increased likelihood that a new contender may attempt to send a packet to deaf or muted nodes as illustrated in Figure 8.8(b). This is a classic misdirected contention that could very easily result in a retry count out. Using directional antennas throughout the 802.11 DFWMAC exacerbates this failure mode.

An alternative is to combine omnidirectional and directional transmissions in the contention exchanges. Under this approach, omnidirectional transmissions separate the source destination (SD) pairs and the directional transmissions attempt to reduce the interference region to be within the omnidirectional virtual sensing region (i.e., cause $(A_{I\text{-}D} \cap (A_{PS\text{-}S} \cup A_{PS\text{-}D}) = A_{I\text{-}D})$ and $(A_{I\text{-}S} \cap (A_{PS\text{-}S} \cup A_{PS\text{-}D}) = A_{I\text{-}S}))$. Thus, this technique does not increase the density of SD pairs but certainly prevents some collisions.

Nevertheless, the majority of current work is directed at trying to employ antennas using DFWMAC like access schemes. They range in aggressiveness. In the most simple, the omnidirectional RTS–CTS exchanges remain unchanged and directional antennas are only used with PDU and ACK exchanges [16]. This scheme pulls the interference region to within the virtual sensing region thus mitigating hidden terminal and deafness type failures. Knowing where to point can be determined within the contention. A shortcoming of this approach is that it still suppresses new exchanges from originating or terminating within the virtual sensing regions. The virtual sensing mechanism must be relaxed to achieve higher capacity. An option is to eliminate virtual sensing and rely on carrier sensing. Huang et al. [17] propose using an omnidirectional RTS followed by directional CTS–PDU–ACK packets. Direction between the nodes is learned in the contention. During the transmission and reception of these latter packets both nodes simultaneously transmit busy tones. New exchanges are suppressed by sensing these tones. A second option is to simply use directional antennas for all the exchanges thus reducing the size of the virtual sensing region. Experiments in [18] show that this improves capacity. An alternative way to relax virtual sensing is to do it directionally as proposed in the directional MAC (DMAC) protocols described in [19]. Neighbors that fall within the virtual sensing region of an exchange can contend or receive in a new exchange so long as they confine their transmissions to directions away from the ongoing exchange. Various versions of this story are proposed in [20–22]. Deviations in [22] include sending information on the position and orientation of nodes in the RTS/CTS transmissions, information on transmit power settings, and receive thresholds in all packet headers. Direction and the potential to cause interference are considered before new transmissions. The motivation for power control is to reduce the overall interference in the network which was found beneficial in [18]. Regardless, directional virtual sensing exacerbates the occurrence of muteness because it is intentional to cause new transmissions in the range of busy nodes. Li et al. [23] propose that, to prevent the muteness problem, new exchanges be permitted during ongoing exchanges only if they can be completed prior to the conclusion of those ongoing exchanges.

Some additional variations are made to the four-way handshake scheme to support long-range exchanges that require directional antennas at both ends. A multi-hop RTS is proposed in [21,22] to coordinate a long-range directional link. The source sends the RTS and remains idle a short time pointing in the direction of the destination, waiting for the RTS to reach that destination, and for it to respond with a directional CTS.

Despite being well-studied, CSMA schemes will likely remain poorly suited for use with smart antennas.

8.3.4.2 Aloha

Aloha protocols will obviously benefit from using directional antennas. Anything that reduces the footprint of transmissions will reduce the occurrence of collisions. However, there are no specific mechanisms available to create any of the antenna use conditions. Slotted Aloha, however, by default creates conditions 3 and 6 (see Section 8.2.4). The potential capacity that smart antennas can create is explored in [24], where the capacity of slotted Aloha is evaluated when nulls can be steered to resolve collisions. It shows that capacity is a function of the number of antenna elements because the number of elements determines the number of possible nulls. More interesting descriptions of the feasibility of exploiting smart antennas with slotted Aloha are found in [7,25]. In [7], the authors propose a packet frame with the two part acquisition preamble that we use as our adaptation model. In [7], a Barker code (i.e., the same type of code used to spread the 1 Mbps direct sequence spread spectrum (DSSS) signal of 802.11) is used to support signal capture. So long as a destination receives the first part prior to any other transmission, and all interfering transmissions arrive during the second part, it can resolve collisions. The same authors in [25] propose a more ambitious use of smart antennas, where they not only resolve collisions but also allow simultaneous reception of packets from multiple sources. Unfortunately, adaptation only occurs at the receivers and slotted Aloha schemes provide no mechanisms to achieve conditions 4 and 5.

8.4 SYNCHRONOUS COLLISION RESOLUTION (SCR) MAC PROTOCOLS

Synchronous collision resolution (SCR) is a new family of contention MAC protocols for ad hoc networks that are especially well-suited for use with smart antennas because they can create all the use conditions. In this section we will describe the general protocol approach and protocol design, and then variations that are tuned to support switched beam, steered beam, and environmentally adaptive antenna technologies.

8.4.1 Overview

Synchronous collision resolution is a broad MAC definition and is best viewed as an access framework in which there are many possible designs. The basic implementation of the SCR MAC is illustrated in Figure 8.9. SCR has four key characteristics:

1. The wireless channel is slotted.
2. All nodes with packets to transmit attempt to gain access to every transmission slot.
3. Contending nodes use signaling to arbitrate their access.
4. All packet transmissions that occur during a transmission slot are sent simultaneously.

Design choices that determine capabilities of SCR are the size and framing of transmission slots, the use of handshake packets, and the specific details of signaling.

Access arbitration consists of collision resolution signaling (CRS) and, optionally, an RTS–CTS handshake. CRS selects a subset of contenders that are good candidates for sending packets at the same time. The RTS–CTS handshake reduces the SD pairs to those that can exchange packets simultaneously.

CRS consists of a series of signaling slots organized into groups of slots called phases in which contending nodes may send very short signals. These signaling slots should not be confused with the longer transmission slots of Figure 8.9. Rather, they occur within a transmission slot during a short period at the very beginning. There are numerous ways to design signaling. The simplest and generally most effective at arbitrating contention is illustrated in Figure 8.10, and consists of one signaling slot per phase. In this design,

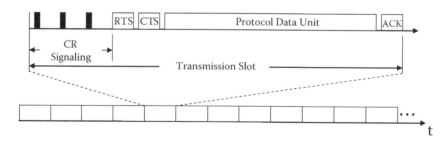

Figure 8.9   Basic Implementation of the Synchronous Collision Resolution MAC Protocol.

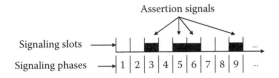

Figure 8.10   Collision Resolution Signaling Using Single Slot Phases.

a probability is assigned to each signaling slot and a contending node will signal in that slot with that probability. There are two assumptions that apply to signals and signaling slots.

1. Signals superimpose such that a receiver that hears multiple signals will still detect a signal.
2. Signaling slots and signals are sized to account for synchronization accuracy, propagation delay for the maximum range, detection time, and receive-to-transmit transition time such that the slot in which a transmitter sends and a receiver detects the signal is unambiguous.

The rules of signaling in this design are as follows.

1. At the beginning of each signaling phase a contending node determines if it will signal. It will signal with the probability assigned to the slot of that phase.
2. A contender survives a phase by signaling in a slot or by not signaling and not hearing another contender's signal. A contender that does not signal and hears another contender's signal loses the contention and defers from contending any further in that transmission slot.
3. Nodes that survive all phases win the contention.

The performance of CRS can be measured in two ways: how well does it resolve contentions locally and how well does it separate survivors spatially. CRS's ability to resolve contentions locally depends on the number of signaling phases used and the assignment of probabilities to the signaling slots of those phases. Our design algorithm is described in Section 8.4.2. Using this design algorithm, 9 phases of signaling can be made > 99% effective at resolving contention to just one survivor with more than 1000 nodes contending for access in range of each other. In multi-hop environments, the synchronized implementation of CRS (i.e., SCR) spatially separates survivors such that the probability that a survivor is in range of another is equivalent to the signaling design's contention resolution probability. Figure 8.11 uses a series of panels to illustrate a series of signals using an 8-phase design which demonstrates SCR reducing 172 contenders to 23 survivors that are spatially separated from each other by at least the range of their signals. A more thorough empirical study of the distribution of the separation of contenders can be found in [26].

At the conclusion of signaling, surviving contenders are separated, but this is not necessarily true for their destinations where interference occurs. The RTS–CTS exchange mitigates this concern. Figure 8.12 illustrates the RTS–CTS exchange. As demonstrated, the role of the RTS–CTS exchange is neither to limit collisions to smaller packets nor to extend channel use detection two hops for hidden terminal protection as is its purpose in the 802.11 MAC protocol. Rather, the RTS–CTS exchange verifies that source-destination pairs can "close" a connection and provides a feedback mechanism to support link

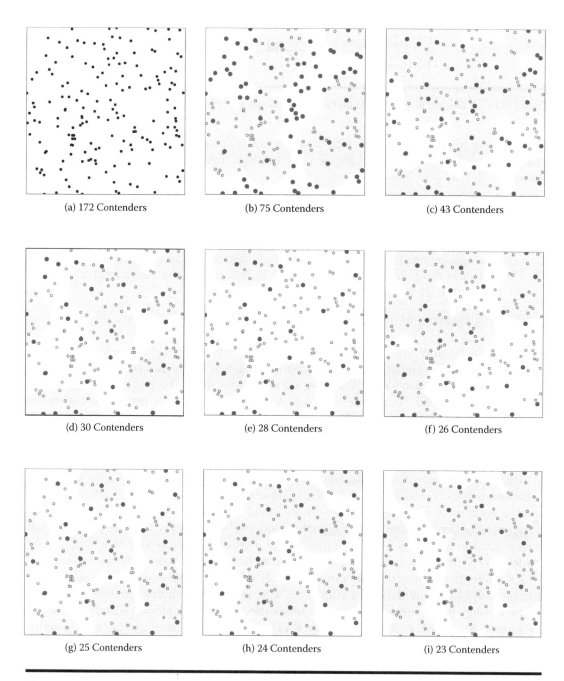

Figure 8.11 Example of Collision Resolution Signaling. Panel (a) Starts With 172 Nodes, All Are Contenders. Panels (b) Through (i) Show the Signals of Each Phase and Their Effect on Which Nodes Remain Contenders. The Larger Dots Are the Contenders at the Conclusion of Each Phase and the Circles Illustrate the Area Covered by the Signals that Are Sent During the Phase. Note that in the Final Result All Contenders Are Separated by at Least the Range of Their Signals.

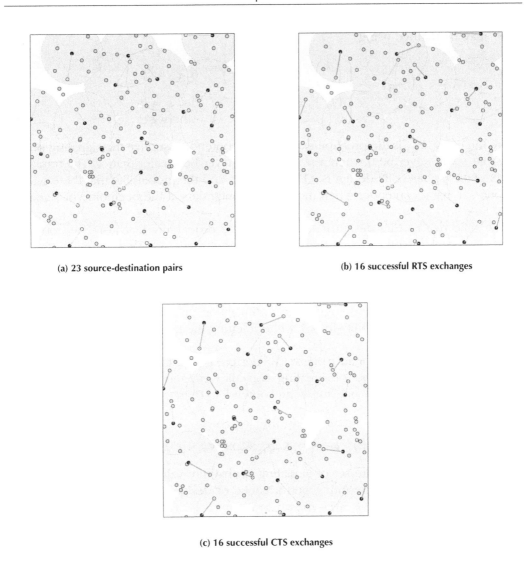

(a) 23 source-destination pairs

(b) 16 successful RTS exchanges

(c) 16 successful CTS exchanges

Figure 8.12 **Example of the RTS–CTS Exchange. Panel (a) Shows the Signaling Survivors From the Example in Figure 8.11, the Destinations for Their Packets, and the Area that Will Be Covered by the Packet Transmission Originating from These contenders. Panel (b) Illustrates the Node Pairs that Successfully Exchange RTS Packets Using a Wider Line Between These Pairs. Panel (b) Shows the Pairs that Complete the RTS–CTS Handshake. Only the Contenders of These Pairs Attempt to Send the PDU Packets.**

adaptation, including antenna adaptation. If a source sending a unicast transmission does not receive a CTS, it loses the contention and foregoes transmitting the PDU. In the case of broadcasts and multicasts, destinations do not send CTS and ACK packets.

8.4.2 The Design of Collision Resolution Signaling Using Single-Slot Phases

Let p^x be the probability that a contender will signal in phase x, and let $\mathbf{P}^x$ be the transition matrix of phase x and $\mathbf{Q}^n$ the transition matrix of the CRS design. The elements of $\mathbf{P}^x$ may be defined using

$$\mathbf{P}^x_{k,s} = \begin{cases} \binom{k}{s} (p^x)^s (1 - p^x)^{k-s} & 0 < s < k \\ (p^x)^k + (1 - p^x)^k & 0 < s = k \\ 0 & \text{otherwise,} \end{cases} \tag{8.2}$$

where the entry $\mathbf{P}^x_{k,s}$ is the probability that s of k contenders survive the signaling phase. Note that s will never be 0 when $k > 1$ and will never be greater than k. The transition matrix of an n phase CRS design is $\mathbf{Q}^n = \Pi^n_{x=1} \mathbf{P}^x$ and the probability that there will be just one surviving contender when there are k contenders at the beginning of signaling is $\mathbf{Q}^n_{k,1}$.

The objective of CRS design is to optimize the probability that just one node will survive the signaling by selecting the signaling probabilities, p^x. Designing CRS to maximize the probability that just one node survives when $k1$ nodes contend is relatively simple; however, a characteristic of CRS is that this maximum may result in a lower resolution probability when $k2$ nodes contend, $k2 < k1$. Figure 8.13 illustrates the effect. We define the design methodology as an optimization problem that will maximize the single node survivor probability for $k1$ without letting the probability for all $k2 < k1$ be less than that at $k1$.

Let q^n be the set of p^x for an n phase CRS design, k_t be a target density of contending nodes, m be the total number of signaling slots allowed, and $S(q^n, k_t, m)$ be the probability that there will be only one surviving contender. Then the optimization problem is

$$\begin{aligned} \max_{q^n} \quad & S\left(q^n, k_t, m\right) \\ \text{s.t.} \quad & S\left(q^n, k, m\right) \geq S\left(q^n, k_t, m\right) \; \forall k, 0 < k < k_t. \end{aligned} \tag{8.3}$$

The best solution for a finite set of signaling probability values can be found through an exhaustive search. The resulting performance of designs using 4 through 9 phases and a design density of $k_t = 50$ is shown in Figure 8.14. Figure 8.15 shows the effect of changing the design density with this design algorithm using 9 phases. The 9-phase design has better than 0.99 probability that just one node will survive signaling for all practical densities of contenders.

8.4.3 Echoing

Because CRS is equally effective with most practical densities of contending nodes, it makes capacity a function of space. There is no congestion collapse. Also, because the

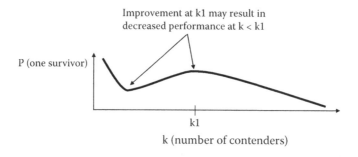

Figure 8.13 The Effect of Optimizing a Signaling Design for a Single Contender Density.

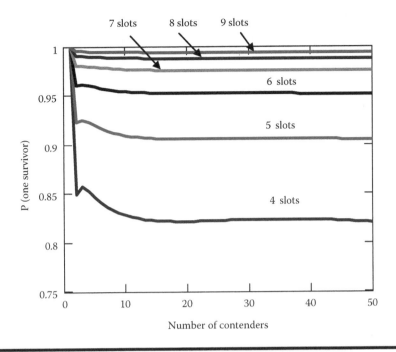

Figure 8.14   CRS Designs Holding the Design Density Constant at $k_t = 50$ and Changing the Number of Signaling Slots.

protocol is synchronous, it does not suffer from hidden terminals, exposed terminals, deafness, or misdirected contentions. However, the separation above does not prevent collisions. This is intentional, so the protocol can benefit from using smart antennas, channelization, or other adaptation techniques. In some cases when neither capture nor other contenders can resolve this conflict, a blocking condition may occur. Blocking

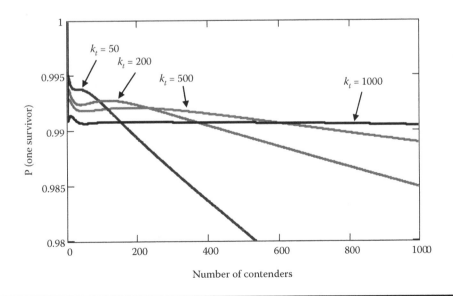

Figure 8.15   CRS Designs Holding the Number of Signaling Slots Constant at 9 and Changing the Design Density.

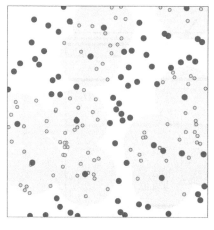

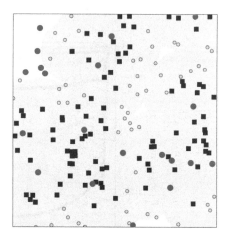

(a) 75 contenders after contention   (b) 19 contenders after echoing

Figure 8.16 Example of Echoing. Nodes that Hear a Contention Signal Echo them in the Next Signaling Slot. Panel (a) Shows the Contention Signal and Panel (b) Shows the Subsequent Echo. The Square Dots in Panel (b) Identify the Nodes that Transmit Echoes. Panel (a) is the Same Initial Set of Signals as Transmitted in Figure 8.11(b) But the Echo Sharply Reduces the Number of Surviving Contenders.

can be resolved by a simple signaling technique called echoing. Signaling phases are designed with two slots. Contenders signal in the first as described earlier, but then neighbors who hear the signals echo them in the second slot. Figure 8.16 is an example. The first signal shown in Figure 8.16(a) is identical to Figure 8.11(b), but, with echoing, all the nodes that hear these first signals echo them. Figure 8.16(b) shows the result. There is a much more sparse set of survivors and this will become sparser still in subsequent signaling phases.

Echoing can be implemented in two ways. First, the entire signal design can use echoing phases. The synchronous unscheduled multiple access (SUMA) protocol is an example [27]. SUMA does not use the RTS–CTS exchange since the echoing phases separate contenders by more than two hops. However, as demonstrated in Figure 8.16, the thinning of the set of contenders will be very aggressive. For this reason, the SUMA protocol divides CRS into two reduction periods with a promotion period between them that provides a method for some of the contention losers in the first reduction period to be promoted back to contenders for the second reduction period (see Figure 8.17). At the conclusion of the first reduction period, all contention winners signal in the promotion slot and all nodes that hear these signals echo them. A former contender that hears neither the assertion signal nor the echo is promoted back to being a contender and then contends in the second reduction period as before. All survivors from the two reduction periods transmit their PDU which is followed by an ACK for unicast transmissions.

The alternative echoing scheme is to conditionally invoke echoing. Figure 8.18 illustrates a 9-"single slot" phase design that can be dynamically converted to a 4-phase echoing design. If a contender detects the condition that a possible block is occurring (e.g., multiple successful contentions, but failed handshakes to the same destination) or if the contender is sending a broadcast where the CTS and ACK responses are not feasible, it invokes echoing by signaling in the echo invoke (EI) slot. The signaling design in Figure 8.18b is the design used by all nodes that send or hear the EI signal.

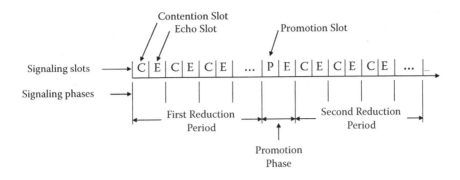

Figure 8.17 The CRS of Synchronized Unscheduled Multiple Access with its Echoing Phases, Promotion Phase, and Two Reduction Periods.

SCR provides many additional features to support quality of service [28], energy conservation [29], and channelization [30, 31]. However, we curtail our discussion at this point because it provides sufficient background to understand its suitability for smart antenna use in MANETs.

8.4.4 Antenna Use Strategies

SCR provides multiple approaches to enable antenna adaptation. In the first approach, a directional CRS is used to support antenna adaptation. In the remaining approaches, adaptation occurs after an omnidirectional CRS together with the handshake packets.

Directional synchronous unscheduled multiple access (DSUMA) [33] was designed for use with switched beam antennas. It is an extension of SUMA where some signals are sent and received directionally. It assumes that contending nodes have learned the beams which best point toward the destinations. So these nodes transmit their assertion signals on those beams. All nodes that do not send signals in the contention signaling slot listen omnidirectionally and those that hear assertion signals echo them omnidirectionally. The contending nodes, however, only listen for echoes in the direction they are

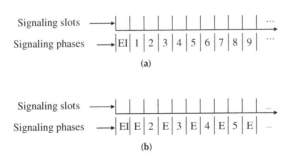

Figure 8.18 A Signaling Design to Selectively Use Echoing: In Most Contentions, Nodes Use the Signaling Design Shown in (a). If the Source Detects a Blocking Condition, Knows the Source to Be an Exposed Node, or Wants to Broadcast a Packet, it May Invoke Echoing. If a Node Signals in the Echo Invoke (EI) Slot then that Node and All of its Neighbors Use the Echoing Design of (b).

contending. At the conclusion of the contention, destinations listen only on the beam they heard the last assertion. This signaling approach could also be implemented with DOA algorithms and arrayed antennas; however, the processing time for each signal may make the approach impractical.

There are four different approaches to exploiting smart antennas at the conclusion of the CRS, together with the RTS–CTS–PDU–ACK exchanges.

1. **Simple pointing**: Assumes nodes know their location and the location of their neighbors and only use this information to point their antennas. A contender starts out by pointing its antenna toward the destination. The destination receives the RTS using an omnidirectional antenna and with the information in the RTS identifies the source, looks up the direction to the source, and points its antenna toward that direction for all subsequent exchanges in the transmission slot.

2. **Adaptive pointing**: A contender starts out by transmitting omnidirectionally or pointing its antenna toward where it thinks the destination is. The destination receives the RTS using a phased array antenna and uses a DOA algorithm to determine the direction to the transmitter. It points its antenna as soon as possible in that direction for the remainder of the transmission slot. The source uses the destination's CTS transmission and the same DOA method to refine its pointing toward the destination.

3. **Environmentally Adapted Reception (EAR)**: A contender starts out by transmitting omnidirectionally or pointing its antenna toward where it thinks the destination is. The destination uses an adaptation technique to optimize reception of the arriving signal. However, the destination transmits using simple pointing. The source adapts to the destination's signal.

4. **Environmentally Adapted Reception and Transmission (EART)**: A contender starts out by pointing its antenna toward where it thinks the destination is. The destination uses an adaptation technique to optimize reception of the arriving signal. The destination uses these weights in transmitting its CTS. The source optimizes its reception of the destination's signal and uses these weights to transmit the PDU. Note that in the derivation of the weights, at each step there are signals from the subsequent receivers and so there is information on where not to point.

In all of these approaches, the basic access mechanism remains the same. Any one of the technologies can be employed at any node without any requirement for all nodes to have the capability. This is so because adaptations do not require the exchange of information between nodes, but occur in the physical layer and because each adaptation contributes to the collective benefit of reduced interference. Nodes with omnidirectional antennas can run SUMA while nodes with switched beam antennas can run DSUMA. Similarly, nodes with omnidirectional, directional, and environmentally adaptive antennas can operate together in the same network using the basic SCR contention mechanism.

8.4.5 Exploitation Conditions

SCR, as described, does not support the creation of condition 1, know where to point; however, it can be built into the protocol where detected directions to previous transmissions from distant nodes are used to estimate the direction in future contentions. Other techniques for this pointing may be built into routing protocols that track node locations [28]. The benefits of SCR are seen in its ability to resolve the latter 5 conditions (numbering refers to the corresponding condition listed in Section 8.2.4):

2. In these protocols there is no ambiguity where not to point. There is no risk that ongoing exchanges will be interrupted by new contentions because SCR is synchronous.
3. SCR enables the adaptive antennas to acquire the conditions for determining the weighting of the antenna elements. With SCR, weights can be determined with each reception. This condition is most important for environmental adaptive transmission (EAT). Weights determined during the reception of the RTS, CTS, and PDU can be applied to each of the subsequent transmissions. The nice feature of SCR is that these weights are derived in the presence of interfering signals from the transceivers that will be receivers during the subsequent transmission. SCR creates the conditions that allow EAT solutions to point nulls toward these receivers.
4. The CRS prevents congestion. At the conclusion of CRS receivers will be in range of no more than two or three interfering transmitters.
5. CRS prevents coincident transmission. CRS causes separation of contenders that results in a nice angular separation between transmitters.
6. SCR preserves the condition. The synchronous nature of SCR does not prevent movement of nodes and of objects in the environment but it prevents new interference from within the network.

8.5 MODELING SMART ANTENNAS IN MANET SIMULATIONS WITH SCR

Modeling of antenna effects in MANETs can have quite different meanings depending on the objective of the simulation study. Those interested in the adaptation algorithms used by the antennas may want a very detailed model of the environment, the propagation that may occur in that environment, and the antenna geometry and radiation patterns. The modeling goal is to support the study and evaluation of the antenna technology itself. These types of models are usually impractical for MANET simulations. Those interested in protocol methods of tracking direction may have a very simple sector model of an antenna and no environmental detail. The modeling goal here is to assess whether the antenna is pointed in the right direction so that the tracking mechanism might be evaluated. Finally, those interested in assessing the antenna technology contribution to a networking solution may abstract the capabilities of the antenna to some lower resolution model that captures the essence of their contribution. Here, the modeling goal is to determine the antenna technology capability that best improves the network's performance. It is this latter modeling approach that we discuss.

8.5.1 Modeling Smart Antenna Processes

The different levels of antenna intelligence have different modeling requirements. Switched beam technologies are the easiest to model. Higher level protocols direct the radios where to point their antennas. Modeling only requires consideration of the power pattern of the antenna, the direction it is pointed, and then its relative direction to the transmitter or receiver of interest. At the time antenna gain is assessed, the relative direction from the mainbeam is used to look up or calculate the gain from the antenna power pattern. The intelligence to point the antenna resides in the protocols, and the only requirement to model the gain at the physical layer is to provide a means for the protocol pointing the antenna to communicate the pointing direction to the part of the model where antenna gain is assessed.

The modeling task for simulations assessing adaptive antenna technologies does not require the modeling of information exchange between protocols and the physical layer, but requires that an assessment be made to determine whether the smart antenna is able to adapt to a desired incoming signal in the presence of interfering signals. The timing model of Figure 8.4 that is discussed in Section 8.2.3 may be used to make this assessment. Two additional assessments may be appropriate. If the access mechanism does not prevent coincident transmissions, then an assessment must be made whether interferer directions are too close to the incoming signal direction. If the access mechanism does not prevent congestion, then an assessment must be made to determine if the antenna can adapt to the quantity of interfering transmitters. Finally, if the adaptation is successful, the model must assess the effect of the adaptation.

8.5.2 Modeling Smart Antenna Effects

The performance of directional antennas is quantified by the directivity of the mainbeam, i.e., its beamwidth. Beamwidth can be specified between half power points, HPBW, or the BWFN. Both are illustrated in Figure 8.1. The second measure of performance is selectivity. This measures the average gain in out-of-beam directions. The specific patterns can be quite complex and will vary based on the specific design of an antenna. We use a more generic model in our simulation called a beam and ball, which is illustrated in Figure 8.19. It accurately models the characteristics of a mainbeam, but blurs the variation of the sidelobes as a single gain in all directions. This model is very general and provides a computationally simple means to access gain in a direction. The calculation

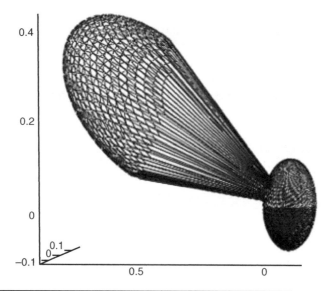

Figure 8.19 Beam and Ball Model for Antenna Gain.

occurs in two steps. First, the angle between the mainbeam and the direction to a receiver is determined and then this angle is used to assess the gain. Let $\mathbf{m}$ be the unit vector that points in the direction of the mainbeam. Let $\mathbf{r}$ be the unit vector toward the receiver. The vector $\mathbf{r}$ can be calculated using

$$\mathbf{r} = \frac{\mathbf{d} - \mathbf{s}}{|\mathbf{d} - \mathbf{s}|}, \tag{8.4}$$

where $\mathbf{s}$ and $\mathbf{d}$ are the coordinates of the source and destination locations. The angle between these directions, θ, is easily calculated using the inner product

$$\theta = \arccos\left(\mathbf{m}^{\mathrm{T}}\mathbf{r}\right) \tag{8.5}$$

and the gain is calculated using

$$g = \left| \begin{array}{ll} g_{mb}\frac{\sin(\theta)}{\theta} & \theta \leq \frac{BWFN}{2} \\ g_{mb} - g_{oob} & \theta > \frac{BWFN}{2} \end{array} \right., \tag{8.6}$$

where g_{mb} is the maximum gain in the mainbeam direction and g_{oob} is the difference in gain from the mainbeam in the ball directions. In the beam and ball model $g_{oob} = $ MSLL.

If a more detailed model of the antenna technology is desired, say the actual antenna pattern of an array antenna, then the gain calculation would need to be replaced by a two-step process. First, the antenna gain pattern must be calculated based on the orientation of the antenna and the direction of pointing. Figure 8.2 illustrates why this is so, the power patterns vary by the relative direction of the pointing to the orientation. In the second step, the actual gain would be determined by the direction to the distant end using this power pattern.

Adaptive antennas are modeled in a softer sense. If the criteria for adaptation are not met then no changes in gain are made at the antenna. If the criteria are met, then, depending on the technology, an adaptation effect is implemented. We model three different methods of adaptation.

AP: If the criteria for adaptation are met, then the antenna for the receiver is pointed in the direction of the captured signal for the remainder of the packet exchange. The gain is assessed as described above. This direction might also be stored as the direction to the source node for the receiver to use in subsequent contentions to send traffic to this node.

EAR: The simplest way to model the environmentally adaptive antenna effect is, if the adaptation criteria are met, to attenuate interfering signals by some level of gain. The specific level of gain should be representative of the combined capability of the adaptation algorithm and the design of the antenna. If the actual attenuation varies with the quantity and specific directions of the interferers, it may be appropriate to model the attenuation with some distribution representative of how the attenuation varies statistically. Creating the statistical distribution requires knowing the antenna technology first. Interfering signals that arrive after the adaptation are not likely to be attenuated by an earlier adaptation, and as shown in Figure 8.3, may more likely require amplification instead. So, in a similar manner, these signals may be amplified by a gain that may also have a distribution representative of the technology being used. Higher fidelity models are feasible if the technology is well-understood. The simulation can both check that the adaptation conditions are met and then use the collection of arriving signals that arrive during the adaptation window to create the inputs for the adaptation algorithm. Gains can then be determined using the adapted antenna patterns.

Environmental Adaptive Transmission (EAT): Assumes that if a radio has adapted to receive a signal, it can use the same information to transmit and thus reduce the gain in the direction of the previously interfering transmitters. Transmissions received at these interfering nodes from the adapted transmitters are attenuated using the same attenuation used in the previous EAR. This mechanism assumes the access protocol caused all the receivers to be transmitters during the receive adaptation, which is the case when using SCR. This model would not be appropriate for a CSMA style protocol.

In this sort of model, the ability to capture one signal in the presence of many is enhanced by using unique training sequences for each destination. If the access approach makes this possible and these sequences are used by the antennas, then the model above would have lower values for the parameters SIRc and SIRa. Other critical characteristics that differentiate antenna technologies are the actual times it takes for adaptations to take place. The duration of the training sequences used in adaptation also affects the required timing between MAC events in a protocol implementation.

8.6   EVALUATION OF SCR

The simulation approach above was used to evaluate the effects of smart antenna technologies on the performance of the standard SCR protocol. The study follows.

8.6.1   Simulation Scenario

Our model of each node included an explicit representation of the SCR protocol together with a perfect router. All transmitters used the same transmit power. The perfect router assumes links exist between pairs of nodes if the arriving signals can achieve a specified SNR when there is no interference. Routes were minimum hop. Path loss was determined using the 2-ray propagation model with vertical polarization on flat earth without terrain features. A total of 156 nodes were randomly placed on a square surface, seven transmission ranges [2] on a side, which was wrapped toroidally. This results in an average node density of ten nodes per transmission area. Nodes were stationary throughout the simulation. Packet arrivals at each node were exponentially distributed at the same rate and each arrival was randomly routed to one of the other nodes in the network. The radio is assumed to have transmission capabilities similar to those of an 802.11 modem using its 1 Mbps DSSS modulation scheme, so we use the bit error rate curves of binary phase shift keying. We sized the transmission slots to send 512 byte payload packets, assume header sizes, RTS, CTS, and ACK packet sizes the same as those used in the 802.11 MAC. Signaling, handshake packets, headers, and interframe spaces account for 34% of a transmission slot's duration and there were approximately 163 transmission slots per second. We used a single scenario, i.e., identical node placement and traffic,

---

[2] We define the transmission range as the distance that a signal has propagated when its strength drops to 10 dB above the thermal noise.

and observed the effects of changing smart antenna techniques and their performance parameters. This network was fully connected with a 10 dB SNR criteria for links.

The best measure of the MAC performance in this scenario and the measure that we use, is MAC throughput, which is the rate packets are exchanged with neighbors. All other performance measures are correlated with this rate. The following information is provided to help the reader interpret the results. The spatial reuse of the channel in the scenario is the MAC throughput (pkts/s) divided by the slots in a second, $\approx$ 163. The total area of the network is 15.6 transmission areas, so a MAC throughput of 2543 pkts/s corresponds to a throughput of one packet per transmission slot per transmission area.

8.6.2 Experiments

We conducted several sets of experiments comparing the effects of varying the directivity (i.e., BWFN) and selectivity (i.e., MSLL) of antennas modeled as beam and ball, and the effectiveness of the adaptation techniques. The standard experiment used a 10 dB SNR for signal and link detection. Table 8.1 lists the details of the modifications for each experiment. The ID numbers in this table are used to identify the experiment performances in the graphs.

Figure 8.20 compares the MAC throughput when using simple pointing. Initially, the selectivity rather than the directivity had the greater effect on capacity, but once the MSLL was below -20 dB, selectivity became more important. We validated this observation

Table 8.1 Experiment Settings

ID	Tech	BWFN	MSLL (dB)	SIR_c (dB)	t_s (μs)	SIR_a (dB)	t_f-t_{sm} (μs)	AG (dB)
1	omni		0					
2	SP	60	-12					
3	SP	30	-12					
4	SP	10	-12					
5	SP	60	-20					
6	SP	30	-20					
7	SP	10	-20					
8	SP	60	-30					
9	SP	30	-30					
10	SP	10	-30					
11	EAR	60	-12	6	1	3	100	-12
12	EAR	60	-12	6	1	3	100	-20
13	EAR	30	-12	6	1	3	100	-12
14	EAR	30	-12	6	1	3	100	-20
15	EAR	10	-12	6	1	3	100	-12
16	EAR	10	-12	6	1	3	100	-20
17	EAR	60	-12	3	1	1	100	-12
18	EAR	60	-12	3	1	1	100	-20
19	EAR	30	-12	3	1	1	100	-12
20	EAR	30	-12	3	1	1	100	-20
21	EAR	10	-12	3	1	1	100	-12
22	EAR	10	-12	3	1	1	100	-20

2 through 10 use adaptive pointing with the same directivity, selectivity, and with SIr_c = dB, t_s = 1 μs, SIR_a = 3 dB, t_f = 100 μs. 11 through 22 use the EART technique.

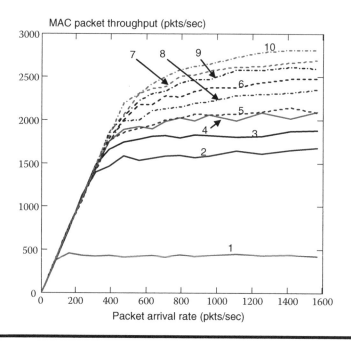

Figure 8.20   Results of Simple Pointing Experiment.

by executing additional simulations holding the load constant and varying BWFN and MSLL. The results are shown in Figure 8.21. MSLL rapidly increases throughput down to −15 dB where it starts to level off. BWFN has a linear effect on throughput. These results indicate that selectivity should take precedence in antenna design until most sidelobes are 15 dB below the mainbeam gain.

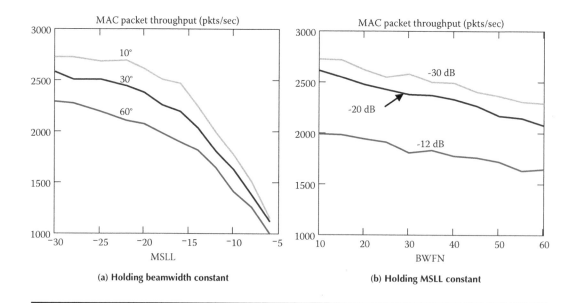

Figure 8.21   Evaluation of the Effect of Directivity and Selectivity on Capacity (Simulations Use a Common Network Traffic Load of 1100 pkts/sec).

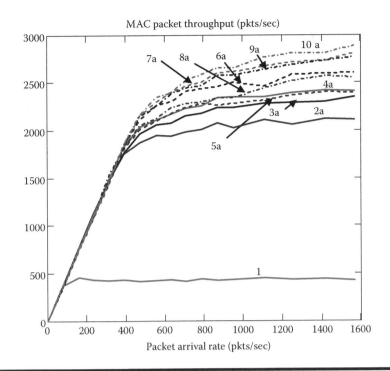

Figure 8.22   Results of Adaptive Pointing Experiments.

Figure 8.22 compares the MAC throughput when using adaptive pointing. There is an improvement over simple pointing due to the adaptation that occurs before receiving the first packet; because the network is stationary all subsequent pointing is the same. We would expect a greater difference in performance if this were a mobile net because the simple pointing techniques are less effective at knowing where to point.

Figure 8.23 illustrates the performance of using EAR. There is little difference between when the EAR gain is −12 dB and −20 dB, indicating that a large gain envisioned in pointing nulls is not necessary. Adaptation effectiveness is most dependent on the robustness of the adaptation (i.e., adaptation can occur in a lot of interference.) Performance also improved with the selectivity of the initial antenna pointing. This may be an unrealistic result as highly directional transmissions can reduce the multipath that enables some EAR techniques to work. The take away observation remains that anything that enables the initial adaptation is good.

Figure 8.24 illustrates the performance of using EART, the combined use of EAR and EAT models. There is little improvement, which further indicates the significance of acquiring the first RTS. There is no difference in the conditions for the success of this first packet between EAR and EART. EAT does not kick in until the CTS and subsequent transmission.

All of our antenna simulations were performed without any processing gain. As we described earlier, one of the techniques that allows a receiver to capture a specific signal among interferers is to use unique training sequences for each receiver. One type of sequence is the spreading code that is used with direct sequence spread spectrum. SCR supports the use of codes in ad hoc environments because it solves the two hard problems of knowing which channel receivers should listen and preventing the near–far

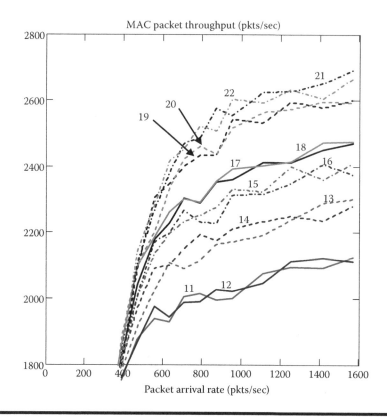

Figure 8.23   Results of Environmentally Adaptive Reception Experiments.

effect (see [30]). We conducted further experiments to determine if the smart antennas would provide an increased throughput when used with processing gain. We repeated all of the experiments described above with a 15 dB processing gain and found that with this large of a processing gain, smart antennas provided little benefit. This does not indicate that there are no benefits with combining these technologies. One of our objectives is to increase the number of exchanges. The combined use of code division multiple access (CDMA) with smart antenna technology promises to support the resolution of primary collisions at receivers. The receivers can recover multiple packets that arrive concurrently. Another advantage to using smart antennas is that they harden transmissions to detection and interception which are very valuable features for tactical military systems.

8.7   OPEN ISSUES AND RESEARCH OPPORTUNITIES

The most challenging open issues with using smart antennas in MANETs concern building the nodes. Most other issues are less issues than they are research opportunities to gain further benefits from antenna technologies. We divide these into two groups, those that expand the role of SCR in exploiting smart antennas and antenna technologies that could dramatically affect the types of protocols that can be used in MANETs.

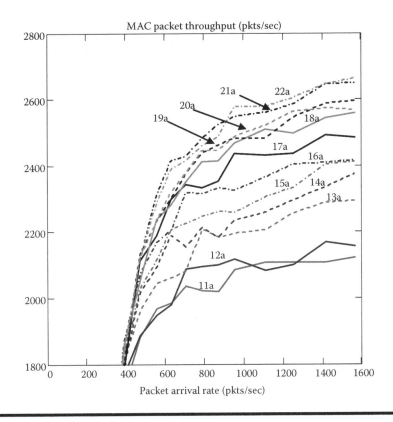

Figure 8.24   Results of Experiments When Using Both Environmentally Adaptive Reception and Environmentally Adaptive Transmission.

8.7.1   Building MANET Nodes with Smart Antennas

The typical approach to building communications systems separates the radio from the antenna so that antennas might be interchanged based on the platform. Clearly adaptive antennas require antennas and radios to be designed as a unit. This commitment can be a disincentive to their development because it requires a commitment to a configuration. However, radio configurations can vary dramatically based on the platform they are placed, meaning a single configuration is not practical. An advantage of the SCR protocol is that systems with different antenna technologies can still work in the same network because the access mechanism does not change. Assuming the modulation remains the same, antennas can be introduced gradually into an existing network as they are developed.

Nevertheless, there are some platforms that may not support smart antennas. The suitability of a platform for smart antenna technology will depend on the frequencies used and the resulting antenna sizes. Smart antennas are unlikely to be appropriate for small platforms and handheld units when low frequencies are used due to the larger antennas required. Then, even if smaller antennas are feasible, some platforms may not be suitable for the antenna technology. Handhelds are particularly suspect due to their small size and the variability of their orientation.

In the design of a MANET system in which exploitation of smart antennas is intended, there must be serious consideration on what modulation schemes to use. Although SCR allows multiple antenna technologies in the same network, they must use the same modulation scheme to interoperate or the communications will default to the lowest common capability. The experience with the 802.11 modems in the 2.4 GHz band gives evidence. The more efficient modulation schemes of the 802.11g and 802.11n must yield to the modulation of the 802.11b when 802.11b devices are part of the network. If the long term intent is to incorporate smart antennas, then the initial design should choose a modulation scheme that will ultimately support the targeted antenna technologies and remain compatible with nodes that do not have the technology.

8.7.2 Advanced Concepts Using SCR

The physical separation between contenders that the CRS of SCR causes is similar to that of a cellular network. With high probability, greater than 0.99, all nodes in range of a contender that survives CRS are idle and ready to receive a packet. Traditionally, each contender sends one packet either to a single destination or broadcasts it to all local nodes. Through the use of antennas and channelization, it becomes possible for a surviving contender to send different packets to multiple destinations in the same transmission. Concepts described in [25,32] may also allow the reception of multiple packets at destinations. Thus, there are a number of new requirements for the protocols used in networks with these capabilities. These protocols will need to schedule packet transmissions in a manner that is compatible with the antenna technology. For example, it may not be possible to send different packets to destinations in a coincident direction and there may be a limit to the number of packets that may be sent. There may also be opportunities to manage spectrum capacity more flexibly. For example, the modulation and channelization approaches may be able to arbitrarily give different proportions of capacity in a transmission to the different downlinks based on demand or a quality of service requirement. Protocols would need to manage this resource allocation.

8.7.3 New Technologies

An exciting new approach in transceiver design is to use block-based processing. With block-based processing, the stream of symbols that are transmitted to communicate a packet are received in blocks and then these blocks are independently optimized; thus, adaptation is done continuously during a packet reception rather than just at the beginning. There is still a need to characterize the channel at the start of the communication, but with this characterization the receivers can not only separate out new interfering transmitters in subsequent blocks they can characterize the channels from these new sources and receive their packets as well [33]. Although this capability does not eliminate the requirements to provide for acquiring the channel, preventing congestion and coincident transmission, it does eliminate the requirement to preserve the condition and so asynchronous protocols may be used. The ramification of this capability on access protocols has not been studied. Further, block-based processing together with frequency duplexing can enable physical layer relaying where transceivers retransmit packets as they are still receiving them, i.e., a previously received and processed block is retransmitted on the relay while subsequent blocks of the same packet are being received [34]. The impact on contention MAC protocol design could be quite dramatic.

8.8 CONCLUSION

This chapter has reviewed the issues in exploiting smart antennas in mobile ad hoc networks and identified the conditions that MAC protocols must create for their exploitation. It described recent proposals for contention MAC protocols and reviewed their potential to exploit directional antennas. Asynchronous access protocols, such as Aloha and the 802.11 DFWMAC, are especially handicapped. One MAC approach, however, synchronous collision resolution, is particularly well-suited for this task. SCR is a contention-based access protocol that creates the conditions necessary for using all types of smart antennas. This chapter described techniques to use in modeling smart antennas in simulations of mobile ad hoc networks. These techniques were then used to compare the effectiveness of various antenna technologies with SCR. There are three main lessons. First, directionality is not as important as selectivity when using directional antennas until the MSLL is 15 dB below the mainbeam gain. Second, small amounts of improvement from adaptation, 10 dB, will get the lion's share of possible improvement over omnidirectional access. And third, the best smart antenna technologies are those that must improve the reception of the RTS packets. SCR supports some very different approaches to increasing the capacity of ad hoc networks. The combined use of SCR, CDMA, and smart antennas could create the foundation of a one-to-many and many-to-one packet exchange paradigm. These new capabilities will place new demands on networking protocols to schedule packet transmissions locally at a node in a manner supportable by the antenna technologies.

REFERENCES

[1] P. Lehne and M. Pettersen, "An overview of smart antenna technology for mobile communications systems," *IEEE Commn. Surveys*, vol. 2, no. 4, pp. 2–13, 4th quarter, 1999.

[2] J. Capon, "High-resolution frequency-wavenumber spectrum analysis," *Proc. IEEE*, vol. 57, Aug. 1969, pp. 1408–1418.

[3] R. Schmidt, "Multiple emitter location and signal parameter estimation," *IEEE Trans. on Antennas and Propagation*, vol. 34, Mar. 1986, pp. 276–280.

[4] R. Roy and T. Kailath, "ESPRIT – estimation of signal parameters via rotation invariance techniques," *IEEE Trans. Aerosp. Electron. Sys.*, vol. 19, Jan. 1983, pp. 134–139.

[5] B. Van Veen and K. Buckley, "Beamforming: a versatile approach to spatial filtering," *IEEE ASSP Mag.*, April 1988, pp. 4–24.

[6] R.T. Compton, Jr., *Adaptive Antennas: Concepts and Performance*, Prentice-Hall, Upper Saddle River, NJ, 1988.

[7] J. Ward and R.T. Compton, Jr., "Improving the performance of a slotted ALOHA packet radio network with an adaptive array," *IEEE Trans. Commn.*, vol. 40, Feb. 1992, pp. 292–300.

[8] L. Kleinrock and F. Tobagi, "Packet switching in radio channels: part I-carrier sense multiple access and their throughput delay characteristics," *IEEE Trans. Commn.*, vol. 23, no. 12, Dec. 1975, pp. 1400–1416.

[9] ANSI/IEEE Std 802.11, Wireless LAN Medium Access Control (MAC) and Physical Layer (PHY) Specifications, IEEE Press, New York, 1999.

[10] ETSI, EN300 652 V 1.2.1, Broadband Radio Access Networks (BRAN); High Performance Radio Local Area Network (HIPERLAN) Type 1; Functional Specification, Sophia Antipolis, France, July 1998.

[11] F. Tobagi and L. Kleinrock, "Packet switching in radio channels: part II–the hidden terminal problem in carrier sense multiple-access and the busy-tone solution," IEEE Trans. Commn., vol. 23, no. 12, Dec. 1975, pp. 1417–1433.

[12] F. Tobagi and L. Kleinrock, "Packet switching in radio channels: part III–polling and (dynamic) split-channel reservation multiple access," *IEEE Trans. Commn.*, vol. 24, no. 8, Aug. 1976, pp. 832–845.

[13] P. Karn, "MACA – a new channel access method for packet radio," *Proc. ARRL/CRRL Amateur Radio 9th Computer Networking Conference*, New York, April 1990.

[14] V. Bharghavan, A. Demers, S. Shenker, and L. Zhang, "MACAW: a media access protocol for wireless LAN's," *Proc. ACM SIGCOMM*, London, Aug. 1994, pp. 212–225.

[15] J.J. Garcia-Luna-Aceves and C.L. Fullmer, "Floor acquisition multiple access (FAMA) in single-channel wireless networks," *Kluwer Academic Publishers Mobile Networks and Applications*, vol. 4, no. 3, 1999, pp. 157–174.

[16] A. Nasipuri, S. Ye, J. Yon, and R. Hiromoto, A MAC protocol for mobile ad hoc networks using directional antennas," *Proc. IEEE Wireless Communications and Networking Conf.*, Chicago, Sept. 2000, pp. 1214–1419.

[17] Z. Huang, C. Shen, C. Srisathapornphat, and J. Jaikaeo, "A busy tone based directional MAC protocol for ad hoc networks," *Proc. of IEEE MILCOM*, Anaheim, CA, Oct. 2002.

[18] R. Ramanathan, "On the performance of ad hoc networks with beamforming antennas," *Proc. ACM MOBIHOC*, Long Beach, CA, Oct. 2001, pp. 95–105.

[19] Y. Ko, V. Shankarkumar, and N. Vaidya, "Medium access control protocols using directional antennas in ad hoc networks," *Proc. of IEEE INFOCOM*, Tel Aviv, Israel, March 2000, pp. 13–21.

[20] M. Takai, J. Martin, A. Ren, and R. Bragodia, "Directional virtual carrier sensing for directional antennas in mobile ad hoc networks," *Proc. ACM MOBIHOC*, Lausanne, Switzerland, June 2002, pp. 183–193.

[21] R. Choudhury, X. Yang, R. Ramanathan, and N. Vaidya, "Using directional antennas for medium access control in ad hoc networks," *Proc. ACM MOBICOM*, Lausanne, Switzerland, June 2002, pp. 59–70.

[22] R. Ramanathan, J. Redi, C. Santivanez, D. Wiggins, and S. Polit, "Ad hoc networking with directional antennas: a complete system solution," *Proc. Of the IEEE Wireless Comm. and Networking Conf.*, Atlanta, GA, March 2004, pp. 375–380.

[23] Z. Li, P. Zhou, and J. Hou, "Fragmentation based D-MAC protocol in wireless ad hoc networks," *Proc. IEEE ICDCS*, Providence, RI, May 2003.

[24] L. Thomson, "Performance of systems using adaptive arrays in mobile ad hoc networks," *Proc. IEEE MILCOM*, Boston, MA, Oct. 2003, pp. 1450–1455.

[25] J. Ward and R.T. Compton, Jr., "High throughput slotted ALOHA packet radio networks with adaptive arrays," *IEEE Trans. Commn.*, vol. 41, pp. 460–470, March 1993.

[26] J.A. Stine, G. de Veciana, K. Grace, and R. Durst, "Orchestrating spatial reuse in wireless ad hoc networks using synchronous collision resolution," *J. Interconnect. Netw.*, vol. 3, no. 3 & 4, Sept. and Dec. 2002, pp. 167–195.

[27] K. Grace, "SUMA – the synchronous unscheduled multiple access protocol for mobile ad hoc networks" *Proc. 11th International Conf. Computer Communications and Networks*, Miami, FL, Oct. 2002, pp. 22–28.

[28] J.A. Stine and G. de Veciana, "A paradigm for quality of service in wireless ad hoc networks using synchronous signaling and node states," *IEEE J. Sel. Areas Commn.*, vol. 22, no. 7, Sept. 2004, pp. 1301–1321.

[29] J.A. Stine and G. de Veciana, "A comprehensive energy conservation solution for mobile ad hoc networks," *IEEE Int. Communication Conf.*, New York, 2002, pp. 3341–3345.

[30] J.A. Stine, "Exploiting processing gain in wireless ad hoc networks using synchronous collision resolution medium access control schemes," *Proc. IEEE WCNC*, New Orleans, LA, 2005, pp. 612–618.

[31] J.A. Stine, "Enabling secondary spectrum markets using ad hoc and mesh networking protocols," *Academy Publisher Commn. J.*, vol. 1, no. 1, April 2006, pp. 26–37.

[32] J. Fite, "Acquiring CDMA packets in an ad hoc network using adaptive arrays," *Proc. IEEE MILCOM*, Boston, MA, 2003, pp. 654–657.

[33] K.H. Grace, J.A. Stine, R.C. Durst, "An approach for modestly directional communications in mobile ad hoc networks," *Springer Telecommn. Sys. J.*, March/April 2005, pp. 281–296.

[34] R. Ramanathan, "Challenges: a radically new architecture for next generation mobile ad hoc networks," *ACM MOBICOM 2005*, Cologne, Germany, Aug. 2005, pp. 132–139.

9

CROSS-LAYER DESIGN FOR WIRELESS SENSOR NETWORKS WITH VIRTUAL MIMO

Yong Yuan and Min Chen

Contents

Energy efficiency, reliability, and quality of services (QoS) provisioning are the main concerns in the design of wireless sensor networks (WSNs) to support the diverse applications. However, the design issues of energy efficiency, reliability, and QoS provisioning in WSNs are multifaceted problems jointly influenced by the physical, medium access control (MAC), network and transport layers. Recently, some virtual multiple-input

241

multiple-output (MIMO) schemes based on single antenna sensors have been proposed and studied to improve the energy efficiency of wireless communication schemes. Though the initial proposal of virtual MIMO schemes focuses on the physical layer design, the adoption of this novel technology also provides a wider design space for the schemes in the upper layers. In this chapter, a cross-layer design scheme based on a virtual MIMO scheme is proposed for WSN. In the design, the sensor nodes form a cooperative node set to transmit data according to the virtual MIMO scheme. Then, the virtual MIMO scheme, multi-hop routing scheme, and hybrid ARQ (HARQ)-based retransmission schemes are jointly designed to improve the performance of energy efficiency, reliability, and QoS guarantees, in terms of delay and throughput. Based on the design, we also developed the model for end-to-end QoS and overall energy consumption of the design in terms of the bit error rate (BER) performance in each link. Then, the energy saving performance and QoS provisioning ability of the scheme are demonstrated through comprehensive simulations. At last, the chapter is concluded by identifying some open research issues on this topic.[1]

9.1 INTRODUCTION

Recent years have witnessed a growing interest in deploying a sheer number of microsensors that collaborate in a distributed manner on data gathering and processing. Sensors are expected to be inexpensive and can be deployed in a large number to harsh environments, which implies that sensors are typically operating unattended. In addition, sensor networks are also subject to high fault rates; connectivity between nodes can be lost due to environmental noise and obstacles; nodes may die due to battery depletion, environmental changes or malicious destruction. In such an environment, reliable and energy-efficient data delivery is crucial because sensor nodes operate with limited battery power and an error-prone wireless channel. On the other hand, wireless sensor networks are expected to be used in a wide range of applications, such as target tracking, habitat sensing, and fire detection, etc. The data gathering in such applications is often required to be timely. For example, when a target enters an area of interest, it may be critical to reduce the delay of sensor reports. If the reported event is not received by the sink node within a certain deadline, reported information may be obsolete and useless. Therefore, end-to-end quality of services (QoS) provisioning is important for such applications in WSNs. In addition, different applications have different end-to-end transmission quality requirements in terms of latency and throughput.

Due to these characteristics, energy efficiency, reliability, and end-to-end QoS provisioning should be jointly considered in the design of wireless sensor networks (WSN). However, these design issues are multifaceted problems influenced by the physical, MAC, network, and transport layers. The energy efficient wireless communication schemes, routing schemes, power conservation schemes and reliable transportation schemes should be jointly considered to maximize the performance in terms of energy efficiency, reliability, and end-to-end QoS.

Among all the related schemes, energy efficiency is deemed as a necessity for the wireless communication scheme in a WSN, because wireless communication has been identified as the dominant power-consuming operation. In addition, the harsh working

---

[1] This is a major revision of the work published in *IEEE Trans. Veh. Technolo.*, vol. 53, no. 3 (May 2006).

environments, channel fading, interference, and radio irregularity further pose challenges on the design of energy-efficient wireless communication schemes for WSNs. In the wireless communication schemes, multiple input multiple output (MIMO) techniques have been studied intensively in recent years [1,2] due to their effectiveness for enhancing reliability, energy, and bandwidth efficiency and the ability to deal with fading phenomena. The characteristics of MIMO techniques make them desirable for WSNs. However, it is difficult to directly apply MIMO techniques in the low-cost small-sized sensors. Some virtual MIMO schemes based on single antenna sensors have been proposed and studied to improve energy saving and spectral efficiency [3–11,13]. In such schemes, multiple individual single-antenna nodes will form a virtual antenna array, and each node will be viewed as an antenna in the array. These nodes will cooperate in the MIMO manner on information transmission or reception. Based on the virtual MIMO design, the advantages of the MIMO scheme will make the physical layer of a WSN more reliable and energy efficient. On the other hand, the adoption of the virtual MIMO scheme in the physical layer also opens a wider design and optimization space for the schemes in the upper layers, such as the retransmission, distributed operation, multi-hop routing, and QoS provisioning schemes.

In this chapter, the state-of-the-art related schemes are summarized and compared, including the virtual MIMO schemes, reliable data transmission schemes, and QoS provisioning schemes. Then, a cross-layer design based on the virtual MIMO scheme is proposed to improve the performance of WSN in terms of energy efficiency, QoS provisioning and reliability. In the cross-layer design, radio irregularity of wireless communications, multi-hop routing, retransmissions, and end-to-end QoS provisioning are jointly considered with the virtual MIMO scheme.

Firstly, we design a single-hop transmission scheme, where an adaptive cooperative nodes selection strategy is proposed to find the optimal set of cooperative nodes to minimize the energy cost. To improve the reliability of the data transmission, three hybrid ARQ (HARQ)-based retransmission schemes are considered to incorporate into the virtual MIMO scheme. The average energy consumption for a successful packet transmission by the virtual MIMO scheme under three retransmission schemes is analyzed and compared. Then the retransmission scheme, using hop-by-hop recovery, is incorporated into the virtual MIMO scheme due to its efficiency. In analysis, the overall energy consumption for a successful packet transmission is found to depend on the average retransmission times and the energy consumption per time transmission, which can be traded off by the bit error rate (BER) performance in transmission. Therefore, an optimal set of transmission parameters, including the BER performance, the number of cooperative nodes, and the number of hops, can be found to minimize the overall energy consumption.

Based on the single-hop transmission scheme, an end-to-end transmission scheme is designed. To simplify the procedure of forming a set of cooperative nodes for the virtual MIMO scheme, the concept of clustering is adopted to organize the sensor nodes into multiple clusters and form the cluster heads as a multi-hop backbone. During the transmission, each cluster head will transmit data to its neighbor cluster through the set of cooperative nodes by the virtual MIMO scheme. The energy cost for the virtual MIMO communication will be defined as the routing cost between two cluster heads in the multi-hop backbone. Then, the shortest path tree (SPT) will be constructed by finding the path with minimum overall energy cost for each cluster head to transmit data to the sink. On the other hand, because the retransmission scheme is considered in each single-hop transmission, the throughput and energy consumption for packet transmission on each link in the SPT will be dependent on the BER performance. Accordingly, the queuing latency and throughput on the link will also be dependent on the BER performance, which in turn impacts the end-to-end latency and throughput. Therefore, the low-level communication parameter, BER performance P_b, will determine the high-level QoS performance in terms of end-to-end latency and throughput. Based on this observation, the end-to-end QoS performance and the overall energy consumption are modeled by the queuing theory in terms of the BER performance of each link in the SPT. The search for the optimal BER performance for each link is modeled as a nonlinear constrained optimization problem to minimize the overall energy consumption without violating the end-to-end QoS requirements. The particle swarm algorithm (PSO) is employed in this chapter to solve the problem.

The remainder of the chapter is organized as follows. Section 9.2 describes the related work of the design. In Section 9.3, the proposed cross-layer design scheme based on virtual MIMO is discussed in detail. Then, in Section 9.4, the energy consumption and QoS performance of the scheme are analyzed and an optimization model is developed to find the optimal parameters. Section 9.5 presents the simulation results. Then, Section 9.6 provides some conclusions and points out aspects that will be the subject of future research.

9.2   RELATED WORK

Our work is closely related to the virtual MIMO scheme design in WSNs, the reliable data transfer in WSNs, and end-to-end QoS provisioning in WSNs. We will give a brief review of the work in these three aspects.

9.2.1   The Related Work in the Virtual MIMO Design in WSNs

The basic idea of the virtual MIMO scheme is extended from the *virtual antenna arrays* (VAA) in the design of ad hoc oriented 4G networks [3,4]. M. Dohler proposed the system capacity analysis, resource allocation strategy, and related protocols about the application of VAA to cellular networks in [3,4]. As for the work of virtual MIMO scheme design in WSN, X. Li [5] proposed a virtual MIMO scheme using two transmitting sensors and space–time block code (STBC) to provide transmission diversity in a WSN with neither antenna array nor transmission synchronization. The author argued that according to the scheme, the full diversity and full rate are achieved which enhances power/bandwidth efficiency and reliability. X. Li also extended the scheme for using any number of transmission sensors by the distributed STBC in [6,7]. In [8], X. Li also proposed a blind channel estimation and equalization scheme in such virtual MIMO schemes. B. Azimi-Sadjadi et al. [9] proposed a method in code division multiple access (CDMA) wireless multi-hop networks which groups transmission nodes into cooperative clusters to reduce the total power expenditure of transmitting nodes. S. Cui [10] analyzed a cooperative MIMO scheme with the Alamouti code for single-hop transmission in a WSN. S. K. Jayaweera considered the training overheads of such schemes in [11], and found that the training overheads can be modeled as proportional to the number of cooperative nodes. S. K. Jayaweera also proposed a virtual MIMO communication architecture based on the vertical Bell Laboratories layered space time architecture (VBLAST) processing [12]. J. N. Laneman also did research work on the system capacity analysis of the virtual MIMO scheme in [13]. S. Jagannathan et al. [14] investigated the effect of time synchronization errors on the performance of the cooperative multiple-input single-output (MISO) systems, and concluded that the cooperative MISO scheme has a good tolerance of up to 10% clock jitter. The previous work of virtual MIMO schemes focuses on the MIMO schemes design in WSNs and the analysis of system capacity and energy consumption. However, the previous work did not consider the impacts of the specific issues of multi-hop networking, reliable transmission, and end-to-end QoS provisioning on the virtual MIMO scheme, which may result in suboptimal system performances in terms of energy efficiency, reliability, and end-to-end QoS. Our work differs mainly with the previous work in that the cross-layer design of the virtual MIMO scheme is considered, which integrates the virtual MIMO scheme with the multi-hop routing scheme, retransmission scheme, and end-to-end QoS provisioning.

9.2.2 The Related Work in the Reliable Data Transfer in WSNs

As for the aspect of reliable data transfer in WSNs, since the pioneer work on reliable transport protocol, pump-slowly, fetch-quickly (PSFQ), presented in [16], there are increasing research efforts on studying the issue of reliable data transfer in WSNs [16–21]. PSFQ works by distributing data from source nodes in a relatively slow pace and allowing nodes experiencing data loss to recover any missing segments from immediate neighbors aggressively. PSFQ employs hop-by-hop recovery instead of end-to-end recovery. In [17], the authors proposed reliable multi-segment transport (RMST), a transport protocol that provides guaranteed delivery for applications requiring them. RMST is a selective negative acknowledgment (NACK)-based protocol that can be configured for in-networking caching and repair. An event-to-sink reliable transport (ESRT) protocol is presented in [18]. In ESRT, the sink adaptively achieves the expected event reliability by controlling the reporting frequency of the source nodes. In [19], a protocol called ReInForM is proposed to deliver packets at desired reliability by sending multiple copies of each packet along multiple paths from sources to sink. Several acknowledgment-based-end-to-end-reliable event transfer schemes are proposed to achieve various levels of reliability in [20]. C. Taddia [21] also proposed and compared four information delivery methods by different retransmission schemes in a WSN. End-to-end or hop-by-hop recovery, forward error correction codes and multipath forwarding are the major approaches to achieve the desired reliability by previous work. However, the reliable data transfer in a WSN is a multifaceted problem influenced by multiple protocol layers. In our work, the retransmission schemes in MAC layer and the virtual MIMO scheme in the physical layer are jointly designed to improve the system performance in terms of energy efficiency and data reliability.

9.2.3 The Related Work in the QoS Provisioning in WSNs

End-to-end QoS provisioning in WSNs has so many applications, such as real-time target tracking in battle environments, emergent event triggering in monitoring applications, etc. The applications often have the QoS requirements in terms of end-to-end latency and end-to-end throughput. There are increasing research efforts on studying the issue of QoS provisioning in a WSN. Sequential assignment routing (SAR) is the first routing protocol for WSN that includes a notion of QoS in its routing decisions [22]. Tian He, et al., proposed a stateless protocol for real-time communication in sensor network, called SPEED [23], to reduce the end-to-end deadline miss ratio in WSNs. K. Akkaya proposed an energy-aware QoS routing protocol to support both best effort and real-time traffic at the same time [24]. The purpose is to meet the end-to-end delay constraint of the real-time traffic and maximize the throughput of the best-effort traffic at the same time. K. Akkaya also used a weighted fair queuing (WFQ) based packet scheduling to achieve the end-to-end delay bound in [25]. We also proposed an integrated energy and QoS-aware wireless transmission scheme for a WSN [26], in which the QoS requirements in the application layer, the modulation and transmission schemes in the data link layer and physical layer are jointly optimized for end-to-end QoS provisioning. In this chapter, we consider the problem of energy-aware QoS provisioning in another way, that is, to model the end-to-end QoS performance and overall energy consumption in terms of the BER performance of each link according to the cross-layer design of the multi-hop virtual MIMO transmission scheme. Then, the search for the optimal BER performance of each link is modeled as a nonlinear constrained optimization problem.

9.3 CROSS-LAYER DESIGN BASED ON THE VIRTUAL MIMO SCHEME

In this section, we will describe the proposed cross-layer design scheme based on virtual MIMO in detail. First, the system architecture of the scheme is described. Then, the design of the single-hop transmission is discussed. Based on the single-hop transmission scheme, the end-to-end cross-layer design is proposed.

9.3.1 System Architecture

The reference system architecture of the proposed cross-layer design based on virtual MIMO is demonstrated in Figure 9.1. In the proposed architecture, the data bits collected by multiple source nodes will be transmitted to a remote sink by multiple hops. During the transmission, the sensor nodes will be organized into multiple clusters. The transmission in each hop can be divided into two main operations. First, the cluster head will broadcast the data bits to the cooperative nodes in the local cluster. We assume an additive white Gausian nodes (AWGN) channel with squared power path loss in such transmission due to the short intracluster transmission range. Then, the cooperative nodes will encode and transmit the data bits to the cluster head in the next hop according to the orthogonal STBC. For the intercluster communications, we assume the transmission from each cooperative node experiences frequency nonselective and slow Rayleigh fading.

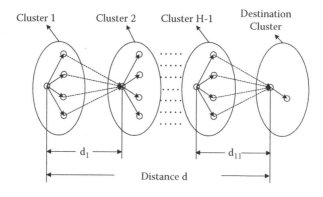

Figure 9.1   Multi-Hop Virtual MIMO Scheme.

Furthermore, the long distance between any two nodes in the network, with respect to the wavelength, gives rise to independent fading coefficients for the cooperative nodes. The rationale behind such channel assumptions is that the intercluster transmission distance is much larger than the intracluster transmission distance and the transmission environments are more complex in the intercluster communication. In the design, the distance between the source cluster to the destination cluster is denoted as d, the number of hops is denoted as H, and the number of cooperative nodes in each single-hop transmission is denoted as J. Because binary phase shift keying (BPSK) can achieve very close performance as the variable-rate modulation scheme with optimal rates, such as multiple quadrature amplitude modulation (MQAM) [12], it is used as the modulation scheme, and the bandwidth is denoted as B. The cluster containing the data source nodes is denoted as S, and the destination cluster containing the sink is denoted as D.

9.3.2 Single-Hop Transmission Scheme Design

During each single-hop transmission, several cooperative nodes will be chosen to communicate with the next cluster head by the virtual MIMO scheme. In order to maximize the performance of single-hop communication between cluster heads, an appropriate strategy should be designed to choose the cooperative nodes. The strategy will be discussed in this subsection. On the other hand, though the virtual MIMO scheme can obtain good BER performance in an energy aware manner, the residual BER will also reduce the reliability of the transmission. To improve the reliability further, the HARQ scheme in the data link layer is incorporated into the hop-by-hop and end-to-end transmission.

9.3.2.1 Strategy to Choose Cooperative Nodes

To maximize the performance of single-hop communications between cluster heads, an appropriate strategy should be taken to choose the optimal cooperative nodes. Suppose that the current cluster head will use J cooperative nodes to transmit data to its neighboring cluster head t by the cooperative MIMO scheme.

Denote the distance between node j and its current cluster head by d_{j1}. Also, denote the distance and path loss for node j to communicate with cluster head t as d_{jt} and k_{jt}, respectively. For each single-hop transmission, the current cluster head will broadcast a data packet to the cooperative nodes. Then, the cooperative nodes will encode and transmit the transmission sequence according to the orthogonal STBC to t toward the sink node. The energy consumption for these two operations in the single-hop transmission will be modeled in the remainder of this section. Then, a novel strategy will be developed to find the optimal set of cooperative nodes to minimize the overall energy consumption.

Let $E_{bt}(1)$ denote the energy consumption for the current cluster head to broadcast one bit to the cooperative nodes. $E_{bt}(1)$ can be broken down into two main components, the transmit energy consumption, $E_{btt}(1)$, and the circuit energy consumption, $E_{btc}(1)$.

The BER performance for BPSK is $P_b = Q(\sqrt{2r})$. Here, r is the signal-to-noise ratio (SNR), which is defined as $r = \frac{P_r}{2B\sigma^2 N_f}$ [27] under the assumption of an AWGN channel, where P_r is the received signal power, σ^2 is the power density of the AWGN and N_f is the receiver noise figure.

In the high SNR regime, we can approximate the BER performance as $P_b = e^{-r}$ by the Chernoff bound [27]. Hence, we obtain $P_r = -2BN_f\sigma^2 ln(P_b)$. Recall that B is the bandwidth of the BPSK modulation scheme. As the assumption of squared power path loss, $E_{bt}(1)$ can be modeled by

$$E_{bt}(1) = E_{btt}(1) + E_{btc}(1)$$

$$= -2(1 + \alpha) N_f \sigma^2 ln(P_b) G_1 d_{max}^2 M_l + \frac{P_{ct} + J P_{cr}}{B} \tag{9.1}$$

where d_{max} is the maximum distance from the cooperative nodes to the cluster head, α is the efficiency of the radio frequency (RF) power amplifier, G_1 is the gain factor at $d_{max} = 1m$, M_l is the link margin, N_f is the receiver noise figure, P_{ct} and P_{cr} are the circuit power consumption of the transmitter and receiver, respectively [10].

Let $f_1(P_b) = -2N_f \sigma^2 ln(P_b)$ and $H(d_{max}) = G_1 M_l d_{max}^2$. Then, (9.1) can be rewritten as (9.2).

$$E_{bt}(1) = (1 + \alpha) f_1(P_b) H(d_{max}) + \frac{P_{ct} + J P_{cr}}{B} \tag{9.2}$$

According to the definition, $H(d_j)$ can be measured as follows. Let the current cluster head transmit a signal with transmit power P_{out}. Then, the power of the received signal at its cluster member, node j, is $P_{j1} = \frac{P_{out}}{H(d_j)}$. Therefore, $H(d_j)$ can be measured as (9.3).

$$H(d_j) = \frac{P_{out}}{P_{j1}} \tag{9.3}$$

From (9.2), we can find that the energy consumption in the intracluster transmission, $E_{bt}(1)$, can be reduced by choosing the nearer cooperative nodes.

To analyze the energy consumption for intercluster transmissions based on the cooperative scheme, denoted by $E_{bt}(2)$, we refine the results in [10]. In [10] an equal transmit power allocation scheme is used as the channel state information (CSI) is not available at the transmitter. If the average attenuation of the channel for each cooperative node pair can be estimated, we can use an equal SNR policy [28] to allocate the transmit power for its effectiveness and simplicity. The average energy consumption per bit transmission by BPSK in such a scheme can be approximated by

$$E_{bt}(2) = (1 + \alpha) \frac{N_0}{P_b^{\frac{1}{J}}} \sum_{j=1}^{J} \frac{(4\pi)^2 d_{jt}^{k_{jt}}}{G_t G_r \lambda^2} M_l N_f + \frac{(J P_{ct} + P_{cr})}{B} \tag{9.4}$$

where N_0 is the single-sided noise power spectral density, P_b is the desired BER performance, G_t and G_r are the transmitter and receiver antenna gains, respectively, and λ is the carrier wavelength [10]. Equation (9.4) is extended from the result in [10] with the settings of different distance and path loss for each cooperative node. The training overhead and transmission rate are not considered in (9.4), which will be considered in Section 9.3.

The average attenuation of the channel for node j can be estimated as follows. Assume the channel is symmetric, and t transmits a signal with transmit power P_{out}, then the power of the received signal at node j, P_{jt}, can be given by

$$P_{jt} = P_{out} \frac{G_t G_r \lambda^2}{(4\pi)^2 d_{jt}^{k_{jt}} M_l N_f} = \frac{P_{out}}{G(d_{jt}, k_{jt})} \tag{9.5}$$

where $G(d_{jt}, k_{jt}) = \frac{P_{out}}{P_{jt}} = \frac{(4\pi)^2 d_{jt}^{k_{jt}}}{G_t G_r \lambda^2} M_l N_f$. Therefore, (9.4) can be reformulated as

$$E_{bt}(2) = (1 + \alpha)\frac{N_0}{P_b^{\frac{1}{J}}} \sum_{j=1}^{J} G(d_{jt}, k_{jt}) + \frac{(J P_{ct} + P_{cr})}{B}$$

$$= (1 + \alpha) f_2(P_b) \sum_{j=1}^{J} G(d_{jt}, k_{jt}) + \frac{(J P_{ct} + P_{cr})}{B} \qquad (9.6)$$

According to (9.6), the transmit power of node j to communicate with cluster head t can be described by

$$P_{outjt} = G(d_{jt}, k_{jt})\frac{N_0 B}{P_b^{\frac{1}{J}}} \qquad (9.7)$$

Based on (9.2) and (9.6), the overall energy consumption for the single-hop transmission can be written as

$$E_{bt} = E_{bt}(1) + E_{bt}(2)$$

$$= (1 + \alpha)[f_1(P_b) H(d_{max}) + f_2(P_b) \sum_{j=1}^{J} G(d_{jt}, k_{jt})] + \frac{(J + 1)(P_{ct} + P_{cr})}{B} \qquad (9.8)$$

From (9.8), the energy consumption for the intracluster transmission, $E_{bt}(1)$, and intercluster transmission, $E_{bt}(2)$, should be traded off to minimize E_{bt}. E_{bt} can be minimized by choosing an appropriate set of cooperative nodes, which can minimize $f_1(P_b) H(d_{max}) + f_2(P_b) \sum_{j=1}^{J} G(d_{jt}, k_{jt})$. To simplify the distributed strategy design, the cooperative nodes should be chosen as the nodes whose $f_1(P_b) H(d_{j1}) + f_2(P_b) G(d_{jt}, k_{jt})$ are minimal. In addition, to balance the energy consumption, the selection criterion is defined as

$$\beta_{jt} = \frac{E_j}{f_1(P_b) H(d_{j1}) + f_2(P_b) G(d_{jt}, k_{jt})} \qquad (9.9)$$

where E_j is the remaining energy in the current round for node j. The rationale behind the definition of β_{jt} is that the node, which has a good tradeoff between $E_{bt}(1)$ and $E_{bt}(2)$ and has more remaining energy, should have a larger chance to be selected as the cooperative node. Therefore, J nodes with maximum β_{jt} will be chosen as the cooperative nodes to communicate with cluster head t.

9.3.2.2 To Incorporate HARQ-Based Retransmission Schemes

Though the virtual MIMO scheme can obtain good BER performance in an energy aware manner, the residual BER will also reduce the reliability of the transmission. To improve the reliability further, the HARQ scheme in the data link layer is incorporated into the data transmission among cluster heads. HARQ is the widely accepted technique to mitigate the link error. The basic idea of HARQ is to combine the automatic repeat request

(ARQ) schemes and forward error correction (FEC) to reduce the average retransmission times for a successful packet transmission. In a HARQ scheme, a FEC code is used to detect and correct the bit errors in the packet [29,30]. If the number of bit errors surpasses the error-correcting capability of the FEC code, a request is sent to the sender to retransmit the packet. Currently, most HARQ schemes can be classified into two types. In HARQ-I schemes, all the transmission attempts of a packet are identical FEC code words containing redundant bits for both error detection and error correction. The error-correcting capability of the FEC part of the scheme can be designed so that most of the erroneously received packets can be corrected, which reduces the number of retransmissions. Generally speaking, HARQ-I schemes are best suited for channel environments where the level of noise and interference is fairly constant. On the other hand, HARQ-II schemes rely on the basic concept of incremental redundancy [31]. In HARQ-II schemes, the parity bits for error correction are sent only when they are needed. On the first transmission attempt, only parity bits for error detection are appended to the message. If errors are detected in the received packet, it is stored in a buffer and a retransmission is requested. The retransmission is not the original packet but a block of parity-check bits formed based on the original message and an error-correcting code. When this block is received, it is used to correct the errors in the previously stored erroneous packet. Many proposed HARQ techniques belong to this type, such as diversity combining [32], code combining [33], and code puncturing [34], etc. Because the HARQ-II schemes have the flexibility in adapting the additional parity bits to changing channel conditions, they are more suitable for the time-varying channels.

As for the scenario of WSNs, the sensor nodes are too function limited to carry out the HARQ-II scheme. In addition, the positions of the sensor nodes are fixed, the level of noise and interference is relatively constant, so the HARQ-I scheme is more suitable for WSNs. To improve the reliability, a simplified HARQ scheme is considered in our design, where the linear block code and stop-and-wait ARQ scheme are combined together to correct the errors and reduce the average retransmission times. Then, the HARQ scheme is incorporated into the following retransmission schemes, similar to the information delivery methods in [21], for the packet transmission among cluster heads.

1. The intermediate cluster heads only perform as digital repeaters and the packet is decoded only at D, retransmissions are requested to S, which is just the end-to-end recovery scheme.
2. The intermediate cluster heads decode the packet and stop a further forwarding of a wrong packet, retransmissions are requested to S.
3. The intermediate cluster heads decode the packet and stop a further forwarding of a wrong packet, retransmissions are requested to the previous cluster head, which is just the hop-by-hop recovery scheme.

To compare the performance of the three retransmission schemes, the amount of energy consumption per successful packet transmission is defined as the criterion for comparison.

In the rest of this section, the amount of energy consumption per successful packet transmission by the virtual MIMO scheme under these three retransmission schemes is modeled and compared.

Denote E_{code} as the energy consumption of the baseband signal processing to perform encoding process, E_{enc}, and decoding algorithm, E_{dec}. The E_{code} of different BCH codes can be found in [35]. Other energy consumption in baseband signal processing is ignored.

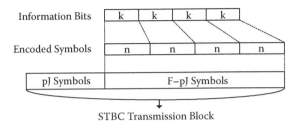

Figure 9.2   The Framework of the Transmission Symbols.

Denote the employed linear block code as (n, m, n_1), in which m information bits will be encoded into a symbol word with n bits, and the word error probability can be computed as in [21].

$$P_w(P_b) = \sum_{i=n_1+1}^{n} \binom{n}{i} P_b^i (1 - P_b)^{(n-i)} \tag{9.10}$$

where P_b is the BER performance.

We denote the encoded symbol word as a packet, so the packet size is just n bits.

On the other hand, as training overhead will be introduced by a MIMO for channel estimation and the number of required training symbols is proportional to the number of transmit antennas [11], we suppose that the block size of the STBC code is F symbols and in each block we include pJ training symbols. According to these assumptions, the framework of the data transmission can be shown in Figure 9.2.

As shown in Figure 9.1, the main operations in each hop include the transmission in a local cluster and the transmission between clusters by the virtual MIMO scheme.

Under the assumption of the AWGN channel with squared power path loss, the average transmit energy consumption per bit in a local cluster can be described by

$$E_{b0} = rd_0^2 + \frac{P_{ct}}{B} \tag{9.11}$$

where d_0 is the transmission distance in the local cluster, P_{ct} is the transmit circuit power consumption and r is a constant based on the circuit design, which can be calculated by (9.1). Because there are J cooperative nodes receiving the bit at the same time, the average receive energy consumption per bit can be described as $\frac{JP_{cr}}{B}$, where P_{cr} is the receive circuit power consumption.

Therefore, the overall energy consumption per packet transmission in a local cluster can be described by

$$E_0(d_0, J) = nrd_0^2 + \frac{nP_{ct}}{B} + \frac{nJP_{cr}}{B} \tag{9.12}$$

According to [10], the average transmit energy consumption per bit transmission by the STBC-encoded virtual MIMO scheme can be described by

$$E_{b1} = (1 + \alpha) \frac{JN_0}{P_b^{\frac{1}{J}}} \frac{(4\pi)^2 d_j^k M_l N_f}{G_t G_r \lambda^2} + \frac{JP_{ct}}{B} \tag{9.13}$$

where α is the efficiency of the power amplifier, N_0 is the single-sided noise power spectral density, d_j is the intercluster distance of the jth hop, M_l is the link margin, N_f is the receiver noise figure, G_t and G_r are the transmitter and receiver antenna gains, respectively, and λ is the wavelength.

Denote $C_2 = \frac{(1+\alpha)(4\pi)^2 N_0 M_l N_f}{G_t G_r \lambda^2}$, then $E_{b1} = \frac{1}{P_b^{\frac{1}{J}}} C_2 d_j^k + \frac{J P_{ct}}{B}$. The cluster head in the next hop also consumes $\frac{P_{cr}}{B}$ energy to receive the bit. Taking into account the training overhead, the total energy consumption per packet transmission in the jth hop can be described by

$$E_1(P_b, d_j, J) = \frac{nF}{F - pJ} \left(\frac{J}{P_b^{\frac{1}{J}}} C_2 d_j^k + \frac{J P_{ct}}{B} + \frac{P_{cr}}{B} \right) \quad (9.14)$$

Based on (9.12) and (9.14), we can model the overall energy consumption of the multi-hop virtual MIMO scheme under the three retransmission schemes.

In retransmission scheme (1), the intermediate cluster heads only repeat the packet, and the packet is only decoded at D. Therefore, the energy consumption for one time packet transmission can be described by

$$E = E_{code} + H E_0(d_0, J) + \sum_{j=1}^{H} E_1(P_b, d_j, J) \quad (9.15)$$

During the transmission, a bit arrives wrong at D if an odd number of errors occurs in the path, then the end-to-end BER performance can be described by [21].

$$P_{bd}(P_b) = \sum_{i=0}^{\lfloor \frac{H-2}{2} \rfloor} \binom{H}{2i+1} P_b^{(2i+1)} (1 - P_b)^{(H-2i-1)} + (H \bmod 2) P_b^{H} \quad (9.16)$$

The word error probability can be computed as [21],

$$P_{e1} = P_w(P_{bd}) \quad (9.17)$$

Then, the average retransmission times for a successful packet transmission can be described as $\frac{1}{1-P_{e1}}$.

Therefore, the overall energy consumption per packet transmission by retransmission scheme (1) can be described by

$$E_{s1} = \frac{1}{1 - P_{e1}} \left[E_{code} + H E_0(d_0, J) + \sum_{j=1}^{H} E_1(P_b, d_j, J) \right] \quad (9.18)$$

In retransmission scheme (2), the intermediate cluster head will decode the packet, the wrong packet will be dropped, and the source cluster head will be requested to retransmit. Then, the word error probability per hop can be described as $P_w(P_b)$. The end-to-end word error probability can be described by

$$P_{e2} = 1 - [1 - P_w(P_b)]^{H} \quad (9.19)$$

Denote P_b as the probability for the packet transmitted b hops before being dropped. Then, P_b can be described by [21],

$$P_b = P_w(P_b)[1 - P_w(P_b)]^{(b-1)} \qquad (9.20)$$

The energy consumption in the jth hop transmission can be described by

$$E_{code} + E_0(d_0, J) + E_1(P_b, d_j, J)$$

Therefore, the overall energy consumption per packet transmission by retransmission scheme (2) can be described by

$$E_{s2} = \sum_{b=1}^{H} \left\{ b[E_{code} + E_0(d_0, J)] + \sum_{j=1}^{b} E_1(P_b, d_j, J) \right\}$$

$$\times P_b + (1 - P_{e2}) \times \left[HE_{code} + HE_0(d_0, J) + \sum_{j=1}^{H} E_1(P_b, d_j, J) \right] \qquad (9.21)$$

In retransmission scheme (3), the intermediate cluster head will decode and buffer the packet, the wrong packet will be dropped, and the previous cluster head will be requested to retransmit. Then, the word error probability per hop can be described as $P_w(P_b)$. The average transmission time for each hop can be described as $\frac{1}{1-P_w(P_b)}$. The energy consumption in the jth hop transmission can be described by

$$E_{code} + E_0(d_0, J) + E_1(P_b, d_j, J)$$

Then, the overall energy consumption per packet transmission by retransmission scheme (3) can be described by

$$E_{s3} = \frac{1}{1 - P_w(P_b)} \times \left\{ H[E_{code} + E_0(d_0, J)] + \sum_{j=1}^{H} E_1(P_b, d_j, J) \right\} \qquad (9.22)$$

As shown in (9.18), (9.21), and (9.22), the overall energy consumption per packet successful transmission should be traded off between the energy consumption per time transmission and the average transmission times. So an optimal BER performance, P_b, should be found to minimize the overall energy consumption per packet successful transmission.

Figure 9.3 shows the relationships between the overall energy consumption per packet transmission and BER performance by the three retransmission schemes, in which the distance of each hop is assumed to be the same. The investigated system parameters are shown in Table 9.1. The employed linear block code is BCH(63, 39, 4).

From Figure 9.3, we can see that there exists an optimal P_b for three schemes with minimum overall energy consumption. The optimal P_b for retransmission scheme (1) is the least one, and the optimal P_b for retransmission scheme (3) is the largest one. For scheme (1), only D will decode the packet and retransmissions are requested to S. So one transmission will start from S and end at D, with tremendous energy cost for retransmission. So the optimal P_b will be small to reduce the number of retransmission

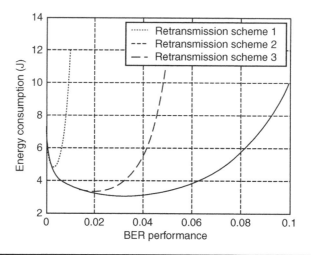

Figure 9.3 **Overall Energy Consumption Versus BER Performance.**

times, which is at the cost of more energy consumption per time transmission. For scheme (3) the retransmission will be requested to the previous cluster head; one retransmission will not cost so much energy. So the optimal P_b will be large to reduce the energy consumption per hop transmission, which is at the cost of more retransmission times. We also can find at the corresponding optimal P_b, retransmission scheme (3) will cost the minimum energy consumption. Figure 9.4 and Figure 9.5 show the optimal P_b and minimum overall energy consumption varying with distance under the different settings of H and J by the three schemes.

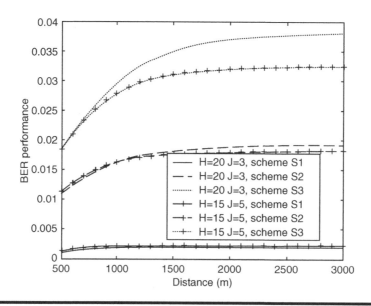

Figure 9.4 **The Optimal BER Performance Varying with Distance by Three Schemes.**

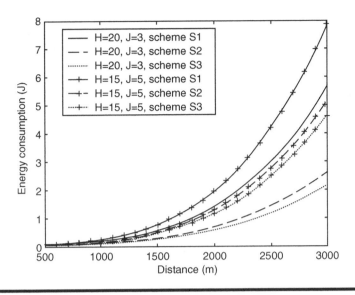

Figure 9.5   The Minimum Overall Energy Consumption Varying with Distance by Three Schemes.

From Figure 9.4 and Figure 9.5, we can draw the conclusion that retransmission scheme (3) can achieve the minimum energy consumption under different distances in three schemes. The optimal P_b for scheme (3) under different distances is about 0.03 in the investigated system parameters. Figure 9.6 also shows the optimal number of hops varying with distance under the different settings of J by the three schemes.

From Figure 9.6, we can find the optimal H by the three schemes is almost linearly increasing with the total distance, which makes the one-hop distance by the three schemes almost fixed. The optimal H by scheme (3) makes the one-hop distance almost fixed as 40 m. Figure 9.7 also shows the optimal J varying with distance under the

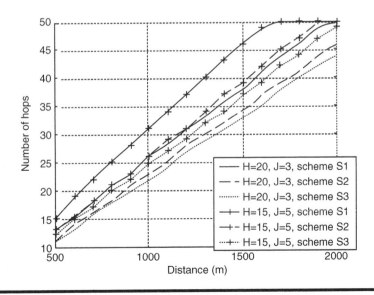

Figure 9.6   The Optimal Number of Hops Varying with Distance by Three Schemes.

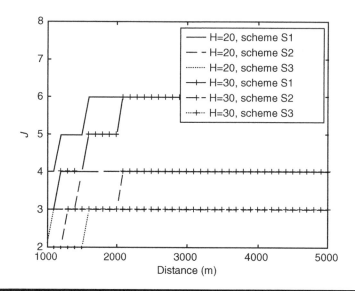

Figure 9.7 The Optimal _J_ Varying with Distance by Three Schemes.

different settings of H by three schemes. From Figure 9.7, we can find the optimal J is almost fixed as 3 by scheme (3).

Based on these analyses, we adopt the retransmission scheme (3) and the joint optimal parameters $(H, J, P_b) = (\frac{d}{40}, 3, 0.03)$ in design.

9.3.3 End-to-End Transmission Scheme Design

Based on the design of the single-hop transmission, the end-to-end transmission scheme will be designed. In the end-to-end transmission scheme, radio irregularity of wireless communications, multi-hop routing, retransmissions, and end-to-end QoS provisioning are jointly considered with the virtual MIMO scheme. To simplify the procedure of forming cooperative nodes set in the virtual MIMO scheme, the low-energy adaptive clustering hierarchy (LEACH) protocol is used to organize the sensor nodes into multiple clusters. Then, the LEACH protocol is extended to enable cluster heads to form a multi-hop backbone, and the single-hop design in the previous section is incorporated into each hop transmission. As assumed in the LEACH protocol, each node has a unique node ID. The transmit power of each node can be adjusted, and the nodes are all time synchronized. Similarly, the operations of the proposed scheme are broken into *rounds*. Each round consists of three phases: cluster formation phase, during which the clusters are organized and the cooperative MIMO nodes are selected; routing phase, during which the routing table is constructed; transmission phase, during which data is transferred from the nodes to the cluster head and forwarded to the sink according to the routing table.

9.3.3.1 Cluster Formation Phase

In this phase, each node will elect itself to be a cluster head with a probability p as specified in the original LEACH protocol. After the cluster heads are elected, each cluster head will broadcast an advertisement message (ADV) by transmit power P_{out} using a nonpersistent carrier sensing multiple access (CSMA) MAC protocol. The message contains the head's ID. If a cluster head receives the advertisement message from another head t and the received signal strength (RSS) exceeds a threshold tb, it will take cluster head t as a neighboring cluster head and record t's ID. As for the non cluster head, node j, it will record all the RSSs of the advertisement messages, and choose the cluster head whose RSS is the maximum. Then, it will calculate and save $H(d_j)$, $G(d_{jt}, k_{jt})$, β_{jt}, and P_{outjt} by (9.3), (9.5), (9.7), and (9.9). Then, node j will join the cluster by sending a join-request message (Join-REQ) to the chosen cluster head. This message contains the information of the node's ID, the chosen cluster head's ID, and the corresponding values of β_{jt}. After a cluster head has received all join-request messages, it will set up a time division multiple access (TDMA) schedule and transmit this schedule to its members as in the original LEACH protocol. If the sink receives the advertisement message, it will find the cluster head with the maximum RSS, send the sink-position (Sink-POS) message to the cluster head, and mark the cluster head as the target cluster head (TCH).

After the clusters are formed, each cluster head will select corresponding optimal J cooperative nodes for cooperative MIMO communications with each of its neighboring cluster heads. As stated in Section 9.2.1, J nodes with maximum β_{jt} will be chosen to communicate with a neighboring cluster head t. If no such J nodes can be found for t, t will be removed from the neighbor list because too much energy is consumed for communicating with t. After selecting the cooperative nodes, the total energy per bit transmission for communications with t, E_{bt}, can be derived by (9.4). Then, E_{bt}, the ID set of the cooperative nodes for each neighboring cluster head, will be stored. At the end of this phase, the cluster head will broadcast a cooperate-request message (COOPERATE-REQ) to each cooperative node, which contains the ID of the cluster itself, the ID of the neighboring cluster head t, the IDs of the cooperative nodes, and the index of the cooperative nodes in the cooperative nodes set for each cluster head t. Each cooperative node that receives the COOPERATE-REQ will store the ID of t, the index and the transmit power P_{outjt}, and send back a cooperate-acknowledge message (COOPERATE-ACK) to the cluster head.

We assume the nodes are locally time synchronized in each cluster at the end of this phase. This could be achieved by having each cluster head transmit a reference carrier and all its cluster members lock to this reference carrier using a phase locked loop. In fact, the clock jitter at the transmit nodes in transmission will cause intersymbol interference (ISI). An accurate synchronization algorithm should be implemented to have very fine synchronization within each cluster, which will cost significant energy consumption. However, as stated in [14], a clock jitter as large as 10% of the bit time does not have much effect on the BER performance for the cooperative MISO scheme. So we do not implement the accurate synchronization algorithm to save energy.

9.3.3.2 Routing Table Construction

To construct the routing table, the basic ideas of distance vector-based routing will be used. Each cluster head will maintain a routing table, in which each entry contains *destination cluster ID, next hop cluster ID, IDs of cooperative nodes, and mean energy consumption per bit*. Initially, only the neighbor cluster heads will have a record in the routing table. Then each cluster head will simply inform its neighboring cluster heads of its routing table. After receiving route advertisements from neighboring cluster heads, the cluster head will update its routing table according to route cost and advertise to its neighboring cluster heads the modified routes. After several rounds of route exchange and update, the routing table of each cluster head will converge to the optimal one. Then, the TCH will flood a target announcement message (TARGET-ANNOUNCEMENT) containing its ID to each cluster head to enable the creation of the paths to it.

9.3.3.3 Data Transmission

In this phase, cluster members will first transmit their data to the cluster head by multiple frames. In each frame, each cluster member will transmit its data during its allocated transmission slot, specified by the TDMA schedule in *cluster formation phase*, and then sleep in other slots to save energy. The duration of a frame and the number of frames transmitted to the cluster head in a slot are the same for all clusters. Thus, the duration of each slot depends on the number of nodes in the cluster. After a cluster head receives data frames from its cluster members, it will perform data aggregation to remove the redundancy in the data. After aggregating received data frames, the cluster head will forward the data packets to the TCH by multiple hops. In each single-hop communication, if there exists J cooperative MIMO nodes, the cluster head will add a packet header to the data packet, which includes the information of source cluster ID, next-hop cluster ID, and destination cluster ID. The cluster head will buffer and encode the data packet according to the linear block coding. Then the encoded data packet is broadcast. Once the corresponding cooperative nodes receive the data packet, they will encode the data packet by orthogonal STBC, and transmit the data to the cluster head in the next hop as an individual antenna, with transmission power P_{outjt} in the MIMO antenna array. In the cooperative MIMO scheme, the transmission delay and channel estimation scheme proposed in [5] can be used in decoding. After receiving the packet, the cluster head in the next hop will decode it and correct the bit errors by the linear block coding. If a word error occurs after decoding, it will send a negative acknowledgment (NACK) message to the previous cluster head to retransmit the packet, otherwise it will send an ACK message to the previous cluster head to remove the buffered packet. The stop-and-wait ARQ scheme is used for the retransmission requirements. The reason not to use other ARQ schemes, such as selective repeat ARQ scheme, is that the transmission distance is so near in WSN that the propagation latency can be ignored. So the throughput by the stop-and-wait ARQ scheme is almost the same as other ARQ schemes. Due to its simplicity, it is more suitable to use in a WSN.

To improve the energy efficiency of the protocol, the communication parameters, including k_c, J, and P_b, should be chosen as the joint optimal ones. The choice of k_c should make the intercluster distance approximately $40\ m$ in the system parameters settings in Table 9.1.

Table 9.1   The System Parameters

$r = 10\ pJ/bit/m^2$	$P_{ct} = 98.2\ mw$	$P_{cr} = 112.6\ mw$
$d_0 = 10\ m$	$d = 2\ km$	$H = 10$
$C_2 = 4.0605e - 12$	$F = 200$	$p = 2$
$J = 5$	$E_{code} = (445 + 752) \times 39\ nJ$ [35]	

9.4 THEORETICAL ANALYSIS OF THE CROSS-LAYER DESIGN

In this section, we will analyze the energy consumption and end-to-end QoS performance of the scheme. Then, an optimization model is developed to find the optimal parameters based on these analyses.

9.4.1 Energy Consumption and End-to-End QoS Performance Analysis

As in the scheme design in Section 9.3, during transmission each cluster head will find the minimum energy consumption of relaying data packets among other cluster heads to the sink. The multi-hop data transmission topology among cluster heads can be treated as a SPT, which is shown in Figure 9.8. The BER performance, P_b, on each link will determine the packet throughput, which can be treated as the packet service rate μ for the sender of the link. Therefore, the queuing latency and throughput of the node can be modeled by the BER performance, P_b, according to the queuing theory. Based on the result, the end-to-end latency and throughput can also be modeled in terms of the $P_b s$ of all links.

According to the assumptions in Section 9.3, the mean time to transmit a packet can be described as $t_f = \frac{nF}{B(F-pJ)}$. Because the stop-and-wait ARQ scheme is used for the retransmission requirements, the throughput can be described as (9.23) if the propagation latency, the processing latency, and the ACK packet transmission latency are ignored [36].

$$G = \frac{1 - P_w(P_b)}{t_f} \tag{9.23}$$

From the view point of the network layer, the throughput of the stop-and-wait ARQ scheme can be viewed as the packet service rate, which can be described as $\mu = \frac{1-P_w(P_b)}{t_f}$ packets/s. Suppose the packets arrive according to the Poisson process and the packet arrival rate is denoted as λ. Let $P_0(\mu, \lambda)$, $P_N(\mu, \lambda)$, and $W(\mu, \lambda)$ denote the probabilities of the queue being empty, being full, and the mean sojourn time for a packet including queuing and servicing, respectively. The existing solutions by the queuing theory can be used directly to compute P_0, P_N, and W [37].

Now we are ready to model the end-to-end QoS performance in terms of the BER performance of each link in the SPT. The SPT can be represented by $T = <V, E_T>$, where node set V is the set of all cluster heads in the SPT, and edge set E_T denotes the set of directed communication links between each pair of cluster heads in the SPT.

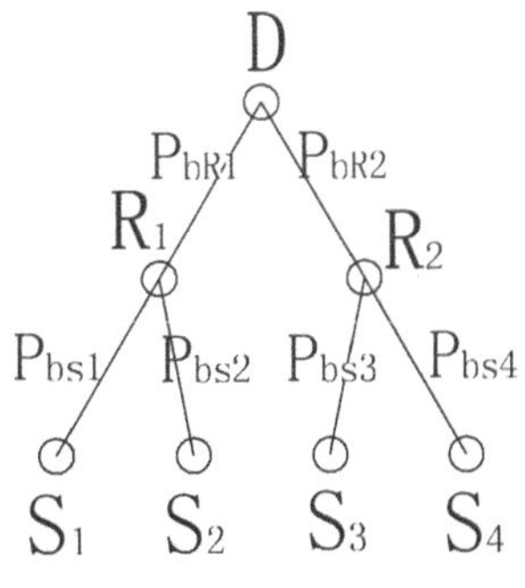

Figure 9.8   The Short Path Tree of the Cluster Heads.

V can be grouped into two subsets, the set of leaf cluster heads (denoted as V_s) and the set of internal cluster heads (denoted as V_r). As for the leaf cluster head, such as S_1 in Figure 9.8, it only receives the packets from its cluster members. However, the internal cluster head, such as R_1 in Figure 9.8, not only receives the packets from its members but also the packets from its children in the SPT. Then, the packet arrival rate can be described by

$$\lambda_i = \begin{cases} \lambda_c, \ (i \in V_s) \\ \sum_{j \in N_{si}} \mu_j(P_{bj})(1 - P_0(\mu_j(P_{bj}), \lambda_j)) + \lambda_c, \ (i \in V_r) \end{cases} \tag{9.24}$$

where N_{si} is the set of children cluster heads in the SPT, λ_c is the intracluster packet arrival rate, and $\mu_j(P_{bjk}) = \frac{1 - P_w(P_{bj})}{t_f}$. To simplify the analysis, we assume λ_c are the same for all clusters, which can be estimated by the number of nodes and the desired number of clusters. However, the extension to the scenario with different λ_c is simple.

Therefore, the probabilities of the queue being empty, being full, and the mean sojourn time of a packet transmission for cluster head j can be described as $P_0(\mu_j(P_{bj}), \lambda_j)$, $P_N(\mu_j(P_{bj}), \lambda_j)$, and $W(\mu_j(P_{bj}), \lambda_j)$.

Denote L_j as the path from cluster head j to the sink in the SPT, the end-to-end latency and throughput for j can be described by

$$La_j = \sum_{i \in L_j} W(\mu_i(P_{bi}), \lambda_i)$$

$$Tb_j = \prod_{i \in L_j} (1 - P_N(\mu_i(P_{bi}), \lambda_i)) \tag{9.25}$$

The mean end-to-end latency and throughput for the whole network can be described by

$$La(\{P_{bj}\}) = \frac{\sum_{j \in V_s \cup V_r} \lambda_j La_j}{\sum_{j \in V_s \cup V_r} \lambda_j}$$

$$Tb(\{P_{bj}\}) = \frac{\sum_{j \in V_s \cup V_r} \lambda_j Tb_k}{\sum_{j \in V_s \cup V_r} \lambda_j} \tag{9.26}$$

Strictly speaking, we only considered the QoS performance of the intercluster communication in (9.26). We have considered the QoS performance of the intracluster communication in [25], which will not be discussed here due to the limited space.

On the other hand, the overall energy consumption of all cluster heads can be described by

$$E_a(\{P_{bj}\}) = \sum_{j \in V_s \cup V_r} \frac{1}{1 - P_w(P_{bj})} [E_{code} + E_0(d_0, J) + E_1(P_{bj}, d_j, J)] \tag{9.27}$$

9.4.2 Parameters Optimization

Based on the above analysis, we developed a model to find the optimal $\{P_{bj}\}$ to minimize the overall energy consumption under the application's end-to-end QoS requirements, which are shown below:

Objective: $minE_a(\{P_{bj}\})$. Refer to (9.27) for the expression of $E_a(P_{bj})$.
Subject to:

■ The requirement on mean end-to-end latency, $La(\{P_{bj}\}) \leq \tau_{app}$.
■ The requirement on mean end-to-end packet loss ratio, $Tb(\{P_{bj}\}) \leq tb_{app}$.
■ $P_{bmin} \leq P_{bj} \leq P_{bmax}$.

Expected solution: Find the optimal $\{P_{bj}\}$.

By solving the optimization model, we can obtain the optimal $\{P_{bj}\}$ to provide the end-to-end QoS requirements by minimum overall energy cost. However, the problem shown above is a nonlinear constrained optimization problem, which is difficult to solve especially when the number of P_{bj} is large. Due to its efficiency in solving such optimization problems, we use the PSO algorithm to find the optimal solution. The PSO algorithm was proposed by J. Kennedy and R. C. Eberhart in 1995 [38], motivated by the social behavior of organisms such as bird flocking and fish schooling. In the PSO algorithm, the local search method and global search method are combined to find the optimal solution. In using the PSO algorithm to solve our problem, we define the particle as the vector containing the $\{P_{bj}\}$. A population with N_p particles is generated. The PSO algorithm is iterated for N_{iter} times to find the optimal solution. Also because our problem is a constrained optimization problem, we convert it to an unconstrained one by the punish function.

During transmission, the sink node will determine the optimal P_{bj} for each link in the SPT and transmit P_{bj} to the related cluster head via the control packet. After receiving the control packet, within its cluster, the cluster head will broadcast a transmit power adjustment packet including P_{bj} and the ID of its parent cluster head in the SPT. After receiving the adjustment packet, the cooperative nodes corresponding to the ID of the parent cluster head will adjust the transmit power by the P_{bj} according to (9.7). This procedure requires knowledge of the topology information of the SPT and the channel gains of each link. We assume that each cluster head will report the following information to the sink, such as the ID of its parent cluster head and the channel gains between itself and its parent cluster head in the SPT during the routing table construction phase. To implement the optimization procedure in a more distributive manner is one of our research interests in the future work.

9.5 SIMULATION AND NUMERICAL RESULTS

In this section, we evaluate the energy saving performance and QoS provisioning of the proposed cross-layer design based on virtual MIMO. Our experiments are organized as follows: First, we demonstrate the energy saving performance of the proposed scheme in the phenomena of fading and radio irregularity; second, we investigate the QoS provisioning performance of the design based on the optimization of $\{P_{bj}\}$ of each link by the optimization model proposed in Section 9.4. In the simulations, the related system parameters are the same as shown in Table 9.1.

9.5.1 Energy Saving Performance of the Cross-Layer Design

In order to evaluate the energy saving performance of the proposed cross-layer design, we simulate the operation of the cross-layer design in multiple rounds, record its energy consumption and compare it to other schemes. The procedure of this simulation will be discussed in this section.

In the simulations, 400 nodes are randomly deployed on a 200×200 field. The location of the sink is randomly chosen in each round. Each node begins with $400J$ of energy and an unlimited amount of data packets to send to the sink. When the nodes use up their limited energy during network operation, they cannot transmit or receive data any longer.

During the simulation, we tracked the accumulated number of packets transferred to the sink, the amount of energy and duration required to deliver the data to the sink, and the percentage of nodes alive. We are interested in the transmission quality and energy saving performance of the proposed scheme. The performance of the proposed multi-hop MIMO-LEACH scheme is compared with the original LEACH and the multi-hop LEACH scheme, in which cooperative MIMO communication is not implemented. The optimal value of k_c for the original LEACH is determined by the model in [15]. We also develop a similar model to find the optimal k_c for the multi-hop LEACH scheme, which will not be discussed here due to the limited space. In the investigated scenario, the optimal k_c for the original LEACH protocol, the multi-hop LEACH scheme, and the proposed scheme are found and set to 3, 41, and 27, respectively. The optimal J for the proposed scheme is found to be 3.

Due to the aggregation operation, the number of effective received packets by sink [15] is deemed as the number of "real" packets after aggregation. Specifically, if no aggregation carries out, the number of effective received packets equals the number of successfully received packets. If the aggregation operation in transmission is *information lossless*, the number of effective received packets is just the number of total packets transferred by the source nodes. We believe that the number of effective received packets is a good application-independent indication of the transmission quality.

Figures 9.9 and 9.10 show the total number of effective packets received at the sink over time and the total number of effective packets received at the sink per a given amount of energy.

Figure 9.9 shows that during its lifetime, the LEACH protocol can obtain better latency performance compared to the multi-hop LEACH scheme and the proposed MIMO LEACH scheme. The reason is that the multi-hop operation in the multi-hop LEACH scheme and the multi-hop MIMO-LEACH scheme will increase the latency, and thus result in less data packets sent to the sink for a given period of time. However, the better latency performance of the LEACH protocol comes from more energy consumption compared to the other two schemes, especially in the fading channel environment. LEACH protocol will consume much more energy due to its single-hop transmission from the cluster heads to the sink, which will result in less network lifetime and less total numbers of transmitted packets. Figure 9.10 shows that, with the same amount of energy consumption, the multi-hop MIMO-LEACH scheme can transmit many more data packets compared to the LEACH protocol and the multi-hop LEACH scheme. From these simulation results, we find that

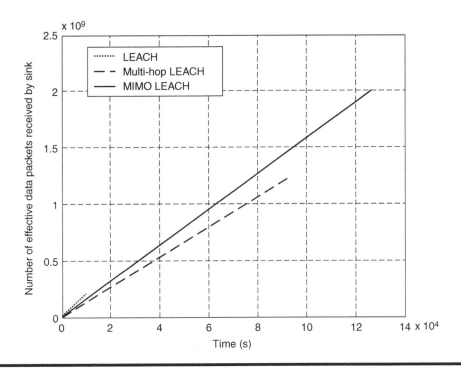

Figure 9.9   Total Amount of Effective Packets Received at the Sink Over Time.

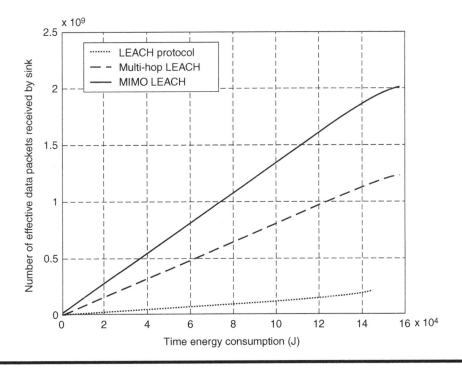

Figure 9.10   Total Amount of Effective Packets Received at the Sink Per Given Amount of Energy.

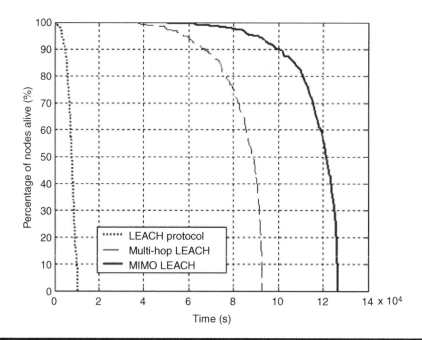

Figure 9.11   Percentage of Nodes Alive Over Time.

the multi-hop MIMO-LEACH scheme is more suitable for the application scenarios which have high requirements on network lifetime but low requirements on latency.

Figure 9.11 shows the percentage of nodes alive over time. From Figure 9.11, we find that the proposed multi-hop MIMO-LEACH scheme can improve the network lifetime greatly. If we define the network lifetime of a WSN as the duration of more than 70% of network nodes remaining alive, it can be observed that the network lifetime of WSN with the original LEACH protocol, the multi-hop LEACH scheme, and the proposed multi-hop MIMO-LEACH scheme is about 0.7×10^4, 8.2×10^4, and 11×10^4 s, respectively. The improvement on network lifetime obtained by the multi-hop MIMO-LEACH scheme is significant.

However, the percentage of nodes alive over time is not always a good indication of the energy saving performance of a protocol. For example, during the same time, though one protocol is worse than other protocols in terms of the energy saving performance, it will still consume less energy if it transmits fewer packets than other protocols. Thus, its lifetime is likely longer. To further investigate the energy saving performance, we also simulate the performance in terms of the percentage of nodes alive per amount of effective data packets received at the sink, which is shown in Figure 9.12. From Figure 9.12, we find that the proposed multi-hop MIMO-LEACH scheme needs significantly less energy to transmit the same amount of data packets. Therefore, the improvement on network lifetime obtained by the multi-hop MIMO-LEACH scheme is significant.

On the other hand, the impacts of the parameters, including the number of cluster heads, k_c, and the number of cooperative nodes, J, are also investigated in the simulation. Figures 9.13 and 9.14 show the percentage of nodes alive over time in different settings of k_c and J.

From the simulation results, including those shown in Figures 9.13 and 9.14, we find that the energy saving performance of the proposed scheme is impacted by the

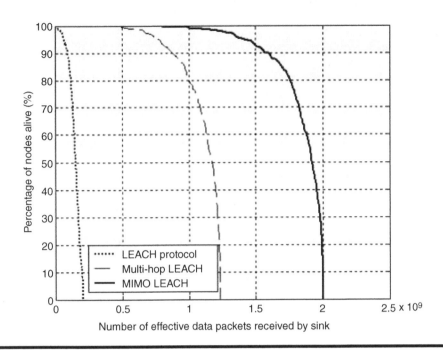

Figure 9.12   Percentage of Nodes Alive Per Amount of Effective Data Packets Received at the Sink.

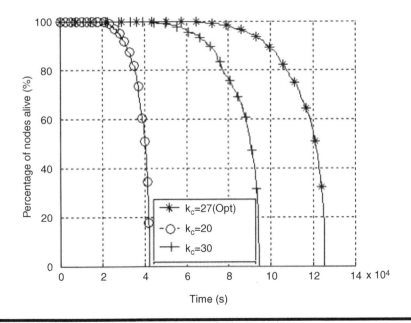

Figure 9.13   The Impact of the Number of Cluster Heads on Energy Saving Performance.

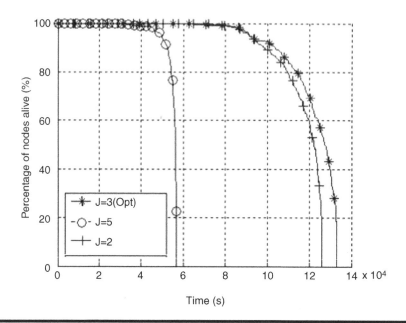

Figure 9.14 **The Impact of the Number of Cooperative Nodes on Energy Saving Performance.**

setting of these parameters. As for the number of cluster heads (*HeadNum*), a large value of *HeadNum* will reduce the distance for each single-hop transmission, which will reduce the transmit energy consumption; a large *HeadNum* also generates a wide search space for the routing table construction, which will also reduce the transmit energy consumption further. However, the larger the *HeadNum*, the more hops in transmission to the sink are needed, which causes more circuit energy consumption for relaying the data packets. Therefore, the number of cluster heads should be chosen to trade-off the transmit energy consumption and circuit energy consumption. As for the number of cooperative nodes, a certain number can form the effective independent multi-path transmission to energy efficiently combat the fading effects. However, too many cooperative nodes will result in large circuit energy consumption, which will cause large overall energy consumption. Therefore, the number of cooperative nodes should also be chosen to trade-off the transmit energy consumption and the circuit energy consumption.

9.5.2   QoS Provisioning Performance of the Cross-Layer Design

In Section 9.4, we proposed an end-to-end QoS model in terms of the BER performance of each link in the SPT. We also proposed to use the PSO algorithm to find the optimal BER performance of each link to minimize the overall energy consumption without violating the end-to-end QoS requirements. In this section, the numerical results will be presented. The structure of the SPT in experiments is shown in Figure 9.15, where S_1, S_2, S_3, and S_4 are the source cluster heads; R_1 and R_2 are the internal cluster heads; the number shown on each link is the distance of the link; P_{bi} is the BER performance of link i, and the intracluster packet arrival rates λ_c for all cluster heads are 40 pps. N_p is set to be 10,000 and N_{iter} is set to be 100 in the simulation. In the experiments, we

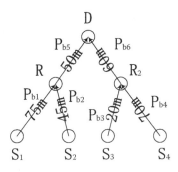

Figure 9.15 The Topology of the SPT in Experiments.

search the optimal BER performance of each link by the PSO algorithm to minimize the overall energy consumption with varied end-to-end QoS requirements.

Figures 9.16 to 9.19 show the convergence of the minimum overall energy consumption, end-to-end latency, end-to-end throughput, and P_bs during the search process of the PSO algorithm. The desired end-to-end latency and end-to-end throughput are 0.04 s and 0.80 s, respectively. From Figures 9.16 to 9.19, we find the algorithm can converge in about 30 iterations, so the convergence speed is fast. Therefore, the PSO algorithm is efficient to solve our problem. The optimal P_{b5} and P_{b6} are less than P_{b1} to P_{b4}, because the internal links should have smaller optimal BER to make the throughput larger than the throughput of the children links in the SPT for end-to-end QoS provisioning.

We also did the experiment to search the minimum energy consumption and optimal P_bs with time-varying end-to-end QoS requirements. In the experiment, the desired end-to-end latency is varied from 0.008 s to 0.038 s, and the desired end-to-end throughput

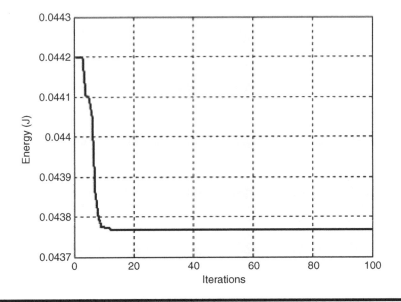

Figure 9.16 The Minimum Energy Consumption Versus Iterations.

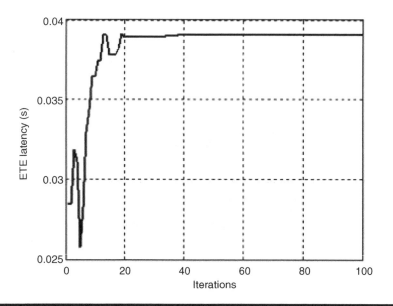

Figure 9.17   The Acquired End-to-End Latency Versus Iterations.

is fixed at 0.8, then the optimal P_bs and minimum overall energy consumption are found by the PSO algorithm. The energy saving performance, by employing the optimal P_bs, is defined as $\eta_E = \frac{E_{ref} - E_{opt}}{E_{ref}} \times 100\%$, where E_{ref} and E_{opt} are the overall energy consumptions by a random setting and optimal setting of P_bs. Figures 9.20 and 9.21 show the actual end-to-end latency and end-to-end throughput acquired by the algorithm.

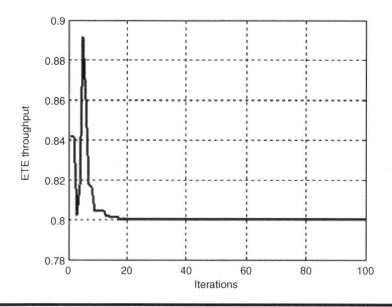

Figure 9.18   The Acquired End-to-End Throughput Versus Iterations.

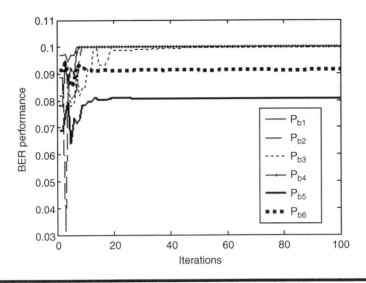

Figure 9.19 The BER Performance of Each Link Versus Iterations.

Figure 9.22 shows the energy saving performance varied with the desired end-to-end latency. Figure 9.23 shows the optimal P_bs varied with the desired end-to-end latency.

Furthermore, we investigated the energy saving performance and end-to-end QoS provisioning of the protocol in a large-scale network. We also did the simulation to search the optimal P_bs by the PSO algorithm in the scenario of a large-scale network, in which 400 sensor nodes are randomly deployed over a $200\,\text{m} \times 200\,\text{m}$ area and the nodes

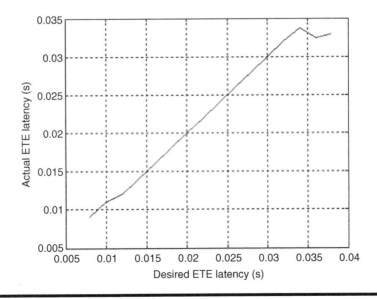

Figure 9.20 Actual End-to-End Latency Versus Desired End-to-End Latency.

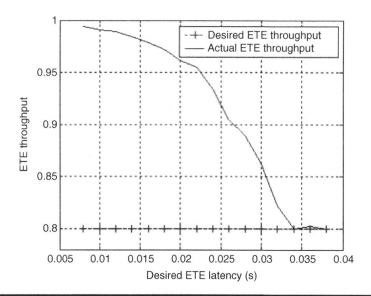

Figure 9.21 Actual End-to-End Throughput Versus Desired End-to-End Latency.

are clustered into 22 clusters. The intracluster packet arrival rates λ_c for all cluster heads are 60 pps. The topology of the network is shown in Figure 9.24.

Figures 9.25 to 9.27 show the convergence of the minimum overall energy consumption, end-to-end latency, and end-to-end throughput during the search process of the PSO algorithm.

For the network shown in Figure 9.24, we also did the experiments to search the minimum energy consumption with time varying end-to-end QoS requirements. In the experiment, the desired end-to-end latency is varied from 11.8 ms to 14.8 ms, and

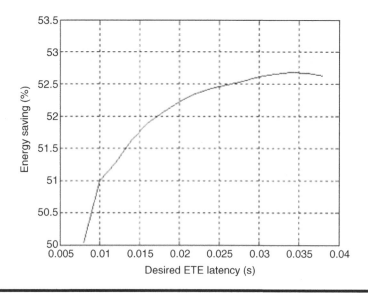

Figure 9.22 Energy Saving Performance Versus Desired End-to-End Latency.

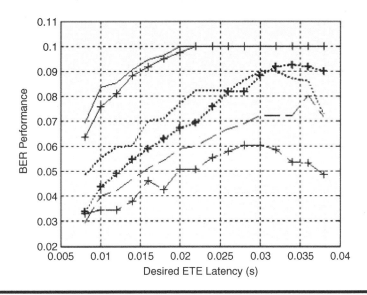

Figure 9.23 The Optimal BER Performance of Each Link Versus Desired End-to-End Latency.

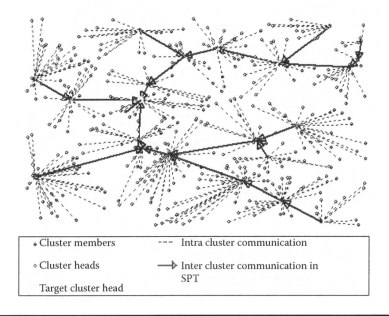

Figure 9.24 The Topology of the Large-Scale Network.

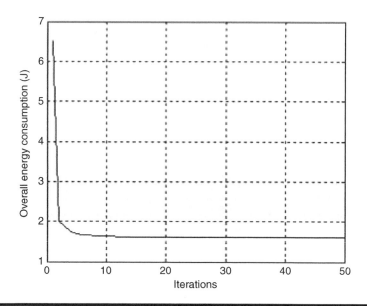

Figure 9.25   The Minimum Energy Consumption Versus Iterations.

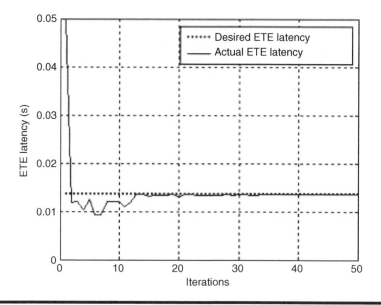

Figure 9.26   The Acquired End-to-End Latency Versus Iterations.

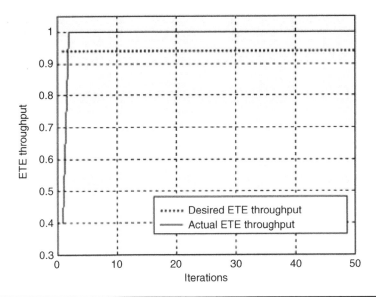

Figure 9.27   The Acquired End-to-End Throughput Versus Iterations.

the desired end-to-end throughput is fixed at 0.78. Figures 9.28 and 9.29 show the actual end-to-end latency and energy saving performance varied with the desired end-to-end latency.

From the experimental results, it can be seen that by adjusting the P_b of each link, the actual end-to-end QoS performances are varied with the end-to-end QoS requirements. And the significant energy saving performance can be acquired by adjusting the optimal BER performance.

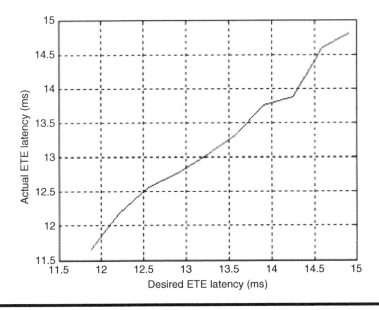

Figure 9.28   Actual End-to-End Latency Versus Desired End-to-End Latency.

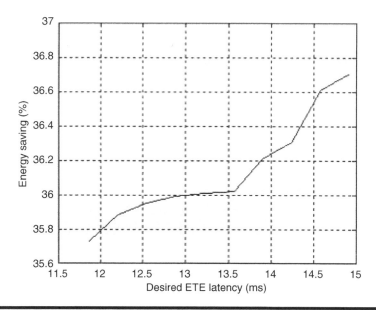

Figure 9.29 Energy Saving Performance Versus Desired End-to-End Latency.

9.6 CONCLUSION AND OPEN ISSUES

In this chapter, we propose a cross-layer design based on the virtual MIMO scheme to increase the energy efficiency and provide the end-to-end QoS guarantee. In the scheme, radio irregularity of wireless communications, multi-hop routing, retransmissions, and end-to-end QoS provisioning are jointly considered with the virtual MIMO scheme. In the cross-layer design, the concept of clustering is adopted to organize the sensor nodes into multiple clusters and form the cluster heads as a multi-hop backbone. Then, the virtual MIMO scheme is incorporated into each single-hop transmission between each pair of cluster heads. In each single-hop transmission, three HARQ-based retransmission schemes are considered to incorporate into the virtual MIMO scheme. The average energy consumption for a successful packet transmission by the virtual MIMO scheme under the three retransmission schemes is analyzed and compared. Then the retransmission scheme, using hop-by-hop recovery, is incorporated into the virtual MIMO scheme due to its efficiency. Then, an adaptive cooperative nodes selection strategy is also designed in the protocol. Based on the single-hop transmission scheme, the end-to-end transmission scheme is designed, which jointly integrates the virtual MIMO scheme, multi-hop routing scheme, and retransmission scheme to improve the performance of energy efficiency, reliability, and QoS guarantees. Based on the cross-layer design, we also developed the model for end-to-end QoS and overall energy consumption of the design in terms of the BER performance in each link of the SPT. A nonlinear constrained programming model is also designed to find the optimal BER performance for all the links in the SPT. The PSO algorithm is employed to solve the programming problem. Simulation results show the effectiveness of the proposed protocol to achieve the goals of minimizing energy consumption. The numerical results show that by adjusting the BER performance of each link, the actual end-to-end QoS performance can be varied with the requirements, and energy can be significantly saved.

In future work, we are interested in incorporating the network layer retransmission schemes into the multi-hop virtual MIMO protocol. In addition, a distributed protocol will be developed to adjust the BER performance of each link in the SPT to provide the end-to-end QoS guarantee while minimizing overall energy consumption.

REFERENCES

[1]   S.M. Alamouti, "Simple transmit diversity technique for wireless communications," *IEEE Journal of Selected Areas in Communications*, vol. 16, pp. 1451–1458, 1998.

[2]   V. Tarokh, H. Jafarkhani, and A. Calderbank, "Space-time block codes from orthogonal design," *IEEE Trans. Inform. Theory*, vol. 45, no. 5, pp. 1456–1466, July 1999.

[3]   M. Dohler, A. Gkelias, and H. Aghvami, "A resource allocation strategy for distributed MIMO multi-hop communication systems," *IEEE Communications Letters*, vol. 8, pp. 99–101, Feb. 2004.

[4]   L. Musavian, M. Dohler, R. Nakhai, and H. Aghvami, "Closed form capacity expressions of orthogonalised correlated MIMO channel," *IEEE Communications Letters*, vol. 8, pp. 365–367, June 2004.

[5]   X. Li, "Energy efficient wireless sensor networks with transmission diversity," *IEE Electronics Letters*, vol. 39, pp. 1753–1755, Nov. 2003.

[6]   X. Li, "Space time coded multiple transmission among distributed transmitters without perfect synchronization," *IEEE Signal Processing Letters*, vol. 11, pp. 948–951, Dec. 2004.

[7]   X. Li, M. Chen, and W. Liu, "Application of STBC-encoded cooperative transmissions in wireless sensor networks," *IEEE Signal Processing Letters*, vol. 12, pp. 134–137, Feb. 2005.

[8]   X. Li, "Blind channel estimation and equalization in wireless sensor networks based on correlations among sensors," *IEEE Transactions on Signal Processing*, vol. 53, pp. 1511–1519, April 2005.

[9]   B. Azimi-Sadjadi and A. Mercado, "Diversity gain for cooperating nodes in multi-hop wireless networks," in *IEEE VTC 2004 Fall*, Los Angeles, vol. 2, pp. 1483–1487, Sept. 2004.

[10] S. Cui, A.J. Goldsmith, and A. Bahai, "Energy-efficiency of MIMO and cooperative MIMO techniques in sensor networks," *IEEE Journal of Selected Areas in Communications*, vol. 22, pp. 1089–1098, Aug. 2004.

[11] S.K. Jayaweera, "Energy analysis of MIMO techniques in wireless sensor networks," in *38th Annual Conference on Information Sciences and Systems (CISS 04)*, Princeton, NJ, Mar. 2004.

[12] S.K. Jayaweera, "An energy-efficient virtual MIMO architecture based on V-BLAST processing for distributed wireless sensor networks," in *2004 First Annual IEEE Communications Society Conference on Sensor and Ad Hoc Communications and Networks*, Santa Clara, CA. pp. 299–308, Oct. 2004.

[13] J.N. Laneman and G.W. Wornell, "Distributed space-time-coded protocols for exploiting cooperative diversity in wireless networks," *IEEE Transactions on Information Theory*, vol. 49, pp. 2415–2425, Oct. 2003.

[14] S. Jagannathan, H. Aghajan, and A. Goldsmith, "The effect of time synchronization errors on the performance of cooperative MISO systems," in *IEEE Global Telecommunications Conference Workshops*, Dallas, pp. 102–107, Dec. 2004.

[15] W.R. Heinzelman, A. Chandrakasan, and H. Balarislman, "An application-specific protocol architecture for wireless microsensor networks," *IEEE Transactions on Wireless Communications*, vol. 1, no. 4, pp. 660–670, Oct. 2002.

[16] C.Y. Wan, A.T. Campbell, and L. Krishnamurthy, "Pump-slowly, fetch-quickly (PSFQ): a reliable transport protocol for sensor networks," *IEEE Journal of Selected Areas in Communications*, vol. 23, pp. 862–872, April 2005.

[17] F. Stann and J. Heidemann, "RMST: reliable data transport in sensor networks," in *First IEEE Int. Workshop on Sensor Network Protocols and Applications*, Anchorage, AK, pp. 102–113, May 2003.

[18] Y. Sankarasubramaniam, O.B. Akan, and I.F. Akyildiz, "ESRT: Event-to- Sink Reliable Transport in Wireless Sensor Networks," in *ACM MobiHoc*, Annapolis, MD, pp. 177–188, June 2003.

[19] B. Deb, S. Bhatnagar, and B. Nath, "ReInForM: reliable information forwarding using multiple paths in sensor networks," in *IEEE LCN'03*, New York, pp. 406–415, Oct. 2003.

[20] N. Tezcan, E. Cayirci, and M.U. Caglayan, "End-to-end reliable event transfer in wireless sensor networks," in *IEEE PIMRC'04*, Barcelona, Spain, vol. 2, pp. 989–994, Sept. 2004.

[21] C. Taddia and G. Mazzini, "On the energy impact of four information delivery methods in wireless sensor networks," *IEEE Communications Letters*, vol. 9, pp. 118–120, Feb. 2005.

[22] K. Sohrabi, J. Gao, V. Ailawadhi, and G.J. Pottie, "Protocols for self-organization of a wireless sensor network," *IEEE Personal Communications*, vol. 7, pp. 16–27, Oct. 2000.

[23] T. He, J.A. Stankovic, L. Chenyang, and T. Abdelzaher, "SPEED: a stateless protocol for real-time communication in sensor networks," in *Proc. IEEE ICDCS'03*, Providence, RI, pp. 46–55, May 2003.

[24] K. Akkaya and M. Younis, "Energy-aware routing of time-constrained traffic in wireless sensor networks," *Journal of Communication Systems, Special Issue on Service Differentiation and QoS in Ad Hoc Networks*, vol. 17, pp. 663–687, 2004.

[25] K. Akkaya and M. Younis, "Energy and QoS aware Routing in Wireless Sensor Networks," *Journal of Cluster Computing Journal, Special Issue on Ad Hoc Networks*, vol. 8, pp. 179–188, July 2005.

[26] Y. Yuan, Z.K. Yang, Z.H. He, and J.H. He, "An Integrated Energy Aware Wireless Transmission System for QoS Provisioning in Wireless Sensor Network," *Elsevier Journal of Computer Communications, Special Issue on Dependable Wireless Sensor Networks*, vol. 29, pp. 162–172, Feb. 2006.

[27] S. Cui, A.J. Goldsmith, and A. Bahai, "Energy-constrained modulation optimization for coded systems," *Proc. IEEE GlobeCom'03*, San Francisco, vol. 1, pp. 372–376, Dec. 2003.

[28] C.S. Park and K.B. Lee, "Transmit power allocation for BER performance improvement in multicarrier systems," *IEEE Transactions on Communications*, vol. 52, pp. 1658–1663, Oct. 2004.

[29] A.R.K. Sastry, "Performance of hybrid error control schemes on satellite channels," *IEEE Transactions on Communications*, vol. 23, pp. 689–694, July 1975.

[30] A.R.K. Sastry and L.N. Kanal, "Hybrid error control using retransmission and generalized burst-trapping codes," *IEEE Transactions on Communications*, vol. 24, pp. 385–393, April 1976.

[31] D.M. Mandelbaum, "An adaptive-feedback coding scheme using incremental redundancy," *IEEE Transactions on Information Theory*, vol. 20, pp. 388–389, May 1974.

[32] A. Banerjee, D. Costello, and T. Fuja, "Diversity combining techniques for bandwidth-efficient turbo ARQ systems," *Proceedings of IEEE International Symp. on Inform. Th.*, Washington, DC, June 2001.

[33] D. Chase, "Code-combining - a maximum likelihood decoding approach for combining an arbitrary number of noisy packets," *IEEE Transactions on Communications*, vol. 33, May 1985.

[34] J. Hagenauer, "Rate-compatible punctured convolutional codes and their applications," *IEEE Transactions on Communications*, vol. 36, pp. 389–400, April 1985.

[35] R. Min, M. Bhardwaj, S.H. Cho, et al., "Low-power wireless sensor networks," in *14th International Conference on VLSI Design (VLSI DESIGN 2001)*, Bangalore, India, pp. 205–210, Jan. 2001.

[36] X.R. Xie, *Computer Networks, 2nd Version* (in Chinese), Publishing House of Electronics Industry, China, 1999.

[37] D.J. Lu, *Random Process and Application* (in Chinese), Tsinghua Press, China, 1986.

[38] J. Kennedy and R.C. Eberhart, "Particle swarm optimization," in *Proc. IEEE Int. Conf. Neural Networks*, Perth, Australia, pp. 1942–1948, 1995.

10

DISTRIBUTED ORGANIZATION OF COOPERATIVE ANTENNA SYSTEMS

Wolfgang Zirwas, Jee Hyun Kim, Volker Jungnickel, Martin Schubert, Tobias Weber, Andreas Ahrens and Martin Haardt

Contents

Cooperative antenna (COOPA) systems have recently gained a lot of interest, as they promise significantly higher performance compared to conventional cellular mobile systems like global system for mobile communication (GSM) or 3G. COOPAs are multi-user MIMO (MU-MIMO) systems which inherently combine features like intercell interference (ICI) cancelation, macrodiversity, spatial multiplexing, and exploit the rank enhancement of the channel matrix. From a practical point of view mobile stations (MS) with low complexity and base stations (BS) with and one or two antenna elements per radio station might be used.

Conventional COOPA systems require a central unit (CU) for pre- or postprocessing of data signals of all involved BSs. Future core networks will be flat, i.e., almost without any central functionalities, which led to the distributed COOPA concept.

In this chapter, besides theoretical analysis of COOPA systems, measurement results will be presented and distributed downlink (DL) and uplink (UL) concepts will be investigated as well. The chapter is concluded by addressing topics and challenges for further studies.

10.1   INTRODUCTION

Similar to fixed networks or personal computers, mobile users also expect significantly increasing data rates for the next generations of cellular mobile networks beyond 3G or 4G. There exist many approaches, like improved coding, adaptive multi-user scheduling, and multi-hop, etc., which have already increased throughput or capacity to some extend. Multiple-input multiple-output (MIMO) antenna techniques reuse the freely available spatial dimension and are very promising as they overcome the fundamental Shannon limit of a single-input single-output (SISO) link. For good radio channel conditions, as provided indoors and at short range, the performance gains of MIMO have been proven, but in cellular outdoor systems, often low signal-to-interference and noise ratio (SINR) due to strong shadowing and multi-cell interference at the cell edge limits the performance. The comparison of single with multiple cell performance shows a significant degradation of spectral efficiency. Hence, interference mitigation has become one of the most promising enhancement techniques for the future.

Therefore, COOPA systems, where suitable processing combats ICI, have become a hot research topic. It is basically a multi-user MIMO(MU-MIMO) approach, where several adjacent BSs are tightly coupled over the backbone network. Each BS/MS can be seen as one or several antenna elements (AE) of the transmit/receive (Tx/Rx) side. Joint transmission/joint detection (JT/JD) algorithms located in a central unit (CU) calculate a common weighting matrix for all BSs, canceling ICI and allowing the system to serve several MSs simultaneously on the same time and frequency resource. This results in a real frequency reuse (f-reuse) equal to 1 or close to 1.

Cooperative antenna systems are interesting for several reasons:

- The system exploits the freely available spatial dimension, which is a well-known feature of any MIMO system. In the case of spatial multiplexing, the capacity may be enhanced by factors compared to that of a single user served by a single link.
- MU-MIMO systems can be deployed with low cost MSs, which might be equipped with only one or two AEs.
- Due to the distributed nature, several adjacent — but geographically distributed — BSs are used as transmit antennas, and full macrodiversity gains are available.
- Maybe the most important advantage for cellular radio systems — as explained above — is the avoidance of ICI between those BSs, which are cooperating due to the common processing in the CU.
- As will be shown, ICI, in conjunction with cooperation, may increase the spectral efficiency even beyond that of a single isolated cell due to rank enhancement.

In spite of these many advantages there is still some reluctance to apply such system concepts. One of the main reasons is the required CU, which would require a hierarchical network structure, principally available in today's 2G or 3G networks, with the radio network controller (RNC). But the vision of network planners, e.g., 3rd generation partnership project (3GPP) universal terrestrial radio access (UTRA) long term evolution (LTE), is a flat hierarchy, which allows for economy of scale for the BSs hardware and is fast and easily deployable with or without simple network planning.

To take care of these requirements a decentrally distributed cooperative radio concept organized over the air has been developed, minimizing the requirements for the backbone network and avoiding several issues that arise when the CU is replaced by decentralized processing. Issues like independently scheduling media accesss control (MAC) units at the BSs, sophisticated channel estimation, and the calculation of the precoding weight matrix $\mathbf{W}$ in a distributed manner, etc. have to be solved.

10.2 MOTIVATION FOR COOPERATIVE ANTENNAS

In the following, the benefits of cooperative antenna systems are described in more detail.

10.2.1 Theoretical Performance Limits and Algorithms

10.2.1.1 Coherent Combining Gain

Consider a system with L BSs in the uplink. Each mobile can be connected simultaneously with multiple synchronized BSs, as illustrated in Figure 10.1. By coherently combining the received signals, the signal-to-noise ratio (SNR) is improved, the transmission becomes more robust toward fading effects, and interference can be reduced. Note that most of the following results can be transferred to the downlink by using the duality between uplink and downlink.

Let us start characterizing the coherent combining gain for a single user scenario. All other interfering users are treated as noise. A signal $x(t)$, with power $p = \mathbb{E}[|x(t)|^2]$, is transmitted to the lth base over a vector channel $\mathbf{h}_l$. The signal at the output of the antenna array is

$$\mathbf{y}_l(t) = \mathbf{h}_l \cdot x(t) + \mathbf{n}_l(t) \tag{10.1}$$

where the vector $\mathbf{n}$ contains uncorrelated noise with variance σ^2, i.e., $\mathbb{E}[\mathbf{n}\mathbf{n}^H] = \sigma^2\mathbf{I}$. The dimension of the vectors $\mathbf{h}_l$ and $\mathbf{n}_l(t)$ is determined by the number of BS antennas, which is assumed to be arbitrary.

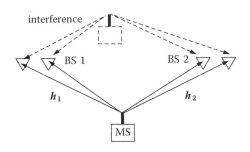

Figure 10.1 Coherent Multi-Base Transmission.

At the lth base, the signal $\mathbf{y}_l(t)$ is received by a linear spatial filter $\mathbf{w}_l$, which combines the outputs of the array antennas. Such spatial filtering with closely spaced array antennas is referred to as *farfield beamforming* [1,2]. Without loss of generality we can assume $\|\mathbf{w}_l\|_2 = 1$.

The beamformer output $\mathbf{w}_l^H \mathbf{y}_l(t)$ of the lth base can be combined with the signals received at other bases. This further improves the SNR, especially for users at the cell edge which might otherwise suffer from excess path loss and shadowing. Introducing weights $\alpha = [\alpha_1, \cdots, \alpha_L]^T$, the output signal of the multi-base combiner can be written as

$$r(t) = \sum_{l=1}^{L} \alpha_l \mathbf{w}_l^H \mathbf{y}_l(t) \tag{10.2}$$

Again, we can assume a normalization $\|\alpha\|_2 = 1$, which has no effect on the achievable SNR, because the desired signal and noise are scaled in the same way. Thus, $\Sigma_l \|\alpha_l \mathbf{w}_l\|^2 = 1$.

The actual combining gain depends on the channel coefficients. An absolute upper bound is obtained by using the following Cauchy–Schwarz-type inequality. The SNR is given as

$$\mathbf{SNR} = \frac{p}{\sigma^2} \left| \sum_l \alpha_l \mathbf{w}_l^H \mathbf{h}_l \right|^2 \leq \frac{p}{\sigma^2} \sum_l \|\mathbf{h}_l\|^2 \tag{10.3}$$

This inequality is fulfilled with equality if and only if $[\alpha_1 \mathbf{w}_1^T, \cdots, \alpha_L \mathbf{w}_L^T]$ and $[\mathbf{h}_1^T, \cdots, \mathbf{h}_L^T]$ are proportional. This solution is known as *maximum ratio combining* (MRC). The normalized MRC beamformers $\mathbf{w}_l = \mathbf{h}_l / \|\mathbf{h}_l\|$[1] lead to effective channel attenuations $\mathbf{w}_l^H \mathbf{h}_l = \|\mathbf{h}_l\|$. The BSs are combined with coefficients $\alpha_l = v\|\mathbf{h}_l\|$, where the scaling factor v is chosen such that $\|\alpha\| = 1$.

It can be observed from (10.3) that the combining gain strongly depends on the factor $\Sigma_{l=1}^{L} \|\mathbf{h}_l\|^2$, which is the sum of the individual channel gains. For a distributed system, the gains $\|\mathbf{h}_l\|^2$ can largely differ due to pathloss and shadowing effects. Thus, combining of multiple BSs is only attractive if all their individual contributions are sufficiently strong. This can be ensured by adaptive assignment strategies. Likely candidates for multi-base combining are users at the cell edge, which would otherwise suffer from a low SNR.

But even when multi-base combining is not feasible, the users still benefit from the SNR improvement due to beamforming. This so-called *array gain* depends on the spatial multipath structure of the propagation channel. For line-of-sight (LoS) channels, the SNR can be improved by a factor M, which is the number of antennas at the BS. For nonLoS channels, beamforming makes the transmission more robust in regard to small-scale fading.

Notice that only MRC can achieve the upper bound (10.3). Any other combining strategy performs worse in terms of SNR, but a disadvantage of MRC is its lack of interference suppression capability. Thus, it is generally suboptimal in the context of a multi-user system with mutually interfering users. The aspect of interference suppression will be discussed in the next section.

---

[1] $\|\cdot\|$ without subscript denotes the Euklidean norm $\|\cdot\|_2$.

10.2.1.2 Multi-User Combining

The traditional way of handling interference is to assign orthogonal resources to all links by time division multiple access (TDMA), frequency division multiple access (FDMA), or code division multiple access (CDMA). This considerably simplifies the system design because the links are no longer coupled by interference. However, reserving a fixed resource for each link can be a waste of spectral efficiency. The available bandwidth is generally best exploited by letting the transmitted signals interfere with each other in a controlled way (see, e.g., [3,4]). Instead of regarding the system as a collection of point-to-point communication links, a joint optimization strategy is preferable.

Consider K users sharing the same resource. Each user is possibly subject to interference caused by other users. The interference depends on the power allocation

$$\mathbf{p} = [p_1, \cdots, p_K]^T \tag{10.4}$$

as well as on the combining strategy chosen at the receiving BS. For a multi-base system, the interference also depends on the BS assignment. In the following, let b_k be the index set of the BSs assigned to the kth user. The channel between the ith transmitter and the BS antennas associated with bases from b_k is denoted as $\mathbf{h}_{i,b_k}$. This vector contains all complex channel coefficients stacked in a single vector.

By coherently combining the antenna outputs of the BSs b_k, an estimate $\hat{x}_k(t)$ of the kth signal $x_k(t)$ is obtained. The weighted receivers $\alpha_l \mathbf{w}_l$, $l \in b_k$ are stacked in a single vector $\mathbf{u}_k$, with $\|\mathbf{u}_k\| = 1$. Then,

$$\hat{x}_k(t) = \mathbf{u}_k^H \mathbf{h}_{k,b_k} x_k(t) + \sum_{i \neq k} \mathbf{u}_k^H \mathbf{h}_{i,b_k} x_i(t) + \mathbf{n}(t) \tag{10.5}$$

The desired signal $\mathbf{u}_k^H \mathbf{h}_{k,b_k} x_k(t)$ is corrupted by interference and noise. The interference cross talk can be characterized by a $K \times K$ nonnegative *coupling matrix* $\mathbf{V}$, with components

$$V_{ki} = \begin{cases} \left| \mathbf{u}_k^H \mathbf{h}_{i,b_k} \right|^2 & k \neq i \\ 0 & k = i \end{cases} \tag{10.6}$$

Then, the interference power caused by the lth transmitter to the kth receiver is given as $V_{kl} p_l$. The effective path gain between the kth transmitter and the kth receiver is $g_k = |\mathbf{u}_k^H \mathbf{h}_{k,b_k}|^2$. Thus, we can write the SINR ratio of the kth user as

$$\mathbf{SINR}_k = \frac{p_k \cdot g_k}{\sum_{l \neq k} p_l V_{kl} + 1} \tag{10.7}$$

Without loss of generality we have assumed additive noise with variance one. Any other noise variance simply corresponds to a scaled version of the power allocation.

It can be observed from (10.7) that the kth SINR is increasing in its own power p_k and decreasing in all other powers. Thus, all K SINR are coupled with each other. Increasing one user's power can lead to a decrease of another user's performance. In addition, the performance depends on the vectors $\mathbf{u}_k$, which in turn depend on the chosen assignment. The set of jointly achievable SINR values is referred to as the SINR feasible region.

The assessment of the performance of such a highly coupled system is difficult; thus, it is assumed that each user is assigned to all BSs, i.e., $b_k = \{1, 2, ..., L\}$. Also, let M denote the total number of antennas in the system.

One possible way to analyze the system is the assumption of *zero-forcing* reception [3]. If $K \leq M$, then interference can be completely removed. This strategy is optimal in the high SNR regime. Without interference, i.e., $\mathbf{V} = \mathbf{0}$, the system is not limited by interference anymore, i.e., any combination of SINRs can be jointly achieved as long as no constraint on the transmission powers is assumed.

For $K > M$, zero forcing is not possible and the system can be interference limited, i.e., not every SINR can be achieved, even with arbitrarily large powers. This effect is due to the mutual interference, which increases when the power is increased. In this case, the SINR feasible region is bounded and depends on the propagation channels. A coarse upper bound is obtained by the following inequality [5].

$$\sum_{k=1}^{K} \frac{1}{1 + \text{SINR}_k} > K - M \tag{10.8}$$

From (10.8), a necessary condition for the achievability of a common target γ by all K users can be derived:

$$\gamma < \frac{M}{K - M} \tag{10.9}$$

This absolute bound holds irrespective of the chosen powers or channel conditions. It can be observed that for $K \to \infty$, the achievable capacity per user goes to zero (as for TDMA). It can also be observed that for a given common SINR target γ, the maximum number of supportable users is upper bounded: $K < M(1 + 1/\gamma)$.

It should be emphasized that these bounds hold for linear combining. When successive interference cancelation (SIC) is used in addition, the system is no longer interference limited [6]. This assumption was also used in [7], where the information-theoretical throughput of a multi-cell system was analyzed asymptotically.

The assumption that each user is connected with all BSs is of theoretical interest and provides a useful benchmark. But from a practical point of view, we are more interested in the case where each user is connected with a subset of BSs. This case will be studied later in the following section.

10.2.1.3 Iterative Resource Allocation

The optimization of a multi-user system can be performed with respect to various design goals, like overall system efficiency, max-min fairness, proportional fairness, network utility, etc. There is no such thing as "the" optimal transmission strategy, so to find a good set of targets $\Gamma = \text{diag}\{\gamma_1, \ldots, \gamma_K\}$ is a nontrivial task.

Suppose that feasible targets Γ are given. Then an interesting question is how to achieve them with minimum transmission power. This fundamental problem was already studied in various contexts [8–17]. A special aspect of the distributed cooperative system is the choice of the assignment strategy and beamforming [8,9,17]. For the uplink, the optimal strategy can be found for each user independently. For a given power allocation

$\mathbf{p}$, the minimum interference power is

$$I_k(\mathbf{p}) = \min_{b_k \in B_k} \left(\min_{\mathbf{u}_k : \|\mathbf{u}_k\| = 1} [\mathbf{V}(u, b)\mathbf{p} + \mathbf{1}]_k \right) \tag{10.10}$$

The inner minimization is carried out for a fixed assignment b_k from a set of possible assignments B_k. The beamformers can be optimized based on the interference-plus-noise spatial covariance.

$$\mathbf{Z}_{k,b_k} = \sum_{l \neq k} p_l \mathbf{h}_{l,b_k} \mathbf{h}_{l,b_k}^H + \mathbf{I} \tag{10.11}$$

The optimizer is

$$\mathbf{u}_{k,b_k} = \alpha(\mathbf{Z}_{k,b_k})^{-1} \mathbf{h}_{k,b_k} \tag{10.12}$$

where α is chosen such that $\|\mathbf{u}_{k,b_k}\| = 1$.

Among all possible assignments from the set b_k, we choose the one that is associated with the minimum interference. This requires an exhaustive search over all combinations; thus the set b_k needs to be kept small.

This choice of beamformers and assignments is only optimal for the given power allocation $\mathbf{p}$, and for another allocation, other strategies can be better. The jointly optimal strategy can be found iteratively, as proposed in [8,12]. This is an extension of results in [11,17] where no beamforming was assumed. It is also a special case of the general framework [16]. The algorithm is the following fixed-point iteration.

$$p_k^{(n+1)} = \gamma_k I_k(\mathbf{p}^{(n)}), \quad \mathbf{p}^{(0)} = [0, ..., 0]^T \tag{10.13}$$

If the targets Γ are feasible, then the iteration (10.13) converges componentwise to the unique power allocation which achieves the targets Γ with equality. Because the optimization is performed with respect to the minimum interference (10.10), the assignment and beamforming strategy is optimally adjusted. There is no other strategy which achieves Γ with less power, so the iteration is globally optimal.

The global optimum can be achieved even faster when knowledge of the coupling matrix $\mathbf{V}$ is exploited [13,14]. Then, for each assignment and beamforming strategy, the optimal power allocation can be computed directly. The following steps are repeated until convergence:[2]

1. Given $\mathbf{p}^{(n)}$, compute beamformers for all possible assignments from the set B_k.

$$\mathbf{u}_k^{(n,b)} = \arg \min_{\mathbf{u}_k : \|\mathbf{u}_k\| = 1} [\mathbf{V}(u, b)\mathbf{p} + \mathbf{1}]_k \tag{10.14}$$

2. Given beamformers $\mathbf{u}_1^{(n,b)}, \cdots, \mathbf{u}_K^{(n,b)}$, choose the assignment which minimizes the interference for each user.

$$b_k^{(n,b)} = \arg \min_{b_k \in B_k} [\mathbf{V}(\mathbf{u}^{(n,b)}, b)\mathbf{p}^{(n,b)} + \mathbf{1}]_k \tag{10.15}$$

---

[2] Here $\mathbf{1}$ is the all-one vector that contains the normalized noise powers.

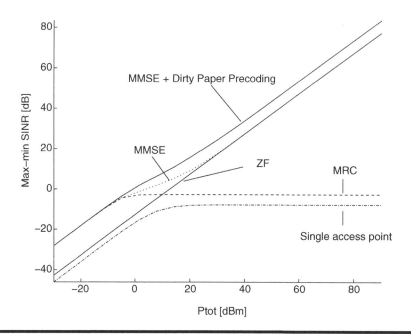

Figure 10.2   Achievable SINR Margin vs. Total Transmit Power for a Multi-Base System with Coherent Base Station Cooperation. Parameters: 7 Base Station, 7 Users Sharing the Same Resource, 100m Cell Radius, Hata-Urban Path Loss.

3. Given assignments $b^{(n,b)}$ and beamformers $\mathbf{u}^{(n,b)}$, compute the power allocation which achieves targets $\gamma_1, \cdots, \gamma_K$.

$$\mathbf{p}^{(n+1)} = \left(\mathbf{I} - \Gamma\mathbf{V}(\mathbf{u}^{(n,b)}, b^{(n,b)})\right)^{-1} \Gamma\mathbf{1} \tag{10.16}$$

This iteration has superlinear convergence. Loosely speaking, this means that the iteration "accelerates" when approaching the optimum. This excellent convergence behavior is enabled by the centralized power update (10.16). If the powers can only be updated in a decentralized manner, the fixed-point iteration (10.13) can be used. Notice that the optimization steps (10.14) and (10.15) are the same for both iterations. They still require a certain degree of cooperation (global channel knowledge) in the system. Typically, the performance degrades when cooperation is restricted. This is illustrated in Figure 10.2. The important outcome is that without cooperation — MRC and single access point — the max-min SINR is limited despite increasing P_{tot} to a low value of about 0 dB due to the interference between the BSs. In contrast, for the cooperative schemes — minimum mean square error (MMSE), MMSE with dirty paper precoding, and zero forcing (ZF) — max-min SINR increases linearly with P_{tot} and is therefore no longer interference limited. Smaller performance differences arise from the applied algorithms like ZF, MMSE or MMSE with nonlinear precoding (MMSEDPC) but they do not change the main behavior.

10.2.2   Outdoor Measurements

In addition to the more theoretical system analysis, a sophisticated measurement campaign has been conducted in Berlin within a campus-like area having a size of about $200 \times 200\,$m with four BS locations.

10.2.2.1 Measurement Setup

Measurements were taken with the RUSK HyEff channel sounder[3] at 5.2 GHz in a bandwidth of 120 MHz, matched to the frequency response of the antennas. The HyEff receiver has the capability to store real-time measurement data continuously onto a huge hard disk array. The single measurement chain of the HyEff system is alternately paused in the radio frequency (RF) domain to allow a consecutive switching onto all Tx and Rx antenna elements with the radio channel in between. Effective Tx power at the antennas was about +33 dBm, and together with an improved link margin the performance is improved by roughly 20 dB, compared to earlier measurements [18] to bridge distances up to about 200 m.

Measurements have been taken in a typical European metropolitan area at the campus of the Technical University of Berlin. Four different receiver positions (acting as BS sites) are located either at the wall of the HHI building or on top of the HFT, MATH, or ILR buildings. The antennas are placed as close as possible to the edges of the roofs. Base station locations and orientations are indicated in Figure 10.3.

The data have been recorded while moving the transmitter on five short tracks indicated by arrows and corresponding numbers in Figure 10.3.

10.2.2.2 Multi-Cell Results

The multi-cell scenario is investigated to demonstrate the potential effects of using both MIMO and joint intercell interference cancelation.

10.2.2.2.1 Enhanced Channel Rank

First, we demonstrate the synergy between the use of MIMO at link level and multi-cell interference cancelation at system level. For an isolated two-user, two-cell scenario, the channel matrices are assembled from individual links. The user at Pos 5 is assigned to the BS on the MATH building and the user at position (Pos) 2a[4] to the BS on the ILR building. Both link scenarios have LoS and, due to polarization, two dominant singular values. The interference link ILR-Pos 5 has LoS as well, while the LoS is blocked in the second interference scenario MATH-Pos 2.

The singular value distributions are shown in Figure 10.4. When the interference is avoided (i.e., interference links are switched off using different resources in each cell, Figure 10.5, left), the joint multi-user multi-cell channel matrix has four dominant singular values. When the interference is included (see Figure 10.5, right), there are two additional degrees of freedom, i.e., altogether six singular values are significantly larger than the others.

This observation is coined *enhanced channel rank* and explained in detail in Figure 10.5. When the two adjacent cells use different resources and the propagation is mostly based on the LoS (Figure 10.5, left), there are two spatial degrees of freedom in both cells due to the polarization, i.e., altogether there are four. When the same resource is used in both cells (Figure 10.5, right), the intercell interference signals add more spatial degrees of freedom because their spatio-temporal properties are statistically independent of the intracell signals. The LoS between ILR and Pos 5 is free and this component actually

---

[3] See www.channelsounder.de

[4] Pos 2a means that only the first part of the track is investigated because properties have changed afterwards.

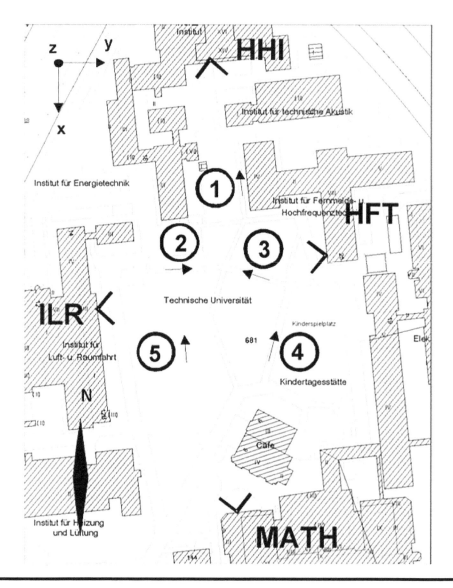

Figure 10.3   Base Station Locations Sector Angles in the Campus of TU Berlin. All Channels Have Been Recorded Consecutively, from Each BS Location to Each of the Five Terminal Tracks.

adds the two more spatial degrees of freedom. On the other hand, the LoS between MATH and Pos 2 is blocked and the transmission distance is very long. Hence, the attenuation is too high to generate two further significant singular values.

As a result, the terminal at Pos 5 can be served with four data streams by jointly transmitting from MATH and ILR BSs, while the terminal at Pos 2 can be served with only two streams because the LoS in the interference link is blocked. Note that this type of macromultiplexing must be supported by a sufficient number of antennas at the terminal. Furthermore, the channel rank is not enhanced but the interference paths can still be used for the purpose of macrodiversity. This means that a single wireless link is established to a given terminal and that it is stabilized by optimally transmitting the same information from two or more BSs.

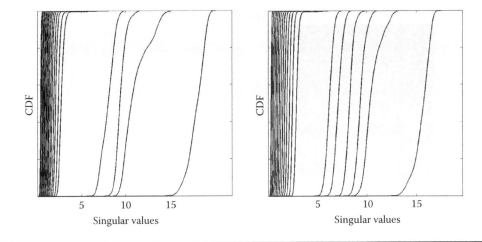

Figure 10.4   Singular Value Distribution for Orthogonal Scheduling of Resources (Left) and for the Cooperative Scheme (Right).

10.2.2.2.2   Enhanced Cell Capacity

There are two major benefits of cooperation in the multi-user multi-cell MIMO scenario. The first comes from the inherent interference cancelation of the joint detection and transmission, enabling the reuse of resources in adjacent cells. The second benefit comes from the enhanced channel rank. The combined effect is demonstrated in the four-cell five-terminal scenario.

Results are shown in Figure 10.6. The gain matrix in Figure 10.6 (left) reveals substantial differences in the mean channel gains for different terminal-to-base-station links. Note that in only a few cases two terminals have similar gain at a single BS (Pos 2 and Pos 3 at HFT and Pos 2 and Pos 5 at ILR) or that a single terminal is similarly received by two BSs (Pos 5 at MATH and ILR). One might conjecture from these results that the ICI is a more or less localized phenomenon, and this observation supports the decentralized approach proposed in this chapter.

In the following, we compare uplink channel capacities without power control. In Figure 10.6 (right), the capacity distributions are plotted for three cases. The mean cell capacity for using different resources in each cell, see curve (a) in Figure 10.6, is only

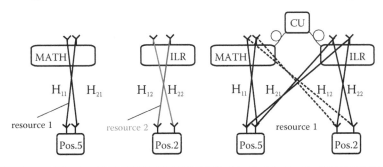

Figure 10.5   Intuition for the Enhanced Channel Rank. Left: Orthogonal Resources in Adjacent Cells Give Two Dominant Singular Values per Cell. Right: When Cooperation Is Enabled in a Central Unit and the Same Resource Is Used, the Interference Link ILR-Pos 5 Gives Two More Degrees of Freedom, Altogether One Gets Six.

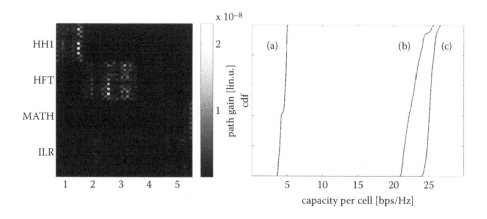

Figure 10.6 **Left: Four-Cell Five-User Channel Gain Matrix Including the Interference Channels Between All Users and All Cells. Right: Distributions of Capacity for the Traditional Deployment with Different Resources in Each Cell (a), Mean Capacity of Isolated Cells (b) and Capacity with BS Cooperation Using a Common Resource (c), All for a Mean SNR of 10 dB Per BS Antenna.**

about 5 bps/Hz, although multiple antennas are used extensively in each MIMO link. The major reasons are that the assignment of different resources in each cell reduces the spectral efficiency per cell due to the frequency reuse factor, and that the individual links are frequently dominated by the LoS (and hence the channels are rank deficient).

For comparison, the mean capacity of the isolated cells (b) is shown in Figure 10.6 as obtained from the capacities of the subchannel matrices between BSs and assigned terminals. It is about four times higher (22 bps/s) because with fictive isolation between the cells, the frequency reuse factor could be unity.

The fictive isolation becomes reality when joint processing is used, including the signals received by all BSs. Then there is an additional gain due to the enhanced channel rank. Altogether, the cell capacity with cooperation is approximately five times higher in the measurement scenario than by using different resources in each cell as shown in Figure 10.6, curve (c).

Altogether, most of the gain comes from interference reduction while a smaller fraction is due to the enhanced channel rank. These huge gains are valid for perfect interference cancelation. It should be noted here that interference in a full-coverage cellular system might be stronger than in the measurement, because signals from more cells are involved, generating interference which cannot be canceled perfectly in practice. But the principal effects should be the same, and significant gains are already possible by using partial interference cancelation, as shown by the multi-cell simulations below.

10.3 DISTRIBUTED SYSTEM CONCEPTS

As shown above, cooperative antenna systems promise the highest capacity and spectral efficiency not only theoretically, but also for real outdoor scenarios. With a CU, a straightforward implementation would be possible. For the distributed approach — here the distributed calculation of the weighting matrices is meant — novel system concepts are required. As illustrated in the sequel, two different concepts are required for the DL and UL, resulting in different challenges.

10.3.1 Downlink

For conventional joint transmission, a CU receives the data for all connected MSs, which should be supported simultaneously over the air from the backbone network. The CU calculates dependent on the overall channel matrix $\mathbf{H}$, an appropriate according weighting matrix $\mathbf{W}$. In the simplest case of ZF, the pseudoinverse of $\mathbf{H}$ is denoted as $\mathbf{H}^+$ and used for preprocessing of the data signals. A ZF scheme that achieves a better performance (also for users with multiple antennas) was presented in [32]. The CU includes the common medium access control (MAC) as well as the MU-MIMO preprocessing in a single point.

Here — to avoid the CU — the cooperation is organized decentrally over the air. The following issues have to be taken into account for the development of a distributed cooperation:

- In real mobile radio systems the scheduling by the MAC entity in each BS is quite unpredictable, as the MAC has to take into account the buffer length of different quality of service (QoS) classes, actual channel conditions, automatic request protocol (ARQ)-messages, etc. For this reason in each radio frame the scheduling might change completely.
- Future backbone networks will be based on the Internet protocol (IP), so the delays over different routes in the backbone network are quite unpredictable, even in managed networks.
- For the highest benefit of cooperative transmission, all BSs need full channel knowledge and free access to all data for all MSs.
- For simultaneous transmission in an orthogonal frequency division multiplex (OFDM) system all BSs and MSs have to be synchronized in frequency and time.

Figure 10.7 shows the distributed cooperative system proposal. The system consists of 3 BSs, BS1, BS2, BS3, and 2 MSs, MS1 and MS2. The BSs are connected to the backbone

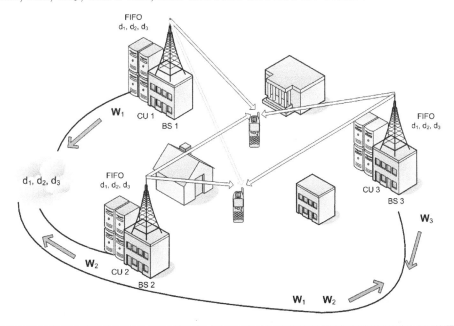

Figure 10.7 Distributed Cooperative System Architecture.

network over cable, optical fiber, or their replacements (microwave or free-space optics) exhibiting different delays, which are assumed to be limited, e.g., up to 20 ms, but unpredictable. So as a first step, all BSs store the data for all MSs in according first in first out (FIFO) buffers. For this purpose a multicast connection has to be established to all involved BSs. As there is no longer a single CU, now 3 distributed "CUs," CU1,...,CU3 calculate their weight matrices $\mathbf{W}_1, \ldots, \mathbf{W}_3$ independently. The name CU is kept despite the fact that the unit is no longer central, and the same processing has to be done as before by a centralized CU.

$$x_{1,s} = w_{11} \cdot x_1 + w_{12} \cdot x_2; \quad x_{2,s} = w_{21} \cdot x_1 + w_{22} \cdot x_2 \tag{10.17}$$

$$r_1 = b_{11} \cdot x_{1s} + b_{12} \cdot x_{2s}; \quad r_2 = b_{21} \cdot x_{1s} + b_{22} \cdot x_{2s} \tag{10.18}$$

$$\begin{pmatrix} r_1 \\ r_2 \end{pmatrix} = \begin{pmatrix} (b_{11} \cdot w_{11} + b_{12} \cdot w_{21}) & (b_{11} \cdot w_{12} + b_{12} \cdot w_{22}) \\ (b_{21} \cdot w_{11} + b_{22} \cdot w_{21}) & \underline{(b_{21} \cdot w_{12} + b_{22} \cdot w_{22})} \end{pmatrix} \cdot \begin{pmatrix} x_1 \\ x_2 \end{pmatrix} \tag{10.19}$$

$$w_{22} = -\frac{b_{11}}{b_{12}} \cdot w_{12}; \quad w_{21} = -\frac{b_{21}}{b_{22}} \cdot w_{11} \tag{10.20}$$

Formulas (10.17) to (10.20) are the main formulas for a 2×2 joint transmission system with ZF (all variables complex), where $x_{i,s}$ are the transmit signals of the BS i, x_j are the data signals for the MS j, and w_{11} to w_{22} and b_{11} to b_{22} are the elements of $\mathbf{W}$ and $\mathbf{H}$, respectively. Here, r_j are the correspondent receive signals at MS j, for simplicity without noise. The goal of the ZF joint transmission is to maximize for $(r_1 \quad r_2)^T$ the underlined terms and to set the interference terms to zero. Formulas (10.20) are required for this. To calculate $\mathbf{W}$ each BS has to estimate all radio channels from each BS to each MS. As can be seen from Figure 10.8, there are three different types of radio channels: those with a direct connection between BS and MS (circles in $\mathbf{H}$ matrices), those where the BS has no connection but its own MS has direct access to the radio channel coefficients (squares in matrices), and those where only the MS from the other BS has direct access.

The difference in the last two cases is that the radio link between the MS and the associated BS will most likely be good — otherwise it would not be associated — while the link quality to another BS is unpredictable and might be poor.

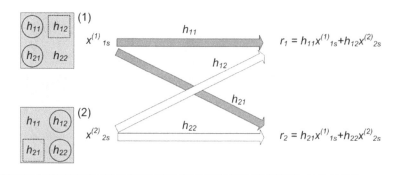

Figure 10.8   Over the Air Estimate of Radio Channels.

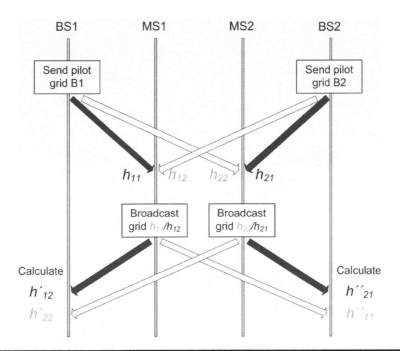

Figure 10.9   Sequence Chart for Distributed Channel Estimation.

The general scheme for channel estimation is shown in Figure 10.9. In the first step, the BSs broadcast orthogonal pilot grids, e.g., on different subcarriers of an OFDM symbol [19]. This allows MS1 to estimate the channel coefficients h_{11} and h_{12} while MS2 can estimate h_{21} and h_{22}. In the following step, the MSs broadcast their estimated radio channel coefficients together with their own pilots over orthogonal pilot grids. From received signals in uplink both BSs can perform an estimation of all channel coefficients. The main difference compared to a system with one single CU is that there will now be different — and therefore diverging — estimates $\mathbf{H}'$ and $\mathbf{H}''$ at each BS, resulting in an additional performance degradation.

It should be mentioned that this proposal can be used in time division duplex (TDD) as well as in frequency division duplex (FDD) systems. For TDD the uplink pilots can be used to directly estimate the circled coefficients while for FDD the pilots are used as a reference signal when estimating the downlink channel. More details about channel estimation in FDD can be found in [19].

Besides channel estimation, the autonomous scheduling at the different BSs is a major challenge of the distributed approach. To accomplish this, the MAC and the physical layers are separated. First, each BS schedules its MSs independently, dependent on channel quality, buffer length, etc. (see message sequence chart (MSC) in Figure 10.10). Next, the BSs broadcast the MAC IDs together with the pilots as described above. The corresponding MSs receive the MAC IDs and rebroadcast the MAC IDs together with their pilot grids and measured downlink channel coefficients so that all BSs are informed about the actual MAC IDs in the neighboring cells.

In principle, each BS knows the data sequence in the FIFO, which has to be transmitted in the next time step. But to improve robustness from time to time, sequence numbers should be inserted into the data queues and transmitted together with the MAC IDs for proper synchronization of the MAC schedulers.

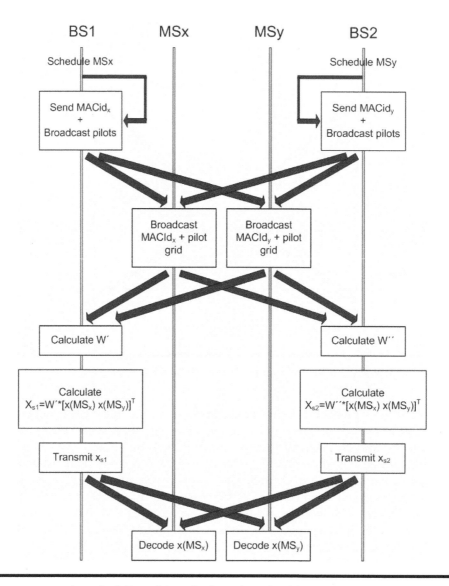

Figure 10.10   MSC for Scheduling of the Distributed JT Algorithm.

Each BS calculates the corresponding weighting matrices $\mathbf{W}'$ and the resulting transmit signals $x_{i,s}$. In the final step the MSs decode the receive signals r_j.

The physical layer part, i.e., mainly the distributed channel estimation, has to be done very fast otherwise the time variant radio channels may have changed. Pilots should be retransmitted in a predefined time slot as soon as channel state information (CSI) estimation has been performed at the MS. In case of analogue CSI estimation, very fast processing is possible, while in case of interpolation, more processing time might be required. Ideal interpolation for radio channels in OFDM systems has been analyzed in [20] for downlink and an improved version in [20a] which is applicable to down and uplink.

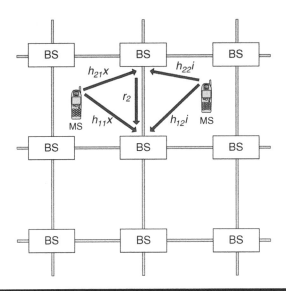

Figure 10.11 Using a Meshed Backbone Network, Each BS Gets Access to the Received Signals from Other BSs. Each BS Has the Capability of Multi-Stream Separation for the Purpose of Interference Reduction.

10.3.2 Uplink

The fundamental question behind reducing the complexity by the distributed COOPA approach is how much cooperation is actually needed to combat not all, but a large fraction of the interference. In other words, there is a certain tradeoff between the number and selection of BSs involved in the cooperation and the residual interference which cannot be canceled. The more cooperation is used, the more interference can be canceled. When just as much interference is canceled as needed, the complexity can be reduced and the system performance is still significantly improved.

The effect of limited interference cancelation has led to the decentralized UL concept proposed in the following. In the centralized approach, the underlying backbone network has the structure of a fixed star with a central intelligence at the CU. In contrast, a meshed backbone network with decentralized intelligence is proposed here (see Figure 10.11).

The signal processing intelligence is now — similar to the DL concept — distributed across the network and not centralized in a single CU. Each BS has the capability to spatially separate signals from multiple cochannel terminals. But the number of locally separable streams can be much larger than the number of physical antennas at the BS. For instance, in the present universal mobile telecommunications system (UMTS) network, two cross-polarized antennas are employed per BS, but each BS could have the signal processing capability to separate eight or more terminals. This capability is used to reduce the intercell interference for those terminals assigned to the BS that is created by terminals assigned to adjacent BSs (see Figure 10.11).

Consider two MSs assigned to two adjacent BSs. The two terminals are assigned to the same resource in both cells and hence there is significant cochannel interference. Now, together with the pilot signals needed to identify their channels, each terminal transmits an identifier to which BS it is assigned. The identifier might be included inherently in the pre-defined orthogonal pilot grids for adjacent BSs e.g. by modulating onto the pilots a cell-specific scrambling code along the frequency and time axes. Based on this information, the serving BS of the desired terminal can estimate the channel coefficients of the interfering terminals and it can identify the particular BS to which each interfering terminal is assigned.

To remove the interference, the serving BS involves the received signals from the so identified adjacent BSs and separates the desired terminal signal from the interference channels using the multistream separation capability of the local signal processing, such that the strongest interferers are canceled.

The central BS detects the signal x of the left terminal having the channel b_{11} which is being disturbed by another terminal with signal i being served by the BS in the north direction. The received signal neglecting the noise is given as

$$r_1 = b_{11} \cdot x + b_{12} \cdot i \tag{10.21}$$

where b_{12} is the channel through which the interference i is received. The BS now estimates the interference channel b_{12} and receives the additional information that this terminal is assigned to the northern BS. Consequently, it requests the received signal

$$r_2 = b_{21} \cdot x + b_{22} \cdot i \tag{10.22}$$

from the northern BS in which pilots are embedded to estimate the channel coefficients b_{21} and b_{22}. Equation (10.21) and Equation (10.22) can be written conveniently in matrix form

$$\begin{pmatrix} r_1 \\ r_2 \end{pmatrix} = \begin{pmatrix} b_{11} & b_{12} \\ b_{21} & b_{22} \end{pmatrix} \begin{pmatrix} x \\ i \end{pmatrix} \tag{10.23}$$

Now, the BS has all the information needed to cancel the ICI i out of the two received signals. Equation (10.23) states a conventional MIMO problem which can be solved by a number of well-known algorithms, among which is the zero forcing, using the pseudoinverse of the matrix $\mathbf{H}$ denoted as $\mathbf{H}^+$

$$\begin{pmatrix} x \\ i \end{pmatrix} = \begin{pmatrix} b_{11} & b_{12} \\ b_{21} & b_{22} \end{pmatrix}^+ \begin{pmatrix} r_1 \\ r_2 \end{pmatrix} \tag{10.24}$$

which is the least complex approach but has the worst performance as well. Better algorithms are ordered successive interference cancelation (V-SIC), sphere decoding, and maximum likelihood detection. The desired terminal signal x is picked out of the reconstructed signal vector $(x \quad i)^T$ and detected, while the interference i is discarded.

In principle, the BS serving a given terminal must have access to the received signals from the adjacent BSs serving the interfering terminals. The interference is canceled locally using the multistream signal processing capabilities of the BS.

The more orthogonal the column and row vectors of the channel matrix $\mathbf{H}$ in (10.23) are, and the more independent linear combinations of the same terminal signals are involved in the interference cancelation, the better is the separability of terminal signals. However, it is a complex task to select the right BSs and terminals needed to reduce the interference below a predefined target.

From their own signal, the serving BS identifies, at first, the BSs to which the interfering terminals are connected. Next, the serving BS requests the received signals from those BSs and estimates the channel coefficients (arranged in the matrix $\mathbf{H}$) between the interfering terminals and the candidate BSs. But unfortunately, these signals may contain additional interference from distant terminals not identified as interferers at the serving BS.

Mathematically this means that the matrix **H** may have more columns than rows

$$\begin{pmatrix} r_1 \\ r_2 \end{pmatrix} = \begin{pmatrix} b_{11} & b_{12} & 0 \\ b_{21} & b_{22} & b_{23} \end{pmatrix} \begin{pmatrix} x \\ i_2 \\ i_3 \end{pmatrix} = \begin{pmatrix} b_{11} & b_{12} \\ b_{21} & b_{22} \end{pmatrix} \begin{pmatrix} x \\ i_2 \end{pmatrix} + \begin{pmatrix} 0 \\ b_{23} i_3 \end{pmatrix} \qquad (10.25)$$

i.e., the signal of the cooperating BS is disturbed by the signal $b_{23} i_3$. Moreover, there are now more unknowns than equations and no cooperation between the terminals is allowed in UL that could remove the interference in advance. While the first part of the matrix is still invertible, in general the second term in (10.25) is residual interference which must be considered seriously.

Owing to this additional interference, the interference cancelation is normally imperfect in the decentralized approach. Hence, only a partial interference cancelation is practical and the target is not perfect, but the best possible interference cancelation. In order to reduce the interference in (10.25) partially, one can write[5]

$$\begin{pmatrix} \hat{x} \\ i_1 \end{pmatrix} = \begin{pmatrix} b_{11}^+ & b_{12}^+ \\ b_{21}^+ & b_{22}^+ \end{pmatrix} \begin{pmatrix} r_1 \\ r_2 \end{pmatrix} - \begin{pmatrix} b_{11}^+ & b_{12}^+ \\ b_{21}^+ & b_{22}^+ \end{pmatrix} \begin{pmatrix} 0 \\ b_{23} i_3 \end{pmatrix} \qquad (10.26)$$

where the second term includes the interference that cannot be canceled. The reconstructed signal x now reads

$$\hat{x} = x - b_{12}^+ b_{23} i_3 \qquad (10.27)$$

where the interference is possibly smaller than without partial interference cancellation where the received signal r_1 is in the presence of cochannel interference from i_2, see (10.21), and the signal reconstructed from r_1 only reads

$$\hat{x} = x - b_{11}^{-1} b_{12} i_2 \qquad (10.28)$$

One must always take care that the partial interference cancellation has actually led to improved signal quality. The best way to select candidate base stations may be an iterative approach. In each step, out of the surrounding base stations, always the one is selected which leads to the highest SINR after partial interference cancellation. This is very similar to the conventional SIC, where the order of cancelling the interference has a huge impact on the performance.

The proposal has some additional advantages in addition to high performance gain, distributed processing without a central unit, and autonomous BS scheduling.

■ Due to the geographically distributed location of BSs, the correlation of radio channels will be low. This gives the BSs full freedom for scheduling. Nothing like user grouping — as might be required for MU-MIMO with one BS — is needed.
■ The interference mitigation is MS centric, in the sense that only the strongest interfering signals are included into the cooperation. This is in contrast to an approach where a central unit processes a number of BSs with high processing effort, irrespective of the achievable performance gain.

---

[5] b_{ij}^+ denotes the (i, j) element of the pseudoinverse of the matrix which is composed of b_{ij}.

◼ The system requirements are quite low and involve mainly orthogonal pilot signals for adjacent BSs, something which is also foreseen for conventional systems to avoid degradation of channel estimation by interfering pilot patterns.

◼ Besides orthogonal pilot signals or MAC-IDs no further overhead is introduced for the air interface.

◼ As data and pilot signals are transmitted together in the UL-direction, even if the real processing might be delayed by some time due to data exchange between BSs, high speed of MSs can be supported.

Challenges for mobile network operators (MNOs) might be the increased data traffic on the backbone network. The backbone should support fast data exchange between BSs to minimize the additional latency as a result of the cooperation.

10.3.3 Challenges and Optimizations

In this chapter some of the main challenges and possible optimizations are presented. The DL channel estimation is one of the most important topics and therefore will be treated in detail. Other issues are the additional data overhead and inter-BS delay requirements in the backbone network where several proposals exist to relax the requirements.

10.3.3.1 Channel Estimation for DL Solution

One of the main questions is what happens due to the radio channel estimation errors at the different BSs for the DL proposal as described above. For this purpose we start with an analysis.

10.3.3.1.1 Detailed Analysis of Channel Estimation

As explained above, the distributed DL system concept replaces the necessity of a CU by multicasting of data signals. As a result, the distributed system has to be organized decentrally over the air to acquire the DL channel matrix $\mathbf{H}$. Due to noise, each BS has its own estimated version $\mathbf{H}'$, resulting in different versions of precoding matrices $\mathbf{W}'$.

In the following simulations, the channel estimation method based on Section 10.3 has been applied, i.e., the elements of $\hat{\mathbf{H}}_k$ — the channel matrix estimates for BS k, the hat sign indicates estimated value, not actual value — have been estimated by suitable pilot signals directly or indirectly by a retransmission of the MSs. Based on $\hat{\mathbf{H}}_k$, the corresponding weight matrices $\hat{\mathbf{W}}_k$ have been calculated for ZF and used for precoding of the transmit signals.

Assuming TDD-based systems, the direct path can be directly estimated by exploiting UL/DL channel reciprocity. However, the estimated CSI has to be rebroadcast to the involved BSs for the indirect radio channels.

The estimated CSI (index lm denotes the link between MS l and BS m) for BS k, which is acquired by the method explained herein, is as follows:

$$\hat{b}_{lm,k}^{DL} = \begin{cases} b_{lm}^{DL} + n_{BS} & (m = k) \\ \left((b_{lm}^{DL} + n_{UE}) b_{lk}^{UL} + n_{BS} \right) / \hat{b}_{lk}^{UL} & (m \neq k) \end{cases} \quad \text{where} \quad \begin{array}{l} \hat{b}_{lk}^{UL} = b_{lk}^{UL} + n_{BS} \\ \\ n_{BS/UE} \sim \mathcal{CN}(0, N_0) \end{array} \quad (10.29)$$

Here, N_0 is the power spectral density of white Gaussian noise, and the noise at BS and MS ($n_{BS/UE}$) is assumed to follow the circularly symmetric complex Gaussian distribution.

For example, the channel matrix at BS k can be expressed as (10.30) when the cooperative area consists of 3 BSs and 2 MSs. The channel matrix for BS 1 ($\hat{\mathbf{H}}_1$) has two directly acquired channel coefficients ($\hat{b}_{11,1}^{DL}$ and $\hat{b}_{21,1}^{DL}$) which have been provided by UL channel estimation of the corresponding links (BS1–MS1 and BS1–MS2). Other channel coefficients are acquired by the retransmission scheme, which is more vulnerable to channel estimation errors and noise effects. Therefore, each BS has its own version of the channel matrix due to the indirect channel estimation, and as a result, the weight matrix can be different for different BSs.

$$\hat{\mathbf{H}}_k = \begin{pmatrix} \hat{b}_{11,k}^{DL} & \hat{b}_{12,k}^{DL} & \hat{b}_{13,k}^{DL} \\ \hat{b}_{21,k}^{DL} & \hat{b}_{22,k}^{DL} & \hat{b}_{23,k}^{DL} \end{pmatrix}, \quad \text{where } k \text{ is the BS index} \tag{10.30}$$

The weight matrix in case of ZF can be calculated by taking the pseudoinverse of the channel matrix.

$$\hat{\mathbf{W}}_k = \hat{\mathbf{H}}_k^+, \quad \text{where } k \text{ is the BS index} \tag{10.31}$$

We should note that the weight matrix calculation process (10.31) is to be performed independently for each BS. Each BS has its own version of the DL channel matrix (10.30); therefore, each BS has its own version of weight matrix, which is indicated by subscript k in (10.31). The MSs, which are served by several BSs forming a cooperative area, may locate near a certain BS. This can result in great differences in received signal power strength for the links between BS and MS, provided that BS transmits the signal with its maximum power. The performance of the indirect channel estimation scheme strongly depends on the link strength. We should also note that the error variances of the channel estimates could be taken into account when the weighting matrix is calculated. However, this has not been considered in the following simulation.

The purpose of the simulation is to evaluate the suggested channel estimation scheme (10.29) and its effect on the system performance in terms of SINR and Shannon throughput.

The simulation results are depicted in Figure 10.13, which shows the cumulative distribution function (CDF) of the achievable SINR for four system deployment scenarios of a cooperative area (CA)[6] with 3 BSs and 2 MSs (Figure 10.12). The average throughput has been introduced as the system performance criterion in addition to the achievable SINR. The instantaneous throughput is calculated by the Shannon formula for the 5 MHz bandwidth case, taking instantaneous SINR as its argument. Average throughput is acquired by taking an average of the instantaneous throughput over simulation drops.[7] The simulation scenario is confined for one CA, and only the intra-CA interference has been dealt with. Therefore, the average throughput herein should be considered only as an evaluation metric to compare performances of different system scenarios. MSs are randomly placed within the CA for each drop.

---

[6] CA stands for cooperative area, which is defined in COOPA. In COOPA, CA consists of a group of BSs and MSs, which are cooperating in transmitting and receiving data. Loosely speaking, the CA in a COOPA is similar to the cell in a conventional cellular system.

[7] It should be noted that this throughput value can be different from the actual system throughput, given the fact that our simulation does not consider inter-CA interference. This argument is also applicable to the achievable SINR.

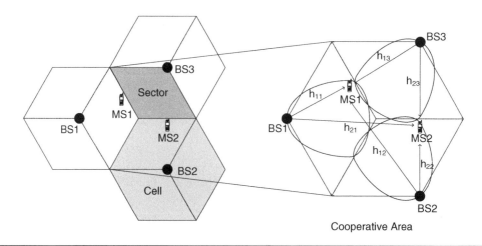

Figure 10.12   CA Topology for 3 BS-2 MS Case, Based on 3 Sector-Cell System.

The following schemes are compared to gain more insight into the system behavior:

■ JT with CU and perfect channel estimation at the CU (pCA)
■ JT with CU and real channel estimation (cCA)
■ Distributed JT with real channel estimation (dCA)
■ Distributed JT with real channel estimation and additional feedback (fb) link (dCAfb)

The CA with CU and perfect channel knowledge (a) provides an upper bound.

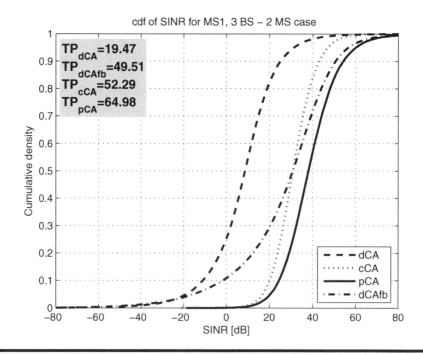

Figure 10.13   Simulation Results of Various CA Schemes.

If real channel estimation, which consults (10.29), is assumed, there is already some degradation. In this case, the channel information has been acquired by the direct channel estimation method only.

In the case of the distributed CA (c), the system throughput reaches only 30% of that of the ideal bound pCA. To improve the performance, an additional feedback link has been added, which proves to be very effective. This increases the throughput by 250% compared to the dCA without feedback case.

10.3.3.1.2 Distributed JT with Feedback

In this section the rationale of the additional feedback is explained. The rationale of the feedback link is:

Even relatively small interference generated from the data of one MS to the other MS will lead to significant performance degradation in the case of the dCA. Due to the sensitivity to estimation errors it will require almost perfect CSI at all BSs to achieve high performance. The MSs are at the best location to measure the interference — either by the help of extra slots dedicated to measurement, i.e., only one MS receives data and the other measures the interference — or by analyzing the variance on the received data symbols.

The interference is fed back regularly in a tracking modus to the BS of each MS, which retains the strongest link quality. Only the BS to which the MS is attached receives the interference value and adjusts the corresponding element of the weight matrix in an effort to remove the interference signal. This will not result in the optimum weight matrix. The change of one weight matrix element, which is designed to remove the interference without considering its effect on the strength of other link, affects the received signal strength for the other MS. However, this simple feedback scheme proves to improve the performance significantly.

10.3.3.2 Combination of Long- and Short-Term CSI Over Backbone Networks

The general scheme for channel estimation is based on orthogonal pilot grids for the BSs and the MSs which are transmitted sequentially as explained above. Here, OFDM is always assumed as the basic transmission format. In this section the analogue feedback concept will be analyzed first. The subspace approach will be explained briefly as the basis for an improved scheme, where the prediction of radio channels at one BS is combined with recent, but low quality CSI estimates at another BS.

Let us assume $p(1)_{1s}$, $p(1)_{2s}$ and $p(2)_{1s}$, $p(2)_{2s}$ are sequentially transmitted pilot symbols from BS1 and BS2, respectively. Then, MS1 estimates the channel coefficients b_{11} and b_{21} based on the receive signal r_1, while MS2 estimates b_{12} and b_{22} at the same time based on the receive signal r_2. In the following step, the MSs broadcast their estimated radio channel coefficients together with their own pilots over orthogonal pilot grids to BS1 and BS2. After that, both BSs can perform an estimation of all channel coefficients.

The general concept for analogue feedback of CSI for a single FDD radio channel is shown in Figure 10.14. The goal is to minimize the overhead for CSI feedback.

The DL channel represented by the vector $\mathbf{h}(f_1)$ within the frequency band f_1 is estimated by a suitable pilot symbol as conventional by maximum likelihood detection, with or without interpolation over different subcarriers. On the UL, intermittently either a pilot symbol for CSI estimation of the UL channel or the analogue DL channel estimates

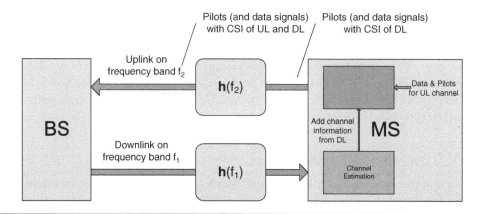

Figure 10.14   CSI Estimation of DL Channel by Adding Analogue CSI on UL Channel.

are added to the UL pilots. As OFDM is assumed this is done for each subcarrier separately. This allows the BS to compare the pilots with and without DL information and to estimate both channels.

For frequency selective radio channels there is a challenge. When the UL channel exhibits a notch, the estimation quality of the DL channel will be very poor. For this reason the mean square error (MSE) for the DL channel after analogue feedback typically will be much higher than for the UL channel.

To avoid fading notches for UL transmission, spreading codes might be used. As a more efficient alternative, the BS informs the MS over the DL channel about a suitable frequency band with a good SNR value in the UL direction. This frequency band with high SINR will be used for feedback of the estimated CSI values of the DL channel from MS to BS (see Figure 10.15).

This results in the following advantages:

■ Fading notches — causing significant degradation due to noise enhancement — are avoided. Typically the SNR at the BS will be even higher than the average value.

■ Chunks,[8] as a consecutive set of subcarriers, exhibit a relatively high correlation in time and frequency direction. Therefore, the radio channel should be relatively flat.

Now we consider the long- and short-term CSI estimates based on subspaces. To improve the channel estimation process for distributed processing, the channel matrix $\mathbf{H}_i$, i indicating the BS number, may be separated into two subspaces containing long-term ($\mathbf{H}_{l,i}$) as well as short-term ($\mathbf{H}_{s,i}$) channel coefficients.

$$\mathbf{H}_i = \mathbf{H}_{l,i} + \mathbf{H}_{s,i} \qquad (10.32)$$

For an IP-based backbone network, it is not possible to guarantee a certain delay $\Delta\tau$ for data exchange of information between different BSs, but it can be expected to be limited

---

[8] Chunk is a block of frequency and time resource, which defines the lowest granularity of resource allocation in OFDMA systems.

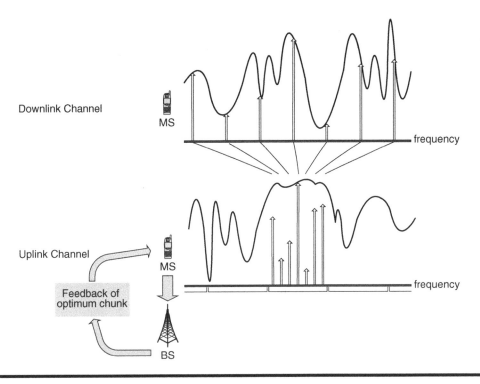

Figure 10.15   Mapping of Estimated DL CSI Values on a Suitable UL Frequency Band.

and should be in the order of some 10 ms, as otherwise, normal latency restrictions for user data could not be fulfilled.

If the observation time t_{ob} for the estimation of the long-term channel subspace $\mathbf{H}_{l,i}$ is larger than $\Delta\tau_{i,j}$, it makes sense to transfer $\mathbf{H}_{l,i}$ to the adjacent BSs. So the idea is to synchronize the long-term part of $\mathbf{H}$ and to update only the short-term part at each BS independently, thereby improving the system performance or reducing the required signal energy of the pilot signals for CSI estimation, as in [19].

For low-dimensional subspaces, there is the advantage that a very accurate prediction over a long time interval is possible.

In Figure 10.16, a simpler linear prediction is shown first. Simulation assumption is a 20 MHz bandwidth with 64 chunks of size 316 kHz each. An artificial elliptical channel model with 10 randomly distributed scatters and shadowing is being used, as this model results in a continuous evolution of the radio channel in contrast to snapshot channel models. The MS is about 500 m apart from the BS and moves with a speed of 3 km/h. The path loss coefficient γ is 2.3.

The best chunk with the highest power on the DL channel is selected and the variation over a time interval of 60 ms is analyzed for this best chunk. Here, the assumption is that the connection of the MS to BS2 (SNR = 6 dB) is significantly worse than to BS1 (SNR = 20 dB in Figure 10.16), but the good CSI estimation of BS1 has to be transmitted over the backbone network with variable delays. For this reason, BS1 has to perform a channel prediction for the time of the data transmission over the backbone network, resulting in an additional prediction error. MRC is used to combine the CSIs from BS1 and BS2.

For prediction, the first 20 samples, i.e., 0 – 10 ms, are used for estimating the slope and offset parameters needed for the linear prediction. The linear prediction fits quite

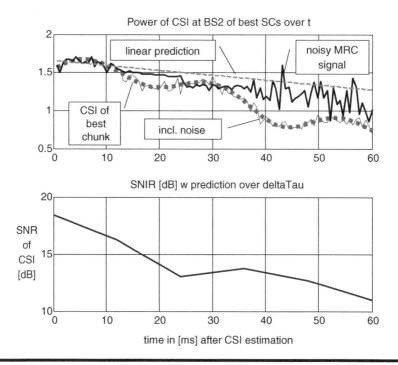

Figure 10.16 Top: Development of Best Chunk Over Time, Noisy Rx Signal at BS1, Linear Prediction and Noisy MRC Signal; Bottom: SNR of CSI Estimation After MRC Over Time Wolfgang Zirwas, Martin Weckerle, Elena Costa, Egon Schulz, 'Channel Estimation for Distributed Cooperative Antenna Systems' *Frequenz Journal of RF-Engineering and Telecommunications, Vol. 60, Edition 506, Germany, 2006 [20a].*

well for a time duration of 20 ms — i.e., for 10–30 ms — while for longer durations the prediction degrades significantly. The resulting mean estimation error has been calculated for many channel realizations. Based on the variance values obtained above, due to MRC, noisy CSI of BS2 will dominate with increasing time over the inaccurate prediction of BS1.

The resulting SNR improvement depends on the channel realization. For very short prediction delays it is around 14 dB. For 40 ms the probability for channel improvement is still about 8 dB, while for higher delays the improvement is reduced.

10.3.3.3 Subspace-Based Channel Estimation

Advanced MIMO schemes like distributed COOPA require accurate CSI either at the Rx [21] or at the Tx side [22] or, even better, at both sides [23]. Unfortunately, the number of channel coefficients grows with the product of the numbers of antennas at the Tx/Rx-side, whereas the channel capacity in the best case grows only linearly with the number of antennas [24]. Consequently, the performance improvement achievable by MIMO transmission will, in practice, be limited by the requirement to estimate the channel coefficients, especially in high mobility scenarios. State of the art channel estimation is based on the transmission of *a priori* known training signals. In this case, the received signal resulting from the training signal transmission is a known linear function of the unknown channel coefficients. The extension to systems with multiple Tx/Rx antennas does not lead to something special, except that the channel coefficients stem from different physical channels. Training signals may now be fed into all the inputs of the MIMO channel at the same time, although this does not offer significant improvements compared to the training signals that are fed into the different inputs at disjoint time intervals [25,26,31]. This chapter introduces a new subspace-based channel estimation technique which exploits long term channel properties to improve the snapshot-based channel estimates, or equivalently, to reduce the number of resources required for training signal transmission to achieve a certain channel estimation performance.

10.3.3.3.1 System Model

Typically, wavefronts impinging at the BS from a distantly moving MS has rather few directions of arrival. Here, let us consider the most simple case of two directions of arrival. The two radio channels corresponding to the two directions of arrival can be described by directional channel impulse responses $\mathbf{h}_d^{(d)}$, $d = 1$, 2. At the reference point, the channel impulse response is just the sum of the two directional channel impulse responses. At a position not too far away from the reference point, i.e., at a position where the wavefronts are still the same as at the reference point, the channel impulse response is a superposition

$$\mathbf{h}(i) = \mathbf{h}_d^{(1)} e^{j\phi_1(i)} + \mathbf{h}_d^{(2)} e^{j\phi_2(i)} \tag{10.33}$$

of the two directional channel impulse responses. The factors $e^{j\phi_d(i)}$, $d = 1$, 2 correspond to the steering factors well-known from the theory of array antennas [27,28]. For the more general case of D directions of arrival, the superposition of (10.33) can be equivalently written as

$$\mathbf{h}(i) = \underbrace{\left(\mathbf{h}_d^{(1)} \cdots \mathbf{h}_d^{(D)} \right)}_{\mathbf{H}_d(i)} \cdot \underbrace{\begin{pmatrix} e^{j\phi_1(i)} \\ \vdots \\ e^{j\phi_D(i)} \end{pmatrix}}_{\mathbf{h}_d(i)} \tag{10.34}$$

It is important to notice that the matrix $\mathbf{H}_d(i)$, made up of the directional channel impulse responses $\mathbf{h}_d^{(d)}(i)$, $d = 1, \cdots, D$, and representing the subspace, does not change if the mobile station only moves in a small area where the wavefronts do not change, whereas the channel impulse response $\mathbf{h}(i)$ and the subspace-based channel vector $\mathbf{h}_d(i)$ change quickly due to fast fading. The channel vector $\mathbf{h}(i)$ typically lies in a rather low-dimensional subspace. In general, using the tall $L \times D$ matrix $\mathbf{H}_d(i)$ spanning the subspace, one can write the channel vector $\mathbf{h}(i)$ of dimension L as a function of the subspace-based channel vector $\mathbf{h}_d(i)$ of dimension D as follows

$$\mathbf{h}(i) = \mathbf{H}_d(i) \cdot \mathbf{h}_d(i) \tag{10.35}$$

Typically, the matrix $\mathbf{H}_d(i)$ changes only slowly over time. Generally, L is much larger than D. The basic idea of subspace-based channel estimation (CE) is that in a certain snapshot i only the subspace-based channel vector $\mathbf{h}_d(i)$ needs to be estimated, which typically results in a significant reduction of the number of unknowns to be estimated as compared to the conventional snapshot-based channel estimator which directly estimates the channel vector $\mathbf{h}(i)$. Finally, the estimate

$$\hat{\mathbf{h}}(i) = \hat{\mathbf{H}}_d(i) \cdot \hat{\mathbf{h}}_d(i) \tag{10.36}$$

of the channel vector $\mathbf{h}(i)$ at time instant i is obtained. For the covariance matrix of the channel vector $\mathbf{h}(i)$

$$\mathbf{R}_h(i) = \mathbb{E}\left[\mathbf{h}(i)\mathbf{h}(i)^{*T}\right] = \mathbf{H}_d(i)\mathbb{E}\left[\mathbf{h}_d(i)\mathbf{h}_d(i)^{*T}\right]\mathbf{H}_d(i)^{*T} \tag{10.37}$$

holds, i.e., the subspace spanned by the columns of the matrix $\mathbf{H}_d(i)$ is equal to the subspace spanned by the covariance matrix $\mathbf{R}_h(i)$. As we are only interested in any set of basis vectors of the subspace, it is sufficient to find one orthonormal basis of the subspace spanned by the covariance matrix $\mathbf{R}_h(i)$.

10.3.3.3.2 Prediction

Movements in mobile environments lead, in general, to Doppler frequency shifts which depend on the mobile's velocity and the directions of arrival described by the factors ν and $\phi_d(i)$, respectively. The assumption of D directions of arrival leads, in general, to D individual Doppler frequencies $f_d(i)$, $d = 1, \cdots, D$. The factors $\phi_d(i)$, $d = 1, \cdots, D$, can be defined as follows

$$\phi_d(i) = 2\pi f_d(i)i\Delta t, \tag{10.38}$$

with $t = i\Delta t$. With the assumption that the mobile moves only in the wavelength range, i.e., $f_d(i+1) \approx f_d(i)$, a prediction of the subspace-based channel coefficients at time interval $(i+1)$ might be possible using the results at time i.

10.3.3.3.3 Results

Subspace estimation is based on estimating the covariance matrix $\mathbf{R}_h(i)$ based on the past I initial channel estimates. To simplify the following basic considerations, it shall be assumed that the influence of noise $\mathbf{n}(i)$ can be neglected, e.g., due to perfect averaging and subtracting its contribution when estimating the covariance matrix $\mathbf{R}_h(i)$. In this case

$$\hat{\mathbf{R}}_h(i) = \frac{1}{I}\sum_{j=i-I+1}^{i}\mathbf{h}(j)\mathbf{h}(j)^{*T} \tag{10.39}$$

holds. A suitable method for determining an orthonormal matrix $\hat{\mathbf{H}}_d(i)$ consists in eigenvalue decomposition [30] of the Hermitian covariance matrix $\hat{\mathbf{R}}_h(i)$. In general, the average energy of the systematic channel estimation error will be small if the nonconsidered eigenvalues are small.

To evaluate the effect of exploiting second-order channel statistics for channel estimation improvements, the eigenvalue profiles of the 3GPP spatial channel models are considered. Channel impulse responses are randomly generated for a SISO system. In Figure 10.17 the simulated, and in descending manner, ordered eigenvalue profiles are shown assuming 3GPP urban macro-, suburban macro-, and urban micro channel models. The eigenvalue profiles of the three scenarios are compared with a scenario with randomly distributed complex Gaussian channel coefficients. There are several dominant eigenvalues in the considered 3GPP scenarios. The noise, in the case of perfect averaging, only leads to an additive term $\sigma^2\mathbf{I}$ in the estimated covariance matrix $\hat{\mathbf{R}}_h(i)$, in the case of white noise ($\mathbf{R}_n = \sigma^2\mathbf{I}$), and optimum training signals. This additive term need not be subtracted as it does not influence the eigenvectors and shifts the eigenvalues only by an additive constant σ^2. One should only consider the eigenvectors corresponding to eigenvalues which are larger than a certain threshold, which will be larger than σ^2.

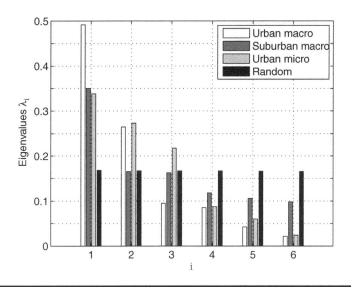

Figure 10.17 Simulated Eigenvalue Profiles Assuming 3GPP Urban Macro and Urban Micro Channel Models.

Furthermore, Figure 10.18 shows the temporal behavior of the subspace channel coefficient assuming the 3GPP urban macro channel model. Thereby, it can be shown that the subspace channel coefficients depend only on the individual Doppler frequencies as stated before.

Summarizing our investigations, the basic principles of a novel subspace-based channel estimation technique are presented, which can improve the performance of channel estimation as compared to conventional channel estimation. The performance gains stem from the exploitation of long term channel properties in the form of the subspace, in which the channel vector lies.

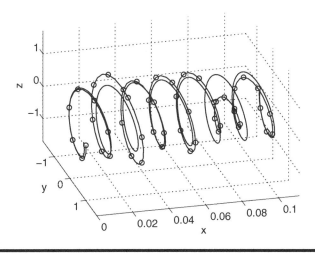

Figure 10.18 Subspace Channel Coefficient Assuming the 3GPP Urban Macro Channel Model (Theory: Solid Line, Simulation: Marked Line, Individual Doppler Frequency: 66.8 Hz).

10.3.3.4 Some Optimizations

A number of advanced schemes for distributed cooperation exist, and some will be addressed shortly in the following:

■ One straightforward solution for reducing the overhead by cooperation is to limit it to those cases where a high gain can be expected. This has several dimensions, i.e., the overhead due to CSI estimation might be reduced by restricting us to low speed MSs which have full queues in their buffers. Thereby, the ratio of overhead to cooperation gain is improved. Contrary, for MSs transmitting only short packages, the overhead for organizing a cooperation is not justified.

■ The overhead in the backbone due to multicasting might be reduced by including only those MSs into multicasting groups which are subject to strong interference, and therefore will profit from cooperation more than others.

■ On the UL, a threshold should be set so that only those interfering signals above that threshold are canceled. Thereby, the backbone network has to transmit data only for these MSs.

■ The combination with HARQ seems promising, as the overhead of cooperation will be reduced to those cases where a first transmission was not successful.

■ The quantization of the data on the UL can be optimized. Typically soft values with, e.g., 14 bit quantization are exchanged over the backbone network, but depending on the maximum intended modulation format, a much smaller quantization like, e.g., 4–8 bits should be sufficient.

10.4 OPEN ISSUES

■ Distributed COOPA systems are based on the cooperation between multiple distributed BSs, so an inter-BS synchronization in frequency and time is required. Corresponding algorithms are well-known and can be found, e.g., in [29]. It should be mentioned that the requirements in the time domain are quite relaxed for OFDM signals due to their cyclic extension or guard interval.

■ As for all MU-MIMO schemes, high precision oscillators with low phase noise, in conjunction with good frequency estimation at the MS, are required to avoid a phase mismatch between the different simultaneously active users. In UL direction, a frequency precorrection might be required depending on the system parameters. With today's available hardware (HW) components, general feasibility has been demonstrated in experimental radio systems for up to 64 QAM modulation, but careful system design is required.

■ Implementation complexity — especially of the MSs — is another issue. At the MS side, interpolation of radio channels based on sparse pilot grids — besides frequency estimation (see above) — has to be processed. This does not add significant overhead compared to conventional systems, as most OFDM systems will require some form of channel interpolation in the future anyway.

■ Channel estimation is one of the most challenging parts and has already been addressed previously in this chapter. As for all cooperative MU-MIMO systems, the required CSI accuracy is typically the same compared to conventional SISO systems in uplink and downlink, respectively. But the net throughput is reduced as more pilot signals are required, while for MSs the higher transmission power of pilot signals might reduce the battery lifetime. For this reason more analysis in the area of channel estimation will be required, in particular for interference limited systems.

■ Another topic demanding more consideration is the optimum design of a relatively high number of orthogonal pilot sequences in cellular systems with good cross correlation in frequency selective radio channels. Different variants and combinations of FDM, TDM, or CDM can be conceived, maybe in combination with frequency hopping and special pilot signal sequences.

10.5 CONCLUSION

The motivation for cooperative antenna systems and distributed cooperation specifically, has been presented based on theoretical considerations and specific system level simulations as well as outdoor measurements. Distributed cooperation concepts for the DL as well as the UL have been explained, proving the feasibility of such systems in the future.

Some advanced schemes — especially for channel estimation — have been proposed, which will be needed for suitable implementations. The simulation results show that the system performance can be greatly improved by adopting a simple feedback scheme. Another interesting option, i.e., the subspace-based channel estimation in combination with suitable prediction, has been given and analyzed. Further research is required to get the full benefit of the described system concepts.

This research project has been supported by the German "Bundesministerium für Bildung und Forschung." The authors alone are responsible for the contents.

REFERENCES

[1] R.A. Monzingo and T.W. Miller, *Introduction to Adaptive Arrays.* Wiley Interscience, New York, 1980.

[2] S.U. Pillai, *Array Signal Processing.* Springer-Verlag, New York, 1989.

[3] S. Verdu, *Multiuser Detection.* Cambridge University Press, Cambridge, UK, 1998.

[4] M.K. Varanasi and T. Guess, "Optimum decision feedback multiuser equalization with successive decoding achieves the total capacity of the Gaussian multiple-access channel," in *Proc. Asilomar Conf. on Signals, Systems and Computers,* Monterey, pp. 1405–1409, November 1997.

[5] P. Viswanath, V. Anantharam, and D. Tse, "Optimal sequences, power control and capacity of synchronous CDMA systems with linear MMSE multiuser receivers," *IEEE Trans. Inform. Theory,* 45, (6) pp. 1968–1983, September 1999.

[6] M. Schubert and H. Boche, "Iterative multiuser uplink and downlink beamforming under SINR constraints," *IEEE Trans. Sig. Processing,* 53, (7) pp. 2324–2334, July 2005.

[7] H. Huang and S. Venkatesan, "Asymptotic downlink capacity of coordinated cellular networks," in *Proc. Asilomar Conf. on Signals, Systems and Computers,* Monterey, vol. 1, pp. 850–855, November 2004.

[8] M. Bengtsson, "Jointly optimal downlink beamforming and BS assignment," in *Proc. IEEE Int. Conf. on Acoustics, Speech, and Signal Proc. (ICASSP),* Salt Lake City, vol. 5, pp. 2961–2962, May 2001.

[9] M. Bengtsson and B. Ottersten. *Handbook of Antennas in Wireless Communications,* chapter 18: Optimal and suboptimal transmit beamforming. Boca Raton, FL: CRC Press, August 2001.

[10] C. Farsakh and J.A. Nossek, "Spatial covariance based downlink beamforming in an SDMA mobile radio system," *IEEE Trans. Commn.*, 46, (11) pp. 1497–1506, November 1998.

[11] S.V. Hanly, "An algorithm for combined cell-site selection and power control to maximize cellular spread spectrum capacity," *IEEE J. Selected Areas in Commn.*, 13, (7) pp. 1332–1340, September 1995.

[12] F. Rashid-Farrokhi, L. Tassiulas, and K.J. Liu, "Joint optimal power control and beamforming in wireless networks using antenna arrays," *IEEE Trans. Commn.*, 46, (10) pp. 1313–1323, October 1998.

[13] M. Schubert and H. Boche, "A generic approach to QoS-based transceiver optimization," *IEEE Trans. Commn.*, 2006 to appear.

[14] M. Schubert and H. Boche, *QoS-based resource allocation and transceiver optimization*, Foundation and Trends in Communications and Information Theory, 2006.

[15] E. Visotsky and U. Madhow, "Optimum beamforming using transmit antenna arrays," in *Proc. IEEE Vehicular Techn. Conf. (VTC) Spring*, Houston, vol. 1, pp. 851–856, May 1999.

[16] R.D. Yates, "A framework for uplink power control in cellular radio systems," *IEEE J. Select. Areas Commn.*, 13, (7) pp. 1341–1348, September 1995.

[17] R.D. Yates and H. Ching-Yao, "Integrated power control and BS assignment," *IEEE Trans. on Veh. Technol.*, 44, (3) pp. 638–644, August 1995.

[18] V. Jungnickel, V. Pohl, H. Nguyen, U. Krüger, T. Haustein, and C. von Helmolt, "High capacity antennas for MIMO radio systems," in Proc. 5th WPMC, Vol. 2, pp. 407–411, 2002.

[19] J. Hahn, W. Baier, M. Meurer, and W. Zirwas, "Spread-spectrum based low-cost provision of downlink channel state information to the access points of FDD OFDM mobile radio systems," *ISSSTA 2006*, Manaus, August 2006.

[20] V. Jungnickel, A. Forck, T. Haustein, S. Schiffermüller, C. von Helmolt, F. Luhn, M. Pollock, C. Juchems, M. Lampe, J. Eichinger, W. Zirwas, and E. Schulz, "1 Gbit/s MIMO-OFDM transmission experiments," *IEEE VTC Fall 2005*, Dallas, 2005.

[20a] Wolfgang Zirwas, Martin Weckerle, Elena Costa, and Egon Schulz, "Channel estimation for distributed Cooperative Antenna Systems," *Frequenz Journal of RF-Engineering and Telecommunications*, Vol. 60, edition 506, Germany, 2006.

[21] P. Wolniansky, G. Foschini, G. Golden, and R. Valenzuela, "V-BLAST: an architecture for realizing very high data rates over rich-scattering wireless channel," in *Proc. URSI International Symposium on Signals Systems, and Electronics (ISSSE'98)*, Pisa, pp. 295–300, September 1998.

[22] T. Weber, A. Sklavos, Y. Liu, and M. Weckerle, "The air interface concept JOINT for beyond 3G mobile radio networks," in *Proc. 15th International Conference on Wireless Communications (WIRELESS 2003)*, vol. 1, Calgary, pp. 25–33, July 2003.

[23] E. Telatar, "Capacity of multi-antenna Gaussian channels," *European Trans. Telecommn.*, vol. 10, no. 6, pp. 585–595, 1999.

[24] M. Meurer and T. Weber, "Imperfect channel knowledge: an insurmountable barrier in Rx oriented multi-user MIMO transmission," in *Proc. 5th ITG Conference on Source and Channel Coding (SCC'04)*, Erlangen, pp. 371–379, January 2004.

[25] I. Maniatis, T. Weber, A. Sklavos, and Y. Liu, "Pilots for joint channel estimation in multi-user OFDM mobile radio systems," in *Proc. IEEE 7th International Symposium on Spread Spectrum Techniques & Applications (ISSSTA'02)*, vol. 1, Prague, pp. 44–48, September 2002.

[26] A. Sklavos, I. Maniatis, T. Weber, and P.W. Baier, "Joint channel estimation in multi-user OFDM systems," in *Proc. 6th International OFDM-Workshop (InOWo'01)*, Hamburg, pp. 3.1–3.4, September 2001.

[27] L.C. Godara, "Applications of antenna arrays to mobile communications, part II: beamforming and direction-of-arrival considerations," *Proceedings of the IEEE*, vol. 85, no. 8, pp. 1195–1245, August 1997.

[28] L.C. Godara, "Applications of antenna arrays to mobile communications, part I: performance improvement, feasibility, and system considerations," *Proceedings of the IEEE*, vol. 85, no. 7, pp. 1031–1060, July 1997.

[29] E. Costa, P. Slanina, V. Bochnicka, and E. Schulz, "Downlink based intra and inter cell time and frequency synchronisation in OFDM cellular systems," *Wireless World Congress*, San Francisco, 2005.

[30] R.A. Horn and C.R. Johnson, *Matrix Analysis*. Cambridge: Cambridge University Press, 1985.

[31] P. Hoeher, S. Kaiser, and P. Robertson, "Two-dimensional pilot-symbol-aided channel estimation by Wiener filtering," in *Proc. IEEE International Conference on Acoustics Speech and Signal Processing (ICASSP'97)*, Munich, pp. 1845–1848, April 1997.

[32] Q.H. Spencer, A.L. Swindlehurst, and M. Haardt, "Zero-forcing methods for downlink spatial multiplexing in multi-user MIMO channels," *IEEE Trans. Sig. Process.*, vol. 52, pp. 461–471, Febuary 2004.

PART III

CASE STUDIES

AND APPLICATIONS

11

EXPERIMENTAL STUDY OF INDOOR
DISTRIBUTED ANTENNA SYSTEMS

Fei Tong and Ian A. Glover

Contents

Distributed antennas can mitigate both fast and slow fading. Few experiments, however, have been made to assess their practical likely performance under realistic channel conditions in realistic environments. In this chapter some results of an indoor distributed antenna system (DAS) experiment are presented. The DAS is emulated using a wideband multipe-input multiple-output (MIMO) channel sounder which provides both spatial and temporal snapshots of the diversity channels.

Assuming the DAS is deployed in the base station while the mobile terminal is equipped with a conventional (single) antenna, then uplink diversity (single input, multiple output) can take advantage of independent reception of the diversity signals. Downlink diversity (multiple-input, single-output), however, is more challenging because the (identical) signals radiated by the individual antenna elements cannot easily be processed separately. Conventional transmit diversity gain is thus limited to that resulting from the incoherent (power) addition of the transmitted signals. The experiment has focused, therefore, on the downlink because the downlink propagation model suggests potentially inferior performance compared to the uplink. The downlink also gives a lower bound on uplink performance.

Selective and multipath antenna diversities are considered first. A detailed examination of measured channel impulse responses (CIRs) shows that the phase variation between components is small within a delay window containing the majority of the CIR energy. Cophasing transmit diversity is therefore possible, in which the delay and phase of the transmitted signal at each base station antenna element are adjusted, such that the peak component of CIRs from each branch align at the mobile terminal both in time and in phase.

Diversity orders of two and three have been compared for all three diversity schemes. The results show that cophasing transmit diversity achieves the greatest power in, and least delay spread of, the received signal. Analysis of the measurements also shows that it achieves best coverage.

Another experimental observation is that disparate path losses between a mobile-terminal and the elements of the distributed antenna lead to unbalanced signal levels in the diversity branches. In a cophasing diversity scheme this results in an unequal downlink signal to interference ratio (SIR). A power control scheme to rebalance the SIR can be devised by solving an eigenvalue problem, thereby maximizing system capacity gain.

11.1 INTRODUCTION

11.1.1 Motivation

Distributed antennas (DAs) can mitigate both fast and slow fading simultaneously. Many studies have demonstrated the potential for both link and capacity gains using theoretical analyses assuming identical independent multiple channels [1–6]. Although these results are sufficient to demonstrate the advantage of DAs, some aspects of their performance, such as downlink diversity gain and the impact of unbalanced link signal-to-noise ratios (SNRs), need to be addressed using a more realistic channel model. A distributed antenna experiment serves this objective. A practical distributed antenna system (DAS) requires a transmission line infrastructure so that the radio frequency (RF) (or baseband signal) can be exchanged between the distributed antenna element and a processing center. This infrastructure is in proportion to the scale of the coverage area, which makes experiments difficult. There are few reports, therefore, describing distributed antenna experiments and assessing their practical performance under realistic channel conditions. Furthermore, those experiments that have been reported are based on narrowband signals [7–12]. The (indoor) experiment described here is broadband, allowing temporal resolution of the channel impulse response down to 10 ns [13] allowing the effect of the frequency-selective channel to be investigated.

11.1.2 Method and Objectives

A central part of this experiment is the capture of temporal and spatial responses of the DA. A wideband multiple-input multiple-output (MIMO) channel sounder is used as the hardware platform. Antenna elements are distributed through an indoor environment to emulate the distributed antenna. These elements are connected back to the processing unit using very low-loss cables. The channel sounder provides vector channel impulse responses (CIRs) by spatial sampling of the coverage area. Postprocessing to simulate any diversity combining technique is then possible. Thus both diversity gain and coverage performance are obtained under realistic channel conditions.

The most likely configuration of a DA will employ multiple antennas for the base-station and a conventional (single) antenna at the mobile (or portable) terminal. On the uplink, each basestation antenna independently receives signals propagating via different paths, thereby benefiting from diversity gain. On the downlink, however, because all basestation antennas transmit the same signal simultaneously (as in a simulcast system [14,15]), signals sum at the single receiving antenna before the mobile station can separate them. The downlink, therefore, cannot easily achieve the same diversity gain as the uplink. In this experiment we have, therefore, payed more attention to the downlink than the uplink.

11.1.3 Outline

First, the DA experiment is introduced. The experiment's objectives are presented, the test environment is described and the antenna configurations are defined. Second, observations relating to the coherence within the CIRs are presented and analyzed, and the consequent suitability of cophasing diversity is commented on. The experimental results are then used to compare the practical performances expected from a range of candidate diversity schemes and orders. Finally, an important issue associated with cophasing diversity is raised and the issue is addressed by solving an eigenvalue problem. Based on the experimental results, conclusions are drawn about the practical performance improvement that DAs might yield.

11.2 EXPERIMENT DESCRIPTION

11.2.1 Measurement Equipment

We use a wideband MIMO channel sounder to emulate a distributed antenna system. The sounder, which has the capability to obtain high resolution, quasi-simultaneous measurements of multiple channels, was developed by QinetiQ (formally the Defence Evaluation and Research Agency of the U.K. Ministry of Defence). It can be divided into transmit and receive subsystems, Figure 11.1 and Figure 11.2. Each subsystem comprises

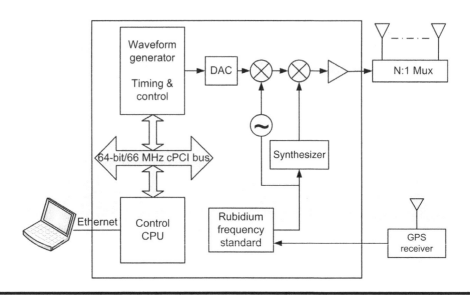

Figure 11.1   Channel Sounder Transmitter.

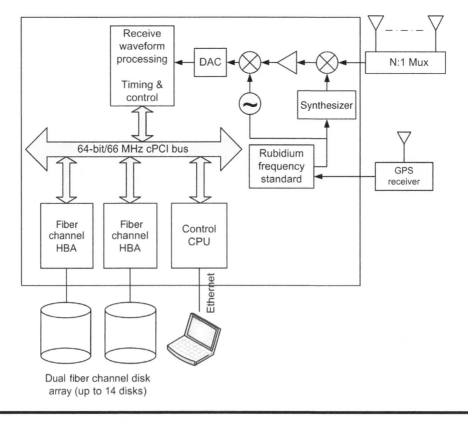

Figure 11.2   Channel Sounder Receiver.

Table 11.1 Channel Sounder Parameters

Band center	2.442 GHz
Sounding bandwidth	250 MHz
Delay resolution	15 *ns* (for 250 MHz bandwidth, equal amplitude paths)
Maximum unambiguous delay	5 μs
Maximum transmitter power	30 dBm
MIMO configuration	3 Tx × 1 Rx

a primary transmit or receive unit, a 16-way antenna multiplexer (connected to the primary transmit/receive unit via coaxial cable) and a laptop PC to provide a user control interface via Ethernet. A fast data storage unit (FDSU) comprising fourteen 73 GB hard drives was used to store raw complex impulse responses. In order for data that was captured by the receive subsystem to be postprocessed for further analysis, an Ethernet interface was used to transfer it to an auxiliary PC network.

The sounder employs pulse compression. The default sounding waveform (a single sideband binary phase-shift keying (BPSK)-modulated, maximal length, pseudonoise sequence with a maximum chip rate of 165 Mchip/s) was used for the measurements. Pulse shaping ensures that a better than 50 dB peak-to-sidelobe ratio was achieved following back-to-back calibration of the transmit and receive units. Tetherless operation is made possible using rubidium frequency standards.

The sounder can measure the wideband, time varying CIR of the equivalent baseband channels between 16 transmitter output ports and 16 receiver input ports. It has been designed, primarily, to allow MIMO characterization of the radio channel using a high-speed, digitally controlled, 16-way antenna multiplexer at both the transmitter and receiver. Measurements are made on each receive antenna in turn before the transmit signal is switched to the next antenna in the series (to minimize RF switching transients). Power is not transmitted during the time that the antenna multiplexer is switched between channels. The multiplexer cycles between channels sufficiently quickly such that, for practical engineering purposes, the set of CIRs comprising a MIMO measurement can be assumed to be simultaneous. (The 16 × 16 cycle time is 12.8 ms.) In this experiment, only 3 transmit antennas are configured.

The sounder's principal operating parameters are listed in Table 11.1.

11.2.2 Measurement Environment and Antenna Deployment

The measurements were made on the fourth floor of the Department of Electronic and Electrical Engineering at the University of Bath. The measured area (approximately 27 m × 11 m) is bounded by the solid line in Figure 11.3 and includes three laboratories and one corridor. The internal walls are constructed, principally, of plasterboard. Laboratories 4.15 and 4.13 (see Figure 11.3) contain desks separated by free-standing partitions. There are some small metallic cabinets placed along the walls.

Three transmit antennas and one receive antenna were used in the measurements. All were sleeve dipoles tuned to the measurement frequency of 2.4 GHz. The antennas were mounted on tripods at a height above the ground of 1.5 m. One transmit antenna is located in each laboratory. The antenna in laboratory 4.13 is identified as 1, the antenna in laboratory 4.15 as 2, and the antenna in laboratory 4.1 as 3. All three transmit, and

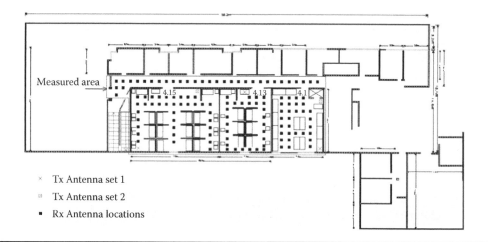

Figure 11.3   Floor Plan of Measurement Area.

one receive, antennas were connected, via low-loss coaxial cable (Table 11.2), to the channel sounder's transmit/receiver multiplexer.

Two sets of transmit antenna locations were used (although in laboratory 4.13 the antenna location remains unchanged, see Figure 11.3). The receive antenna locations formed a 1 m square grid (black squares in Figure 11.3) within the measurement area.

The CIRs of the three channels were measured sequentially, at each receive location, in a period that was short enough to deem them simultaneous. After the measurement, two data sets had been acquired, each of which comprised three groups of CIR measurements corresponding to each transmit antenna unit.

11.2.3   X-dB Window and Data Integrity

For every measured location, there exists a vector of channel responses. The normalized sequence of samples from the ith of K antennas represents a finite impulse response, tapped delayline model:

$$h_i(t) = \sum_{n=1}^{L_i} \alpha_i(n)\delta(t - \tau_i(n)), \ i \in [1, K] \tag{11.1}$$

Table 11.2   Cable Loss at 2.4 GHz

Cable	Length (m)	Cable Loss (dB)
Receiver	2.2	1.8
Transmitter 1	2.2	1.8
Transmitter 2	22.5	16.6
Transmitter 3	22.5	12.8

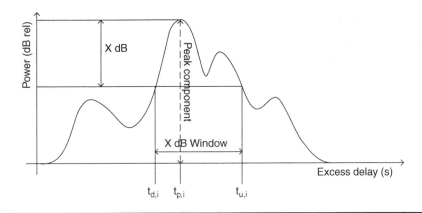

Figure 11.4   X-dB Window.

where L_i is the number of multipath components (or taps) and $\alpha_i(n)$ is the complex gain of each component with relative delay. Then $\tau_i(n)$ is discrete delay with values that are mutiples of the channel sounder delay resolution and n is the component index. We also define $\phi_i(n) = arg[\alpha_i(n)]$ as the phase angle of $\alpha_i(n)$.

To facilitate the study, we define an X-dB window with respect to the CIR peak component, within which most of the channel response power resides. The window is delineated by two cut-off points, both X-dB below the peak (see Figure 11.4).

The two cutoff points have excess delays $t_{d,i} = \tau_i(n_d)$ and $t_{u,i} = \tau_i(n_u)$, and the peak component has excess delay $t_{p,i} = \tau_i(n_p)$. Then, n_d, n_u, and n_p are the indices of the components at first and last cutoff points and CIR peak. The window edges are not necessarily symmetrical about the peak component.

COST 231 [16] has defined a CIR data quality criterion. It states that a valid CIR must have 18 dB clearance between its peak value and the noise floor and, when calculating CIR metrics, only sample points 3 dB above the noise floor should be used (see Figure 11.5). This criterion was adopted to ensure data integrity.

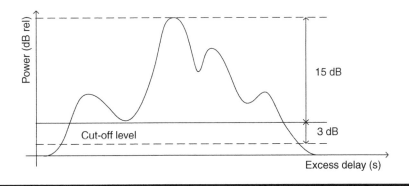

Figure 11.5   CIR Integrity Check.

11.3   POSTPROCESSING FOR TRANSMIT DIVERSITY

11.3.1   Downlink Diversity

The downlink channel is multiple-input single-output (MISO). In contrast to the uplink, where each multipath component can be adjusted in phase, only the initial phase and time delay of the signal transmitted from each antenna unit can be controlled separately. The transmitted downlink signals sum at the mobile station's single antenna. Thus, the composite signal at the mobile station is expressed by:

$$b_c(t) = \sum_{i=1}^{K} b_i(t - \Delta T_i)e^{j\varphi_i} \tag{11.2}$$

where ΔT_i and φ_i are time and phase delay adjustment at the ith transmit antenna and K is the number of transmit antennas. The downlink diversity becomes an optimization problem to determine K pairs of $(\Delta T_i, \varphi_i)$ to maximize the gross power of $b_c(t)$. Because each multipath component has its own phase angle, the components from all antennas arriving at the same time may not be phase aligned after the initial phase adjustment, i.e., the selected phases could result in constructive addition at one delay and destructive addition at another delay. The following two candidate schemes attempt to solve this problem.

11.3.2   Selective Diversity

A simple solution is selective diversity. This scheme avoids destructive summation of multipath components by selecting the antenna with the best channel condition. It also has the advantage of simple implementation. An application of this scheme is fast sector handover.

11.3.3   Multipath Antenna Diversity

In spread spectrum signaling, multipath components with sufficiently different time delay can be isolated by a RAKE receiver. Taking advantage of this feature, [17–22] propose that delay is intentionally introduced between signal branches (i.e., different transmit antennas). These signals then present themselves as multipath components in the RAKE receiver and can be combined independently. We call this multipath antenna diversity.

The basestation can estimate the path length between a mobile station and each basestation antenna and the time dispersion suffered by each signal branch. For each antenna, ΔT_i is chosen to adjust the arriving time instant so that channel responses from different antennas do not overlap. Because this scheme avoids the direct summation of signal branch components, the phase does not need adjustment.

We have used the X-dB window to determine the differential delay between each transmit antenna. To facilitate the derivation of the required time adjustments and without loss of generality, we assume the signal arrives first (via the channel with CIR $b_1(t)$) at antenna 1 and is used as the time reference. (The selection of the reference antenna has an impact only on the absolute time delay at the mobile station.) This scheme is illustrated in Figure 11.6. We define the time adjustment at the other transmit antennas with respect to this and denote them as T_i ($T_1 \equiv 0$). Each channel branch has a different X-dB window and signals from all branches must avoid overlapping of their windows.

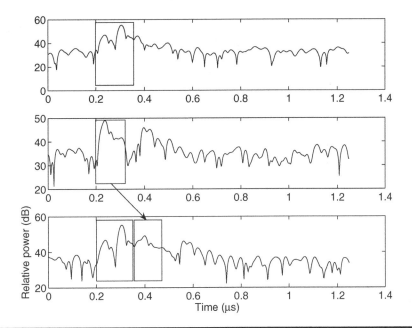

Figure 11.6 Multipath Antenna Diversity Example.

We therefore express the differential delay at each antenna iteratively, the time shift for each signal branch with increasing index being given by:

$$\Delta T_i = t'_{u, i-1} - t_{d,i}, \ i > 1$$

$$t'_{u,i} = t_{u,i} + \Delta T_i$$

$$\Delta T_1 = 0 \tag{11.3}$$

where $t'_{u,i}$ is the new delay of the lower cutoff component after time shifting, Figure 11.6.

By using the X-dB window, we prevent the powers within the windows (from all antennas) from summing with random phase. The signals will overlap outside the window, of course, but the power involved is small. The deeper the window (i.e., the larger X is), the less power overlaps, but this results in long delay with a consequent requirement for long tap delay lines in the RAKE receiver. A tradeoff is possible by adjusting the depth of the window.

11.3.4 Cophasing Transmit Diversity

11.3.4.1 Intra-CIR Phase Coherence Observation

The schemes proposed above essentially prevent multipath components from different antennas arriving at the same time. Therefore, these can, at best, achieve only power-wise summation of diversity branch signals. The full advantage of multiple antennas cannot therefore be realized, resulting in inferior performance compared to the uplink.

The central idea underpinning these schemes is the avoidance of multipath component cancelation. If the delay and phase spread are not too large, however, the majority

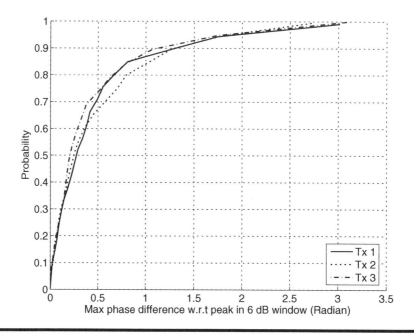

Figure 11.7 **CDF of Maximum Phase Deviation from Peak Component.**

of components could be, at least approximately, cophased. The gain obtained due to coherent summation, even with some components suffering destructive cancelation, may still be greater than the power-wise summation of all components. This possibility motivated our study of the delay variation of the CIR phase described below [23].

We examine the maximum phase difference, ϕ_m, between the peak CIR component and all other components within the X-dB window, i.e.,

$$\phi_m = \max_n |\phi(n) - \phi(n_p)| \tag{11.4}$$

where

$$n_p = \arg[\max_n \alpha(n)], \ n \in [n_d, n_u] \tag{11.5}$$

The cumulative distribution of ϕ_m shows that for a 6-dB window, approximately 90% of samples have a maximum phase difference less than 60°, Figure 11.7.

The results suggest that the variation of phase angles within the X-dB window of the CIR is relative modest. The physical explanation of this is that two components arriving at closely spaced time instants tend to propagate via similar paths. They therefore undergo similar phase shifts during propagation. (Conversely, the probability that two signals propagating over very different paths arrive closely spaced in time is small.)

Absolute phase angle will, of course, follow a uniform distribution. This clearly does not translate, however, into a uniform distribution for phase difference between multipath components within a single CIR.

The temporal coherence of the channel will, of course, impose an upper bound on the time delay that can be tolerated between a channel measurement at the user terminal and the relevant channel parameters becoming available at the base station.

11.3.4.2 Diversity Scheme

If the components can be cophased, greater gain will be achieved than power-wise addition. This requires both delay and phase angle adjustment of the signal transmitted from each antenna.

The observation of maximum CIR phase shift relative to the peak component suggests that if the peaks of two delay profiles are time aligned and cophased, then the corresponding component pairs within the 6-dB windows are likely to have a phase angle difference less than 60°. This, in turn, suggests that a cophasing transmission diversity scheme is possible. Assuming the downlink CIR is known at the base station, the timing delay and phase of the signal at each transmit antenna can be adjusted so that all CIRs, seen at the mobile station, are aligned with respect to their peak components and these peak components will be cophased.

At each base station antenna, each signal is delayed by ΔT_i and adjusted in phase. This scheme is illustrated in Figure 11.8. The rectangles, not shown to scale, indicate the X-dB windows to be aligned. The resulting channel impulse response is:

$$b'_i = b_i(t + \Delta T_i)e^{j\varphi_i}$$

$$\Delta T_i = t_{p,i} - t_{p,1}, \ i > 1$$

$$\varphi_i = \phi_i(t_{p,i}) - \phi_1(t_{p,1})$$

$$\Delta T_1 = 0 \tag{11.6}$$

This scheme only assures precise cophasing of the CIR peak components. However, because the phase angle of the adjacent component changes slowly within the 6-dB

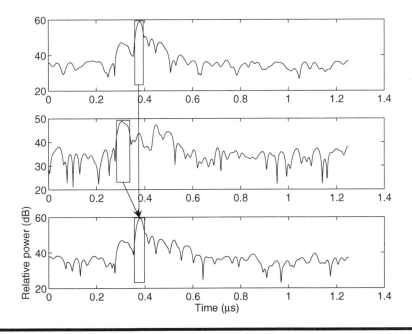

Figure 11.8 Cophasing Transmit Diversity Example.

window, the remaining components will, with high probability, sum approximately constructively. Compared to power-wise combining, summation with a nonzero but small phase angle difference can still achieve improved gain. Given two complex signals A and B with phase angle difference θ, the gain obtained by vector summation relative to power summation is:

$$\frac{|A + B|^2}{|A|^2 + |B|^2} = 1 + \frac{2|A||B|\cos(\theta)}{|A|^2 + |B|^2} \tag{11.7}$$

If the phase angle is less than $60°$ and $|A| = |B|$, the advantage will be greater than $10\log_{10}(3/2) = 1.76\,\text{dB}$.

11.4 EXPERIMENTAL RESULTS AND ISSUES

11.4.1 Diversity Performance

11.4.1.1 Performance Metrics

Transmit diversity schemes change the channel response experienced by the mobile terminal. Two important metrics of the resulting channel quality are gross power and root mean square (RMS) delay spread. The gross power delivered to the receiver is:

$$P_{tot} = \sum_{n=1}^{L} |\alpha(n)|^2 \tag{11.8}$$

RMS delay spread summarizes the severity of time dispersion and is related to the tapped delay line model parameters by:

$$T_{RMS} = \sqrt{\sum_{n=1}^{L} \frac{|\alpha(n)|^2}{P_{tot}} (\tau(n) - \bar{\tau})^2} \tag{11.9}$$

where

$$\bar{\tau} = \frac{1}{P_{tot}} \sum_{n=1}^{L} |\alpha(n)|^2 \tau(n) \tag{11.10}$$

In a wideband system, signal bandwidth is larger than the coherence bandwidth (which is approximately half the reciprocal of the RMS delay spread) and the channel is frequency selective. An ideal RAKE receiver performs optimal reception under these channel conditions. Assuming perfect channel information, the output of the RAKE receiver represents the gross power of the channel response. If multipath components are spaced by more than the reciprocal of the signal bandwidth, each of the paths can be resolved allowing all the power (P_{tot}) to be collected.

11.4.1.2 Diversity Gain Benchmark

We first examine the three diversity schemes, i.e., selective, multiplath antenna, and cophasing transmit diversity, assuming all diversity branches have equal gross power.

Before we use the measured CIR to evaluate these schemes, we develop a benchmark performance criterion using a generic CIR model.

Multipath antenna (MPA) diversity achieves power summation of individual CIRs. For co-phasing transmit (CPT) diversity, we assume the multipath profile contains L components and has normalized (unit) power. (If we assume the total power of the delay profile is constant, then L is trivial to the diversity result.) The upper bound on gain occurs when the phase difference between all signal branches for all values of delay is $0°$. The gain for two diversity branches at power level P_1 and P_2 relative to power-wise addition is given by:

$$G_r = \frac{\sum_{n=1}^{L}(P_1|\alpha_1(n)|^2 + P_2|\alpha_2(n)|^2 + 2\sqrt{P_1 P_2}|\alpha_1(n)||\alpha_2(n)|\cos(\theta_n))}{P_1\sum_{n=1}^{L}|\alpha_1(n)|^2 + P_2\sum_n|\alpha_2(n)|^2}$$

$$\leq 1 + \frac{\sum_{n=1}^{L} 2\sqrt{P_1 P_2}|\alpha_1(n)||\alpha_2(n)|}{P_1 + P_2} \tag{11.11}$$

As stated previously, here $\sum_n|\alpha_i(n)|^2 \equiv 1, \forall i$.

The relative gain depends on the distribution of power across multipath components. Intuitively, a maximum occurs when individual CIRs from both diversity branches have the same impulse response, i.e., $\alpha_1(n) = \alpha_2(n), \forall n$. The second term in (11.11) can be identified as the zero-lag amplitude cross correlation. When the two functions have identical form the cross correlation is maximized. When the two branches also have equal power, the maximum gain is achieved, i.e.,

$$G_r \leq 1 + \frac{2\sqrt{P_1 P_2}}{P_1 + P_2}$$

$$\leq 2 \tag{11.12}$$

The two-branch performance for the ideal case is, therefore, upper bound as above.

11.4.1.3 Diversity Gain Evaluation

We have used the measured CIRs to evaluate the practical performance of the three diversity schemes. Because only the cophasing transmit diversity scheme relies heavily on the detailed CIR, we focus on this scheme and compare it with the MPA scheme. In the comparison, the CIRs are taken from antennas 2 and 3 for every measured location. The total power of all CIRs has been normalized so that all combinations are realized using balanced power branches, i.e., the CIR from each branch has the same total power.

We have examined performance degradation compared to performance benchmark. Performance degradation is due to two factors: nonideal cophasing at multipath components other than peak components and nonidentical channel impulse response magnitudes in different branches. To isolate these two factors, we also simulate a scheme in which we rotate the phase angle of each individual multipath component separately so that all are precisely cophased. We use this as a reference to investigate the loss of gain due to imperfect co-phasing of all other but the peak multipath components.

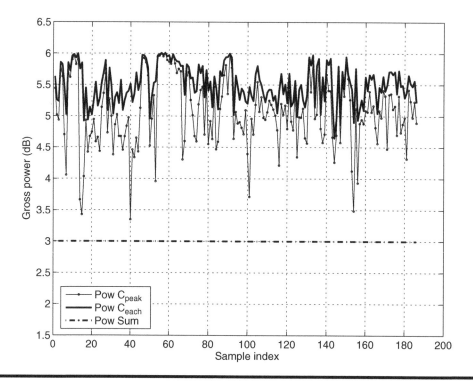

Figure 11.9   Comparison of Combined Signal Power for All Diversity Schemes (Pow C$_{peak}$: Performance of CPT; Pow Sum: Performance of MPA; Pow C$_{each}$: Performance of Perfect Cophasing at Every Delay Instant).

In Figure 11.9, each curve shows the gross power of a different combining scheme at each receiving position. The ideal combining result represents the upper bound.

CPT diversity has better performance than MPA diversity. If we compare the MPA and CPT schemes by examining the relative gain, we obtain the CDF shown in Figure 11.10. Approximately 90% of samples have a relative gain greater than 1.4 dB and 50% have a relative gain greater than 2 dB.

The CDF in Figure 11.11 shows that in more than 90% of locations the gain loss due to imperfect phase alignment of components other than peak components is less than 0.9 dB.

Even for the ideal cophasing case, very few samples have a total combined power 3 dB better than the MPA scheme. This is because the delay profiles of all branches are not necessarily the same.

11.4.2   Coverage

We now examine the coverage performance of diversity schemes using measured CIRs (without normalization). At every receiving location, the simulation generates a CIR using each of the transmission diversity schemes by combining 2 or 3 branches, i.e., selective diversity (SEL(2), SEL(3)), multipath antenna diversity (MPA(2), MPA(3)), and cophasing

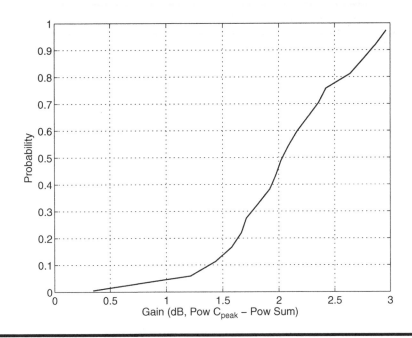

Figure 11.10   CDF of Relative Diversity Gain.

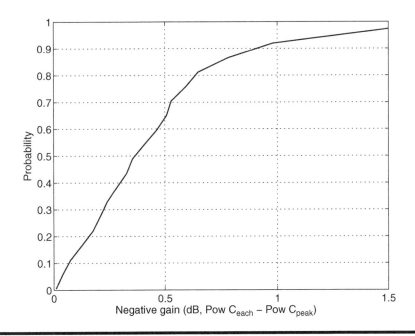

Figure 11.11   CDF of Negative Gain with Respect to C_{each}.

Table 11.3 Mean Gross Power (Data Set I)

Diversity Scheme	Mean Power Ratio W.R.T No Diversity (dB)
None	0
SEL(2)	0.7
SEL(3)	1.9
MPA(2)	2.5
MPA(3)	4.7
CPT(2)	4.8
CPT(3)	8.2

transmit diversity (CPT(2), CPT(3)). We use the performance of the central antenna (Tx 1) as the no diversity reference (referred to as None in Tables 11.3–11.6).

We use the gross power of combined CIRs averaged over the whole measurement area as a coverage metric. To compare the various diversity schemes, we calculate the ratio between the mean gross power for each diversity scheme and the single antenna scheme. We also calculate the variance of received gross power, Table 11.3 and Table 11.4.

It is observed that MPA and CPT schemes result in the largest gross powers. This is because these two schemes make use of signals from all antenna units. The CPT scheme has the best coverage. For data set I, CPT has achieved 2.3 dB and 3.5 dB gain over MPA for 2 and 3 antennas. For data set II, the gain is 0.8 dB and 2.2 dB for 2 and 3 antennas, respectively. When comparing the results from the two data sets, the ratio of different schemes varies. This is due to different transmit antenna locations. In both data sets, the advantage of antenna diversity is apparent. Selective antenna diversity appears to be more sensitive to antenna position than cophasing transmit diversity, at least in the case of these limited measurements.

In addition to gross power, RMS delay spread is an important metric for channel quality. Even in code division multiple access (CDMA) systems, which can isolate multipath components, a small delay spread means that fewer taps are required in the RAKE receiver. The statistics of RMS delay spread are shown in Table 11.5 and Table 11.6. Compared to selective diversity, multipath antenna diversity suffers longer delay spread. This is due to the delay(s) introduced at the transmitter. Co-phasing transmit diversity has a similar (slightly smaller) delay spread. The smaller delay spread is due to the fact that in a selective scheme the CIR with the greatest gross power is chosen, which usually has

Table 11.4 Mean Gross Power (Data Set II)

Diversity Scheme	Mean Power Ratio W.R.T No Diversity (dB)
None	0
SEL(2)	3.7
SEL(3)	4.6
MPA(2)	4.4
MPA(3)	5.9
CPT(2)	5.2
CPT(3)	8.1

Table 11.5 RMS Delay Spread (Data Set I)

Diversity Scheme	RMS Delay (ns)
None	25.0
SEL(2)	41.7
SEL(3)	29.6
MPA(2)	92.8
MPA(3)	128.9
CPT(2)	42.2
CPT(3)	30.0

more multipath components and thus greater delay spread. In CPT diversity, although the delay spread of each branch contributes to the resulting CIR, the cophasing of peak components reduces relatively the other components with longer delay. In contrast to multipath antenna diversity (in which delay spread increases with increasing numbers of antennas) each additional signal in CPT diversity does not necessarily result in extra delay spread. In fact, order 3 CPT diversity achieves reduced delay spread. It is thought that this is because the order 2 scheme employs antenna units located at the extreme ends of the measurement area. The order 3 scheme adds an antenna near the center of the measurement area. This centrally located antenna results in less delay spread because the propagation paths to it are more similar in length than is the case for the antennas at more extreme locations.

11.4.3 Unbalanced Link SNR and Optimal Power Allocation

Because the signals from different transmit antennas propagate via different paths, the SNR of different transmit diversity branches can be very disparate. The diversity gain can, therefore, be a sensitive function of location. (We previously showed that when signal levels are the same, maximal gain is achieved.) The experiment shows that the variance of gross power is greater for the diversity scheme than the nondiversity (single antenna) scheme.

A practical DA system needs to determine the best power allocation between antennas for each different mobile location. In the cophasing diversity scheme shown above,

Table 11.6 RMS Delay Spread (Data Set II)

Diversity Scheme	RMS Delay (ns)
None	25.0
SEL(2)	40.4
SEL(3)	28.4
MPA(2)	77.6
MPA(3)	109.7
CPT(2)	38.4
CPT(3)	28.1

each diversity antenna transmits the same power. This "equal gain" combining scheme does not give optimal diversity gain. An improvement is to weight the transmitted power at each antenna differently. A classic approach is to use the maximal ratio combining principle. The power at each transmit antenna is made proportional to the nominal channel gain between the antenna and mobile station. This weighting results in optimal SNR. To simplify the analysis we assume that the delay profiles for all diversity branches are the same but with different propagation delay and loss. We also assume that cophasing is perfect. At a particular mobile location, the received signal can then be expressed as:

$$r_c(t) = \sqrt{p} \sum_{i=1}^{K} g_i \sum_{n=1}^{L} \alpha_i(n) \delta(t - \tau_i(n)) \qquad (11.13)$$

where g_i is the channel gain, $\alpha_i(n)$ is the normalized multipath gain ($\sum_n |\alpha_i(n)|^2 \equiv 1, \forall i$), and p is total transmitted power for one mobile user. As each antenna is weighted by $\sqrt{g_i}$, the received signal power from each antenna is pg_i^2.

11.4.3.1 Equalization of SIR

Power control is used in CDMA networks to control intercell and intracell interference. In the downlink, because every mobile user receives a composite of signals for all users from the base station antennas, the power ratio of signals from the same base station antenna remains the same as at the transmitter. Interference arises from all unwanted users' signals. For a single receiving antenna and an incoherent (power) diversity combining scheme, an equal power allocation for each user at every diversity antenna can result in equal received signal-to-interference ratio (SIR) for each user after combining. In a cophasing diversity scheme, after combining, the gain of the intended signal is greater than that of the interference. (The time and phase adjustment is performed on each mobile user's signal independently.) Because the gain realized by combining depends on the signal level ratios between diversity branch, the received SIRs at different mobile locations are not necessarily equal. An eigenvalue-based approach can solve this problem [24–26].

If we ignore loss of orthogonality (because the assumption is that this factor will affect all users equally), then the SIR, γ_j, for each user is given by:

$$\gamma_j = \frac{p_j (\sum_{i=1}^{K} g_{ij})^2}{\sum_{m, m \neq j}^{M} p_m \sum_{i=1}^{K} g_{im} g_{ij}} \qquad (11.14)$$

We define a transmission gain matrix, Γ:

$$\Gamma = \begin{bmatrix} g_{11} & \cdots & g_{1M} \\ \vdots & \ddots & \vdots \\ g_{K1} & \cdots & g_{KM} \end{bmatrix}_{K \times M} \qquad (11.15)$$

To equalize received SIR, the total transmitted power for each user p_j, $j \in [1, M]$ needs to be controlled. We define the transmission power vector P as:

$$P = [p_1 \ p_2 \cdots p_M], \ p_j \in \Re_+ \tag{11.16}$$

and the weighting matrix, Q, as:

$$Q = \Gamma' = \begin{bmatrix} g_{11} & \cdots & g_{K1} \\ \vdots & \ddots & \vdots \\ g_{1M} & \cdots & g_{KM} \end{bmatrix}_{M \times K} \tag{11.17}$$

Each row in this matrix determines the power allocation between all antennas for one mobile user. We now define the following quantities:

$$Z = (Q \times \Gamma - \Lambda) \times D^{-1} \tag{11.18}$$

where Λ and D are diagonal matrices with values:

$$D = \begin{bmatrix} (\sum_{i=1}^{K} g_{i1})^2 & \cdots & 0 \\ \vdots & \ddots & \vdots \\ 0 & \cdots & (\sum_{i=1}^{K} g_{iM})^2 \end{bmatrix}_{M \times M} \tag{11.19}$$

and

$$\Lambda = \begin{bmatrix} \sum_{i=1}^{K} g_{i1}^2 & \cdots & 0 \\ \vdots & \ddots & \vdots \\ 0 & \cdots & \sum_{i=1}^{K} g_{iM}^2 \end{bmatrix}_{M \times M} \tag{11.20}$$

The SIR equalization problem can then be expressed as:

$$\frac{1}{\gamma^*} p_i z_{ii} = \sum_{m, m \neq i}^{M} p_m z_{mi} \tag{11.21}$$

The optimum power distribution between users, P^*, is now a solution of the following eigenvalue problem:

$$\lambda^* P^* = P^* \times Z \tag{11.22}$$

We use an inverse square propagation model to simulate the channel gains of mobile stations at different locations to all diversity antennas. Figure 11.12 and Figure 11.13 show the CDF of a downlink SIR for 8 and 16 users for CPT diversity using this power distribution scheme. We can see from the CDF that the SIR is not uniform, i.e., even after the application of power control the SIR is dependent on the spatial distribution of

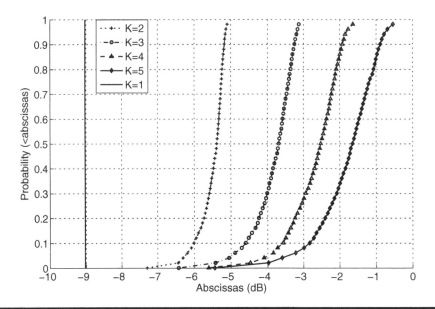

Figure 11.12   CDF of Downlink SIR (8 Users).

mobile users. The slope of the CDF is related to the number of base station antennas. The more antennas, the smaller the slope. This can be explained by the observation that when there are more antenna units, the number of ways in which the mobile users can be clustered around them increases.

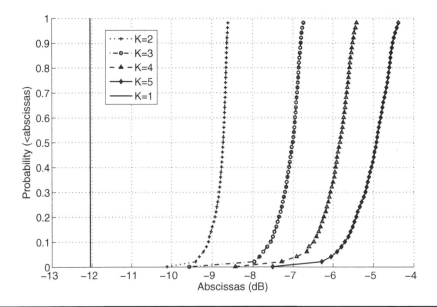

Figure 11.13   CDF of Downlink SIR (16 Users).

11.5 CONCLUSIONS AND FURTHER WORK

11.5.1 Conclusions

Distributed antenna diversity has asymmetric performance on uplink and downlink. Due to MISO channel characteristics in the downlink, its performance is normally inferior to the uplink. In narrowband systems, when time dispersion is negligible, cophasing transmission is possible. In wideband systems, achieving a combining gain for signals dispersed in time is more difficult. Using a wideband MIMO channel sounder, distributed antenna performance, in particular that on the downlink, has been investigated for various transmit diversity schemes.

Observation of channel impulse responses in an indoor environment suggests that phase variation within the impulse response time window containing the majority of power is small. Based on this observation, a wideband CPT diversity scheme has been proposed. An evaluation of diversity performance in terms of both gross power and RMS delay spread after combining shows that CPT diversity is superior to multipath antenna diversity. Both of these schemes have superior performance to simple selective diversity.

The coverage performance of the different diversity combining schemes has also been investigated. Even partial cophasing (practical CPT) achieves better coverage than power-wise combining (MPA). (This applies to both uplink and downlink.) Compared to a single antenna base station, distributed antennas improve radio coverage significantly, irrespective of the combining technique used. This improvement may be especially important at higher frequencies because here propagation loss and shadowing are more serious than at lower frequencies.

11.5.2 Further Work

Some issues related to the study presented in this chapter need further investigation. While the proposed transmit diversity scheme for wideband signals is based on measurements in an indoor environment, it is also applicable, in principle, to the outdoor environment. The outdoor environment, however, typically results in greater time dispersion. This may cause the phase of the CIR to change more rapidly with delay. Because energy is concentrated near the peak CIR component, however, cophasing (the peak component) may still give an advantage over power-wise combining. A separate study is therefore needed for outdoor environments.

The practical issue of obtaining the channel state required to implement transmit diversity also needs addressing. This is particularly critical for cophasing diversity because it relies on the accurate knowledge of phase and delay information. The duplex mode will have an impact in this respect. For time division duplex (TDD) systems, downlink channel information can be obtained from channel estimation of the reciprocal uplink. For frequency division duplex (FDD) systems, however, reciprocity does not apply. One option is to use a reverse signaling channel to transmit the channel information measured at the mobile station back to the base station. An obvious disadvantage is that this scheme sacrifices channel resource due to the signaling overhead. The coding, transmission, and decoding of channel information will also inevitably introduce delay in applying the channel information to the cophasing process. The accuracy and timeliness of the channel state report will have an impact on the resulting performance.

CDMA has been widely adopted in 2G and 3G cellular networks. Before implementing distributed antennas in CDMA systems, however, some practical problems must be solved. One is loss of orthogonality in the simulcast downlink. The degree to which orthogonality is compromised is random, a function of the relative arrival time of other unwanted user multipath signals as well as the multipath dispersion of the wanted user signals. Because the signals targeting the same mobile station are aligned in time, the self interference is smaller for CPT diversity. For the same reason, however, the signals for different mobile stations transmitted from the same antenna unit cannot be aligned at orthogonal code boundaries. The interference arising from unwanted user signals therefore becomes more serious than other schemes which can maintain orthogonal code time alignment. One solution to avoid this performance compromise is to use spreading codes with good auto- and cross-correlation characteristics irrespective of relative delay.

REFERENCES

[1] R. Bernhardt, "Macroscopic diversity in frequency reuse radio systems," *IEEE Journal on Selected Areas in Communications*, vol. 5, no. 5, pp. 862–870, 1987.

[2] A.L. Brandao, L.B. Lopes, and D.C. McLernon, "Base station macro-diversity combining merge cells in mobile systems," *Electronics Letters*, vol. 31, no. 1, pp. 12–13, 1995.

[3] K. Kerpez, "A radio access system with distributed antennas," *IEEE Trans. Vehicular Technology*, vol. 45, no. 2, pp. 265–275, May 1996.

[4] W. Papen, "Uplink performance of a new macro-diversity cellular mobile radio architecture," *6th IEEE International Symposium on Personal, Indoor and Mobile Radio Communications*, Toronto, vol. 3, pp. 1118, 27–29 Sept. 1995.

[5] A.A. Abu-Dayya and N.C. Beaulieu, "Micro- and macrodiversity NCFSK(DPSK) on shadowed Nakagami-Fading channels," *IEEE Trans. on Commnications*, vol. 42, no. 9, pp. 2693–2702, Sept. 1994.

[6] V. Emamian, M. Kaveh, and M.S. Alouini, "Outage probability with transmit and receive diversity in a shadowing environment," *IEEE Wireless Communications and Networking Conference*, Orlando, vol. 1, pp. 54–57, March 2002.

[7] I. Stamopoulos, A. Aragon, and S. Saunders, "Performance comparison of distributed antenna and radiating cable systems for cellular indoor environments in the dcs band," *12th International Conference on Antennas and Propagation*, Exeter, U.K., vol. 2, pp. 771–774, March 2003.

[8] A. Saleh, A. Rustako, and R. Roman, "Distributed antennas for indoor radio communications," *IEEE Trans. on Communications*, vol. 35, no. 12, pp. 1245–1251, Dec. 1987.

[9] A. Arredondo, D.M. Cutrer, J.B. Georges, and K.Y. Lau, "Techniques for improving in-building radio coverage using fiber-fed distributed antenna networks," *45th IEEE Vehicular Technology Conference*, Atlanta, vol. 3, pp. 1540–1543, 28 April–May 1996.

[10] P. Chow, A. Karim, V. Fung, and C. Dietrich, "Performance advantages of distributed antennas in indoor wireless communication systems," *44th IEEE Vehicular Technology Conference*, Stockholm, vol. 3, pp. 1522–1526, 8–10 June 1994.

[11] L.R. Grundmann and S.T. Nichols, "An empirical comparison of a distributed antenna microcell system versus a single antenna microcell system for indoor spread spectrum communications at 1.8GHz," *2nd International Conference on Universal Personal Communications*, Ottawa, vol. 1, pp. 59–63, 12–15 Oct. 1993.

[12] K. Bye, "Leaky-feeders for cordless communication in the office," *8th European Conference on Electrotechnics*, Stockholm, pp. 387–390, 13–17 June 1988.

[13] F. Tong, I.A. Glover, S.R. Pennock, P.R. Shepherd, and N.C. Davies, "Indoor distributed antenna experiment," *14th IST Mobile & Wireless Communications Summit*, Dresden, 19–23 June 2005.

[14] S. Zürbes, W. Papen, and W. Schmidt, "A new architecture for mobile radio with macroscopic diversity and overlapping cells," *5th IEEE International Symposium on Personal, Indoor and Mobile Radio Communications*, The Hague, vol. 2, pp. 640–644, 18–23 Sept. 1994.

[15] S. Ariyavisitakul, T.E. Darcie, L.J. Greenstein, M.R. Phillips, and N.K. Shankaranarayanan, "Performance of simulcast wireless techniques for personal communication systems," *IEEE Journal on Selected Areas in Communications*, vol. 14, no. 4, pp. 632–643, May 1996.

[16] COST231, Urban transmission loss models for mobile radio in the 900- and 1800 MHz bands (Revision 2), *European Commission*, Brussels, Belgium, 1996.

[17] E. Sousa, "Antenna architectures for CDMA integrated wireless access networks," *6th IEEE International Symposium on Personal, Indoor and Mobile Radio Communications*, Toronto, vol. 3, pp. 921, 27–29 Sept. 1995.

[18] H. Yanikomeroglu and E. Sousa, "CDMA distributed antenna system for indoor wireless communications," *2nd International Conference on Universal Personal Communications*, Ottawa, vol. 2, pp. 990–994, 12–15 Oct. 1993.

[19] H. Yanikomeroglu and E. Sousa, "CDMA sectorized distributed antenna system," *5th IEEE International Symposium on Spread Spectrum Techniques and Applications*, Sun City, South Africa, vol. 3, pp. 792–797, 2–4 Sept. 1998.

[20] L. Welburn, J.K. Cavers, and K.W. Sowerby, "Optimizing the downlink power of macrodiversity antennas in an indoor DS-CDMA system," *50th IEEE Vehicular Technology Conference*, Amsterdam, vol. 3, pp. 1905–1909, 19–20 Sept. 1999.

[21] J.H. Lee, J.H. Roh, J.H. Kwun, and C.E. Kang, "A controlled distributed antenna system for increasing the Eb/No in the DS-CDMA System," *9th IEEE International Symposium on Personal Indoor and Mobile Radio Communication*, Boston, vol. 3, pp. 1401–1405, 8–11 Sept. 1998.

[22] J. Yang, "Diversity receiver scheme and system performance evaluation for a CDMA system," *IEEE Trans. on Communications*, vol. 47, no. 2, pp. 272–280, Feb. 1999.

[23] F. Tong, I.A. Glover, S.R. Pennock, and P.R. Shepherd, "Co-phase transmission diversity for distributed antenna," *Antennas and Propagation Society International Symposium*, Washington, D.C., vol. 1A, pp. 60–63, 3–8 July 2005.

[24] S. Grandhi, R. Vijayan, D. Goodman, and J. Zander, "Centralized power control in cellular radio systems," *IEEE Trans. on Vehicular Technology*, vol. 42, no. 4, pp. 466–68, Nov. 1993.

[25] A. Klein, B. Steiner, and A. Steil, "Known and novel diversity approaches as a powerful means to enhance the performance of cellular mobile radio systems," *IEEE Journal on Selected Areas in Communications*, vol. 14, no. 9, pp. 1784–795, Dec. 1996.

[26] L.A.D. Rocha and J. Brandao, "General analysis of downlink power control in CDMA systems," *ITS'98 Proceedings, International Telecommunications Symposium*, San Antonio, vol. 1, pp. 172–76, 9–13 Aug. 1998.

12

A CASE STUDY ON DISTRIBUTED ANTENNA SYSTEMS

Troels B. Sørensen

Contents

Passive distributed antenna systems (DASs) consisting of distributed feeder lines or single point antennas are now often installed in large office buildings where they provide efficient coverage throughout the building. More sophisticated DASs with intelligent reuse and the ability to adapt to changing interference and traffic conditions are less common, despite their potential for increased capacity in comparison with traditional (pico)cellular-based concepts. This chapter explores a case study of one indoor environment where the site-specific propagation characteristics are taken into account in evaluating the capacity potential from *adaptive* DASs. After an introduction, the chapter gives a description of the site-specific environment considered in this study, including models for the large-scale path loss and user deployment that is later assumed for simulations. Radio resource assignment is described in terms of algorithms for power allocation and access port assignment, as well as algorithms for (dynamic) channel assignment. After an outline of simulation assumptions, system capacity comparisons are given between the adaptive DAS and a system with fixed channel and access port assignment. The chapter is concluded with a discussion on the results and some open issues for deploying DASs in the indoor environment.

12.1 INTRODUCTION

The distributed antenna system (DAS) is a system with spatially distributed access ports, possibly simple antennas as the name suggests, connected to a central controller. Conceptually, therefore, it behaves like a single logical cell. In its simplest form a DAS is nothing more than a passive distribution of signals to a set of distributed antennas by coaxial cables and power splitters (a *passive* DAS). The signal distribution may even take place along the (feeding) coaxial cable without any definable single point antennas. This solution is commonly known as the radiating coax (leaky feeder), where controlled slots, or a loosely woven shield in the coax, leaks part of the radio frequency (RF) energy propagating within the wire [1]. To better control the polarization and radiation, and most importantly to have individual control over the power radiated (or received) through the access ports, discrete single point antennas will be assumed here.

In urban areas, a cellular operator may have only a few frequencies available for provision of indoor cellular services to fit the frequency allocation into the operators overall (macro and microcellular) channel plan. Depending on the capacity need and the available frequencies, there will be a need to reuse frequencies inside the building. Traditionally this is achieved by cell-splitting (multi-cell approach) to create dedicated picocells similar to the micro and macrocells in the outside cellular network. The *adaptive* DAS can be seen as a solution which combines the best of a passive DAS and a multi-cell system, in the sense that it retains the trunking efficiency[1] of the centralized single-cell approach and avoids the increased handover rate of the multi-cell. The further capacity advantage comes from the ability to adapt to changing propagation and interference conditions in combination with dynamic channel allocation, which allows for efficient reuse of the available system resources provided that sufficient isolation exists between the access ports. The indoor environment can provide high isolation between access ports due to the screening provided by interior partitions; however, at the same time it can cause subtle propagation and interference conditions as a result of the close proximity of users to the building structure. The close proximity can result in significant interference changes for only a small change in user position.

One of the very initial investigations on the passive DAS was made by Kerpez in [2]. It was shown that the DAS increases capacity as a secondary effect to the increase in signal-to-interference (SIR) ratio resulting from increasing the number of distributed antennas per logical cell. The corresponding reduction in dynamic range was pointed out in the empirical (measurement based) study reported in another early contribution by Saleh [3]. More recent simulation-based studies similar to the one presented here compare the picocell to both passive and adaptive DASs based on voice service provisions in a large office building [4]. Also, the generic comparison to systems with single-input multiple-output processing (SIMO) has been done, e.g., in [5] where power and capacity gains are shown in comparison with an equivalent but centralized antenna array processing, and in [6] where the downlink spectral efficiency is shown to be superior in a multiple-input multiple-output (MIMO) system with distributed as compared to centralized transmit antennas. In [6] specifically, the authors are considering the case with simultaneous transmission from several access ports, in which there is a tradeoff between the improvement

---

[1] Trunking efficiency derives from having a large population sharing a limited, but larger, number of channels in a statistical multiplex; it is usually measured by the population size that can be offered a particular grade of service (GoS).

from diversity and the deterioration from the increased interference in transmitting the same information from several access ports.

In this study, we consider the problem where each user connects to one access port. Largely, this problem can be treated based on the work and algorithms of Hanly [7] and Yates [8] who simultaneously, and independently, derived algorithms for the joint port (in their work base station) assignment and uplink power allocation problem to minimize each individual user's transmission power. For a system with low load, users always connect to the port with the minimum path loss. However, as the load increases some of them are reassigned to other ports to minimize interference within the system, and hence reduce total transmitted power. It turns out that the minimum power allocation can be achieved by having each user minimize their own power, which therefore allows a decentralized implementation despite its centralized, or global, outset [7]. This is possible because there is no penalty in changing the port assignment for a particular user in uplink — it does not change the interference situation of other users — and hence the assignment is decoupled from the power allocation. The situation is quite different for the downlink where the same decoupling cannot be achieved because changing port assignment creates a new interference situation for the other users in the system. However, Farrokhi et al. [9] extended the algorithm of Hanly and Yates [7,8] to the downlink. Farrokhi et al. relied on the equivalence between uplink and downlink for fixed (known) assignment to derive the BUD algorithm. The acronym BUD refers to the three steps of the resulting algorithm:

1. **B**ase station (port) assignment for uplink direction.
2. **U**plink power allocation based on the assignment in step 1.
3. **D**ownlink power allocation based on the same assignment.

Steps 1 and 2 are based on the Hanly and Yates algorithm, which in the following is referred to as the Hanly algorithm. The BUD algorithm can be shown to give the minimum total downlink power allocation.

Based on very ideal assumptions for the operation of the DAS, but realistic site-specific propagation modeling, the aim in this chapter is to see what potential capacity increase adaptive DAS systems may bring to a system with orthogonal channels. This is applicable to many indoor systems which deploy the global system for mobile communication (GSM) or a similar system for voice service provision. The downlink direction of transmisison is considered of most interest due to the coupling between assignment and power allocation.

For simplicity, the capacity is measured by the number of supported voice users at a specified blocking level, and compared between a fixed channel and access port assignment (picocell), and a DAS with centralized and adaptive channel assignment. The site-specific modeling makes sure we include the realistic effects of random large-scale (shadowing) variations in the analysis, and by extension of the above mentioned BUD algorithm, also, in the joint downlink power allocation and access port assignment. Related to the operation of the DAS, there are many practical problems such as signal distribution and availability of radio signal measurements for making port assignment and power allocation. Limitations and constraints in this respect are not included here.

12.2 CASE STUDY: DAS IN AN INDOOR ENVIRONMENT

A representative multistory office building was initially selected for measuring the propagation characteristics within a configuration of distributed antennas. Later, based on these measurements, a site-specific propagation model of the office environment was

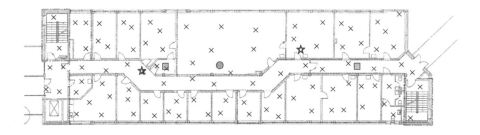

Figure 12.1 Center Floor Plan of the Office Building Showing the Three Different Port Configurations Discussed in the Text (Stars, Squares, and Circle). 'x' Markers Indicate the Possible Random User Locations on This Particular Floor.

developed to generalize the findings for the study of other port configurations as well as for studying different user distributions. This was possible due to the availability of site-specific information in the form of a three-dimensional geometric model for the office building. The office building is a three-story building with an approximate ground coverage of 50×15 m. Outer wall construction is brick, but with large metal-tinted windows covering both sides of the building. Floors are separated by reinforced concrete decks, while interior walls are made of plasterboard on metal studs and extend from one concrete floor to the next through suspended ceiling tiles. The suspended ceiling tiles are a mix of plaster and fibreboard placed at about 2.7 m above floor level, leaving a space of about another 0.7 m above for cable guides and ventilation ducts. The floor plan for the center floor is shown in Figure 12.1, which is typical of this building with its central office corridor flanked by mostly small office cubicles and a few larger meeting rooms or open spaces. Stairways connecting floors are placed at both ends of the central corridor, and on the ground floor there is a big entrance hall at the left end of the building (not shown in the figure).

12.2.1 Indoor Propagation Characterization

A measurement campaign with two omnidirectional antennas placed on each of two floors was conducted to obtain data for the development of a site-specific propagation model. Antenna positions were very similar, but not identical to the configurations shown in Figure 12.1, and with antennas mounted just below the suspended ceiling tiles for vertical polarization in what seemed to be a typical (or likely) deployment. A wideband sounding system, capable of simultaneous sounding to each of the four antennas, was used to sample the large-scale path loss at a number of randomly selected measurement positions over the two floors; in total 87 positions were characterized with about an equal number of positions on each floor. The details of the operation and setup of the measurement system are outside the scope of this chapter, but due effort was paid to spatial averaging and calibration at each position to obtain the path loss including the characteristics of transmitting and receiving omnidirectional dipole antennas from that position to each of the four port (antenna) locations [10].

The site-specific path loss model derived from these meaurements is based on general physical propagation principles, though rightfully it belongs to the class of empirical path loss models. Basically, it is a partition-based model [11], but augmented based on the principle that the path of least attenuation dominates the path loss. The main motivation was to minimize the residual error between the measurement and the path loss predicted by the model — the so-called standard error — and in this process, to get a suitable statistical description of the prediction residuals to allow their inclusion as a random component to the model. One can imagine that the smaller the standard error the less similarity between the random variations in path loss to different port positions over the measurement area; this will be the case because the residuals represent progressively smaller and smaller detail (clutter), which is likely to be uncorrelated between different paths and which will tend toward Gaussian distribution. The exact details of the model are not of importance here; for this the reader is referred to [10]. What is important is the ability to get accurate predictions and, hence, being able to generalize the findings from the measurements. Figure 12.2 shows the distribution of the aggregate prediction residuals after constrained linear least-squares optimization of the resulting model from the calibrated path loss data. The approximate log-normal (Gaussian in dB) behavior facilitates a simple description of the residuals so that the total path loss can be modeled as the sum of a deterministic part and a random normal part.

From further analysis of the residuals, it was noticed that ports located inside office cubicles had slightly reduced standard error compared to ports in the corridor area: 3.9 dB versus 4.7 dB, respectively. Accordingly, the random normal part of the site-specific path loss model was made to apply a similar distinction. It further turned out that, although the standard error obtained was quite small (considering a total dynamic range of almost 60 dB), there was a distinct positive correlation in the path loss between different ports. This was also included to the model so that the random normal part is drawn from a multivariate normal (in dB) distribution with correlation 0.4.

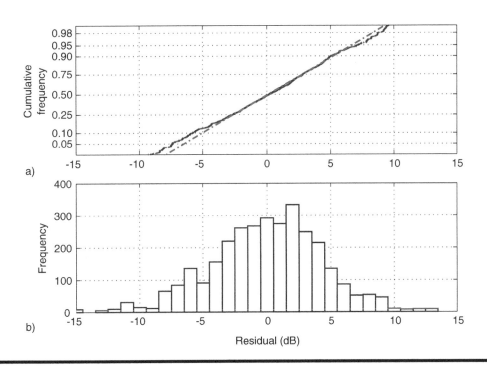

Figure 12.2   Empirical Distribution of Prediction Residuals for the Path Loss Model. a) Normal Probability Plot of the Residuals; b) Frequency Histogram of the Residuals (Standard Deviation is 4.4 dB).

Table 12.1 Location Clusters and Associated Probabilities

Cluster:	Corridors Y_1	Labs Y_2	Stairways Y_3	Offices Y_4
$P[Y_j]$:	0.25	0.25	0.10	0.40

In analyzing predictions versus measurements, it was confirmed that there was no significant difference between predictions on same floor, different floor, or between ports, hence supporting the generalization of the measurements by the described path loss model.

12.2.2 User Deployment Scenario

The office building is composed of different types of rooms, notably: corridors, offices, stairways, meeting rooms, and facility rooms. To study different and realistic user distributions, a probabilistic user deployment model is defined to take into account that users are more likely to be present in one type of room than others. The model is static, in the sense that it does not accommodate for users moving from one location to another, but simply assigns a location to a particular user. The model is probabilistic because location is chosen on the basis of the location cluster probabilities specified in Table 12.1.

A location cluster, Y_j, is a grouping of locations with similar characteristics, for instance, those locations that are within offices (cluster Y_4). Apart from Y_1 to Y_4, one additional cluster, Y_5, is defined as a subgroup of the corridor cluster for the purpose of representing areas where users tend to gather, e.g., outside a meeting room. The individual locations within each cluster are uniformly distributed, and except for Y_5, all clusters are disjoint which makes the model fairly easy to implement. Cluster Y_5 is dealt with by assigning a higher (uniform) probability to locations common to Y_1 and Y_5, in a ratio of 1 to 3. Specifically,

$$P[Y] = \sum_{j=1}^{4} P[Y_j] \equiv 1 \tag{12.1}$$

In the system simulations shown later, Monte Carlo simulation is performed over the user deployment model based on a total of 322 possible user locations defined on the three floors of the office building. These locations have been defined by randomly sampling the floor plans as shown for the center floor in Figure 12.1; this total equals the size of the set Y in (12.1).

12.3 RADIO RESOURCE ASSIGNMENT

The radio resource assignment in an adaptive DAS may be divided between the problem of power allocation and port assignment to minimize the total downlink transmitted power (and interference), and the problem of assigning a channel (the channel reuse problem). In the following, the formulation of both of these problems is outlined together with the algorithms to deal with them; the algorithms are used in the subsequent system simulations to see the potential capacity increase with adaptive DAS systems.

The radio resource assignment will be made to satisfy a certain quality constraint, taken here as a specified user SIR requirement. We are concerned mainly about the effects from large-scale fading, and assume implicitly that the SIR target, averaged over fast fading, is set high enough to be able to cope with fast fading. It is intuitively acceptable that the minimum total allocated power is reached when all users have their SIR requirement satisfied at equality [7]. The notion of minimum total allocated power does not mean that each individual allocation is at its minimum value. Generally, it is possible to reduce the power allocated to a particular user below that given by the power allocation for minimum total transmitted power; necessarily, the power allocation for some of the other users must increase to keep a constant minimum sum. For a component-wise optimum solution this is different. In this case, the total minimum solution is also the minimum for each individual allocation. The latter is possible for the uplink direction, but not for the downlink [8,9], as is considered here. Fortunately, from a practical point of view, this is also the most desirable situation because of the power consumption issues related to the mobile terminal and the public concern of the potentially damaging effects of microwave radiation.

12.3.1 Power Allocation and Port Assignment

Consider the general downlink problem in Figure 12.3 in which the ports can be either a base station in the multi-cell system or a radio port in the adaptive DAS. Each of M users may connect to K different ports, but at any time user i is connected to exactly one of these, b_i; thus, $b_i = k$ if user i is connected to port k. The assignment of users to ports is represented by the vector $\boldsymbol{b} \in \{1, 2, \ldots, K\}^M$, and referred to simply as b. For a particular user there might be restrictions as to the allowable ports (connections), hence the active set $\Psi_i \subseteq \{1, 2, \ldots, K\}$ is the set of permissible ports for user i. The set of users connected to port k under assignment b is $\Theta_k^{(b)} = \{i : b_i = k, i = 1, 2, \ldots M\}$. There is a maximum of K^M possible assignments.

The power allocated to user i's downlink transmission under port assignment b is $q_i^{(b)}$ (for simplicity the (b) notation is omitted in Figure 12.3). The power allocation vector $\boldsymbol{q}^{(b)} = [q_1^{(b)}, \ldots, q_i^{(b)}, \ldots, q_M^{(b)}]^T$ represents the aggregate allocation under assignment b. The path gain between user i and the port assigned to user j, b_j, is denoted by $a_{b_j,i}$; $a_{b_j,i}$ is determined from the path loss model and hence is assumed log-normal. With this notation, the downlink symbol level SIR, ω_i, for user i can be expressed as in

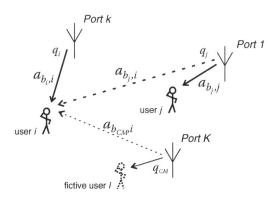

Figure 12.3   Multi-User Interference in the Downlink of an Adaptive DAS.

(12.2) where the multi-user interference from the downlink transmissions of other users is conditioned by the orthogonality, $o_{j,i}$, between multi-user signals.[2] Receiver noise is expressed in terms of additive white Gaussian noise (AWGN) with a power spectral density η_i in a system bandwidth of W Hz. A spreading factor ξ_i has been included in (12.2) to generalize the formulation to the case with quasi-orthogonal channels, e.g., wideband code division multiple access (WCDMA), as it can be shown that the formulation applies equally well in this case also. For this chapter, ξ_i will be taken to be a constant equal to 1 for $i = 1 \ldots M$. The SIR must satisfy the specified SIR requirement $\overline{\omega}_i$.

$$\omega_i = \frac{q_i^{(b)} a_{b_i,i}}{\frac{1}{\xi_i} \left(\sum_{j=1}^{M} o_{j,i} q_j^{(b)} a_{b_j,i} + \eta_i W \right)} \geq \overline{\omega}_i \tag{12.2}$$

Part of the multi-user interference arises from the transmission of common channels, which all users in the coverage area are expected to decode. Each (logical) cell has its own set of common channels transmitted with a certain fraction of the maximum port power. The common channels are allocated within the power vector $q^{(b)}$ with fixed port and power allocation by including a fictive user for each common channel transmitted from the ports. In total we assume L common channels.

Rewriting (12.2) in matrix notation with the power vector $q^{(b)} = [q_1^{(b)}, \ldots, q_i^{(b)}, \ldots, q_M^{(b)}]^T$ we get

$$\frac{q_i^{(b)}}{\overline{\omega}_i/\xi_i} \geq G_{(i,:)}^{(b)} O_i^{(b)} q^{(b)} + \frac{\eta_i W}{a_{b_i,i}^{(b)}} \tag{12.3}$$

In (12.3), $G_{(i,:)}$ is the ith row of the normalized interference matrix with elements $g_{i,j} = a_{b_j,i}/a_{b_i,i}$, $i = 1 \ldots M$, $j = 1 \ldots M$, representing the fractional power (interference) seen by user i due to user j's transmission. Matrix G can be expressed in terms of the path gain matrix A as in

$$G^{(b)} = diag(A^{(b)})^{-1} A^{(b)} \tag{12.4}$$

where $A^{(b)}$ is defined by

$$A^{(b)} = \begin{bmatrix} a_{b_1,1} & a_{b_2,1} & \cdots & a_{b_M,1} \\ a_{b_1,2} & & & \vdots \\ \vdots & & \ddots & \\ & & a_{b_j,i} & \\ a_{b_1,M} & \cdots & & a_{b_M,M} \end{bmatrix} \tag{12.5}$$

The assignment b may change as more users are added to the system and will affect the power allocation $q^{(b)}$ and the interference matrix G; this explains the superscript notation (b).

---

[2] In GSM, for example, users on different time slots, and possibly different frequencies, are fully orthogonal ($o_{j,i} = 0$) when time synchronized, while users on the same time slot but different frequencies have orthogonality representative of the frequency selection in the mobile terminal. Users (re)using the same channel have orthogonality 1.

The orthogonality matrix O_i is a diagonal matrix specific to each user i with element values depending on the channel assignment between users.

$$\mathbf{O}_i^{(b)} = diag([o_{1,i}, \ldots, o_{j,i}, \ldots, o_{M,i}]), \qquad o_{j,i} = \begin{cases} 0 & \text{if } j = i \\ \{0, o_{acr}, 1\} & \text{if } j \neq i \end{cases} \qquad (12.6)$$

In (12.6), o_{acr} is in correspondence to the adjacent channel rejection in the mobile terminal. In the general case for an adaptive DAS, we can have transmissions on the same frequency but different time slots, from different ports. Potentially, if the cell size is large enough, this can lead to a lack of orthogonality due to the difference in propagation delay between ports and users. For the indoor picocells this is not a problem and we can treat these cases as completely orthogonal.

A new effective interference, G', and path gain, A', matrix can be defined by pre-multiplying matrices G and A, respectively, with the orthogonality matrices O_i, $i = 1 \ldots M$.

$$\mathbf{G}'^{(b)} = diag(\mathbf{A}^{(b)})^{-1}\mathbf{A}'^{(b)} = diag(\mathbf{A}^{(b)})^{-1} \begin{bmatrix} \mathbf{A}_{(1,:)}^{(b)}\mathbf{O}_1^{(b)} \\ \vdots \\ \mathbf{A}_{(i,:)}^{(b)}\mathbf{O}_i^{(b)} \\ \vdots \\ \mathbf{A}_{(M,:)}^{(b)}\mathbf{O}_M^{(b)} \end{bmatrix} \qquad (12.7)$$

With this notation, the inequality governing all user $i = 1 \ldots M$ can be obtained from the extension of (12.3)

$$\mathbf{q}^{(b)} \geq \begin{bmatrix} \bar{\omega}_1/\xi_1 & 0 & 0 \\ 0 & \ddots & 0 \\ 0 & 0 & \bar{\omega}_M/\xi_M \end{bmatrix} \left(\mathbf{G}'^{(b)}\mathbf{q}^{(b)} + W \begin{bmatrix} \eta_1/a_{b_1,1}^{(b)} \\ \vdots \\ \eta_M/a_{b_M,M}^{(b)} \end{bmatrix} \right) \qquad (12.8)$$

or equivalently, by introducing matrix D and E for the first and last term.

$$\mathbf{q}^{(b)} \geq \mathbf{D}(\mathbf{G}'^{(b)}\mathbf{q}^{(b)} + W\mathbf{E}^{(b)}) = \mathbf{D}(diag(\mathbf{A}^{(b)})^{-1}\mathbf{A}'^{(b)}\mathbf{q}^{(b)} + W\mathbf{E}^{(b)})$$

$$= \mathbf{D}\, diag(\mathbf{A}^{(b)})^{-1}(\mathbf{A}'^{(b)}\mathbf{q}^{(b)} + W\eta) \qquad (12.9)$$

To minimize downlink interference we need to first consider the equivalent uplink problem, in accordance with the BUD algorithm proposed by Farrokhi et al. [9]. The algorithm relies on the known (joint optimization) convergence results of Hanly [7] and Yates [8] for uplink and the equivalence between uplink and downlink for fixed port assignment as discussed previously. This equivalent uplink problem, where multi-user interference to user i is now in the reception of user i's signal at its assigned port, can be shown to be

$$\tilde{\mathbf{q}}^{(b)} \geq \mathbf{D}\, diag(\mathbf{A}^{(b)})^{-1}(\mathbf{A}''^{(b)}\tilde{\mathbf{q}}^{(b)} + W\tilde{\eta}^{(b)}) \qquad (12.10)$$

$$\boldsymbol{G}''^{(b)} = diag(\boldsymbol{A}^{(b)})^{-1}\boldsymbol{A}''^{(b)} = diag(\boldsymbol{A}^{(b)})^{-1} \begin{bmatrix} (\boldsymbol{A}^{(b)})^T_{(1,:)} \boldsymbol{O}^{(b)}_1 \\ \vdots \\ (\boldsymbol{A}^{(b)})^T_{(i,:)} \boldsymbol{O}^{(b)}_i \\ \vdots \\ (\boldsymbol{A}^{(b)})^T_{(M,:)} \boldsymbol{O}^{(b)}_M \end{bmatrix}$$

The tilde ($\sim$) and the double prime ($''$) designate uplink variables. Note the different interpretation of the receiver noise where $\tilde{\eta}^{(b)}$, $\tilde{\eta}^{(b)}_i = \tilde{\eta}_{b_i}$ is the receiver noise power spectral density at the port to which user i is assigned. The uplink problem has been studied extensively by Yates [8] and Hanly [7] who proposed an algorithm for the joint port assignment and power allocation problem. Additionally, they gave information on the convergence properties of the algorithm and the necessary conditions for a feasible solution.

The minimum uplink power allocation (total and componentwise) is given when the constraints are satisfied at equality in (12.10). For a fixed assignment b, the necessary and sufficient condition for a positive solution (a feasible solution) is that the spectral radius[3] of $\boldsymbol{D}\, diag(\boldsymbol{A}^{(b)})^{-1}\boldsymbol{A}''^{(b)}$ is less than unity [7]

$$\rho(\boldsymbol{D}\, diag(\boldsymbol{A}^{(b)})^{-1}\boldsymbol{A}''^{(b)}) < 1 \tag{12.11}$$

In fact, for the fixed assignment b we can first check for feasibility and then solve explicitly for the unique solution $\tilde{\boldsymbol{q}}^{*(b)}$

$$\tilde{\boldsymbol{q}}^{*(b)} = [\boldsymbol{I} - \boldsymbol{D}\, diag(\boldsymbol{A}^{(b)})^{-1}\boldsymbol{A}''^{(b)}]^{-1}\boldsymbol{D}\, diag(\boldsymbol{A}^{(b)})^{-1} W\tilde{\eta}^{(b)} \tag{12.12}$$

given $\boldsymbol{I}$ as the identity matrix. It can be shown that given an optimum assignment b^* for the uplink, the same (fixed) assignment is feasible also for the downlink where there exists any. The proof is based on the result in [12], which can be stated for the present problem as

$$\rho(\boldsymbol{D}\, diag(\boldsymbol{A}^{(b)})^{-1}\boldsymbol{A}'^{(b)}) = \rho(\boldsymbol{D}\, diag(\boldsymbol{A}^{(b)})^{-1}\boldsymbol{A}''^{(b)}) \qquad \text{if } \boldsymbol{A}''^{(b)} = (\boldsymbol{A}'^{(b)})^T \tag{12.13}$$

The transpose relation between matrices is fulfilled in this case due to the symmetry in orthogonality factors between arbitrary users n and m, $o_{mn} = o_{nm}$. The uplink problem can now be written in a notation similar to the downlink one

$$\tilde{\boldsymbol{q}}^{(b)} \geq \boldsymbol{D}(\boldsymbol{G}''^{(b)}\tilde{\boldsymbol{q}}^{(b)} + W\tilde{\boldsymbol{E}}^{(b)}) = \boldsymbol{D}(diag(\boldsymbol{A}^{(b)})^{-1}\boldsymbol{A}'^{(b)}\tilde{\boldsymbol{q}}^{(b)} + W\tilde{\boldsymbol{E}}^{(b)}) \tag{12.14}$$

$$= \boldsymbol{D}\, diag(\boldsymbol{A}^{(b)})^{-1}((\boldsymbol{A}'^{(b)})^T\tilde{\boldsymbol{q}}^{(b)} + W\tilde{\eta}^{(b)})$$

with optimum solution

$$\tilde{\boldsymbol{q}}^{*(b^*)} = [\boldsymbol{I} - \boldsymbol{D}\, diag(\boldsymbol{A}^{(b^*)})^{-1}(\boldsymbol{A}'^{(b^*)})^T]^{-1}\boldsymbol{D}\, diag(\boldsymbol{A}^{(b^*)})^{-1} W\tilde{\eta}^{(b^*)} \tag{12.15}$$

With the same assignment b^*, we may write the (feasible) solution $\boldsymbol{q}^{*(b^*)}$ for the downlink problem in (12.9)

$$\boldsymbol{q}^{*(b^*)} = [\boldsymbol{I} - \boldsymbol{D}\, diag(\boldsymbol{A}^{(b^*)})^{-1}\boldsymbol{A}'^{(b^*)}]^{-1}\boldsymbol{D}\, diag(\boldsymbol{A}^{(b^*)})^{-1} W\eta \tag{12.16}$$

---

[3] Largest (positive) eigenvalue.

While in general the optimum solution $\tilde{q}^{*(b^*)}$ yields the minimum (componentwise and, therefore, minimum total) transmitted power among all feasible port assignments b, this is not the case for the downlink solution $q^{*(b^*)}$; $q^{*(b^*)}$ is optimum only for assignment b^* and not among all possible downlink assignments. However, in the special case where we can assume same receiver noise for all ports, and similarly for all mobile terminals, it can be shown that $q^{*(b^*)}$ gives the minimum total downlink transmitted power solution among all feasible port assignments [9]. Mathematically, this is

$$q^{*(b*)} = \arg \min_{q^{(b)} \geq \mathbf{0}} \mathbf{1}^T q^{(b)}, \forall b \qquad \text{and} \qquad W_\eta = W_\eta \mathbf{1}, \; W_{\tilde{\eta}} = W_{\tilde{\eta}} \tilde{\mathbf{1}} \qquad (12.17)$$

$$\mathbf{1} = [1, 1, \ldots, 1]^T, |\mathbf{1}| = M$$

$$\tilde{\mathbf{1}} = [1, 1, \ldots, 1]^T, |\tilde{\mathbf{1}}| = K$$

If we neglect noise sources other than thermal noise, this assumption is reasonable for receivers operating within the same bandwidth W, and it is still possible to have different noise figures for port and mobile terminal receivers.

The proof of (12.17), which is not elaborated here, is based on the iterative algorithms proposed by Hanly [7] and Yates [8] to solve for $\tilde{q}^{*(b^*)}$, and subsequently $q^{*(b^*)}$, which proves to be a more practical approach to the solution of (12.14) and (12.9) than the explicit expressions in (12.12) and (12.16). These algorithms, at iteration step $t + 1$, are based on the assumption that the other users remain fixed at their step t power levels, and will converge from the all zero vector $q^t|_{t=0} = \mathbf{0}$ [7]. With the assumption of the noise constraints in (12.17), the uplink iteration at step $t + 1$ is given by

$$\tilde{q}_i(t + 1) = \min_{k \in \Psi_i} \frac{\overline{\omega}_i}{\xi_i} \frac{1}{a_{b_i,i}^{(b_i=k)}} \left(\left[A_{(:,i)}^{(b_i=k)} \right]^T O_i^{(b_i=k)} \tilde{q}(t) + \tilde{\eta} W \right) \qquad (12.18)$$

$$b_i(t + 1) = \arg \min_{k \in \Psi_i} \frac{\overline{\omega}_i}{\xi_i} \frac{1}{a_{b_i,i}^{(b_i=k)}} \left(\left[A_{(:,i)}^{(b_i=k)} \right]^T O_i^{(b_i=k)} \tilde{q}(t) + \tilde{\eta} W \right)$$

$$i = 1 \ldots (M - L)$$

Concerning the common channels, the uplink optimization problem is defined for a reduced problem with only real users because there is no sensible way to define the path gain between the fictive users and the ports in the uplink. Hence in (12.18), only real users $i = 1 \ldots (M - L)$ are considered.

Note the column indexing subscript on matrix A to comply with (12.10). For each user, at iteration step $t + 1$, an active set Ψ_i of candidate ports is considered. The port allowing user i to transmit at the lowest possible power level is assigned to user i, thus $b_i(t + 1)$ and the power allocation become $\tilde{q}_i^{b(t+1)}(t + 1)$. Having done this for all real users we get uplink assignment $b(t + 1)$ and power allocation $\tilde{q}^{b(t+1)}(t + 1)$, which is subsequently used to update the downlink power allocation for users $i = 1 \ldots (M - L)$, now including the common channels with fixed assignment $\boldsymbol{b}_{CM}$ and constant power allocation $\boldsymbol{q}_{CM}$ in $\boldsymbol{q}^{(b)}$. Based on (12.3) to (12.9) the update is

$$q_i(t+1) = \frac{\overline{\omega}_i}{\xi_i} \frac{1}{d_{b_i,i}^{(b(t+1))}} \left(\begin{bmatrix} \boldsymbol{A}_{(i,:)}^{(b(t+1))} & \boldsymbol{A}_{CM(i,:)} \end{bmatrix} \begin{bmatrix} \boldsymbol{O}_i^{(b(t+1))} & \boldsymbol{0} \\ \boldsymbol{0} & \boldsymbol{O}_{CMi} \end{bmatrix} \begin{bmatrix} \boldsymbol{q}(t) \\ \boldsymbol{q}_{CM} \end{bmatrix} + \eta W \right)$$

$$i = 1 \ldots (M - L) \qquad (12.19)$$

Eventually, the algorithm BUD converges to the optimum power allocation vector $\boldsymbol{q}^{*(b*)}$. If the user configuration is feasible (cf. the eigenvalue constraint in (12.11)) the algorithm is guaranteed to converge, likewise from the all zero power vector. However, starting from an infeasible power allocation, $\boldsymbol{q}$, can complicate convergence. In a practical implementation, algorithm BUD is iterated a maximum number of T times to reach the optimum allocation. According to [8], $T = 20$ is nearly as effective as $T = 100$ (for a balanced SIR system), and therefore many iterations do not necessarily improve the solution. If the user configuration is infeasible, the algorithm does not converge to any fixed point [7] and the power allocation in the iterative loop will increase (without bound) while each user's ω_i converges to the maximum attainable [8].

The power update in (12.18), with the uplink as an example, can be expressed in terms of the so-called *standard* interference function, $I(\boldsymbol{q})$, of Yates [13].

$$\tilde{q}_i(t+1) = I_i(\tilde{\boldsymbol{q}}(t)) = \frac{\overline{\omega}_i}{\xi_i} \frac{[\boldsymbol{A}_{(:,i)}^{(b_i=k)}]^T \boldsymbol{O}_i^{(b_i=k)} \tilde{\boldsymbol{q}}(t) + \eta W}{d_{b_i,i}^{(b_i=k)}} = \frac{1}{\omega_i(t)\overline{\omega}_i} \tilde{q}_i(t) \qquad (12.20)$$

According to the convergence criterion of Yates, if the interference function is standard (positivity, monotonicity, and scalability properties are satisfied) the iteration will converge. In the interpretation of (12.20), convergence of the power update means that the normalized SIR, $\omega_i/\overline{\omega}_i$, converges to unity. The power is updated in inverse proportion to this quantity.

All the interference terms in (12.20) refer to the large-scale variation, where at each step the power allocation and the orthogonality matrix are considered constant. The large-scale variations are determined from the path loss model: this includes the deterministic path loss prediction of the model plus a random variation, cf. matrix A. The random variation, characterized over many different paths and combinations of materials, was identified to have an approximate log-normal distribution, correlated between the ports. We will think of it as the unpredictable effect of clutter as it occurs over a small area, local to any fixed prediction point (user location). This variation relates to the fact that users are never completely stationary or the effects of structural changes to the environment, e.g., when doors open or close in the path between transmitter and receiver. Based on these assumptions, the (cochannel) interference is modeled as a composition of individual interfering signals whose large-scale power level follows a log-normal distribution with logarithmic mean predicted by the (deterministic) path loss model and logarithmic standard deviation equal to that of the (random) residuals. The resulting signals are correlated between ports due in part to correlated residuals and also correlation in the deterministic predictions. It is reasonable to assume that the phase shift observed in each individual signal varies significantly owing to scattering, such that we can assume the signals to add incoherently (power combining) when averaged over the small area local to any fixed prediction point.

Accordingly, the composite interference is a sum of log-normally distributed signals which can be approximated by another log-normal distribution. The average noise power, ηW, is a characteristic of the receiver and its long-term value is a constant; as such, it may be assumed log-normal with standard deviation of 0 dB. The assumption can be extended in the presence of external noise by increasing the mean and standard deviation. This simplification means that we can treat the whole numerator in (12.20) as an approximate log-normal *random variable* (r.v.).

The SIR, ω_i being the difference between the desired and the sum of interfering signals, is therefore also (approximately) a log-normal r.v. whose mean and variance can be calculated by the Yeh and Schwartz algorithm. Schwartz and Yeh [14] proposed an algorithm to calculate the mean and standard deviation of a sum of uncorrelated log-normal r.vs. This algorithm was later extended to include correlated signals [15] and is therefore applicable for the calculation of the mean and standard deviation of the SIR given the individual (logarithmic) signal mean and standard deviation, and the correlation between them.

Consequently, when we allocate a transmit power of q_i for user i the resulting SIR, ω_i, will vary around the mean as a log-normal r.v. due to the random, and correlated, variation between their own signal and interfering signals, and between interfering signals themselves. To account for this variability in the update of the (mean) transmit power allocation, (12.20) is modified to include a SIR protection ratio to maintain a sufficiently low probability that the SIR falls below the specified requirement $\overline{\omega}_i$. The protection ratio, or critical value δ^p, will change with the standard deviation of the SIR for a fixed (outage) probability p as illustrated in Figure 12.4. As shown, for the top SIR target the allocated power must be increased while less power is needed to meet the bottom target. With this modification (12.20) becomes

$$\tilde{q}_i(t+1) = I'_i(\tilde{\boldsymbol{q}}(t)) = \frac{1}{\mathrm{E}[\omega_i(t)]/\overline{\omega}_i} \tilde{q}_i(t)\, \delta_i^p \tag{12.21}$$

The iterative equations form the basis for the system simulations given later. Before going into simulation details we first need to address the problem of channel assignment.

12.3.2 Dynamic Channel Assignment

The dynamic channel assignment differs between the multi-cell and the adaptive DAS. In the multi-cell, users connect to the strongest base station and are allocated a channel if one is available; if not, the user is blocked and no attempt is made to access other base stations (the active set size is 1). Initially, the same happens in the adaptive DAS, but here, users may connect to any port; therefore the active set size is K. Also, when the common pool of channels is exhausted an attempt is made to reuse currently active channels.

The decision as to which channels are reused is based on an orthogonal signal representation of the signal constellation received on the K ports from transmitting a known pilot signal. In this chapter, we assume this constellation to be described by a multivariate normal distribution with known mean and variance, in accordance with the predictions of the path loss model. This is of course a very idealistic assumption because in practice, such information and assumptions may be difficult to fulfill. However, it serves to investigate the potential capacity gain from the adaptive DAS. The orthogonal

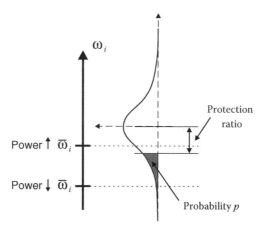

Figure 12.4   Power Adjustment in Relation to SIR Protection Ratio.

representation is based on a method similar to *principal components analysis* [16], which selects components $a_j \in \Re^K$ of the orthogonal basis $A = [a_1, a_2, \dots, a_K]$ so as to maximize the power of the signal constellation vector projected in the direction of a_j. In matrix notation this is

$$\text{Maximize } Q_v = diag\left(A^T \Gamma A\right) \qquad \text{subject to } A^T A = I \qquad (12.22)$$

where the maximization is to be in an ordered sequence and Γ is the sample product, or sample variance, matrix of the signal constellation. This optimization problem is exactly what is solved by the eigendecomposition of the real symmetric sample variance matrix. The decomposition computes matrices U and S so that Γ can be represented by $\Gamma = USU^T$. Specifically, $Q_v = diag\left(U^T\left(USU^T\right)U\right) = diag(S)$ is a descending sequence of positive eigenvalues and $U^T U = I$. Therefore, U is the desired orthogonal basis A with element vectors $a_j = u_j$ as the principal component directions, and corresponding eigen-values $\lambda_j = Q_{v_j}$ which account for the power explained by that component. To see how to obtain the components of Γ, we let the signal constellation be represented by a logarithmic vector y as in (12.23), where y_k is the signal received at port k.

$$V = \sqrt{e^y} = \sqrt{[e^{y_1}, \dots, e^{y_K}]^T} \qquad (12.23)$$

Component $\gamma_{i,j}$ of Γ is given by the statistical expectation operation in (12.24) in which y is expressed in natural logarithms, $Y_k = y_k/\beta$, $\beta = \ln(10)/10$ (Y_k is in dB).

$$\gamma_{i,j} = \mathrm{E}[\sqrt{e^{y_i}}\sqrt{e^{y_j}}] = \mathrm{E}[e^{(y_i + y_j)/2}] \qquad (12.24)$$

The right-hand expression can be viewed as a function $g(y_i, y_j)$ operating on the r.vs and whose expectation can be approximated by [17]

$$\mathrm{E}[g(y_i, y_j)] = g + \frac{1}{2}\left(\frac{\partial^2 g}{\partial y_i^2}\sigma_{y_i}^2 + 2\frac{\partial^2 g}{\partial y_i \partial y_j}\rho\sigma_{y_i}\sigma_{y_j} + \frac{\partial^2 g}{\partial y_j^2}\sigma_{y_j}^2\right) \qquad (12.25)$$

σ_{y_i} and σ_{y_j} are the logarithmic standard deviation of the normal variates, and ρ is the correlation between them. The function g and its derivatives are evaluated at the (logarithmic) mean values μ_{y_i} and μ_{y_j}. With this approximation (12.24) evaluates to

$$\gamma_{i,j} = E[g(y_i, y_j)] = g(\mu_{y_i}, \mu_{y_j})\left(1 + \frac{1}{8}(\sigma_{y_i}^2 + 2\rho\sigma_{y_i}\sigma_{y_j} + \sigma_{y_j}^2)\right) \tag{12.26}$$

where $g(y_i, y_j) = e^{(y_i + y_j)/2}$. Once all the components of Γ have been calculated, the eigen-decomposition produces the a_j and the corresponding eigen, or latent values, λ_j. From these values, the channel assignment procedure calculates two variables, b_r and b_t, which indicate the discrimination that the new (reuse) user can obtain against the existing (test) user (b_r) and similarly for the test user against the reuse user (b_t). Alternatively, they can be seen as evaluating the power intercepted when the other user is transmitting along its *maximum power dimension*, compensated for their different channel power gain. This evaluation accounts for the total signal constellation (on all K ports), their correlation, different mean and variance, and assumes no actual assignment. The assignment necessarily requires that a favorable reuse is decided upon, that is, channel reuse between users with sufficient mutual isolation. Otherwise, the procedure in the previous section for port assignment and power allocation will be unable to find a feasible solution, which allows all ports to stay within the maximum power handling capability. Assuming that the two users require the same power allocation, the decision to reuse the channel of an existing user is based on the simple threshold rule in (12.27). The channel that we assign to the new user is the channel currently assigned to user t^*.

$$t^* = \arg \min_{t \in \Theta^{(b)}} (\max(b_t, b_r)) \wedge \max(b_t, b_r) < H \tag{12.27}$$

H is a discrimination threshold parameter which basically determines how aggressive we should be in reusing channels. Variable t refers to the test user among the set of all currently active users $\Theta^{(b)}$ assigned according to b. Note that the set $\Theta^{(b)}$ possibly includes other reuse users, and that the assignment b may change as more users are introduced to the system. The parameter H sets the minimum acceptable discrimination (maximum of b_t, b_r) and hence determines which (test) users are considered for reuse. Among those, if any, the threshold rule selects in favor of the test user which shows the best mutual discrimination. As such, there is no direct relation between the parameter H and the discrimination, or rejection, which is required to make two users coexist. The influence of H will be demonstrated by simulation.

12.4 SYSTEM SIMULATIONS

Based on the previous algorithms we compare the system capacity, measured by the number of supported voice users, between a system with fixed channel and access port assignment and an adaptive DAS. The capacity is measured based on two blocking mechanisms, referred to respectively as channel and power blocking: Channel blocking is due to a lack of channels while power blocking is due to an insufficient port power handling capability; the latter refers to the case when the total allocated power on a port exceeds the maximum port power handling capability. Either mechanism can be categorized as hard blocking, in the sense that it leads to service denial in the following system simulations. Nominally, the capacity is measured at a total blocking level of 2%

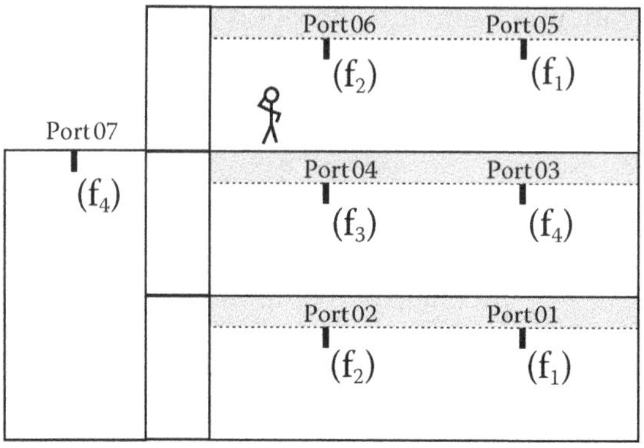

Figure 12.5 Multi-Cell Frequency Assignment in the Site-Specific Setting. The Frequency Assignment Is Shown in Brackets Next to the Port Symbol.

and obtained by Monte Carlo simulation over the user deployment model. It is noted that channel blocking, as it is applied in this study, is not necessarily constrained by the number of channels allocated to the system, but implicitly includes the reuse (strategy) of a particular system configuration.

12.4.1 Simulation Assumptions

Figure 12.5 shows a simplified drawing of the office building. For the comparison of the two systems, a seven port reference configuration has been selected as shown in the figure; the two port positions on the center floor can also be seen in Figure 12.1 in two different configurations. The simulation assumptions are summarized in Table 12.2.

For the fixed reference system, a conventional reuse has been applied based on the assumed total of 4 carriers available for in-building coverage. The GSM-like carriers, with 8 time slots on each, are time slot synchronized so that the only (intrinsic) interference between channels results from finite filtering attenuation. It is assumed that the frequency selection in the mobile terminals can attenuate neighboring frequencies by a minimum of 30 dB. For the multi-cell system, each base station needs to transmit its own cell

Table 12.2 Settings for the System Simulations

	Multi-cell		DAS
Total number of carriers		4	
Total number of common channels	K		1
Total number of traffic channels	$K(8-1)$		$8 \cdot 4 - 1$
SIR target (dB)		9 to 18	
SIR tolerance (dB)		0.25	
Port power handling (dBm)		20	
Relative level for common channels	1		$1/K$
Attenuation between carriers (dB)		> 30	
Active set size	1		K
Noise power level (dBm)		-90 (standard deviation 5 dB)	

Note: The number 8 refers to the number of time slots available on each carrier.

broadcast, while the adaptive DAS uses one common cell broadcast for the whole service area. The multi-cell transmits the common channel at the full power handling capability, while the DAS reduces the transmission power by K on each port; basically, this is the equivalent of the power reduction from cell splitting under free space propagation. The frequency reuse in Figure 12.5 may be considered a rather progressive strategy when only little a priori site information is available. A more conservative layout is possible with a larger frequency allocation where none of the frequencies are reused. The present minimum frequency allocation results in a realistic number of users for an office building of this size and at the same time stresses the limitation on the channel resources.

All systems are interfered by transmissions from external cells. For convenience, this interference is modeled similar to the signals inside the building as a log-normal varying signal. The median external noise level is set at −90 dBm, a number which is in reasonable agreement with measurements taken in a similar office building, while the variability is specified by a standard deviation of 5 dB.

One of the advantages in increasing the number of ports is to get the size and cost benefit of low power port devices. The power handling capability has been set accordingly at a level of 100 mW (20 dBm) per port, or base station. With the available frequencies up to four carriers can be transmitted simultaneously in the DAS, and at different power levels. Only the average power increase from transmitting several simultaneous carriers has been considered in the simulations. The maximum power constraint is handled explicitly after the power allocation and assignment by calculating the total power on each port or base station. If the power handling capability is exceeded on port k due to the introduction of a new user to the system there are several possibilities. Simply, this could be defined as power blocking (multi-cell); alternatively for DASs, a reassignment is attempted of one or several users $i \in \Theta_k$. The reassignment is done by excluding port k from the active set Ψ_i of these users. If, after running a new port assignment and power allocation, the new (and any other existing) allocation stays within the maximum power constraint, the user is allocated to the system and the reassignment has succeeded in "diverting" power to less loaded ports. Otherwise, a power blocking takes place. Depending on the discrimination threshold parameter H, the channel assignment may decide prior to this that there are no more channels available; in this case we get channel blocking. Potential reassignment candidates are ranked according to their absolute uplink power change when moved from the optimum assignment to the second best assignment. The rationale is that the one with the smallest change also generates the smallest increase in the total transmitted downlink power, in accordance with the operation of the BUD algorithm.[4]

In the system simulation, the calculated SIR, ω_i, for a user i, is compared against the corresponding SIR requirement ω_i: $\omega_i(dB) \geq \omega_i(dB) - \delta_i^p(0, std(\omega_i))$ as per Figure 12.4. In the specification of the outage probability, it has been assumed that there are suitable mechanisms to sustain or prevent SIR outage conditions, as they cannot be avoided in a probabilistic way of speaking. For the present study it was decided to use a protection ratio equal to one standard deviation, corresponding to a probability $p = 0.159$. The SIR calculation uses the correlation coefficient established for the residuals in the path loss model, $\rho = 0.4$. This value applies when signals are received from different ports, but not from the same port where they are fully correlated. For the algorithmic implementation, a value close to 1 ($\rho = 0.9$) served to represent this latter case. Different SIR targets are

---

[4] Note that the uplink power is obtained as part of the BUD algorithm.

studied, e.g., the default of 18 dB was chosen because GSM does not typically benefit from microdiversity in the indoor environment, hence amble fading margins must be reserved to cope with small-scale fading.

Apart from the stopping criterion discussed in the previous section, ($T = 40$), an additional criteria based on SIR tolerance has been applied so that the port assignment and power allocation iterations are stopped when the SIR is within tolerance of the target. Initially, a (feasible) number of users is added to the system where later, users are added one-by-one. At each system load approximately 5000 users have been considered, so that many more Monte Carlo simulations are done for low load conditions. It has been verified that this number is large enough to get blocking probabilities within 0.25%. At each Monte Carlo step the user deployment changes according to the model, and so does implicitly the order of appearance of users. In addition to the blocking statistics, statistics relating to channel and port assignment are collected during simulations.

As an alternative to the adaptive DAS and the multi-cell there is a third possibility, namely, a central coverage single-cell system. This system can be configured with transmissions from a single antenna, or alternatively, with transmissions from multiple antennas (passive DAS). In terms of capacity the two are similar, but the passive DAS has the potential to reduce the total transmitted power and the dynamic range. Actual improvements will be commented on in the following section for the selected case study.

12.4.2 System Capacity Results

Initially, simulations are shown for the port configuration indicated by the star marker in Figure 12.1. In this configuration, ports are aligned vertically on the three floors of the building as recommended in [18] for a fixed frequency assignment with reuse between floors. The motivation put forward in [18] was the observation of a strong positive correlation between signals from vertically aligned ports compared to the moderately strong negative correlation between horizontally transposed ports (a similar behavior is observed from the path loss model). Strong positive correlation between desired and interfering signals is desirable from a SIR point of view.

Figure 12.6 shows the blocking performance of the multi-cell reference system. In this case, the blocking performance is determined solely by channel blocking. The result is compared to the multi-cell Erlang-B blocking under uniform load. This assumes the Erlang capacity of a 7 channel trunk (1 carrier with 8 time slots less one reserved for the common channels) multiplied by K. Also shown is the Erlang-B blocking for an equivalent single-cell system with a $7 \times K$ channel trunk. It can be seen that the simulation result emanates from the uniform load blocking curve and ends up along the equivalent single-cell curve. At the blocking levels of interest (1 to 2%) performance is somewhere in between. The user deployment model described in the previous section is not exactly uniform load, but close, and for very high load, users effectively access a trunk of $7 \times K$ channels. Crucial for the deviation from Erlang-B blocking at low load is the fact that Erlang-B assumes an infinite user population with an average offered traffic corresponding to the abscissa in Figure 12.6, wheras the simulations account for a finite user population of M users.

In total, the multi-cell system is allocated 4×8 channels. Hence, as shown in Figure 12.6, at a 2% blocking level it is possible to make use of a significant part of the total channel allocation (27 users corresponding to 85% load in this definition). This is possible of course due to the fixed frequency reuse of the multi-cell system. The multi-cell has, similar to the passive DAS single-cell, a considerable coverage improvement

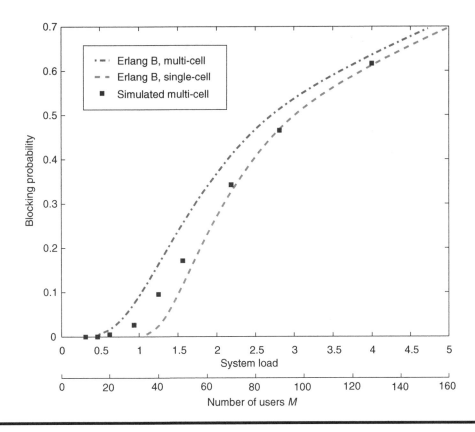

Figure 12.6   Multi-Cell Blocking Behavior for the Frequency Reuse in Figure 12.5. The System Load Is Relative to the Total System Capacity (4 Carriers with 8 time Slots, thus 32 Erlangs).

compared to the single-cell, single-antenna system. Based on the port location indicated on the center floor in Figure 12.1 (circle marker), and a total power handling capability equal to K times the capability of the distributed ports, the single-antenna system transmits at a mean power level 24.8 dB higher than the distributed systems. Furthermore, the dynamic range characterized in terms of the mean standard deviation of the power transmitted from the (distributed) ports is more than 100 times higher in the single-antenna system. From considering idealistic cells under a simple path loss model, the dynamic range improvement is $K^{-\alpha/2}$, where α is the path loss exponent: For $K = 7$ this would imply a path loss exponent in excess of 4.7, a value which is quite reasonable and in the range often quoted for indoor propagation.

The multi-cell hard blocking result is reproduced in Figure 12.7 to compare with the simulation result for the adaptive DAS system. The discrimination threshold parameter H has been set at -10 dB. At a 2% blocking level there is a 35% improvement in the system capacity. Consequently, the (IDAS) system is making an even more efficient use of the total channel allocation (116% load). At 1% blocking, the improvement is 51%. In terms of reuse factor we can define an ideal adaptive DAS system to have a reuse of 1. For the current scenario this implies that all time slots on the four frequencies are reused on each and every port. Accounting for the common channel this makes a total of 217 channels. Clearly, the isolation between ports is not large enough to achieve such a situation. In the multi-cell, because only channel blocking is in effect, we may simply count the number of times a time slot is reused in the fixed frequency assignment; for the reuse

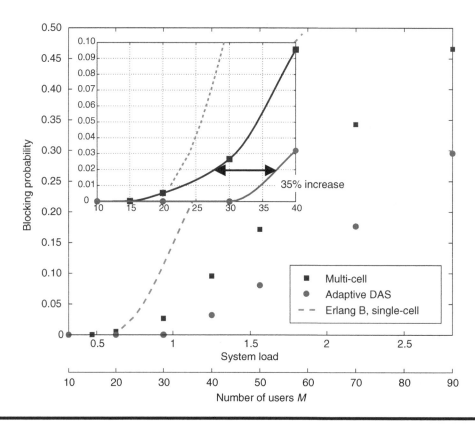

Figure 12.7   Blocking Performance of the Adaptive DAS in Comparison to Multi-Cell. Discrimination Threshold $H = -10$ dB. The Insert Is a Zoom of the 1-2% Blocking Range.

factor this makes a total of 49 channels relative to the 217, which is 0.23. Hence, with a 35% improvement, the DAS achieves a reuse of approximately one-third in this particular case study ($0.23 \times 1.35 = 0.31$). Statistics for the number of simultaneously active carriers show that this result is achieved with only a few carriers transmitted simultaneously: At an offered load of 40 users, 25% of the carried calls require two simultaneously active carriers, and less than 5% calls for three carriers.

Another comparison to make is how effective the systems are in increasing capacity above the trunking capacity. For the multi-cell, the trunking capacity is represented by the Erlang-B, multi-cell curve in Figure 12.6, and for the adaptive DAS by the Erlang-B, single-cell curve in Figure 12.7. At 2% blocking, the multi-cell achieves a 35% increase and the DAS a 65% increase. Part of the difference is from increased trunking efficiency in going from $8 - 1$ to $4 \times 8 - 1$ channels per port, and part is attributed to increased channel reuse. Figure 12.8 shows how the reuse of channels progresses with the system load. For 30 users there is no need to reuse and every channel is used only once. Increasing the offered load beyond 31 users necessitates reuse, and at 50 users, approximately half of the carried calls coexist with cochannel interference from another user; about 20% must cope with 2 cochannel users. These numbers change against more cochannel users as the load is increased, but stabilizes for an offered load somewhere between 70 and 90 users, where the blocking is excessively high. The numbers change somewhat with the discrimination setting as will be seen—the more aggressive the higher reuse.

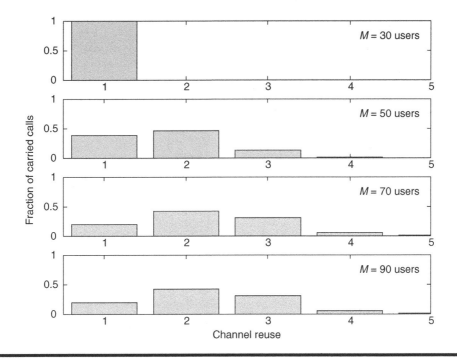

Figure 12.8 Progression of Channel Reuse with System Load in the Adaptive DAS.

The port assignment may change from the (initial) minimum path loss assignment to minimize the total downlink power allocation when more users are introduced to the system. The simulations show that it does not happen that often — only for very high load. Figure 12.9 shows the progression of reassignments with system load. At an offered load of 50 users, approximately half of the Monte Carlo runs had users who stayed with the minimum path loss assignment, while the other half had one or more reassignments of users, most of them with only one reassignment. A partial explanation for the small number of reassignments is the rather small number of reassignment candidates; in each time slot there are only four. Similar to channel reuse, the number of reassignments changes somewhat with the discrimination setting — the more aggressive, the more reassignments.

Whereas the blocking for the multi-cell system is dominated by channel blocking, in the DAS it is a combination of channel and power blocking. Figure 12.10 shows how the blocking in the DAS is partitioned between channel and power blocking for two different values of the discrimination threshold parameter H. Power blocking is shown explicitly by the diamond ($\diamond$) marker while channel blocking results as the difference to the total blocking. At the -10 dB setting used in Figure 12.7 (solid markers in Figure 12.10), channel blocking is the dominant mechanism, though naturally its influence diminishes with system load: the more users that are introduced to the system, the more likely to have admitted (by channel reuse) users for which there is no feasible power allocation. It is possible to trade off between channel and power blocking by being more or less aggressive in reusing channels, that is, increasing or decreasing H. For $H = -7$ dB (open markers in Figure 12.10) power blocking is now dominant, except at low load ($M = 40$ users). There is a decrease in total blocking at 40 users, but an increase at 50 users. This suggests, and is evidenced by results further on, that an aggressive setting can

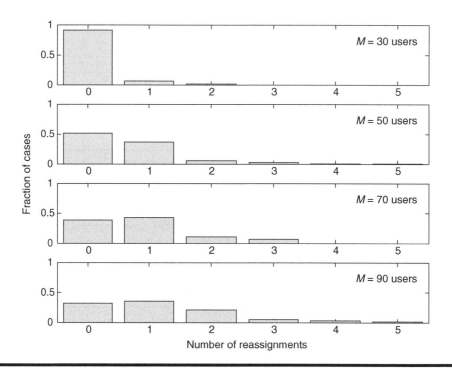

Figure 12.9 Progression of Reassignment from Minimum Path Loss Assignment in the Adaptive DAS.

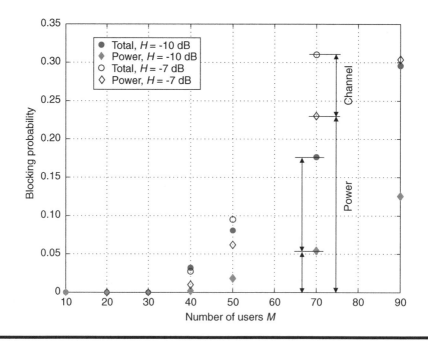

Figure 12.10 Influence on the Blocking Behavior from the Discrimination Threshold Parameter *H*.

change the shape of the blocking curve so as to serve more users at the 1 to 2% blocking level, a ("knee-point") behavior which is well-known from dynamic channel assignment systems [19]. At the same time, this behavior can make the system more sensitive to the number of users in the system. In this case, decreasing H below -10 dB does not add to the capacity because power blocking is almost absent for this setting.

In the following, results will be shown for the alternative port configuration indicated by the square markers in Figure 12.1. Compared to the first, this second configuration increases the isolation between ports, as evidenced from the results in Figure 12.11. The multi-cell performance is identical to the first configuration, except for a slight increase ($< 0.5\%$) from power blocking in the range of 30 to 50 users. In comparison, the capacity gain of the adaptive DAS is now 50%, compared to previously 35%, at the 2% blocking level. With a change of the discrimination threshold to $H = -7$ dB, the gain can be increased by another 5% in the tradeoff between code and power blocking.

It was already mentioned that the user deployment model is close to a uniform distribution of users over the three floors of the building. To see the effect from a nonuniform deployment, users were confined mainly to location cluster Y_2 ($P[Y_2] = 0.8$ and $P[Y_4] = 0.2$). This situation causes the capacity to drop. The multi-cell, in particular, will suffer from increased channel blocking, but also for the DAS there will be less isolation to separate users to reuse channels. The simulated blocking curves are shown in Figure 12.12 along with the corresponding results from Figure 12.11. The nonuniform load is handled satisfactorily in the DAS, partly due to the common channel pool (trunking efficiency). In the extreme case where all users access the same port, the multi-cell will suffer an 89% capacity reduction at 2% blocking compared to 45% for the DAS. For the case depicted in Figure 12.12 capacity drops by 8.3% for the DAS and 14.5% for the multi-cell. The "knee-point" behavior of the DAS curve is clearly

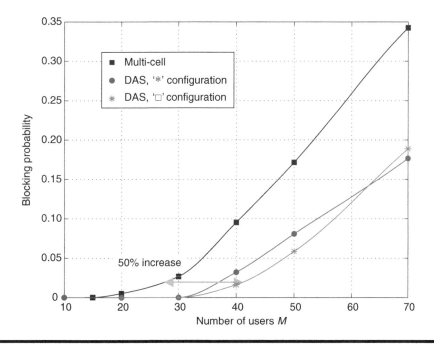

Figure 12.11 Blocking Performance of the Adaptive DAS for Two Different Port Configurations. Discrimination Threshold $H = -10$ dB.

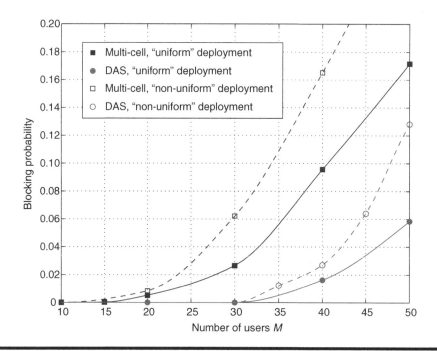

Figure 12.12   Blocking Compared Between Uniform and Non-Uniform User Deployment. Discrimination Threshold $H = -10$ dB.

distinguishable from the multi-cell for the nonuniform user deployment, and can be even further emphasized by increasing H: At 40 users, power blocking is only 0.5% compared to 2.2% for channel blocking; increasing H to -7 dB triples the power blocking, but lowers the total blocking just below the result in Figure 12.12. At an offered load of 35 users, total blocking is more than halved to 0.7%, though the net effect at the 2% level is but small. Overall, Figure 12.12 demonstrates that the DAS system is more robust to changes in the user deployment.

The SIR target selected for the simulations will have an influence on the achievable improvement in the system capacity because, as we have seen, the blocking in the DAS system is a combination of channel and power blocking, where the latter is related to the SIR. Figures 12.13 and 12.14 show the trend with progressively decreasing SIR target down to 9 dB, that is, eight times lower than the default setting of 18 dB. For increased readability, the curves show the result of polynomial interpolation within the range of simulated values (10 to 70 users in steps of 5). In each case, the result is given for different values of the discrimination threshold parameter H to add further to the observations made so far. It can be seen that there is an optimum setting for the threshold parameter, in which the capacity gain relative to the multi-cell system reaches a maximum; for both less and more aggressive settings the gain is lower. We can infer that the optimum setting corresponds to the point in Figure 12.10 where channel blocking has been reduced to its minimum value at the desired blocking level. Furthermore, this optimum setting is biased toward more aggressive settings with decreasing value of the SIR target: $H_{opt} = -10$ dB for SIR target of 15 dB and $H_{opt} = -5$ dB for SIR target of 9 dB. The figures also confirm the tradeoff between offered quality and system capacity, in that a lower SIR requirement leads to higher system capacity.

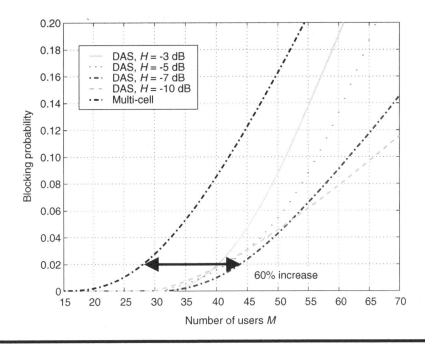

Figure 12.13 Blocking Performance of the Adaptive DAS in Comparison to the Multi-Cell System for a Target SIR of 15 dB and Different Values of the Discrimination Threshold Parameter _H_.

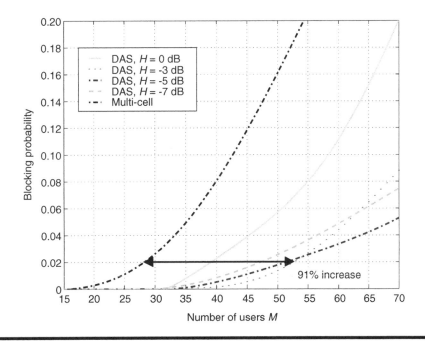

Figure 12.14 Blocking Performance of the Adaptive DAS in Comparison to the Multi-Cell System for a Target SIR of 9 dB and Different Values of the Discrimination Threshold Parameter _H_.

12.5 CONCLUDING REMARKS AND SOME OPEN ISSUES

From the simulation results, there is a clear benefit from an adaptive DAS over a conventional multi-cell system with fixed reuse. The capacity gains achieved are in the same range, but generally lower, than those reported in [4] for a different and less elaborate case study.

It is quite clear that performance is dependent on the port configuration. From a capacity point of view, it is therefore important to plan with maximum isolation between ports, but dynamic effects, such as handover between subcells, can change the situation in favor of configurations with a larger overlap between cells, that is, less isolation. This is a tradeoff to be made in each specific situation; a tradeoff which will likely call for accurate site-specific prediction tools such as those discussed in [20]. It seems however, that because of the trunking efficiency of the DAS, designers will need to pay less attention to changing user distributions.

Implementation issues relating to the distribution of multicarrier signals seem manageable. Each port should have the capability to process four carriers in this case study, but most of the time no more than two carriers are transmitted simultaneously on each port, and at low power levels. It is expected that a limitation to two carriers per port will not significantly change the simulated capacity gains. The low power levels are mainly achieved at the expense of having many ports within the system for which a suitable distribution system is required. While interesting solutions are available (fiber-radio and distributed (synchronized) pico base stations to mention a few) these technological challenges are outside the scope of the present chapter. For future research activities, it is of particular interest to study the distributed pico base station solution in view of the many practical problems that will limit the capacity potential of the adaptive DAS. Particular attention should be given to the availability of measurements - that is, how many, at what update rate, and how accurate - to operate the adaptive DAS with sufficient performance gain compared to other alternative solutions.

12.6 SUMMARY

The simulation results in this chapter were based on a site-specific case study involving a three-story office building. The primary aim was to see to what extent an adaptive DAS system could increase the downlink capacity in a representative office environment under the assumption that only a few carriers were allocated to the system. A GSM-like system was assumed, with users deployed according to a probabilistic user deployment model. The evaluation was based on a generic, but not necessarily practical, framework for the port assignment and downlink power allocation on one side, and dynamic channel assignment on the other.

The results quite clearly demonstrate that the adaptive DAS is superior to a conventional multi-cell system where a fixed frequency assignment has been made. The DAS makes more efficient use of the frequency allocation and can increase capacity by at least 55% compared to the multi-cell at a 2% hard blocking level. The exact configuration of the ports has an influence on the achieved capacity, so that increased isolation between them leads to higher capacity. Nonuniform user deployment is handled satisfactorily in the DAS system, primarily due to its trunking efficiency.

In the analysis, the reuse in the DAS is controlled by a single discrimination threshold parameter, which basically determines how aggressive the system is to be in reusing channels. The results indicate that a very aggressive setting can change the shape of the blocking curve to serve more users at the blocking levels commonly of interest. This is increasingly true for lower and lower SIR targets, which, together with the fact that the DAS is limited by both channel and power blocking (blocking due to insufficient port power handling capability), leads to higher and higher system capacity gains compared to the multi-cell. For a SIR target of 9 dB the capacity gain is as high as 91%.

An extension of the study to include the more practical aspects relating to the implementation of the adaptive DAS was briefly discussed.

REFERENCES

[1] A.J. Motley and D.A. Palmer, "Directed radio coverage within buildings," *Radio Spectrum Conservation Techniques Conference*, Birmingham, U.K., Sept. 6–8, 1983.

[2] K.J. Kerpez, "A radio access system with distributed antennas," *IEEE Trans. on Vehicular Technology*, vol. 45, no. 2, May 1996.

[3] A.A.M. Saleh, A.J. Rustako, and R.S. Roman, "Distributed antennas for indoor radio communications," *IEEE Trans. on Communications*, vol. 35, no. 12, Dec. 1987.

[4] A. Åslundh, M. Frodigh, and K. Wallstedt, "A Performance Comparison of Three Indoor Radio Network Concepts Based on Time Dynamic Simulations," in *Proceedings of Personal Indoor and Mobile Radio Communications (PIMRC) Symposium*, Helsinki, Finland, Sept. 1–4, 1997.

[5] M.V. Clark, T.M. Willis, L.J. Greenstein, A.J. Rustako, V. Erceg, and R.S. Roman, "Distributed Versus Centralized Antenna Arrays in Broadband Wireless Networks," in *Proceedings of IEEE Vehicular Technology Conference (VTC)*, Rhodes, Greece, Spring 2001.

[6] H. Zhuang, L. Dai, L. Xiao, and Y. Yao, "Spectral efficiency of distributed antenna system with random antenna layout," *IEE Electronics Letters*, vol. 39, no. 6, March 2003.

[7] S.V. Hanly, "An algorithm for combined cell-site selection and power control to maximize cellular spread spectrum system capacity," *IEEE Journal on Selected Areas in Communications*, vol. 13, no. 7, Sept. 1995.

[8] R.D. Yates and C.-Y. Huang, "Integrated Power Control and Base Station Assignment," *IEEE Trans. on Vehicular Technology*, vol. 44, no. 3, Aug. 1995.

[9] F. Rashid-Farrokhi, K.J. Ray Liu, and L. Tassiulas, "Downlink power control and base station assignment," *IEEE Communications Letters*, vol. 1, no. 4, July 1997.

[10] T.B. Sørensen, "Intelligent Distributed Antenna Systems (IDAS); Assessment by Measurement and Simulation," Ph.D. dissertation, Aalborg University, ISBN 87-90834-36-4, April 2002.

[11] M. Keenan and A.J. Motley, "Radio coverage in buildings," *British Telecom Technology Journal*, vol. 8, no. 1, Jan. 1990.

[12] J. Zander and M. Frodigh, "Comments on performance of optimum transmitter power control in cellular radio systems," *IEEE Trans. on Vehicular Technology*, vol. 43, no. 3, Aug. 1994.

[13] R.D. Yates, "A framework for uplink power control in cellular radio systems," *IEEE Journal on Selected Areas in Communications*, vol. 13, no. 7, Sept. 1995.

[14] S.C. Schwartz and Y.S. Yeh, "On the distribution function and moments of power sums with log-normal components," *Bell Systems Technical Journal*, vol. 61, no. 7, Sept. 1982.

[15] A. Safak, "Statistical analysis of the power sum of multiple correlated log-normal components," *IEEE Trans. on Vehicular Technology*, vol. 42, no. 1, Feb. 1993.

[16] J. Edward Jackson, *A User's Guide to Principal Components*, New York: John Wiley & Sons, 1991.

[17] A. Papoulis, *Probability, Random Variables, and Stochastic Processes*, New York: McGraw-Hill, 1991.

[18] K.S. Butterworth, K.W. Sowerby, A.G. Williamson, and M.J. Neve, "Influence of Correlated Shadowing and Base Station Configuration on In-Building System Capacity," in *Proceedings of IEEE Vehicular Technology Conference (VTC)*, Ottawa, Canada, May 18–21, 1998.

[19] J. Zander and S.-L. Kim, *Radio Resource Management for Wireless Networks*, Norwood, MA: Artech House, 2001.

[20] A. Aragon-Zavala, B. Belloul, V. Nikolopoulos, and S.R. Saunders, "Accuracy Evaluation Analysis for Indoor Measurement-Based Radio Wave Propagation Predictions," *IEE Proceedings- Microwaves, Antennas and Propagation*, vol. 153, no. 1, Feb. 2006.

13

RF SYSTEM ENGINEERING FOR A CDMA DISTRIBUTED ANTENNA SYSTEM

Craig J. Stanziano and Dang-Jye Shyy

Contents

This chapter describes radio frequency (RF) system engineering, design principles, and field results for a code division multiple access (CDMA) cellular distributed antenna system (DAS). The CDMA DAS has many advantages compared to the traditional macro-cellular system. The advantages include higher capacity, better performance, and a faster and potentially more cost-effective deployment. We present descriptions of DAS characteristics in terms of transport mechanisms from the base station to the distributed antennas, and the system specifications of a DAS. We present the theoretical analysis on propagation modeling, link budget, and capacity budget. We address how to tune the propagation models using drive test results. We also validate our theoretical performance analysis using real field performance measurement results. Also discussed is the design guideline of how to deploy a CDMA DAS based on real-life experiences.

13.1 INTRODUCTION

The CDMA cellular DAS solutions have existed in cellular networks since the early years of cellular telecommunications. As operator networks mature and traditional site selection becomes an increasingly difficult task, the role of DAS is becoming an attractive alternate design solution. Performance benefits from the DAS delivery solution could create the service differentiator from its competitor. Field results depicted in Section 13.8 indicate the potential for enhanced capacity and utilization of deployed resources through the implementation of DAS.

The transport medium for a DAS can take on many forms, from traditional off-air repeaters to a tethered architecture consisting of a coaxial, fiber optic, or hybrid fiber/coaxial based solution. The ability to leverage these outside plant (OSP)/tethered assets allows for the ability to rapidly and cost effectively deploy wireless services.

The hybrid solution with a CDMA protocol will be the focus of this chapter. Today, you find DAS being operated on cable television services (i.e., CATV) OSP, defined as hybrid fiber to coaxial (HFC) networks. The delivery of wireless services on these networks has been fielded since 1997.

The elements attached to the HFC network are defined as donor and remote units. Certain vendors have unique identifiers for these elements. The identifiers specific to the CATV HFC network were referred to as headend interface converter (HIC) (i.e., donor units) and cable microcell integrator (CMI) (i.e., remote units). DAS design is different from macro-cell design in the following aspects: (a) CMI cells represent distributed antenna elements and are interfaced to the base transceiver station (BTS) through the CATV plant and (b) the propagation characteristics of CMI cells are dictated by both building and terrain because the antenna height is at or below the clutter. This characteristic will also be apparent with any remote unit whose antenna heights are mounted similarly.

Field results in Section 13.8 present a compelling case that performance benefits from large scale DAS deployments can offset higher initial build cost expectations.

13.2 DAS CELL CHARACTERISTICS

DAS cells have advantages over traditional cell builds in the following aspects:

- High potential for no zoning permits
- Minimal site acquisition
- Minimal site leasing
- Minimal construction (other than installation)
- Expedited time to market
- Flexibility in adding coverage and capacity where overhead or underground OSP reside
- Cost savings (significant when leveraging existing infrastructure)
- Maximum control of interference levels to in-band microwave links

Typical applications of DAS cells are highway, downtown, residential or zoning avoidance, remote area, coverage extension or infill, and indoor scenarios.

CMI cell coverage is economically difficult to predict due to antenna height being at or below building clutter. Also, coverage of cross streets is traffic dependent due to diffraction caused by various vehicle types (i.e., trucks versus sedans). During RF design, a fair amount of margin must be allocated for the system to absorb these dynamic changes. Margin can be accommodated using overlapping regions between CMI cells, soft/softer handoff between CMI sectors, as well as soft handoff between a CMI sector and a macrocell sector.

Several trends of radio propagation can be observed for DAS cells. Radio propagation specific for CMI cells will be addressed in Section 13.3.

To determine the initial coverage area for DAS cells, the following steps need to be followed:

- To establish an ubiquitous coverage solution between the DAS and traditional (macro) deployed sites a service boundary will need to be established. This boundary is recommended to be referenced to the networks link budget. The link budget thresholds which reflect a service level of in-building performance levels will be used in establishing the service boundary.
- Identify residual coverage deficient areas that require service to be provided.
- Overlay the CATV HFC network OSP onto both of these coverage areas. The coverage areas that reside within the perimeter of the OSP are capable of being serviced with a CATV HFC DAS solution. Fringe coverage area to the OSP can be evaluated for possible extension.
- Engage the Sales and Marketing Team to identify key areas of interest (AOI) within the CATV HFC network OSP and consider these AOI as anchor points for the design.
  - Key areas may include major highways or roads as well as important landmarks such as shopping centers, convention centers, arenas, stadiums, golf courses, gated communities, and airports.
  - Identify in-building coverage and in-vehicle coverage objectives for the area.
- Derive a coverage contour which will be considered the DAS solution boundary (to cover the key areas).
- Submit a composite layout (CATV HFC network OSP, traditional cell build coverage contour, and AOI) for review and concurrence to a network coverage objective.
- Project Management then coordinates with the RF design group, CATV HFC Network OSP group to schedule field survey/continuous wave drive testing of the coverage objective for placement of DAS elements.

The downstream or forward link of CATV is from 54 MHz to 750 MHz and the upstream or reverse link is from 5 MHz to 42 MHz. The allocation of spectrum for different services is described in Table 13.1.

The HIC is an interface between the BTS and CATV plant; it converts the BTS frequency to a downstream CATV transport frequency (54 MHz to 750 MHz) and converts the upstream CATV transport frequency to PCS receive frequency.

The CMI is an interface between CATV plant and PCS antenna; it converts the downstream CATV frequency to a personal communication service (PCS) transmit frequency and converts a PCS receive frequency to an upstream CATV frequency. The CMI is a simplex device and emits its forward link signal through one transmit antenna. The CMI receives its reverse link signal through two spatially separated antennas. In other words, each CMI transmits only one carrier and receives one carrier's diversity pair. The CMI power amplification gain and frequency assignment are controlled by the HIC. The coverage of a sector is determined by simulcasting a cluster of CMI. Co-located at the headend are HICs, headend control unit (HECU), and BTSs.

Table 13.1 CATV Spectrum Allocation

5–15 MHz	155–42 MHz	545–450 MHz	450–500 MHz	500–750 MHz
PCS and Telephony Services	Multimedia Services	Cable Services	PCS and Telephony Services	Multimedia Services

In traditional cellular design, BTS location determines the sector coverage because BTSs and antennas are typically co-located. With the HIC/CMI technology, the sector coverage is extended to where the CATV plant resides and is not constrained by where the BTS is located. Today, terminology used to describe this deployment configuration is known as a "BTS Hotel."

A majority of CMIs will be mounted on the aerial strand that is present in land easements (rights-of-way (ROW)) granted to utility companies. CMI installation needs to adhere to guidelines established by the governing Public Utility Commission or other identified organization. For example, in California there is the California Public Utility Commission (CPUC) General Order No. 95: Rules for Overhead Electric Line Construction.

A CATV node, also referred to as HFC, is an interface between optical fiber and coaxial cable. In the downstream (forward link), the node serves as an optical to electrical converter (for all downstream frequencies) as well as an amplifier. In the upstream (reverse link), the node combines the electrical signals from different coaxial cable branches and converts them into an optical signal with a laser diode.

Due to the tree-and-branch architecture of the CATV plant, the return path is shared among different CMIs. The noise from different CMIs and ingress from the plant is summed at each node. The ingress contains the following noise components: narrowband interference from the air, the return-path impulse noise, and the common mode distortion due to the nonlinearity of the CATV plant.

The CMI maintains the following characteristics, such as power amplifier output at transmit port (34 dBm, i.e., 2.5 W), number of carriers per CMI (1), one transmit antenna and two receive antennas per CMI, nominal CDMA traffic loading (50 %,[1]) maximum number of CMI per node (4), and one CMI noise figure (5 dB).

The HIC maintains the following characteristics, such as number of carriers per HIC (1), maximum number of CMI per one-carrier HIC (24), nominal number of CMI per one-carrier HIC (12), number of sectors per HIC (3), maximum number of HIC per node (3), and total one-CMI/HIC/BTS noise figure should be less than or equal to 8.0 dB (5 dB for CMI, 0 dB for HIC, and 4 dB for BTS).

The following factors determine the number of CMI per node:

- *Ingress Noise:* Different nodes have different ingress noise characteristics. Ingress noise can be managed through preventative maintenance and OSP modification, but both have economical concerns to address.
- *Node Density in an Area:* If node density in an area is high, there is no reason to reuse the same node unless one particular node has better ingress noise characteristics. Design should minimize multiple CM, per node where possible.
- *Differential Delay:* When reusing the same node for multiple simulcasting CMIs or using simulcasted nodes, measurement of the differential delay between these CMIs is required. The minimum differential delay between CMIs is 1 chip, but BTS manufacturers recommend that the differential delay should be greater than 2.5 chips duration. One chip duration is defined as $1/1228800 Hz = 0.814$ μs. (For the RAKE receiver to be able to distinguish different Rayleigh multipaths, the differential delay between two multipaths must be greater than one chip

---

[1] Due to the CATV laser dynamic range, a 50% capacity loading is used to be conservative. Higher loading (85 to 95%) can be realized once multimedia operations are confident of experiencing no impact.

Table 13.2   CDMA Reverse Link Linear Dynamic Range Requirement

Noise Components	Requirement	Residual Dynamic Range
Minimum receive signal level above ingress noise	10 dB	25.0 dB
User loading factor or noise rise (50%)	3 dB	22.0 dB
Waveform factor (PA peaking)	10 dB	12.0 dB
No. of receive antenna per CDMA carrier (2)	3 dB	10.0 dB
Max. no. of CDMA carriers (4)	6.02 dB	3.98 dB

duration.) If this condition cannot be met, one of the following approaches needs to be taken: place the CMI site at a different node or insert artificial cable delay for the CMI site. It is also recommended the *maximum differential delay* for a simulcast CMI be less than 90 μs; the *maximum differential delay* for simulcast CMI of different sectors be less than 240 μs.

■ *Dynamic Range:* A node has typically 50 dB of dynamic range. Actual dynamic range has been observed to be 35 dB (above the ingress noise floor). The dynamic range is shared among the following noise components: currently, the minimum CDMA reverse link receive signal level is set 10 dB above ingress noise floor (to maintain 1% frame error rate (FER) under no loading condition), the sector loading margin, CDMA waveform peak to average factor[2] (10 dB) to absorb the spikes when they occur, number of receive antenna per CDMA carrier, and maximum number of CDMA carriers. The number of CMIs per node depends on the remaining dynamic range after all the noise components have been considered. An example is given in Table 13.2 to illustrate this concept. From the above analysis, there is 3.98 dB left for the noise generated from the number of CMI per node if 10 dB above ingress noise floor is assumed. The noise generated by n CMIs per node is $10\log_{10}(n)$. Therefore, the maximum number of CMI per node is 4. The residual dynamic range is shared over all CATV services but potential for additional capacity loading is available.

A diagram showing simplified HIC/CMI cell configuration is shown in Figure 13.1.

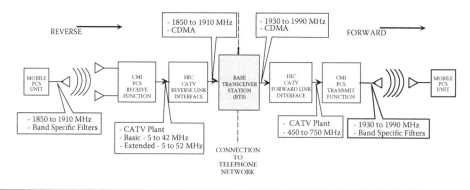

Figure 13.1   HIC/CMI Cell Configuration.

---

[2] This 10 dB is a conservative application and improved CATV plant ingress management will allow for a reduction once confidence is established that wireless operations have no impact on Wireline and Multimedia services.

13.3 DAS CELL OUTDOOR PROPAGATION AND LINK BUDGET

13.3.1 DAS Cell Outdoor Propagation

This subsection presents a DAS propagation model where the major obstruction to wave propagation is due to building clutter. (In this case, orientation and height of buildings, as well as their associated street layout, dictate propagation characteristics because buildings reflect, diffract, and shadow radio waves.) Drive test data is used to tune the coefficients associated with the model. The model is used for two purposes in the design process. First, it can be used to compute average cell radius based on the maximum allowable path loss determined from the link budget. Second, it can guide the RF engineer in placement of the initial CMI sites by predicting CMI cell coverage. Propagation characteristics for a DAS cell must be addressed from two aspects: line-of-sight (LoS) and nonline-of-sight (NLoS) propagation. The LoS condition is defined as point A (e.g., BTS antenna) and can visually see point B (e.g., mobile subscriber (MS) antenna).

13.3.1.1 Line-of-Sight (LoS) Propagation

There are two typical antenna locations: at an intersection or mid-street. If the antennas are mounted at the intersection of two streets, the antennas are considered to be LoS to those two streets. In the LoS condition, when the MS is close to the antennas, the path loss exponent is approximately two (free space loss). A path loss exponent of two means that there is a 20 dB loss whenever the path distance (between transmitter and receiver) is increased ten-fold. When the MS is away from the antennas, the path loss exponent becomes four (and as high as eight). (The LoS propagation is determined by the direct path and reflected path from the ground.) There is a break point which separates these two propagation path loss exponents. The break point is either based on where the first Fresnel zone touches ground or touches terrain. This distinctive phenomenon shows that it is not appropriate to perform a regression fit using only one slope and one intercept point because the standard deviation would be large. A two slope regression fit should be implemented to make sure that the two best-fit lines intercept at the break point. The formula for the break point distance using the first Fresnel zone is

$$D_{break} = 4 \ h_{CMI} \ h_{MS}/\lambda \tag{13.1}$$

where h_{CMI} is the effective antenna height of a CMI cell (in meters), h_{MS}: effective antenna height of MS (1.5m), and λ is the wavelength (0.156 m).

For example, for a 7.31 m (24 feet) antenna height of a CMI cell, the distance of the break point is 281.15 m. Break point distance versus CMI cell antenna height curve is shown in Figure 13.2.

As for the LoS propagation model, the lower bound for the loss is:

$$Prop\_Loss = L_b + 20 \ \log_{10}(d/D_{break}), \quad \text{if } d \leq D_{break} \tag{13.2}$$

$$Prop\_Loss = L_b + 40 \ \log_{10}(d/D_{break}), \quad \text{if } d > D_{break} \tag{13.3}$$

$$L_b = -20\log_{10}(\lambda^2/(8\pi h_{CMI}h_{MS}) \tag{13.4}$$

The upper bound for the loss is:

$$Prop\_Loss = L_b + 20 + 25 \ \log_{10}(d/D_{break}), \quad \text{if } d \leq D_{break} \tag{13.5}$$

$$Prop\_Loss = L_b + 20 + 40 \ \log_{10}(d/D_{break}), \quad \text{if } d > D_{break} \tag{13.6}$$

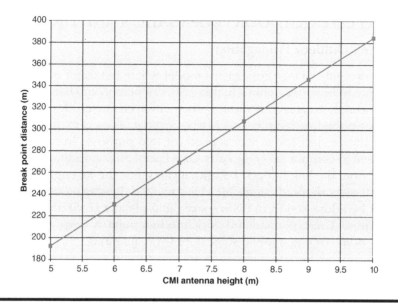

Figure 13.2 Break Point Distance (m) vs. CMI Antenna Height (m).

For example, for a 7.31 m (24 feet) antenna height ($L_b = 81.1$ dB), the lower bound and upper bound propagation loss versus distance curves are shown in Figure 13.3.

13.3.1.2 Nonline-of-Sight (NLoS) Propagation

The NLoS propagation exhibits the following characteristics:

■ There is a sudden drop of the signal strength (20 to 25 dB loss) when the MS turns onto a NLoS street.

■ Loss due to corner turning has a linear effect, i.e., if you turn two corners into a street (which is not LoS with the transmitter), the loss will be 40 to 50 dB.

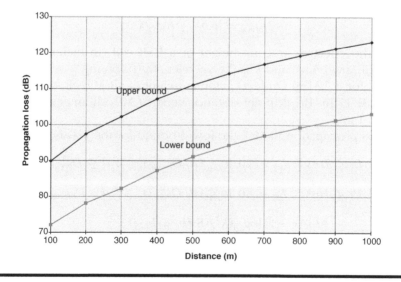

Figure 13.3 The Lower Bound and Upper Bound of Propagation Loss (dB) vs. Distance (m) for LoS Propagation.

- Path loss exponent for NLoS street is higher than that for the first half (d < D_{break}) of LoS street.
- Power drop and path loss exponent depends on street widths and distance from the transmitter to the turning corner. When distance to the turning corner increases, the level of power drops due to corner turning and the path loss exponent for NLoS increases.
- Signal strength is higher for a wider street than for a narrower street.
- Signal strength also depends on the amount of traffic at intersections.

NLoS propagation loss can be specified as follows [2]:

$$Prop\_Loss = L_f + A + 10 B \log_{10}((d_1 + d_2)/d_1) + correction\ factor\ (cf), \qquad (13.7)$$

where:
$L_f = -20 \log_{10}(\lambda/(4\pi d_1))$
A: Loss due to corner turning within 10 to 15 m (20 to 25 dB)
B: Slope of loss in NLoS street [referenced to $(d_1 + d_2)/d_1$]
d_1: Distance from transmitter to corner
d_2: Distance from corner to receiver along NLoS street
cf: Used to tune the model for different morphologies

The above analysis tells us that to have proper coverage for a NLoS street, the selection of d_1 (the distance from the CMI site to the corner) is very important.

An example using a perfect grid street layout is shown in Figure 13.4 and Figure 13.5. The street block is assumed to have a square shape and the length is 200 m. In the previous equation: A is assumed to be 20 dB and B is 4. Assume the effective isotropic radiated power (EIRP) is 39 dBm. Based on the link budget result in Section 13.3.2, the signal level threshold for suburban in-building is −76.9 dBm (using 4 CMIs simulcasting and 6.0 dBi antennas). The dotted line shows the contour for the in-building signal level threshold. LoS radius is about 0.9 km and the NLoS is about 0.25 km. The general conclusion for DAS cell design is that the MS should be, at most, one turn away from the LoS street.

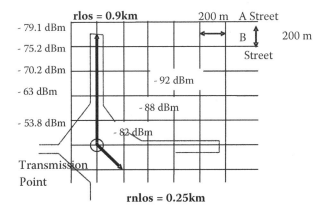

Figure 13.4   Street Grid with Signal Strength at LoS and NLoS for 6 dBi Antenna at Intersection.

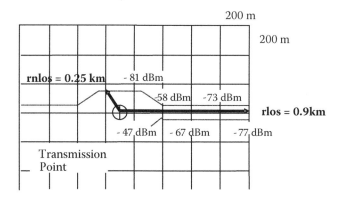

Figure 13.5   Street Grid with Signal Strength at LoS and NLoS for 6 dBi Antenna at Non-intersection.

13.3.1.3   Design Procedures

The procedure of estimating DAS cell coverage is as follows:

- Examine street map (or examine aerial photography) and pick a potential CMI cell site within the CATV node coverage ring[3] (of 0.5 km) applying the following coverage prediction rules:
  - If there are two LoS streets (the antenna is mounted at an intersection), the coverage shape is concave diamond. If there is one LoS street (the antenna is mounted between intersections), the coverage area has a bat shape.
  - Estimate cell radius (r_{los}) for the four LoS tails.
  - Estimate cell radius (r_{nlos}) between two LoS tails. Estimate A and B for NLoS propagation loss equation (based on street layout) and adjust the correction factor based on the morphology type.[4]
  - Estimate concave diamond or bat shape using r_{los} and r_{nlos}.
- Perform a field survey and drive test for each CMI site (create a form to manage this data). If necessary, readjust the site location during field survey based on the following factors:
  - Would the terrain affect the predicted coverage?
  - The CMI site should not be obstructed by any large structure (e.g., tall buildings or water tanks) or vegetation to the desired coverage area.
  - The CMI site should not be LoS with any other CMI site[5] (to avoid excessive overlap area).

---

[3] There are two reasons for using 0.5 km as the node coverage ring. First, the typical CMI cell radius is about 0.3 km. Second, when CMI is colocated with a fiber node, ingress noise can be reduced; consequently, reverse link dynamic range requirement of laser diode can be maintained as well as filtering of ingress noise can be maximized and need for minimal OSP modifications.

[4] A and B are calculated using MATLAB to process CMI measurement data (for different combinations of street width and street orientation). Effect of different morphologies will be used to adjust the correction factor. Morphology types are categorized based on average building height, flat area, hilly area, and vegetation.

[5] Due to the concave diamond shape of CMI cells, two neighbor CMI cells should not reside on the same street; they should be offset by at least one street. However, two CMI neighbor cells can be close diagonally.

- The CMI site should not be LoS with any other macrocell site (to avoid excessive interference).
- Make sure that the CMI site can see the expected handoff buildings which are covered by neighbor CMI sectors or neighbor macrocells.
- Make sure the CMI site has a good near-field characteristics (free of obstruction).

■ Use the following guidelines when processing and analyzing drive test data:
- Any holes in CMI coverage must be in areas with no in-building traffic zone (such as streets or outdoor).
- CMI cell location must be adjusted to provide enough simulcasting region (30%) with neighbor CMI cells. CMI cell location must be adjusted to provide enough (30%) handoff area with neighbor macrosectors. To maintain soft handoff area not only guarantees the anticipated diversity gain required at the cell edge, it also prevents excessive interference from the reverse link.
- CMI cells should avoid unwanted handoff areas at the intersection of two LoS tails generated from two CMI sectors.
- While performing these coverage analyses, do not forget to preserve the in-building coverage.

For the r_{los} and r_{nlos}, we have

$$r_{nlos} < (2\ (r_{los})^2)^{0.5} \tag{13.8}$$

It is conjectured that

$$r_{nlos} \geq (2\ (D_{break})^2)^{0.5} \tag{13.9}$$

Aerial photography of DAS cells should be obtained. Analysis of the CMI cell coverage and the corresponding photo intelligence will provide estimation for other CMI cells' coverage if they share similar characteristics. The drive test data can be used for:

■ Propagation model tuning (Drive test data can be used to tune LoS and NLoS propagation models.)
■ CDMA performance analysis

13.3.1.4 *Drive Test Data Samples*

To properly plan for the number of CMI and BTS required for a particular design solution, the area should be divided by the headend OSP boundaries. Drive testing should be performed to characterize the different morphologies. In this subsection, the average CMI coverage area (km^2) for a simulated headend service area is described.

CMI sites will be classified as suburban morphology types. The signal level thresholds for in-building, in-vehicle, and outdoor coverage are listed below. The derivation of maximum propagation loss for different morphologies is discussed in Section 13.3.2. The maximum propagation loss is a function of antenna gain. As discussed previously, three types of antenna gain are available: 5 dBi, 6 dBi and 8 dBi omnidirectional antenna. The length of an 8 dBi antenna does not satisfy the general order (GO) 95 strand mounting standard. The length of 5 dBi and 6 dBi antennas is about the same and they both satisfy the GO 95 strand mounting standard. Therefore, 6 dBi antennas should always be used in CMI design. The following signal level thresholds are derived using 6 dBi antennas.

Table 13.3   Average In-Building CMI Cell Area of a Hypothetical Headend Service Area

Headend	In-Building Area (km²)	In-Building Area with Overlap (km²)	In-Building Theoretical Radius (km)
A	0.704	0.577	0.47

- In-Building signal level threshold $= 39$ dBm $- 121.9 + 10\log_{10}(4) = -76.88$ dBm
- In-Vehicle signal level threshold $= 39$ dBm $- 133.9 + 10\log_{10}(4) = -88.88$ dBm
- Outdoor signal level threshold $= 39$ dBm $- 138.5 + 10\log_{10}(4) = -93.48$ dBm

The results of the sample test that will allow for the average in-building and in-vehicle CMI cell area (using four CMIs per sector) at different headend services are shown in Tables 13.3 and 13.4.

Table 13.5 shows total in-building and in-vehicle coverage area of the hypothetical headend service area. Because the average CMI cell area is known (from Tables 13.3 and 13.4), the number of CMIs and BTS for the hypothetical headend service area can be derived; they are listed in Table 13.5.

13.3.2   DAS Cell Link Budget

The link budget can be used to predict maximum propagation loss for a balanced CMI sector, and consequently, the cell radius can also be estimated. But the cell shape should not be assumed to be circular, because the cell shape is highly dependent on the local environment. The link budget can also be used to obtain signal threshold levels for in-building, in-vehicle, and outdoor coverage for different morphologies. With 2.5 W CMI output power, the CMI EIRPs are 39, 40, and 42 dBm for three types of CMI antenna with gains of 5 dBi, 6 dBi, and 8 dBi, respectively. (*Note:* CMI power amplifier output is a function of simulcasting the CMI.) The formula for computing signal level threshold is listed below.

$$Signal\_Level\_Threshold = EIRP_{cmi} - Max\_Rev\_Prop\_Loss + Ant\_G_m - Line\_Loss_m$$

$$= EIRP_{cmi} - Max\_Rev\_Prop\_Loss \qquad (13.10)$$

Examples of signal level thresholds for different morphologies are shown in Table 13.6. The CMI transmit power is 2.5 W.

$$x = 10\log_{10}(n) \qquad (13.11)$$

where n is the number of simulcasting CMI cells.

Table 13.4   Average In-Vehicle CMI Area of a Hypothetical Headend Service Area

Headend	In-Vehicle Area (km²)	In-Vehicle Area with Overlap (km²)	In-Vehicle Theoretical Radius (km)
A	2.483	2.04	0.89

Table 13.5   CMI and BTS Quantity for a Hypothetical Headend Service Area

Headend	In-Building Coverage Area (km²)	In-Vehicle Area Coverage Area (km²)	Population Covered (Census Data)	No. of CMI	No. of 3-Sector BTS
A	131.79	195.31	301,697	325	28

Derivation of the maximum propagation loss using the link budget is addressed later in this section. Table 13.6 provides this derivation solution with the effects of a varying simulcast solution. The following is an example calculation for a four simulcast suburban scenario:

$$Sig\_Lvl\_Threshold_{suburban} = +40.0\text{dBm} - 119.9\text{dB} + 6.0\text{dBi} - 0.0\text{dB}$$

$$= -81.9\text{dBm} + 10\log_{10}(4)$$

$$= -75.9\text{dBm} \tag{13.12}$$

Before discussing the CMI link budget, a comparison of differences between macrocells and CMI cells is necessary. Empirically, the significance of the maximum mean propagation loss differential would deter a design engineer from pursuing DAS as a solution. However, the effectiveness of strategic CMI placement ensures that the majority of MSs are being served at signal strengths above margins with a reduction in the other cell and MS interference from controlled propagation.

Tables 13.7 and 13.8 illustrate typical CDMA network design link budget criteria in both a macro and CMI scenario. The tables provide the design engineer the ability to understand the differentials that exists between the two architectures.

There are several issues for the CMI link budget:

- The fade margin may be different from that in macrocell because the slow fading characteristics are different. For a macrocell, slow fading is typically caused by terrain. For CMI cells, slow fading is typically caused by buildings.
- Because frequency reuse efficiency is highly dependent on the path loss exponent, the method described in Section 13.4 needs to be followed to derive frequency reuse efficiency for CMI.

Table 13.6   Signal Level Thresholds for In-Building, In-Vehicle, and Outdoor Coverage Per Morphology with Diversity Receive Antennas at 40 dBm EIRP

Morphology	In-Building	In-Vehicle	Outdoor
Dense Urban (dBm)	$-71.9 + x$	$-93.9 + x$	$-100.9 + x$
Urban (dBm)	$-78.9 + x$	$-93.9 + x$	$-100.9 + x$
Suburban (dBm)	$-81.9 + x$	$-93.9 + x$	$-100.9 + x$
Rural (dBm)	N/A	$-93.9 + x$	$-100.9 + x$

Table 13.7   Reverse Link Difference between Macro Cell Link Budget and CMI

Parameters	Macro Cell	CMI Sector
Cell Antenna Gain	16 dBi	5 dBi, 6 dBi, 8 dBi
Cell Loss	2 dB	0.0 dB
Cell Noise Figure	5 dB	8.5 dB + $10 \log_{10}$(No. of simulcasting DAS/CMI)
Interference Margin	−4.6 dB (DU, U, S)	−3dB
	−3 dB (*Rural*)	

Note: Cell loss is to account for cable loss from CMI to antenna and is assumed to be negligible. BTS noise figure is equal to noise figure of CMI plus $10 \log_{10}$(# of simulcasting DAS/CMI).

- Due to constant change of cell shape (resulting from multipath and shadowing caused by moving vehicles), it may be difficult to control the handoff region. More handoff percentage needs to be allocated to CMI cells than specified.
- Because the percentage of soft/softer handoff is larger for CMI cells, the power control factor is larger and the percentage of traffic channel overhead also becomes larger. The sector may become forward link limited.
- Because a CMI cluster (sector) provides higher multipath diversity gain, $(E_b/N_t)_{sp}$ can be reduced, where *sp* stands for set point. In macrocell softer handoff, a single channel element is responsible for simulcasting voice frames to multiple sectors. The softer simulcasting is very similar to CMI simulcasting. It is expected that the forward and reverse link simulcasting diversity gain is the same as the softer diversity gain. However, because the link budget always assumes the worst case, i.e., MSs are at the cell edge, simulcasting diversity gain for MSs at the cell edge may be smaller than those near the cluster center. In addition, $(E_b/N_t)_{sp}$ is assumed to have no improvement in link budget analysis. But in reality, $(E_b/N_t)_{sp}$ is different for MSs at different locations of a simulcasting cluster.

13.3.2.1   Reverse Link Budget

The reverse link budget for one CMI simulcasting sector is discussed below. As shown in Table 13.9, the fade margin for rural morphology is different from that for dense urban,

Table 13.8   Forward Link Difference between Macrocell Link Budget and CMI Sector Link Budget

Parameters	Macrocell	CMI Sector
Total Cell Transmit Power	16 W	2.5 W
Number of Active MS	13	10
Number of Active Channels	24	19
Cell Antenna Gain	16 dBi	5 dBi, 6 dBi, 8 dBi
Cell Loss	2 dB	0.0 dB
Max. Mean Propagation Loss (for *Suburban* In-Building)	137 dB	115.9 dB

Note: Macrocell uses 65% loading and CMI cell uses 50% loading. The maximum mean propagation loss is for 6 dBi antenna and four simulcasting CMI.

Table 13.9 CMI Simulcasting Reverse Link Budget per Morphology

REVERSE LINK		MORPHOLOGY		
Parameter	Descriptions	Dense Urban	Light Urban	Suburban Rural
MS Tx Power	23 dBm	23 dBm	23 dBm	23 dBm
MS Tx Power	0.2 W	0.2 W	0.2 W	0.2 W
MS Net Antenna Gain	0.0 dBi	0.0 dBi	0.0 dBi	0.0 dBi
Body Loss	−2.0 dB	−2.0 dB	−2.0 dB	−2.0 dB
MS EIRP	21.0 dBm	21.0 dBm	21.0 dBm	21.0 dBm
MS EIRP	0.1 W	0.1 W	0.1 W	0.1 W
Fade Margin	−8.0 dB	−8.0 dB	−8.0 dB	−5.0 dB
Soft Handoff Gain	4.0 dB	4.0 dB	4.0 dB	3.0 dB
Sector Loading	50%	50%	50%	50%
Receiver Interference Margin	−3.0 dB	−3.0 dB	−3.0 dB	−3.0 dB
Building Penetration Loss	−25.0 dB	−18.0 dB	−15.0 dB	−5.0 dB
BTS Rx Antenna Gain	5.0 dBi	5.0 dBi	5.0 dBi	5.0 dBi
BTS Cable Loss	0.0 dB	0.0 dB	0.0 dB	0.0 dB
kT	−174.0 dBm/Hz	−174.0 dBm/Hz	−174.0 dBm/Hz	−174.0 dBm/Hz
BTS Noise Figure	8.5 dB	8.5 dB	8.5 dB	8.5 dB
Baud Rate (14.4 kbps)	41.6 dBHz	41.6 dBHz	41.6 dBHz	41.6 dBHz
$(E_b/N_t)_{sp}$	7.0 dB	7.0 dB	7.0 dB	7.0 dB
BTS Rx Sensitivity (14.4 kbps)	−113.9 dBm	−113.9 dBm	−113.9 dBm	−113.9 dBm
Maximum Reverse Prop. Loss	**110.9 dB**	**117.9 dB**	**120.9 dB**	**132.9 dB**

Note: The in-building penetration loss for rural morphology is intended for in-vehicle penetration loss because the 90% area reliability for rural morphology is for in-vehicle, not for in-building. $(E_b/N_t)_{sp}$ of 7 dB is derived considering a receive diversity antenna gain of 4 dB.

light urban, and suburban because the former is used for in-vehicle coverage and the latter is used for in-building coverage. Also, the soft handoff diversity gain is 3 dB for rural morphology and 4 dB for other morphologies. The reason is that the reverse soft handoff diversity gain is a function of standard deviation of the slow fading process. When one traffic channel fades, and if standard deviation of the slow fading process is large, the diversity gain provided by the other nonfaded traffic channel is large compared with another slow fading process with a smaller standard deviation.

The formula for sensitivity (S) used in the reverse link budget is expressed below:

$$S = (E_b/N_t)_{sp} + N_0 W - 10\log_{10}(W/R) - 10\log_{10}(1 - x)$$

$$S = EIRP_{m;max} - Body\_Loss_m + Ant\_G_b - Line\_Loss_b - Fade\_Margin$$

$$+ Soft\_Rev\_HO\_Gain - Max\_Rev\_Prop\_Loss - Building\_Loss \qquad (13.13)$$

$$N_0 W = 10\log_{10}(KTW) + Noise\_Figure + 10\log_{10}(n) \qquad (13.14)$$

where x and n are the sector loading and the number of simulcasting CMIs.

Table 13.10 Values Used for Penetration Loss and Fade Margin for Dense Urban

Dense Urban	In-Building	In-Vehicle	Outdoor
Penetration Loss (dB)	25	5	0
Fade Margin (dB)	8	6	4

The values used in the link budget to compute the in-building, in-vehicle, and outdoor coverage are listed in Tables 13.10 through 13.13.

13.3.2.1.1 In-Building Coverage Maximal Propagation Loss

The formula for maximum reverse path loss for in-building coverage versus number of CMI cells for different morphologies is listed below (6.0 dBi antenna gain):

$$Max\_Rev\_Prop\_Loss(Dense\ Urban) = 111.9 - 10\log_{10}(n)\text{dB} \qquad (13.15)$$

$$Max\_Rev\_Prop\_Loss(Light\ Urban) = 118.9 - 10\log_{10}(n)\text{dB} \qquad (13.16)$$

$$Max\_Rev\_Prop\_Loss(Suburban) = 121.9 - 10\log_{10}(n)\text{dB} \qquad (13.17)$$

where n is the number of simulcasting CMI cells (1–16).

13.3.2.1.2 In-Vehicle Coverage Maximal Propagation Loss

The formula for the maximum reverse propagation loss for in-vehicle coverage versus number of CMI cells for different morphologies is listed below (6.0 dBi antenna gain):

$$Max\_Rev\_Prop\_Loss(Dense\ Urban) = 133.9 - 10\log_{10}(n)\text{dB} \qquad (13.18)$$

$$Max\_Rev\_Prop\_Loss(Light\ Urban) = 133.9 - 10\log_{10}(n)\text{dB} \qquad (13.19)$$

$$Max\_Rev\_Prop\_Loss(Suburban) = 133.9 - 10\log_{10}(n)\text{dB} \qquad (13.20)$$

$$Max\_Rev\_Prop\_Loss(Rural) = 133.9 - 10\log_{10}(n)\text{dB} \qquad (13.21)$$

Table 13.11 Values Used for Penetration Loss and Fade Margin for Light Urban

Dense Urban	In-Building	In-Vehicle	Outdoor
Penetration Loss (dB)	18	5	0
Fade Margin (dB)	8	6	4

Table 13.12 Values used for Penetration Loss and Fade Margin for *Suburban*

Dense Urban	In-Building	In-Vehicle	Outdoor
Penetration Loss (dB)	15	5	0
Fade Margin (dB)	8	6	4

13.3.2.1.3 Outdoor Coverage Maximal Propagation Loss

The formula for the maximum reverse propagation loss for outdoor coverage versus number of CMI cells for different morphologies is listed below (6.0 dBi antenna gain):

$$Max\_Rev\_Prop\_Loss(Dense\ Urban) = 140.9 - 10\log_{10}(n)\text{dB} \qquad (13.22)$$

$$Max\_Rev\_Prop\_Loss(Light\ Urban) = 140.9 - 10\log_{10}(n)\text{dB} \qquad (13.23)$$

$$Max\_Rev\_Prop\_Loss(Suburban) = 140.9 - 10\log_{10}(n)\text{dB} \qquad (13.24)$$

$$Max\_Rev\_Prop\_Loss(Rural) = 140.9 - 10\log_{10}(n)\text{dB} \qquad (13.25)$$

13.3.2.2 Forward Link Budget

The forward link budget is discussed below. The principle is to balance forward and reverse links using the *Max_Rev_Prop_Loss*. Determination of power allocation for pilot, sync, paging, and traffic channels is performed. The forward and reverse links are balanced if

$$(E_b/I_t)_{achieved} = (E_b/I_t)_{sp} \qquad (13.26)$$

$$(E_c/I_t)_{achieved} = (E_c/I_t)_{sp} \qquad (13.27)$$

If $(E_b/I_t)_{achieved} > (E_b/I_t)_{sp}$ or $(E_c/I_t)_{achieved} > (E_c/I_t)_{sp}$, it is reverse link limited. CMI transmission power needs to be reduced to balance the links. If $(E_b/I_t)_{achieved} < (E_b/I_t)_{sp}$ or $(E_c/I_t)_{achieved} < (E_c/I_t)_{sp}$, it is forward link limited. CMI transmission power needs to be increased to balance the links.

The set point values for different forward channels are listed in Table 13.14.

Because reverse link coverage shrinks with more simulcasting CMIs, the forward link transmit power needs to be adjusted accordingly such that forward and reverse links can be balanced.

The actual forward link budget is omitted here due to page limitation.

Table 13.13 Values Used for Penetration Loss and Fade Margin for Rural

Dense Urban	In-Building	In-Vehicle	Outdoor
Penetration Loss (dB)	N/A	5	0
Fade Margin (dB)	N/A	5	4

Table 13.14   Set Point Values for Different Forward Channels

$(E_b/I_t)_{sp}$ for traffic channel	7 dB
$(E_b/I_t)_{sp}$ for pilot channel	−13 dB
$(E_b/I_t)_{sp}$ for sync channel	balanced with pilot (13 dB)
$(E_b/I_t)_{sp}$ for paging channel	balanced with pilot (10 dB)

13.4   DAS CELL REVERSE LINK CAPACITY

The CMI cluster sector capacity needs to be computed for the reverse link such that the capacity requirement for a given area can be determined to support traffic demand in year 2005. There are two representations of CDMA capacity: soft Er capacity and maximum number of active users (N_{max}). The soft capacity formula proposed by Viterbi cannot be used by BTS manufacturers for two reasons. First, Viterbi uses I_0/N_0 as the blocking level, but BTS manufacturers use sector loading as the blocking level. Second, the Viterbi formula only has one blocking level (for new calls) to compute soft capacity, but BTS manufacturers have two soft blocking levels (one for new calls and the other for handoff calls). The N_{max} formula is used to compute the maximum number of active users, and then convert the N_{max} to the number of Er using the Er B model. The grade of service (GOS) is 2%. The formula for the N_{max} (for one carrier) is given below.

$$N_{max} = \eta[1 + F\ PG/((E_b/I_t)_{sp}\ \rho], \text{ where} \qquad (13.28)$$

η: Sector efficiency,
F: Frequency reuse efficiency,
PG: Processing gain (85.33),
ρ: Reverse link voice activity factor (0.403),
$(E_b/I_t)_{sp}$: Reverse link (E_b/N_t) set point.

Unlike in macrocell design, a CMI sector is formed by simulcasting n CMI. The CMI sector efficiency is determined by the overlap area between two CMI cluster sectors, not by the antenna pattern. Therefore, the RF engineer can create any sector efficiency they would like to have as long as a softer handoff condition is appropriate. As an example, for softer handoff of 20% overlapping, the sector efficiency is 80%.

Frequency reuse efficiency is a function of propagation characteristics (such as path loss exponent). The definition of frequency reuse efficiency is:

$$F = (own\ cell\ interference/(own\ cell\ interference + other\ cell\ interference)) \quad (13.29)$$

Typical values of frequency reuse efficiency versus path loss exponent are shown in Figure 13.6.

Theoretically, the number of simulcasting CMIs will not affect frequency reuse efficiency (assuming that propagation characteristics do not change much when all simulcasting CMIs for a sector are in the same morphology). Hence, frequency reuse efficiency is not a function of the number of simulcasting CMIs.

The reverse link (E_b/N_t) set point for different mobility (with receive antenna diversity) is listed in Table 13.15.

Differences between macrocell capacity and CMI sector capacity for reverse links are listed in Table 13.16.

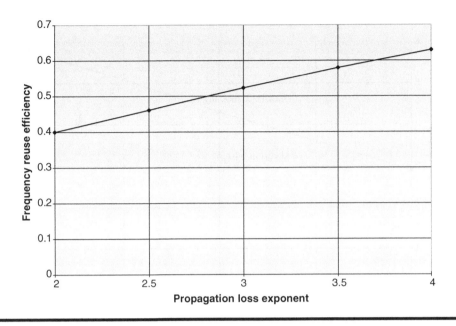

Figure 13.6   Frequency Reuse Efficiency vs. Path Loss Exponent.

**Table 13.15   Reverse Link (E_b/N_t)
Set Point for Different Mobility**

8 km/hr	30 km/hr	100 km/hr
5.2 dB	7.1 dB	5.5 dB

Table 13.16   Reverse Link Differences between Macrocell Capacity and CMI Sector Capacity

Parameters	Macrocell	CMI Sector
Sector efficiency	0.7	Between 0.65 and 0.8
Frequency reuse Efficiency	0.54 (DU), 0.6 (U), 0.65 (S), 0.8 (*Rural*)	Based on path loss exponent and Figure 3.6
Loading	65% (DU, U, S), 50% (*Rural*)	50%
$(E_b/I_t)_{sp}$	6 dB (L), 7 dB (M), 8 dB (H)	6 dB (L), 7 dB (M), 6 dB (H)

CMI sector capacity for low mobility, mixed mobility, and high mobility is given in Figures 13.7 through 13.10. CMI sector capacity for a given morphology depends on the path loss exponent measured for the morphology. An example is given below for computing sector capacity using Figures 13.7 through 13.10.

Assume path loss exponent is 4 for suburban morphology and 30% CMI cell overlapping. The sector efficiency is 70%. The N_{max} can be found using Figure 13.8 by proper scaling. Figure 13.9 predicts N_{max} to be 18 for 0.65 sector efficiency. For 0.7 sector efficiency, N_{max} becomes 19. With 50% loading, the number of active users per sector becomes 9 or 10. Figure 13.7 yields a sector Er capacity boundary of 4.34 Er or 5.08 Er.

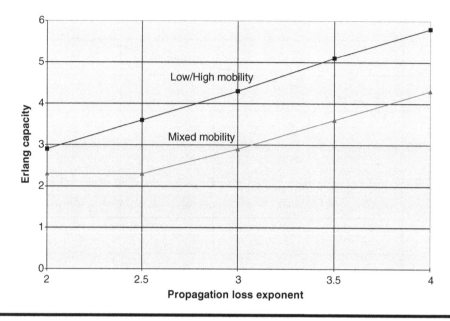

Figure 13.7 CMI Sector Erlang Capacity vs. Path Loss Exponent for Low/High Mobility and Mixed Mobility with 0.65 Sector Efficiency.

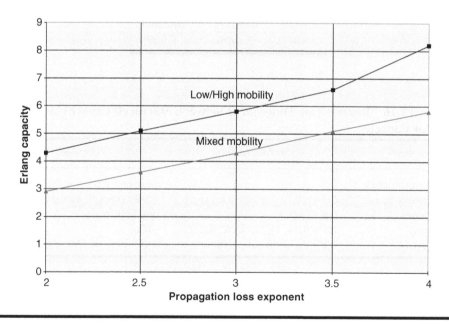

Figure 13.8 CMI Sector Erlang Capacity vs. Path Loss Exponent for Low/High Mobility and Mixed Mobility with 0.8 Sector Efficiency.

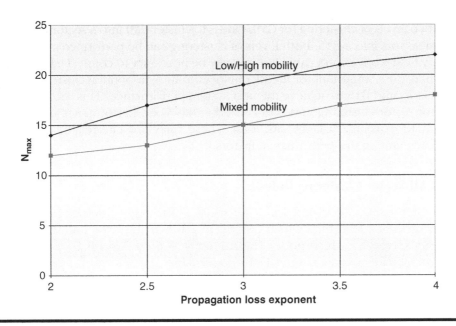

Figure 13.9   CMI Sector N_{max} vs. Path Loss Exponent for Low/High Mobility and Mixed Mobility with 0.65 Sector Efficiency.

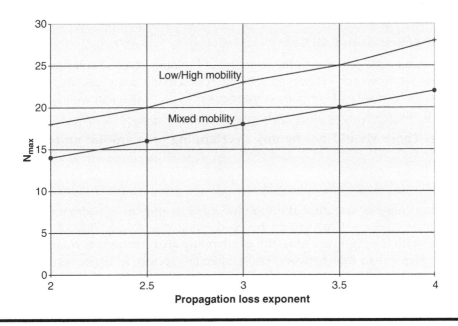

Figure 13.10   CMI Sector N_{max} vs. Path Loss Exponent for Low/High Mobility and Mixed Mobility with 0.8 Sector Efficiency.

13.5 DAS SECTOR CLUSTERING AND BTS CLUSTERING

There are two levels of clustering for CMI. One is to cluster CMI into a sector and the other is to cluster sectors into a BTS. Both levels of clustering can be performed at the headend without any field work except cable lashing may be necessary to connect a CMI to a more desirable node. It is important for the RF engineer to find an optimal clustering topology for both sectors and BTS to achieve the best system performance. This section addresses the rules for sector clustering and BTS clustering based on performance considerations from CATV, RF coverage, CDMA, and cost. These rules are interrelated and tradeoffs must be made among the performance factors.

13.5.1 CMI Sector Clustering Rules

There are three issues associated with CMI sector clustering: to determine the number of CMI per sector, to determine which CMI belongs to which sector, and to know which CMI is being served by which node. The solutions to these issues are based on cable plant constraints, RF coverage, and CDMA performance.

The number of simulcasting CMI cells per cluster (sector) ranges from 1 to 24; it is a tradeoff among local propagation environment, maximum propagation loss derived from link budget, sector coverage, sector capacity, forward link amplifier output power, cost, and simulcasting diversity.

When choosing the size for a sector, a CMI sector should maintain a macrosector size for uniform treatment of cells regardless of their types. More discussion on the number of simulcasting CMI per sector will be presented later, and the standard shapes for different applications of CMI cells are discussed first.

13.5.1.1 CMI Cell Shape

13.5.1.1.1 Downtown CMI Cells

For a downtown CMI cell, due to the structure of buildings, typical cell shape is concave diamond or bat. However, caution must be exercised to eliminate overlapping area from four CMI cells because a MS only has three fingers in the RAKE receiver. Hence, the first rule of CMI clustering:

> **Rule 1: There should not be any overlapping area among more than three CMI.**

13.5.1.1.2 Highway Cells

For highway cells, to maximize the time for a MS to stay in one sector, a rectangular shaped sector (linear sector) should be used where the longer side of the rectangle lies parallel with the highway. Also, the overlapping area between two highway sectors should be larger than that between two residential sectors to make sure that a high-speed vehicle has sufficient time to accomplish handoff signaling in the handoff region, as shown in Figure 13.11.

> **Rule 2: CMI cells along the highway should be clustered into one sector.**
> **Rule 3: Overlap distance between two highway sectors should be larger than 3 dB.**
> **Rule 4: Neighbor list with less than or equal to 20 neighbor sectors must be preserved.**

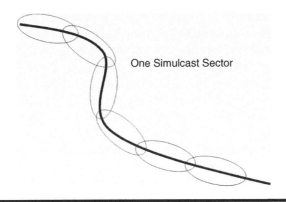

Figure 13.11 CMI Sector Cluster Configuration for Highway.

13.5.1.1.3 Other Types of Cells

For other types of cells, the typical shape should be circular, although the shape is determined by the terrain and the local environment.

If the shape of each CMI cell is circular, attempts should be made to maintain the traditional 3-sector hexagonal shape whenever possible to be consistent with the macrocell design philosophy. Sectors with hexagonal shape make the task of pseudo noise (PN) offset planning much easier.

Rule 5: Hexagonal shape should be maintained for CMI sector.

13.5.1.2 Number of CMI Cells Per Sector

When selecting the number of simulcasting CMIs per sector to achieve a certain coverage and capacity, be aware that the CMI sector size should be consistent with that of a macrocell for uniform treatment of CMI and macrosectors. Table 13.17 explains the factors affecting the decision of the number of CMIs per sector. It is assumed that a 6 dBi antenna is used for CMI design.

The cost includes CMI, HIC, and BTS. To minimize the number of BTSs (as well as HIC), one BTS should support more CMI. However, although more CMI means less BTS, it also means more cost for the CMI themselves. Also, remember that CMI cost is a function of maximum amplifier output power and number of receive antennas required per CMI. The number of receive antennas required per CMI is determined by diversity gain.

The diversity gain needs to consider forward and reverse links as well as number of receive antennas per CMI. For the forward link, at least 3 CMI are required for the MS to receive three direct paths.

A simulated traffic demand for a suburban morphology for years one, five, and ten is 2.43 Erlang/km^2, 16.08 Erlang/km^2, and 36.45 Erlang/km^2, respectively, exists. From the previous sector cluster capacity analysis, for utility pole-mounted CMIs, four CMIs

Table 13.17   Factors Affecting Number of Simulcasting CMI Per Sector

	Cost	Diversity	Capacity	Coverage	Forward Link Power
Wood Pole	4	≥ 3	4	4	≥ 2
Light Pole	≥ 3	≥ 3	≤ 4	≥ 8	≥ 2

per sector require two carriers to support year five demand and three carriers for year ten demand. For light pole-mounted CMI, four CMIs per sector require two carriers to support year one demand and ten carriers for year five demand.

More CMIs do not necessarily provide more coverage; a tradeoff exists between the number of CMIs per sector and the effective CMI cell radius (which is a function of the number of CMIs per sector). Based on the previous analysis, four CMIs per sector provides the maximum coverage for utility-mounted configuration. Because a lightpole-mounted CMI has a larger effective cell radius, eight CMIs can be used to achieve a sector effective radius to be about 1.5 km, which would be similar to typical macrocell solutions.

Based on the link budget, for a 2.5 W output power, at least two CMIs should be used for one sector. Collected field data indicates 1 W output power is feasible for coverage and ease of implementation. Based on the link budget, for a 1 W forward output power device, at least four CMIs should be used for one sector. However, a conservative design approach is recommended and a CMI forward link power of 2.5 W is recommended.

For both utility pole-mounted and light pole-mounted CMIs, the number of CMIs per sector is recommended to be 4.

Rule 6: Nominal number of CMIs per sector (for utility pole) is 4.

13.5.1.3 Cable Constraints

Historically, a node typically serves 1000 homes and the geographical area of these 1000 homes constitutes the node boundary. It is recommended that CMIs served by a node be within that node's boundary (today, 500 homes passed per node is actively pursued).

Rule 7: CMIs served by a node must be within the node boundary.

A HIC has one shared downstream frequency shift key control tone for all CMIs (in 3 sectors). Although there is a separate upstream phase shift key control tone for CMIs within a sector, the control frequency is time shared by CMIs in three sectors. Summarily, CMIs of the same cluster cannot be served by two different HICs.

Rule 8: CMIs of the same sector cannot be served by two different HICs.
Rule 9: It is best a node only assigns to a CMI belonging to the same HIC.
Rule 10: CMIs in one cluster (sector) should form contiguous coverage.
Rule 11: Three sectors assigned to the same HIC should form contiguous coverage.

13.5.1.4 Delay Constraints

Timing delay is a complex topic within traditional wireless networks. The inclusion of DASs requires even further design considerations. CMI cells of the same sector transmit pilots with the same PN offset. The maximum differential delay of CMI cells of the same sector is recommended not to exceed 110 PN chips (90 μs). The minimal differential delay of CMI cells of the same sector cannot be less than 2.5 PN chips (2.0 μs) duration.

Rule 12: Differential delay of simulcast CMI cells is recommended not to exceed 110 PN chips (90 μs).
Rule 13: Differential delay of simulcast CMI cells is recommended to be larger than 2.5 PN chips (2.0 μs).

Differential delay limits are established by both hardware and performance criteria. The hardware is based on interim standard 95 (IS-95) protocol standards. There is a 512

PN chip delay budget (416 μs) limitation that must be preserved under all operating conditions. This 512 PN chip delay is referenced from the channel element cards within the BTS channel element through the transport infrastructure (filtering, HIC, HFC plant, CMI, air propagation) to the MS. The distribution of these delays needs to be understood so that a feasible design is deployed. For example, the ability to support a simulcast CMI sector with a large differential delay at a long distance from the BTS is less probable than at a closer distance.

The following provides a scenario that explains the hardware limitations. The BTS anticipates a power adjustment command on the MS within 3.75 ms. Assume the BTS and MS processing time takes 3.35 ms. The maximum allowable round-trip propagation delay is 400 μs. This results in a one-way propagation delay of 200 μs and fiber velocity (0.6×speed of light) accounting for the majority of delay, the maximum one-way fiber length for CMI cell deployment is 36 km (without considering the HIC, coaxial, CMI, over the air, and differential delays). Increases in differential delay would require a decrease in distance.

There are three BTS translations that need to be defined to better understand the previous statement.

- *Maximum Differential Transmit Delay (μs)*: This is a setting in microseconds which defines the delta between the earliest and latest arriving CMI in the simulcast sector.
- *Transmit Propagation Delay (μs)*: This is a setting in microseconds that defines the minimal (i.e., earliest) one way (CE to CMI antenna) transmit path delay of the simulcast sector.
- *Receive Propagation Delay (μs)*: This is a setting in microseconds that defines minimal (i.e., earliest) one way (CMI antenna to CE) receive path delay of the simulcast sector.

These translations will normalize the delay (i.e., timing advance) incurred from the transport infrastructure and allow for access search window translations to be applied to MSs being served from that simulcast sector.

There are five access search window translations (3 reverse link, 2 forward link) that have performance affecting results and optimizing the values are critical as shown in Table 13.18. The three reverse link (i.e., MS based) are the active, neighbor, and remainder search windows. The two forward link (i.e., BTS) are sector size and cell search window. The definition of these translations is provided below:

- *Active Search Window*: The mobile uses this search window to identify active or candidate set pilots. It is specified in units of PN chips and is centered around the earliest arriving multipath component of the pilot.
- *Neighbor and Remainder Search Window*: The mobile uses these search windows to identify neighbor and remainder set pilots in that order. Window time centering is established by the target pilot PN offset relative to the arrival time of the reference pilot in the active set.
- *Sector Size (miles)*: Establishes the maximum range (one way air propagation delay) this sector is desired to cover. Typically entered in units of miles.
- *Cell Search Window (μs)*: The BTS searches for MSs attempting to access this sector in time, up to the limit entered. Typically this setting should match the sector size value.

Table 13.18 Delay Budget and Search Window Size

Delay Budget (μsec)	Window Size (PN chips)	SRCH_WIN_A SRCH_WIN_N SRCH_WIN_R
$T_d \leq 2 \times 1.64$	4	0
$T_d \leq 2 \times 2.45$	6	1
$T_d \leq 2 \times 3.27$	8	2
$T_d \leq 2 \times 4.09$	10	3
$T_d \leq 2 \times 5.72$	14	4
$T_d \leq 2 \times 8.17$	20	5
$T_d \leq 2 \times 11.44$	28	6
$T_d \leq 2 \times 16.34$	40	7
$T_d \leq 2 \times 24.51$	60	8
$T_d \leq 2 \times 32.68$	80	9
$T_d \leq 2 \times 40.85$	100	10
$T_d \leq 2 \times 53.11$	130	11
$T_d \leq 2 \times 65.36$	160	12
$T_d \leq 2 \times 92.32$	226	13
$T_d \leq 2 \times 130.72$	320	14
$T_d \leq 2 \times 184.42$	452	15

Same sector differential delay limits ensure that the CDMA RAKE receiver search window will not exceed a 130 PN (2×53.11 μs) chip delay spread (translation setting of 11). BTS equipment manufacturers do not endorse operation with a larger search window size due to performance degradation. The degradation is caused by the MS requiring larger search times and reduced revisit opportunities to current pilot selections.

13.5.1.5 Neighbor List

The maximum number of neighbors each sector can resolve in the neighbor list is 20. When performing CMI clustering (especially for highway cells), keep the above constraint in mind.

Rule 14: Each sector can have at most 20 neighbors.

13.5.1.6 Noise Figure

The mean noise figure of combined HIC/CATV/CMI is about 7.2 dB. CMIs exhibit a variance with high (9 dB), average (8 dB), and low (7 dB) noise figures. Because CMI reverse link cell size is a function of noise figure, high noise figure CMIs can be used to reduce the reverse link coverage and low noise figure CMIs can perform the opposite. For example, a two-CMI sector should use high noise figure CMIs while a six-CMI sector should use low noise figure CMIs.

Rule 15: Assign low noise figure CMIs to a cluster with more than four CMIs and assign high noise figure CMIs to a cluster with less than four CMIs.

13.5.1.7 One-CMI Sector

When the number of CMIs (with 2.5 W transmit power) per sector is less than four, the reverse link coverage is larger than that of forward link. There are three approaches to reduce the reverse link coverage: additive Gaussian noise, attenuation, and noise figure impairments.

13.5.1.8 Hole Fill-In CMI for Macrocells

When CMIs are used to fill in the holes of a macrosector, those CMIs cannot reuse the same BTS of the macrosector. The reason is HICs for CMIs are located at headend. There is no easy way of connecting the HIC with the macrosector BTS. Therefore, disjoint CMIs form a sector without contiguous coverage. There are several issues associated with CMIs used to fill in the coverage holes. First, because the CMIs in the same cluster are not next to each other, there is no simulcasting (i.e., no diversity gain). Second, the soft handoff percentage for the macrosector as well as the CMI sector is largely increased.

13.5.2 BTS Clustering

The BTS clustering needs to consider the following factors.

- There should not be any overlapping area among more than 3 CMI.
- Minimize soft handoff events for highway CMI sectors; neighbor CMI sectors on the highway should belong to the same BTS.
- Minimize traffic in handoff region.
- Maintain necessary softer handoff area among three sectors of the same BTS and necessary soft handoff area between two sectors belonging to two (BTS).
- Check differential delay between CMI sectors. CMI sectors transmit pilots with different PN offsets. The differential delay of CMI sectors is recommended not to exceed 226 PN chips (2×92.32 μs) or a translation setting of 13. (The propagation delay contains the delay from channel element to HIC, the delay from HIC to node through fiber, and the delay from node to CMI antenna through coaxial cable.)

13.6 CMI CELL OVERLAPPING CONSIDERATIONS

One of the considerations for CDMA RF design is to maintain a proper percentage of soft handoffs for each sector. Two approaches have been proposed to maintain a required soft handoff region: incremental approach and uniform approach. The pros and cons of these two approaches are discussed below.

13.6.1 Incremental Approach

In this approach, the RF engineers not only select CMI locations, they also cluster CMI cells to a sector at the same time. Advantages of this approach are that overlapping requirements for different types of diversity (simulcasting overlapping, soft handoff, and softer handoff) can be optimized. The disadvantages of this approach are that sector boundaries are hard to control and it seems to be impractical and time consuming to cluster CMIs into a sector consistently. (Note that no field work is required when clustering CMIs into a sector; the cable integration group will perform the necessary labor at the headend.)

13.6.1.1 Overlapping Area between Simulcasting CMI Cells

The CMI overlapping area within a cluster must consider the RAKE receiver characteristics and the number of multipaths. The optimal number of multipaths is three for a MS. The scenario of less than three multipaths results in the optimal diversity gain being reduced for the forward link because the MS cannot catch 100% of the energy. When more than three multipaths are present, the optimal diversity gain is also reduced because extra multipaths create self-interference.

The difference between CMI cells overlapping and macrocells soft handoff overlapping is addressed below. In theory, soft handoff provides a better diversity gain than multipath. However, this conclusion is made under the assumption that the multiple Rayleigh paths come from the same BTS. For simulcasting, the multiple (direct) paths come from different CMI cells. In this case, the simulcasting diversity gain (4 dB) should be about the same as that for soft handoff. For simulcasting, the situation that two (direct) paths have a differential delay less than one chip duration can be completely avoided (by calculating the fiber and coaxial cable delay when performing site selection). For simulcasting, power levels from two (direct) paths are about equal; for nonsimulcasting multipaths, the 2nd and the 3rd resolvable paths are typically 15 dB and 17 dB less than the direct path. For simulcasting multiple direct paths, the cross-correlation coefficients are close to zero; for non-simulcasting multipaths, the cross-correlation coefficient between direct and 2nd path as well direct and 3rd path is about 50%. (Two Rayleigh paths can be considered to be uncorrelated if their cross-correlation is less than 0.1.) All these supporting evidences prove that simulcasting diversity gain is larger than nonsimulcasting diversity gain.

For two CMI simulcasting, there is a region where none of the signals are dominant. The width of this region depends on path loss exponent and the difference between two signal levels ($RSSI_{diff}$) which may result in deep fades. The typical values for receive signal strength indicator ($RSSI_{diff}$) are between 5 dB and 10 dB. Link budget already considers a slow fading standard deviation of 8 dB; it is assumed that to maintain at least two (direct) paths in the simulcasting region, propagation loss across the simulcasting region is 3 dB. Table 13.19 illustrates the width of the simulcasting region for different path loss exponents.

Assume the cell radius is R and the simulcasting width is $2z$, as shown in Figure 13.12. Assume the path loss exponent is γ, then the signal strength at distance $R - 2z$ from transmitter A is proportional to $1/(R - 2z)^\gamma$. The signal strength at distance R from transmitter B is proportional to $1/(R)^\gamma$.

$$(R)^\gamma/(R - 2z)^\gamma = 2 \qquad \text{(in linear scale)} \tag{13.30}$$

$$2z/R = (2^{1/\gamma} - 1)/2^{1/\gamma} \tag{13.31}$$

Table 13.19 Recommended Overlapping Region for Simulcasting CMI Cells for Different Path Loss Exponents

γ	Simulcasting Region (% of overlapping in distance)
2	0.146 km (29.2%)
3	0.103 km (20.6%)
4	0.08 km (16%)

Note: With radius $R = 0.5$ km.

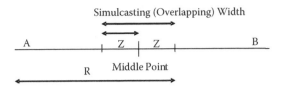

Figure 13.12   Simulcasting Scenario.

For a 3-cell, 4-cell, 5-cell, or 6-cell cluster, the area where three cells overlap should be around 1% to 2% of the cell area. There should not be any overlapping area by more than three CMI cells.

13.6.1.2   Soft Handoff Overlapping Area between CMI Sectors

The nominal soft and softer handoff percentage between CMI sectors should be larger than that in the macrocells to absorb the effect of dynamic change of CMI cell coverage (due to fading). Based on Reference [1], for CMI cells, a soft handoff percentage is 30% and a softer handoff percentage is 35%. The path loss across a soft handoff zone should be 3 dB for 30% soft handoff. The overlapping distance for soft handoff is $r(1 - 2^{-1/\gamma})$. For a sector radius of 0.5 km and a path loss exponent of 4, the overlapping distance is 0.08 km (the number of bins required to be overlapped is one 3-second bin).

Because the soft handoff zone is derived based on a 3 dB criterion, accuracy of signal level plots is very important. The ability to utilize predictive tools for this purpose is limited. Effective use of predictive propagation modeling tools requires high resolution clutter data to ensure resulting standard deviations are applicable. This could be economically ineffective. Predictive propagation models in traditional deployments are challenged in achieving standard deviations of 8 dB or better. This level of standard deviation will not allow for an effective analysis. Use of predictive modeling to estimate a handoff region is not recommended. Drive test data to examine the soft handoff overlap region is a requirement.

13.6.1.3   Soft Handoff Overlapping Area between CMI Sector and Macrosector

To have a proper percentage of soft handoff users (assuming the users are uniformly distributed in the area), path loss across a soft handoff zone between CMI sector and macro-sector should be between 3 and 4 dB.

13.6.2   Uniform Approach

This approach does not perform sector clustering until a large group of CMI sites have been determined. CMI cell overlap is continuous with a predetermined percentage. The advantage is ease of implementation because the RF engineer does not have to spend time in optimizing the overlapping area for soft and softer handoff. The disadvantage is that the network performance is not optimized. The issue of determining the overlapping area for CMI cells is a derivation achieved by considering the following factors.

More overlapping area creates more multipath diversity and lowers the $(E_b/N_t)_{sp}$ requirement. More overlapping area also maintains the soft handoff requirement at the cell edge; consequently, the interference at the cell edge is minimized and the soft diversity gain is also achieved. However, too much softer handoff area reduces the sectorization efficiency and as a result the capacity is also reduced. Based on the above discussion, the overlapping percentage between CMI cells is recommended to be 30% as a compromise. The RF engineer should be aware that this recommendation is not rigid; the engineers should use their CDMA discipline to determine the overlapping percentage based on the local environment. In some cases, sector clustering is obvious (such as the highway sectors) and optimization of the overlapping area to obtain the best performance is clearly understood.

13.7   ANTENNA MOUNTING CONFIGURATION

This section addresses the requirements for distance of the side-mounted antenna: from pole, space diversity, antenna isolation, and antenna desensitization.

13.7.1   CMI Antenna Cable Specification

The specification for the cable used to connect the CMI device and antenna is recommended to be no larger than 1/2 foam dielectric. This will allow for a tight bending radius needed for strand mounted antennas.

13.7.2   CMI Antenna Mounting Configuration on Cable Strand

The recommended mounting configuration for CMIs on cable strand is shown in Figure 13.13. The minimum separation from the nearest obstruction of the transmit antenna is 6 feet (1.83 m). The obstruction can be a utility pole or any device co-located on the cable strand. The minimum separation from the nearest obstruction of the receive antenna is 3 feet. The receive antennas are mounted inverted, with reference to the transmit antenna to achieve isolation between antennas (see Section 13.7.4, Antenna Isolation). To satisfy the GO 95 standard, the bases of receive antennas are actually mounted 3 inches above the cable strand such that there is sufficient separation from the tips of receive antennas and the phone line on the pole. This special arrangement can be achieved using an offset bracket. (Note that the near field zone is about $5 \lambda = 0.8$ m; 5λ corresponds to a free space loss of 36.2 dB.)

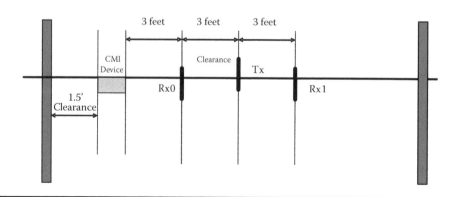

Figure 13.13   CMI Antenna Mounting Configuration on Cable Strand.

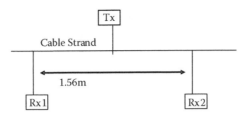

Figure 13.14 Mounting Configuration to Achieve Space Diversity Using Two Receive Antennas.

13.7.3 Space Diversity

Space diversity is achieved by physically separated receive antennas. The principle behind space diversity is that fading on two signals is uncorrelated; the result is that these two signals can be optimally combined or selected to get maximum diversity gain. The space between two receive antennas is $\lambda r/d$, where r is the distance between the MS and receive antenna, and d is the distance of the MS to immediate surroundings (that will cause multipath). Assume $r = 0.5$ km and $d = 0.1$ km (for a suburban environment), the required spacing is $(0.156) \times 0.5/0.1 = 0.78\ m = 5\lambda$.

A conservative recommendation on spacing between two receive antennas is about 10 λ (1.56 m). Each CMI is equipped with a diversity receive capability, as depicted in Figure 13.14. This spatial diversity is secure via a 6 ft. separation of the two receive antennas. This separation will produce ~12 λ factor which exceeds the recommended 10 λ requirement.

13.7.4 Antenna Isolation

The typical isolation requirement between nonduplexed transmit and receiver antennas is 30 dB. Either horizontal isolation or vertical isolation can be used to achieve this requirement.

The formula to achieve vertical isolation given the spacing between nonduplexed transmit and receive antennas is

$$A_v = 28 + 40\log_{10}(h/\lambda). \tag{13.32}$$

$$= 28 + 40\log 10(h/0.156). \tag{13.33}$$

$$= 30\ \text{dB}, \quad \text{given } h = 0.175\ \text{m (6.9 inch)} \tag{13.34}$$

Spacing between nonduplexed transmit and receive antennas using vertical isolation is recommended to be 0.2 m. The 0.2 m separation is between the two antennas; one is mounted straight up and the other is mounted upside down, as depicted in Figure 13.15. (Note 0.2 m distance corresponds to a free space loss of 24.13 dB.)

The formula for horizontal isolation given the spacing between nonduplexed transmit and receive antennas is

$$A_h = 22 + 20\log_{10}(h/\lambda) - (Tx\_G + Rx\_G - Disc\_Loss) \tag{13.35}$$

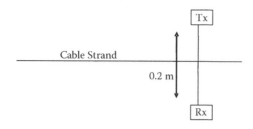

Figure 13.15 Mounting Configuration to Achieve Vertical Isolation between Transmit and Receive Antennas.

where Tx_G is the maximum transmit (Tx) antenna gain (dBd), Rx_G is the maximum receive (Rx) antenna gain (dBd), and $Disc\_Loss$ is the sum of the discrimination loss at both antennas.

Because a CMI uses an omnidirectional antenna, the formula can be simplified to:

$$A_b = 22 + 20\log_{10}(b/0.156) - (Tx\_G + Rx\_G) \tag{13.36}$$

The horizontal separation requirements for three types of CMI antennas are listed in Table 13.20 and depicted in Figure 13.16.

13.7.5 Antenna Desensitization

Because the CMI cell antenna height is about 7 m and is right on the street, the likelihood for the MS to be very close to the CMI cell is very large. Because the MS has a limit on minimum transmission power, to prevent the CMI cell from desensitization, CMI antennas should not be placed directly next to the windows of a building.

13.8 DESIGN PROCEDURES AND RESULTS FOR DAS/CMI FIELD PERFORMANCE MEASUREMENTS

In this section, a summarization of the report is provided and RF design steps for CMI cells are defined. The flow chart of the design process is presented in Figure 13.17.

An effort to assess a PCS DAS network was conducted in February 2000. The goal of this assessment was to determine the potential for a capacity enhancement from the CMI network as compared to its adjacent macro solution. In driving to validate this goal, a number of attributes of the network resulted. Network attributes such as differentiation between the CMI and macro serving areas, and sector assignment of the simulcasting CMIs through exhaustive measurements of the network areas were determined.

Table 13.20 Horizontal Separation Requirements for CMI Antennas

Antenna Gain (dBi)	Spacing (m)
5	0.782
6	1
8	1.56

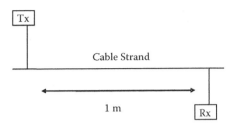

Figure 13.16 Mounting Configuration to Achieve Horizontal Isolation between Transmit and Receive 6 dBi Antennas.

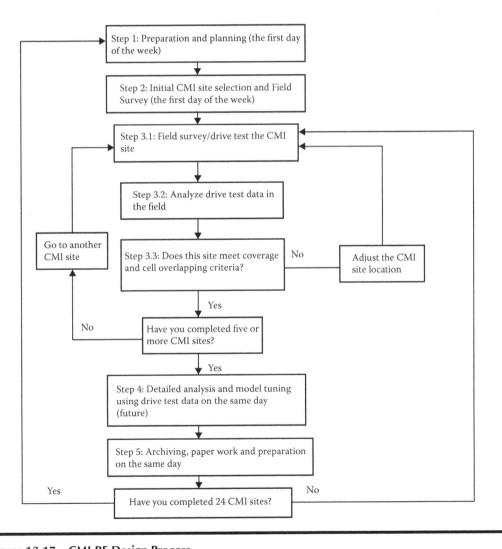

Figure 13.17 CMI RF Design Process.

Wireless carrier RF engineering departments in general establish performance criteria that, for the most part, are based on "call attempts," "percentage of calls," "percentage dropped," and "percentage blocked." These three fundamental components not only provide the "tip off" to engineering staffs on where to focus their efforts, but indicate where the capacity of the network is in demand. It is the efficient use of this capacity that brings a competitive advantage to wireless carriers.

In a PCS CDMA network, "noise" is the factor that limits the capacity yield. It is this factor, addressed in the following subsections, that concludes that the CMI delivery method increases capacity by 22% through limiting the effects of noise within its serving area.

The main objective of this data collection effort was to assess the metric differentials between the CMI and macro network. This differential is then applied to an industry-accepted derivation that will indicate either a capacity gain or reduction. Capacity determination within a CDMA network is a challenging task due to the dynamitic and intrinsic properties that constitute this protocol. These properties such as power control, voice activity, other cell interference, and soft handoff either impair or improve the capacity yield. To better understand the conclusions drawn, effects of these properties are discussed. The following sections establish the bounds of expectations through derivation of theoretical limits.

13.8.1 Theoretical Limitations

A telephonic communications path consists of two links, forward and reverse, that establish the ability to convey information between subscribers. It is the integrity or robustness of these links that establishes the "call quality" the subscriber experiences. Typically the reverse link is the limiting factor. This is due to the limited transmit power available to the PCS MS (+24 dBm) compared to (+39 dBm) from the BTS.

Maintaining a nominal MS SNR ratio will result in an experience of "good" call quality; however, the call quality will decrease as subscribers increase, because each additional subscriber decreases the SNR for all subscribers. Assume that a single cell serves all subscribers and perfect power control maintains the received signal strengths from all MSs equally at the BTS receiver, then determining signal-to-noise as a function of the number of MSs is possible.

$$Signal\_to\_Noise(E_b/N_t) \approx \frac{A_1^2 \times (T_b/2)}{(M-1) \times \frac{A_1^2}{3} \times T_c} = \frac{3\,PG}{2(M-1)} \tag{13.37}$$

The assumptions for the above equation are:

- $A_1 = A_i$ because all users are received at the BTS with the same signal strength (perfect power control)
- All users served by the same cell
- No additive or other interfering sources
- PG = 1.2288 Mbps/14.4 kbps = 85

Figure 13.18 establishes the absolute perfect case scenario for maximum capacity yield in an omniCDMA BTS. Typically, an E_b/N_t of 7 dB is the link budget target to achieve "good" call quality (< 2% FER) and in the perfect world this would result in a simultaneous subscriber usage of 26 at a 100% pole point with an omnicell.

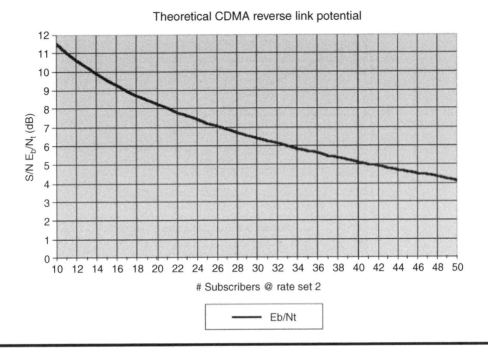

Figure 13.18   Maximum Capacity of Omni BTS.

13.8.2   Real World Factors on Capacity

The actual capacity yield is affected by a number of real world factors. These factors have either an enhancing or degrading effect on the capacity yield curve. It is these factors that are determined for both the CMI and macro networks. Through the implementation of strategic drive routes that attempted to isolate the two RF delivery techniques a comparable set of data was derived. This resulted in the establishment of a proportional value for each network based against typical. To further clarify, if a typical value for a particular factor is one, then the deviation surmised between CMI and macro is normalized to the typical value.

The following real world factors that affect CDMA capacity in either a positive or negative manner are CDMA demodulator, power control limitations, CDMA other cell interference, soft handoff, voice activity, and sectorization.

Of these, only CDMA demodulator and voice activity are factors that remain constant between the comparisons of the two RF delivery techniques. For completeness, a description of these effects on capacity is discussed.

13.8.2.1   CDMA Demodulator

CDMA receiver demodulator technology achieves 0.1% frame error rate (FER) with an $E_b/N_o = 6.8$ dB signal, on an ideal nonfading channel with additive white Gaussian noise (AWGN). Realize N_o represents just the white Gaussian noise and N_t represents white Gaussian noise plus interference. The curve shows the best performance that can be achieved in a perfect laboratory environment. Anticipated technological enhancements in CDMA receiver chipsets could prove to enhance CDMA capacity yields. In regard to the comparison, wireless carriers are driving their link budgets to a 6.8 dB E_b/N_t to

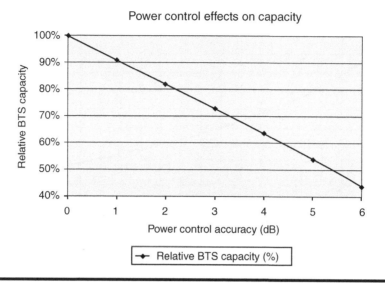

Figure 13.19 Power Control Effects on Capacity.

achieve a 1% FER with a 50% pole point (11 active users/sector). It is this value that can be compared between networks to realize the capacity differential.

13.8.2.2 Power Control Limitations

The power control accuracy has a strong effect on the CDMA channel capacity. Industry reports indicate on average that a 1.5 dB standard deviation on all signals is received by the BTS receiver. This result limits capacity yields to a maximum of 85% as shown in Figure 13.19. The curve shows capacity reduction as a function of power control accuracy and assumes interference from co-channel neighbors. The path loss between neighbors is a function of $1/r^4$ (40 dB per decade), where R is the distance between cochannel interferers.

Composite mobile Tx data collected from the two networks indicated that the CMI network indicated a 1 dB higher mean than the macro network but had a 2 dB improvement in standard deviation. The higher mean was expected due to the additional hardware non-linearity that the reverse link is exposed to in the CMI network. This increase is considered negligible especially because the standard deviation had a pronounced improvement. Figure 13.20 and Figure 13.21 capture the results discussed.

Mobile Tx is perceived as having a strong correlation to the power control accuracy but a differential metric between the two networks was not concluded. What can be surmised is that the CMI network with the tighter standard deviation is experiencing a different propagation coefficient due to the simulcasted environment producing an improved fading characteristic. A common metric (85% efficiency) for both networks was used in the capacity calculation, therefore; the potential for improved capacity yield(s) from those determined in this chapter is possible.

13.8.2.3 Other Cell Interference

It is this real-world factor that was analyzed between the two networks with extreme focus. Co-channel, as defined in this chapter, reduces capacity by increasing the receiver

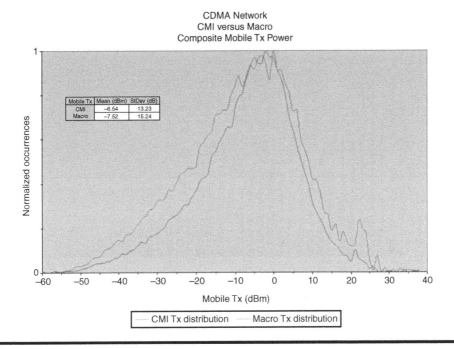

Figure 13.20   Network Composite Mobile Tx.

noise floor. This noise-floor rise is attributed to other cell signal propagation interacting (covering) within the same areas of service as adjoining cells. The CMI network with its low antenna centerline would exhibit improved propagation control to minimize the effects of other cell interference.

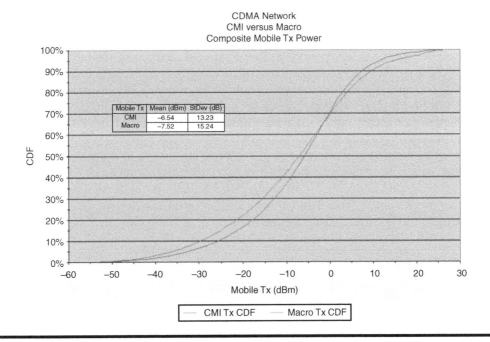

Figure 13.21   Network Composite Mobile Tx CDF.

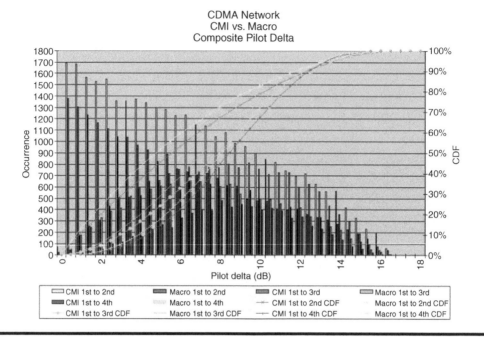

Figure 13.22   CMI versus Macro Composite Pilot Delta.

The subscriber phone has a search window that is constantly scanning for cell sectors (pilots) that could offer the best frame erasure rate (FER) performance during that call. The RAKE receiver that performs the search has limited abilities and in scenarios where excessive cell sectors (pilots) are present, degrading call quality ensues. To address this factor in a comparison between the two networks, a relationship of the serving cell/sector (pilot) to the next three was performed. Figures 13.22 and 13.23 depict the performance variations between the two networks with regard to other cell interference by showing the pilot delta from the primary serving pilot. Figure 13.22 indicates that the CMI exhibits less other cell interference as compared to the macro. This is derived from the cumulative distribution function (CDF) for each primary to N pilot curve. The increased pilot delta from the CMI in all pilot categories gives an indication that the propagation within the CMI is better contained than the macro. Figure 13.23 breaks out the CDF curve(s) to better illustrate this condition.

Table 13.21 extracts the analysis of the pilot relationships at a critical 3 dB design point. This is a critical juncture because designs attempt to achieve a 30% coverage overlap at the 3 dB points. The CMI network indicates an ability to achieve this goal better than the macro by 8%. The capacity reduction factor derived in Table 13.21 is used to determine the final capacity assessment.

13.8.2.4   Soft Handoff

Soft handoff improves overall system performance by controlling the subscriber to the minimum power required by the base station with the lowest path loss. Without soft handoff, a subscriber will be ordered to transmit full power at the edge of the coverage

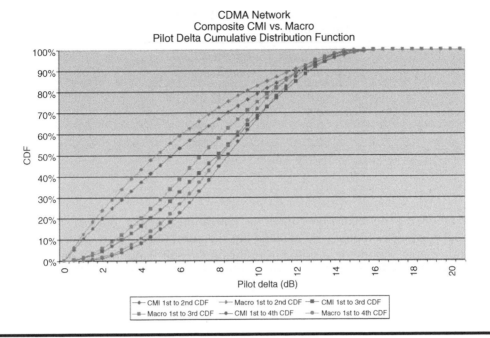

Figure 13.23   CMI versus Macro Network Composite Pilot Delta CDF.

area; with soft handoff, the subscriber will be ordered to transmit the minimum power required for an adequate signal at the BTS with the lowest path loss. Because the MS transmits lower power, on average, interference is reduced to other users in the system, increasing capacity.

The gain attribute from soft handoff has been quantified to be 2 dB with power control invoked [3]. This assumes that the cell edge has coverage that limits powering up subscribers. Typically, a 30% overlap at the 3 dB coverage thresholds is targeted. Information collected in this market on both networks indicates that the CMI achieved a 43% coverage overlap with pilots that were within 3 dB of the server while the macro

Table 13.21   Other Cell Interference Summary

Pilot Delta	Ideal Ideal 3 dB Pt.	CMI 3 dB Pt.	30% Pt. (dB)	Macro 3 dB Pt.	30% Pt. (dB)	30% Pt. Delta (dB)
1 to 2	30%	29%	3	34%	2.5	0.5
1 to 3	0%	10%	5.5	12%	5	0.5
1 to 4	0%	4%	7	5%	6	1
Total	30%	43%	4.21	51%	3.65	0.56
Capacity Reduction Factor	1.40	1.58	—	1.69	—	—

Table 13.22 Voice Activity Factor Distribution

Activity Factor (A)	Percent of Calls With Activity > A
0.4	47.0%
0.5	19.0%
0.6	4.0%
0.7	0.5%

resulted in a 51%. The results both meet the typical design goal, but as is the case, too much is also capacity limiting by maintaining the subscribers in an extended handoff state. The comparison metric resulted in reducing the soft handoff state by 13% for the CMI and 21% for the macro. Table 13.21 captures this information by determining the differential between the percentage area served with pilots equal to or greater than 3 dB. The result alludes to the fact that strategic coverage control is achievable through the implementation of the CMI solution.

13.8.2.5 Voice Activity

Though a factor in assessing capacity, the same criteria is applied to both networks, therefore, no differential can be resolved. Natural speech includes active periods and quiet periods called spurts and pauses, as shown in Table 13.22. Spurts are generally syllables and words while pauses include the times in conversation when the party is listening. In a typical conversation, the speech spurts last between 1 and 2 seconds and the activity factor is about 40%. The average speech time can be modeled as shown in [4].

The CDMA capacity can be increased by taking advantage of the pauses and listening times, because the transmitter does not transmit during these periods, reducing the channel interference, therefore, increasing capacity.

13.8.2.6 Sectorization

The concept of sectorization of a BTS is straightforward. Sectoring a BTS will increase capacity because the directional antenna only receives signals from a fraction of the targeted serving area. The ideal 120° antenna will only receive signals from 1/3 of the cell, reducing the interference by 2/3, and increasing the capacity by a factor of 3. The real-world effect though is that actual antennas do not possess an ideal pattern that only accepts subscriber signals from a 120° sector, and rejects remaining subscribers from areas outside the 120° sector. There is always some overlap of the sector antenna patterns, so interference is not reduced by a factor of 3, but more typically the reduction is 2.55.

The comparison metric derived for this factor resulted in the CMI achieving better than typical results of 2.58 while the macro derived a 2.45 value. The CMI was successful in maintaining better control of its overlap between primary and secondary pilots. This is also indicated in Table 13.21, where the CMI resulted in a 29% overlap of primary and secondary pilots compared to the macro's 34%. Only the primary and secondary pilots

Table 13.23   Overall BTS Sector Capacity Metrics

Variable	Macro	CMI
Processing Gain (PG)	85	85
Power Control Efficiency (PCE)	0.85	0.85
Soft Handoff Gain (SHG)	1.264	1.582
Voice Activity Factor (V)	2	2
Sectorization Factor (S)	2.45	2.58

were considered for this analysis because they best represented the first tier neighbor relationship.

13.8.3   Overall BTS Sector Capacity

The equation presented here takes into account the real-world factors addressed so far. The equation only considers a single carrier and is applicable because the market was not determined from measurement data to support a multicarrier capability in either network. The equation derivation is obtained from Figure 13.24. Typically, an E_b/N_t of 7 dB is practical and the corresponding subscriber/sector is determined. Table 13.23 contains the metrics used in the equation to derive the curves in Figure 13.24.

$$Subscriber\_Sector_{Total} = \frac{PG}{E_b/N_t} \times PCE \times \frac{SHG}{OCI} \times V \times S \qquad (13.38)$$

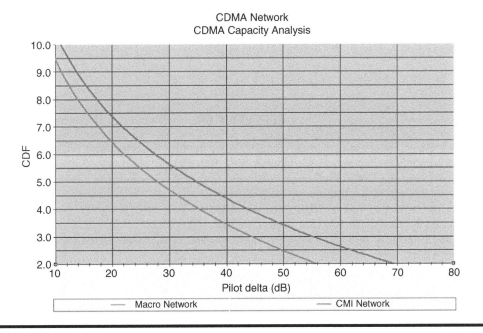

Figure 13.24   Network Capacity Comparison.

13.9 CONCLUSION AND FUTURE RESEARCH DIRECTION

13.9.1 Conclusion

The use of DAS in wireless network deployments is a growing trend. The concept of DAS utilizing existing operator's HFC infrastructures combines both an economical and performance capability enhancement. The empirical data presented describe capacity gains through the use of DAS.

13.9.2 Future Research Direction

The DAS concept is developing into another defined multiple antennas distribution solution. The multiple-input multiple-output (MIMO) technique has been portrayed as the next revolution for wireless technologies. The MIMO technique is an effective mechanism to increase the capacity and reliability of wireless communications by taking advantage of its spatial multiplexing gain or diversity gain. However, conventional MIMO is expensive and has its associated drawbacks: the form factor of the radio is large, multiple RF front ends are required, multiple antennas are difficult to place on a mobile platform especially for lower frequency operation, and the channel may not support multiple uncorrelated streams. Cooperative MIMO offers an alternative way of implementing MIMO where single-antenna devices cooperative with each other to form a virtual MIMO system. In cooperative MIMO, multiple nodes coordinate their transmissions so that cooperative parallel transmissions can be established between multiple source nodes and a destination node. Cooperative MIMO also takes advantage of cross-layer techniques. Some of the functions required at the physical layer are pushed up to the medium access control (MAC) layer. Further, these elevated MAC layer functions are distributed to a number of nodes to jointly, cooperatively forming a virtual MIMO system. These functions include synchronization among different streams (antennas), selection of antennas, and assignment of space–time coding. It would be possible to use DAS architecture to implement cooperative MIMO. The performance of this approach would be left as a future research topic.

REFERENCES

[1] D.J. Shyy, "Impact of deploying cdma2000 data to voice capacity using cellular simulator," *OPNETWORK Conference*, Washington, DC, Aug. 2002.

[2] V. Erceg et al., "Urban/suburban out-of-sight propagation modeling," *IEEE Communications Magazine*, vol. 30, pp. 56–61, June 1992.

[3] J.S. Lee and L.E. Miller, *CDMA System Engineering Handbook*, Mobile Communications Series, Boston A/A:Atech House, pp. 315–316, 1998.

[4] J.M. Fraser, D.B. Bullock, and H.G. Lang,"Over-all characteristics of a TASI system," *BSTJ*, vol 41, pp. 1439–1473, July 1962.

14

MULTI-HOP VIRTUAL CELLULAR NETWORK

Eisuke Kudoh and Fumiyuki Adachi

Contents

Higher speed data transmissions are strongly expected in mobile communication services. However, there will be a serious problem. As data transmission rates increase, a larger transmit power is required to keep the same received bit energy necessary for the required quality of service (QoS). To avoid such large transmit power, the cell size should be significantly reduced, resulting in a nano-cell or pico-cell network. One efficient realization of such networks is the virtual cellular network (VCN), in which each virtual cell (VC) consists of a central port (CP), which is a gateway to the network and many distributed wireless ports (WPs). The signal transmitted from a mobile terminal is received by its surrounding WPs and relayed to the CP using a wireless multi-hop technique. This chapter introduces the wireless multi-hop VCN concept suitable for high-speed packet data services in future mobile communication systems. A novel multi-hop route construction algorithm and a distributed dynamic channel allocation algorithm for multi-hop VCN are presented.

14.1   INTRODUCTION

The 3rd generation (3G) mobile communication systems, known as IMT-2000 systems, have a data transmission capability of up to 2 Mbps [1]. Furthermore, high-speed downlink packet access (HSDPA) [2] aiming to achieve a maximum peak throughput of around

14 Mbps has been developed and will be put into service for the enhancement of the IMT-2000 systems. However, because the information transferred over the Internet is increasingly becoming rich, the wireless transmission capability of the enhanced IMT-2000 will sooner or later become insufficient. There is a strong demand for peak rates of around 100 Mbps ~ 1 Gbps, even in mobile communication systems, to offer mobile users Internet-related broadband multimedia services [3]. This is the task of the so-called 4th generation (4G) mobile communication systems [4].

There may be two important technical issues to realize 4G systems. For such high-speed data transmissions, the channel is severely frequency selective due to the presence of many interfering paths with different time delays, and this significantly degrades the transmission performance. A promising wireless access technique that can overcome the channel frequency selectivity and even take advantage of this selectivity to improve the transmission performance is code division multiple access (CDMA) with frequency-domain equalization [5].

Another issue to realize 4G networks is the power limitation. In a cellular network, a service area is covered with many cells. A cell is illuminated by a base station. The typical cell radius, i.e., the coverage by a base station, may be smaller than a few kms [6]. The cell radius depends on the allowable maximum transmission power and the required QoS. Neglecting the shadowing loss and fading loss, the received bit energy E_b from a mobile terminal (MT) at cell boundary is given by

$$E_b = P_t \cdot T \cdot r^{-\alpha}, \tag{14.1}$$

where P_t is the transmit power of an MT, T is the bit duration, r is the cell radius, and α is the path loss exponent [7]. As the transmission rate becomes higher (the bit duration T becomes shorter), the transmit power P_t of an MT should be increased to keep the required bit energy. For example, assuming that the transmit power is 1 W for 8 kbps data transmission, a transmit power of as large as 12.5 kW is necessary for 100 Mbps data transmission at the same distance r. To avoid such huge transmit power, the cell size should be made significantly small. Assuming a path loss exponent $\alpha = 3.5$ and a cell radius equal to 1 km for 8 kbps data transmission, a cell radius as small as 68 m is necessary for 100 Mbps data transmission in order to keep the transmit power equal to 1 W. However, reducing the cell size requires more base stations. Therefore, the control signal traffic for handover and location registration may increase. The use of the present cellular concept may not be optimal to realize a nano- or even pico-cell wireless network. One efficient realization of such a small cell wireless network is the virtual cellular network (VCN) [8,9]. The VCN is composed of a central port (CP), which is the gateway to the network, and many distributed wireless ports (WPs).

In 3G systems, direct-sequence CDMA (DS-CDMA) is adopted as an access technique [2]. In DS-CDMA, transmit power control (TPC) is indispensable to reduce the adverse effect of fading as well as the well-known near–far problem. DS-CDMA can also be used in the multi-hop VCN. In the wireless multi-hop VCN, stationary WPs are deployed. This is different from the so-called wireless multi-hop network, in which an MT relays the signal to other terminals [10]. The multi-hop technique is also used in wireless mesh networks [11] for wireless local area networks (LANs). Wireless mesh networks are comprised of two types of nodes: routers and clients. Each client node interconnects via wireless multi-hop links. The mesh network can extend the service area of wireless LANs. On the other hand, the multi-hop VCN aims to realize the 4G mobile network. Because the mobile network must accommodate many users, the multi-hop VCN must be a truly frequency-efficient network. To reduce the interference power and increase the frequency efficiency, the total transmit power minimization route construction algorithm is used [8]. To avoid the interference in the multi-hop links, the available frequency band is divided into several frequency channels and different frequency channels are allocated to adjacent multi-hop links. To efficiently allocate the frequency channels, the channel segregation dynamic channel allocation (CS-DCA) algorithm is applied [9]. To support the users, mobility, handover and location registration are necessary. To reduce the control signal traffic for handover and location registration, the distributed mobility management is applied.

After presenting the VCN concept in Section 14.2, Section 14.3 describes the multi-hop route construction algorithm. Section 14.4 describes the multi-hop channel allocation using the CS-DCA algorithm and Section 14.5 presents some concluding remarks.

14.2 MULTI-HOP VCN CONCEPT

Figure 14.1 compares the VCN and the present cellular network. The wireless multi-hop VCN consists of a CP and many distributed WPs. For the VCN, WPs are stationary. However, to change the network topology flexibly according to the traffic distribution or radio propagation environment, the installation or removal of WPs is made whenever necessary. The signal transmitted from an MT is received by its surrounding WPs. The WPs that directly communicate with the MT are called end WPs. Because each end WP can act as a site diversity branch, the transmit power of an MT can be significantly reduced compared to the present cellular network. If all the end WPs communicate with the CP directly, the transmit powers of some of them may become very large due to path loss, shadowing loss, and multipath fading. To avoid this, the wireless multi-hop technique is applied. The signals received at (or transmitted from) the end WPs are relayed to (or from) the CP.

For mobility management, the MT location information updating is essential. To reduce the location information control message traffic, a distributed mobility management scheme is used, in which the CP supports the intravirtual cell (VC) mobility management and the control station in the core network supports the interVC mobility management. The VC control layer, which is inserted between the data link layer and the network layer, is introduced as illustrated in Figure 14.2. The VC control layer manages the route construction, channel allocation, and location registration in each VC.

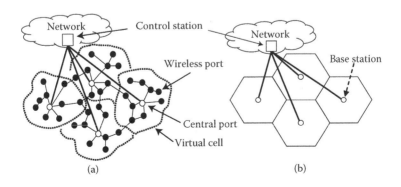

Figure 14.1 VCN and Present Cellular Network, (a) Virtual cellular network, (b) present cellular network.

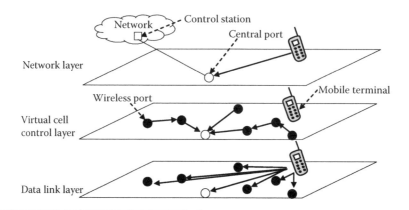

Figure 14.2 Layer Structure.

In the multi-hop VCN, two types of channel are necessary: the communication channel for relaying the data and the control channel for route construction, channel allocation, and location registration. The system bandwidth is divided into several frequency bands with different carrier frequencies.

The features of the multi-hop VCN are summarized here:

1. The control signal traffic for handover and location registration to/from the network increases as the cell size becomes smaller. However, in the VCN, a group of distributed WPs acts as one virtual base station and hence, the control traffic will not increase (however, there exists control signal traffic within each VC for multi-hop route construction and maintenance).
2. Because each end WP acts as a site diversity branch, the transmit power of each MT and the total transmit power of the end WPs can be made significantly smaller than the present cellular network.
3. Reducing the transmit power contributes to the reduction of the interference power to other VCs, and thus the frequency efficiency improves significantly.
4. Grouping of distributed WPs, to construct a VC, may not necessarily be fixed but can be different for each user and the VC size for the uplink may not necessarily be the same as for the downlink.

14.3   WIRELESS MULTI-HOP ROUTING

For uplink (downlink) data transmissions, many WPs can be used to relay the signal transmitted from an MT (the CP) to the CP (an MT). The routing algorithm is an important technical issue to select the relaying intermediate ports until the CP. Routing algorithms proposed for wireless multi-hop networks or ad hoc networks [12–15] can be applied to VCNs. To increase the frequency efficiency, a routing algorithm that minimizes the total uplink transmit power while limiting the number of hops is introduced [8]. However, the carrier frequency of the control channel for route construction is different from the data channels. The fading observed at a different frequency is diverse. For a rich scattering radio propagation environment, the fading correlation decays quickly as the frequency separation increases. This means that the fading observed at the control channel may be different from the data channel. Therefore, the multi-hop constructed route may not necessarily minimize the total transmit power for the data transmission. To reduce the degradation of the transmit power efficiency caused by the fading correlation between the control and the data channels, multi-hop maximal ratio combining (MHMRC) diversity is applied [16,17].

The total transmit power for wireless multi-hop communication is evaluated by computer simulation to show that the wireless multi-hop can considerably reduce the total multi-hop transmit power, while avoiding unnecessary large time delay, irrespective of the path loss exponent and the shadowing loss standard deviation.

14.3.1 Route Construction Procedure

The received signals at all end WPs in the VC, where the MT is located, are relayed to the CP. A multi-hop route connecting each end WP and CP is constructed to minimize the total transmit power of WPs for multi-hop relay. Uplink case (end WP-to-CP) is considered. Figure 14.3 shows the message flow of route construction. The route construct

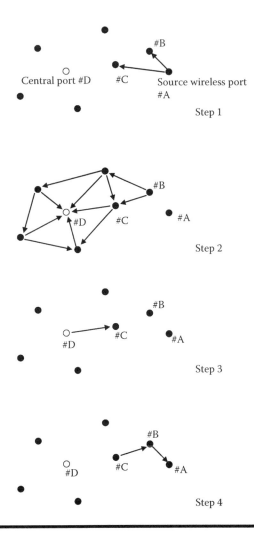

Figure 14.3   Example of Uplink Route Construction Message Flow.

request message is sent periodically from all WPs to the CP via other WPs, and the route notification message is sent back from the CP to each WP via other WPs. To avoid excessive transmission delay, the maximum number of hops is limited. To limit the maximum number of hops, the number of hops is included in the route construct request message. The header of the route construct request message includes: transmit power, source WP address, the number of hops, transmitting WP address, and total required transmit power of WPs along the route. The header of the route notification message includes: destination WP address, transmitting WP address, the source WP address, and required transmit power of the destination WP. As the maximum number of hops increases, the route construction control message traffic may increase. However, because we are assuming that all WPs are stationary, the multi-hop route updating interval may not necessarily be too short and thus, the increase in the route construction control message traffic is not severe.

Let the MT $\sharp A$ be the source WP as shown in Figure 14.3. There are several candidate routes branching from the source WP. The WP of the i-th hop along the candidate route k is represented as $W_k(i)$. The route construction procedure is as follows:

Step 1: Source WP $\sharp A = W_k(0)$ $(k = 0, 1, \ldots)$ (see Figure 14.3) sends the route construct request message with transmit power $P_t(W_k(0))$.

Step 2: Upon reception of the k-th route construct request message from the WP $W_k(j-1)$, the WP $W_k(j)$ checks the number j of hops, and if j is less than the allowable maximum number of hops, the WP $W_k(j)$ computes the required transmit power $P_{t,req}(W_k(j-1), W_k(j))$ of the WP $W_k(j-1)$ using the following equations

$$P_{t,req}(W_k(j-1), W_k(j)) = P_{req} + P_t(W_k(j-1)) - P_r(W_k(j)) \text{ in dB,} \qquad (14.2)$$

where P_{req} is the required received signal power, $P_t(W_k(j-1))$ is the transmit power of the WP $W_k(j-1)$, and $P_r(W_k(j))$ is the received signal power at the WP $W_k(j)$. Then, the total transmit power P_k of the WPs along the k-th route reaching the WP $W_k(j-1)$ is computed using

$$P_k = \sum_{i=1}^{j} P_{t,req}(W_k(i-1)), W_k(i)). \qquad (14.3)$$

If the WP receives more than one route construction request message, the WP selects the route that has the minimum total required transmit power

$$\hat{k} = \arg\min_k \{P_k\}, \qquad (14.4)$$

and multicasts the route construction request message to other WPs. For example, WP $\sharp C (= W_0(1) = W_1(2))$ receives the route construction request messages from WPs $\sharp A (= W_0(0))$ and $\sharp B (= W_1(1))$. Because we assume

$$P_{k=1} = \left(P_{t,req}(\sharp A = W_1(0), \sharp B = W_1(1)) + P_{t,req}(\sharp B = W_1(1), \sharp C = W_1(2)) \right)$$

$$< P_{k=0} = P_{t,req}(\sharp A = W_0(0), \sharp C = W_0(1)), \qquad (14.5)$$

the $k = 1$ route is selected and the WP $\sharp C$ relays the route construction message received from WP $\sharp B$.

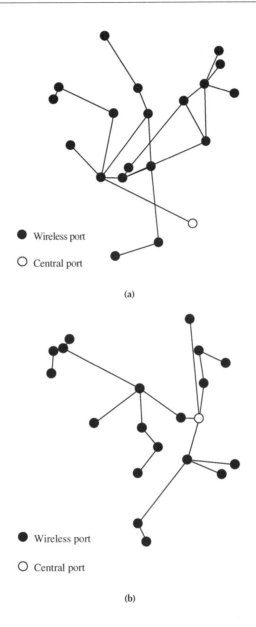

(a)

(b)

Figure 14.4 Examples of Constructed Routes.

Step 3: The CP ♯D chooses the route which minimizes the total required transmit power and multicasts the route notification message that includes the destination WP address (♯$C = W_1(2)$) and the required transmit power of WP ♯C.

Step 4: When the WP finds its address in the received route notification message, it relays the route notification message to the source WP (WP ♯$A = W_1(0)$).

Step 5: The source WP (WP ♯$A = W_1(0)$) receives the route notification message.

Figure 14.4 shows some examples of constructed routes, $K = 20$ and $N = 5$, where N is the maximum allowable number of hops for $\alpha = 3.5$ and $\sigma = 7$ dB. Figure 14.4(a) shows an example in which the multi-hop routes have been constructed, where the

CP transmits (receives) the uplink (downlink) signals via only one WP. Of course, this does not always happen. Sometimes, the CP needs to have multiple connections with surrounding WPs as seen in Figure 14.4(b).

14.3.2 MHMRC Diversity

To reduce the interference power and increase the frequency efficiency, the total transmit power minimization route construction algorithm was introduced in the previous section. However, because the carrier frequency of the control channel for route construction is different from the data channels, the multi-hop route constructed may not necessarily minimize the total transmit power for the data transmission. To reduce the degradation of the transmit power efficiency, caused by the fading correlation between the control and the data channels, multi-hop maximal ratio combining (MHMRC) diversity can be applied [16,17]. In multi-hop connection, as illustrated in Figure 14.5(a), each WP relays the signal to its next port. But, the same signal may be received by multiple ports along the route.

Figure 14.5(b) explains the concept of MHMRC. User link (MT-to-end WP) is also considered in this section. An MT transmits its signal, which is received by port ♯1, but the same signal is received by ports ♯2, ♯3 and ♯4. Port ♯1 relays its received signal to

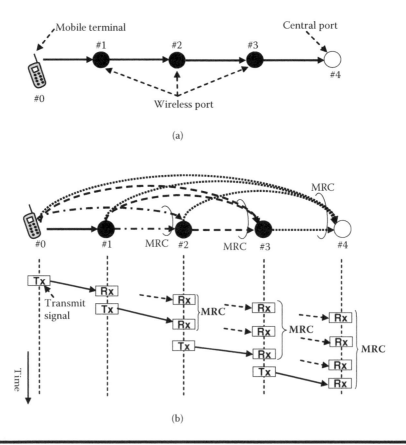

(a)

(b)

Figure 14.5   Multi-Hop Diversity Relay, (a) Multi-Hop Relay, (b) MHMRC Diversity Relay.

port ♯2. Therefore, port ♯2 receives the same signal twice, first from port ♯0 and then from port ♯1. Therefore, port ♯2 can combine them before relaying the signal to port ♯3, which can combine the 3 received signals before relaying the signal to port ♯4. Thus, port ♯4 can receive the same signal four times to combine. During the relaying process, a WP may also receive the signals transmitted from its next ports. However, those signals from the next ports will be received after having sent the signal and therefore cannot contribute to multi-hop diversity combining. For diversity combining, the well-known MRC [7] can be used. Using the MHMRC relay, the port transmit power can be reduced. Because the same signals transmitted from the previous ports have been received before the signal from the immediately previous port was received, the delay time of MHMRC is the same as that of the simple multi-hop relay.

To evaluate the transmit power reduction, the numerical expressions of transmit power are derived below. In DS-CDMA the RAKE receiver can collect the signal energy from all the received signal paths [23]. We assume an ideal RAKE combining and an L-path Rayleigh fading channel. The received signal power $P_r(j)$ at port ♯j from port ♯i is given by

$$P_r(j) = P_t(i) \cdot d_{i,j}^{-\alpha} \cdot 10^{-\frac{\eta_{i,j}}{10}} \cdot \sum_{l=0}^{L-1} |\xi_{i,j}(l)|^2, \qquad (14.6)$$

where $P_t(i)$ is the transmit power from port ♯i, α is the path loss exponent and $d_{i,j}$, $\eta_{i,j}$, and $\xi_{i,j}$ are, respectively, the distance, the shadowing loss (in dB), and the l-th path complex path gain between WPs ♯i and ♯j. Assuming a uniform power delay profile of the multipath channel, $\xi_{i,j}$ are independent complex Gaussian variables with zero mean and $E[|\xi_{i,j}|^2] = 1/L$, where $E[*]$ denotes the ensemble average operation. We assume an ideal TPC based on the signal-to-noise power ratio (SNR) measurement. For a multi-hop relay without diversity, the transmit power $P_t(i)$ is given by

$$P_t(i) = \frac{P_{target}}{d_{i,j}^{-\alpha} \cdot 10^{-\frac{\eta_{i,j}}{10}} \cdot \sum_{l=0}^{L-1} |\xi_{i,j}(l)|^2}, \qquad (14.7)$$

where P_{target} is the target received signal power.

To determine the total transmit power along the MHMRC route, we consider an n-hop connection from the MT to the CP; port ♯$i = 0$ is the MT and port ♯$i = n$ is the CP, whereas port ♯$i = 1 \sim n-1$ is the intermediate port. The received power $P_r(1)$ from the MT ♯$i = 0$ is given by

$$P_r(1) = P_t(0) \cdot d_{0,1}^{-\alpha} \cdot 10^{-\frac{\eta_{0,1}}{10}} \cdot \sum_{l=0}^{L-1} |\xi_{0,1}(l)|^2. \qquad (14.8)$$

Therefore, the MT transmit power $P_t(0)$ is expressed as

$$P_t(0) = \frac{P_t(i)}{d_{0,1}^{-\alpha} \cdot 10^{-\frac{\eta_{0,1}}{10}} \cdot \sum_{l=0}^{L-1} |\xi_{0,1}(l)|^2}. \qquad (14.9)$$

For ports $\sharp i = 1 \sim n - 2$, the received power $P_r(i + 1)$ at port $\sharp(i + 1)$ is the sum of all the received powers from all the previous ports and is given by

$$P_r(i + 1) = \sum_{j=0}^{i} P_t(j) d_{j,i+1}^{-\alpha} 10^{-\frac{\eta_{j,i+1}}{10}} \sum_{l=0}^{L-1} |\xi_{j,i+1}(l)|^2$$

$$= P_t(i) d_{i,i+1}^{-\alpha} 10^{-\frac{\eta_{i,i+1}}{10}} \sum_{l=0}^{L-1} |\xi_{i,i+1}(l)|^2 + \sum_{j=0}^{i-1} P_t(j) d_{j,i+1}^{-\alpha} 10^{-\frac{\eta_{j,i+1}}{10}} \sum_{l=0}^{L-1} |\xi_{j,i+1}(l)|^2.$$

$$(14.10)$$

Because $P_r(i) = P_{target}$ with TPC, the transmit power $P_t(i)$ of the port $\sharp i$ is given by

$$P_t(i) = \frac{P_{target} - \sum_{j=0}^{i-1} P_t(j) d_{j,i+1}^{-\alpha} 10^{-\frac{\eta_{j,i+1}}{10}} \sum_{l=0}^{L-1} |\xi_{j,i+1}(l)|^2}{d_{i,i+1}^{-\alpha} \cdot 10^{-\frac{\eta_{i,i+1}}{10}} \cdot \sum_{l=0}^{L-1} |\xi_{i,i+1}(l)|^2}.$$

$$(14.11)$$

Transmit power $P_t(j)$ is obtained recursively from (14.9) and (14.11). If the received power $P_r(i + 1)$ at the port $\sharp(i + 1)$ from the other previous ports, $\sharp 0 \sim \sharp i$, is larger than the target received power P_{target}, i.e., $P_r(i+1) > P_{target}$, the port $\sharp i$ can be removed from the constructed route, i.e., $P_t(i) = 0$; the transmit power $P_t(i - 1)$ of the port $\sharp(i - 1)$ becomes

$$P_t(i - 1) = \frac{P_{target} - \sum_{j=0}^{i-2} P_t(j) d_{j,i+1}^{-\alpha} 10^{-\frac{\eta_{j,i+1}}{10}} \sum_{l=0}^{L-1} |\xi_{j,i+1}(l)|^2}{d_{i-1,i+1}^{-\alpha} \cdot 10^{-\frac{\eta_{i-1,i+1}}{10}} \cdot \sum_{l=0}^{L-1} |\xi_{i-1,i+1}(l)|^2}.$$

$$(14.12)$$

Using this route modification algorithm, the number of hops decreases and consequently, the delay time also decreases.

14.3.3 Computer Simulation

MTs and WPs are randomly located in each VC. The average total transmit power along the route from an MT to the CP is evaluated by computer simulation. To limit the relay time, the maximum allowable number of hops is limited to N. The number K of WPs in each VC is also an important design parameter. We evaluate the impact of the radio propagation parameters (path loss exponent α, the shadowing standard deviation σ, the number L of propagation paths, and also the fading correlation ρ between the control channel and data communication channel) on the average total transmit power. The total transmit power P_{total} is given by the sum of the transmit powers along the route:

$$P_{total} = \sum_{i=0}^{n-1} P_t(i).$$

$$(14.13)$$

The normalized average power P_{norm} with the MHMRC diversity is defined as the average total transmit power along the route normalized by that of the single-hop case (conventional cellular case), i.e., $P_{norm} = E[P_{total}]/E[P_{single-hop}]$, where P_{total} is given by (14.13) and $P_{single-hop}$ is given by (14.9) with $i = 0$ (MT) and $j = n$ (CP). Therefore, P_{norm} is given by

$$P_t(i) = \frac{E[P_{total}]}{E[P_{single-hop}]} = \frac{E\left[\sum_{i=0}^{n-1}\left(\frac{1 - \sum_{j=0}^{i-1}\frac{P_t(j)}{P_{target}}d_{j,i+1}^{-\alpha}10^{-\frac{\eta_{j,i+1}}{10}}\sum_{l=0}^{L-1}|\xi_{j,i+1}(l)|^2}{d_{i,i+1}^{-\alpha}10^{-\frac{\eta_{i,i+1}}{10}}\sum_{l=0}^{L-1}|\xi_{i,i+1}(l)|^2}\right)\right]}{E\left[\frac{1}{d_{0,n}^{-\alpha}10^{-\frac{\eta_{0,n}}{10}}\sum_{l=0}^{L-1}|\xi_{0,n}(l)|^2}\right]}.$$

(14.14)

The transmit power $P_t(j)$ is given by (14.9) for port $\sharp j = 0$, and is given by (14.11) for ports $\sharp j = 1 \sim n-1$. Because $P_t(0) \propto P_{target}$, it can be easily understood that $P_t(j)/P_{target}$ is not a function of P_{target}. As a consequence, P_{norm} does not depend on P_{target}.

Figure 14.6 plots the normalized average total transmit power as a function of N with ρ as a parameter for $\alpha = 3.5$, $\sigma = 7$ dB, $L = 2$, and $K = 50$. Clearly, multi-hop VCN can significantly reduce the total transmit power. When $N = 4$, with MHMRC (without MHMRC) total multi-hop transmit power can be reduced to -24 dB (-22 dB) of that of the conventional cellular case ($N = 1$). Also, the normalized total multi-hop transmit power is almost the same for $N > 4$. This suggests that the maximum allowable number of hops can be limited to avoid an unnecessary long relay time. In addition, the MHMRC decreases the total transmit power for all values. The power reduction by

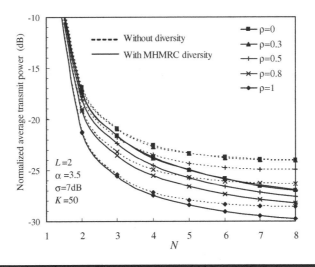

Figure 14.6 Impact of ρ on the Transmit Power.

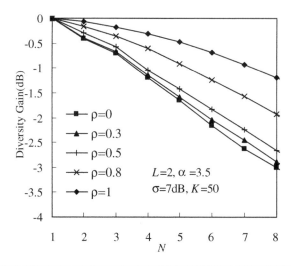

Figure 14.7   Impact of ρ on the Diversity Gain.

MHMRC is larger when ρ becomes smaller. For $\rho \geq 0.5$, the MHMRC diversity gain gets smaller, and becomes very small when $\rho = 1$. This is because, when $\rho = 1$, the route construction channel and the data communication channel experience the same fading, and hence the data communication route is also the minimum transmit power route. The fading correlation property between the control channel and the data communication channel may be given by [18]

$$\rho = \frac{1}{L}\frac{\sin\left(\frac{2\sqrt{3}\pi L}{\sqrt{L^2-1}}\Delta f\tau_{rms}\right)}{\sin\left(\frac{2\sqrt{3}\pi}{\sqrt{L^2-1}}\Delta f\tau_{rms}\right)}\exp\left(j2\sqrt{3}\pi\sqrt{\frac{L-1}{L+1}}\Delta f\tau_{rms}\right), \qquad (14.15)$$

where Δf is the carrier frequency separation between the two channels and τ_{rms} is the root mean square (rms) delay spread of the fading channel. When $L = 2$ and $\Delta f \cdot \tau_{rms} = 0.165$, the frequency separation is $\Delta f = 165$ kHz for $\rho = 0.5$.

To evaluate the MHMRC power reduction, we computed the MHMRC diversity gain as a function of N. The diversity gain is defined as the ratio of the average total transmit powers with and without MHMRC diversity. Figure 14.7 plots the MHMRC diversity gain (in dB) as a function of N with ρ as a parameter for $K = 50$, $\alpha = 3.5$, $\sigma = 7$ dB, and $L = 2$. It is seen from this figure that as ρ decreases, the diversity gain increases; it is about 3.7 dB when $\rho = 0$ and $N = 10$.

As understood from (14.11), the transmit power of each WP is affected by the propagation parameters. Below we evaluate the impact of the radio propagation parameters (path loss exponent α, the shadowing standard deviation σ, the number L of propagation paths, and also the fading correlation ρ between the control channel and data communication channel) on the MHMRC diversity gain.

Figure 14.8 plots the MHMRC diversity gain as a function of N with α as a parameter for $K = 50$, $\sigma = 7$ dB, $L = 2$, and $\rho = 0$. It is seen that as α decreases, the diversity gain increases. A reason for this is discussed below. Comparing (14.9) and (14.11), we can see that the MHMRC diversity gain depends on the second term of the numerator of (14.11), i.e., $\sum_{j=0}^{i-1} P_t(j)d_{j,i}^{-\alpha}10^{-\frac{\eta_{j,i}}{10}}\sum_{l=0}^{L-1}|\xi_{j,i}(l)|^2$, which increases as α decreases, and hence the transmit power reduces. Consequently, the diversity gain increases as α decreases.

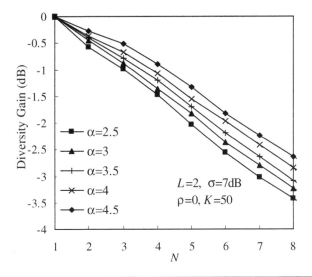

Figure 14.8   Impact of α on the Diversity Gain.

Figure 14.9 plots the MHMRC diversity gain as a function of N with σ as a parameter for $K = 50$, $\alpha = 3.5$, $L = 2$, and $\rho = 0$. As σ increases, the diversity gain decreases. This can be explained below. As σ increases, the route selection diversity effect increases [8]. Therefore, the port transmit power even without MHMRC reduces and hence, the second term of the numerator of (14.11) decreases. As a consequence, the diversity gain decreases as σ increases.

Figure 14.10 plots the MHMRC diversity gain as a function of N with L as a parameter for $\alpha = 3.5$, $\sigma = 7$ dB, $K = 50$, and $\rho = 0$. As L decreases, the MHMRC gain increases. The reason for this is given below. As L decreases, the variations of $\sum_{l=0}^{L-1} |\xi_{j,i}(l)|^2$ in the second term of the numerator of (14.11) increase; therefore, the diversity gain increases as L decreases.

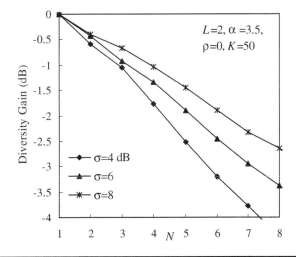

Figure 14.9   Impact of σ on the Diversity Gain.

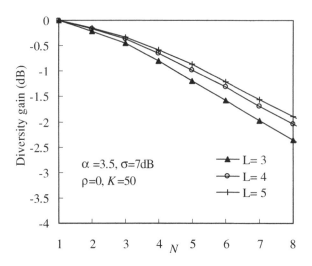

Figure 14.10   Impact of L on the Diversity Gain.

The number K of WPs in each VC is also an important design parameter. Figure 14.11 plots the MHMRC diversity gain as a function of N with K as a parameter for $\alpha = 3.5$, $\sigma = 7$ dB, $L = 2$, and $\rho = 0$. As K increases the MHMRC gain decreases. This is because as K increases, the possibility of choosing a smaller transmit power route increases; therefore, the port transmit power without MHMRC diversity can be reduced; hence, the second term of the numerator of (14.11) decreases also, thereby the diversity gain decreases as K increases.

Using the route modification algorithm with MHMRC diversity, the number of hops reduces, and this reduces the data relay time between the MT and the CP. Figure 14.12 plots the cumulative distributions of the number of hops with and without MHMRC cases with N as a parameter for $L = 2$, $\alpha = 3.5$, $\sigma = 7$ dB, $\rho = 0$, and $K = 50$. When $N = 5$, the number of hops at the probability of 90% is almost the same for both with and without MHMRC; however, when $N = 10$, it is 8 hops using MHMRC, while 9 hops

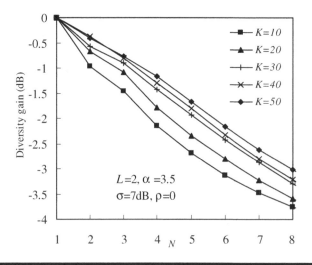

Figure 14.11   Impact of K on the Diversity Gain.

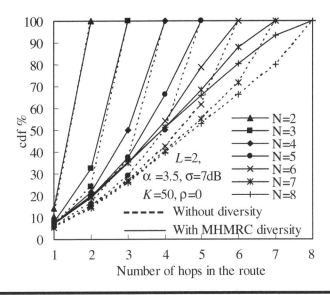

Figure 14.12   Cdf of the Number of Hops in the Route With *N* as a Parameter for With and Without MHMRC Cases.

without diversity. To evaluate the relay time reduction, we compute the distribution of the difference between the number of hops without MHMRC diversity and that with MHMRC diversity. Figure 14.13 plots the distribution of the reduction in the number of hops by MHMRC diversity with N as a parameter for $L = 2$, $\alpha = 3.5$, $\sigma = 7$ dB, $\rho = 0$, and $K = 50$. It can be seen that as the maximum allowable number of hops N increases, more hops can be reduced. This is because as N increases, the number of diversity branches (or the number of the same signals received from previous ports along the route) increases; thereby a larger diversity gain is obtained, and thus, more numbers of hops can be removed by using the route modification algorithm.

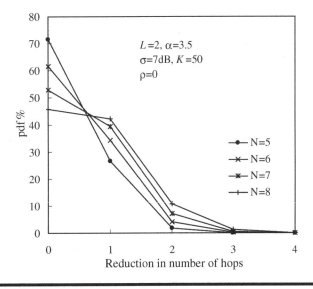

Figure 14.13   Distribution of the Reduction in the Number of Hops by MHMRC.

14.4   WIRELESS MULTI-HOP CHANNEL ALLOCATION

If the same frequency channel is used for multi-hop transmission, it may be a source of high interference causing the degradation of the system capacity. To avoid the interference in the multi-hop links, the available frequency band is divided into several frequency channels and different frequency channels are allocated to adjacent multi-hop links. In this section, a channel refers to a frequency channel. In the wireless multi-hop VCN, an efficient channel allocation algorithm is necessary. Channel allocation schemes are classified into fixed channel allocation (FCA) and dynamic channel allocation (DCA) schemes [19]. Using FCA, predetermined fixed channels are allocated to each WP. FCA cannot adapt to the change in traffic conditions and user distributions. On the other hand, using DCA, all channels are available at each WP and one of the available channels is allocated if the channel meets the required quality. DCA can be implemented either in a centralized or a distributed fashion [19]. The latter is promising for the multi-hop VCN. In this section, we introduce an on-demand strategy to allocate channels to the multi-hop uplinks between the MT and the CP, using a CS-DCA algorithm [20] with some modifications to meet with the wireless constraints in a DS-CDMA VCN [21]. Using computer simulation, comparison between multi-hop VCN and present cellular network blocking probabilities is given.

In high traffic load, due to decreased transmit power in the multi-hop links, the decreased interference leads to less blocking probability compared to the present cellular network. The performance difference between the two networks depends on the number of the channels available. Improving the blocking probability performance leads to an increase in the supportable load of the network for the given blocking probability.

14.4.1   Application of CS-DCA to Uplink Multi-Hop Communication

The problem of dynamic channel assignment has been extensively considered in the context of cellular networks [19]. However, there are significant differences between the two networks. The multi-hop communication imposes additional complexity, as non-conflicting channels must be allocated to the wireless links along the route.

As illustrated in Figure 14.14, we will consider the data transfer on a link, $WP_T - WP_R$, using channel f_i. For this link allocation to be made successfully, the following criteria need to be satisfied:

(a) WP_T must not receive from any other WP using channel f_i. Otherwise, the transmission from WP_T will interfere at WP_T.

(b) WP_R must not be involved in any other call transmission in channel f_i. Otherwise, the transmission from WP_T will interfere at WP_R.

We consider that the system bandwidth is divided into several frequency channels with different carrier frequencies. One of the available channels is allocated to a link between the MT and its nearest WP or between two adjacent WPs along a multi-hop

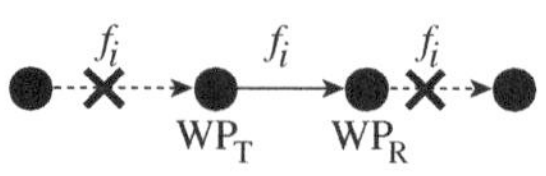

Figure 14.14   Interference Model for an Uplink Multi-Hop Transmission.

Table 14.1   An Example of Channel Priority

Channel Index i	Channel Priority Value P(i)	Number of Times the Channel Is Tested N(i)
0	0.8	20
1	0.6	15
. . .	. . .	. . .
$C-1$	0.1	10

route. Because DS-CDMA is applied, the same channel can be shared by different multi-hop links. In this case, a link ($WP_T - WP_R$) can serve multiple calls in the same frequency channel, using different orthogonal spreading codes, if the wireless constraints are satisfied, resulting in an efficient usage of the limited frequency resource.

As described in the previous section, the construction procedure of the route between each WP and the CP, based on the total transmit power minimization criterion, is carried out using a control channel having a carrier frequency different from the channels for multi-hop communications. After the whole route between the MT and the CP is constructed, the CS-DCA [20], which was proposed for single-hop link calls in the present cellular networks, is applied to assign frequency channels to all the uplinks along the MT-CP path. The single-hop link assignment procedure is repeated in a sequence over the multi-hop path. If at any link of the multi-hop path there is no channel available, the call is blocked.

In CS-DCA, each receiving side WP is equipped with a channel priority table (as in [20]); priority function value and the number of times the channel was checked are listed. Table 14.1 shows an example of a channel priority table. In Table 14.1 and hence forth, C is the number of available channels. In the channel priority table, priority value and the number of times the channel is tested are listed. The channel priority value is defined as: (the number of times the channel has been successfully allocated)/(the number of times the channel has been tested). The channel priority value is updated as follows. When the channel is successfully allocated, the channel priority value P is updated as

$$(NP + 1)/(N + 1) \to P, \tag{14.16}$$

where N is the number of times the channel has been tested. When the channel is not successfully allocated, P is updated as

$$NP/(N + 1) \to P. \tag{14.17}$$

The probability that the channel is usable (i.e., the signal-to-interference plus noise power ratio (SINR) of that channel is larger than the required value) is denoted as q. After the channel priority value is updated many times, P approaches a certain value in a steady state, where the following relationship holds

$$q\frac{NP + 1}{N + 1} + (1 - q)\frac{NP}{N + 1} = P. \tag{14.18}$$

From (14.18), we have $q = P$, i.e., the channel priority value P approaches the probability that the channel is usable. Therefore, the channel whose usable probability q is higher is selected more frequently.

The WP receiver selects a channel among available ones using its channel priority table. The CS-DCA procedure for one link in the multi-hop communication is as follows.

Step 1: For a link ($WP_T - WP_R$), if a channel f_0 is allocated to the same link in another call, WP_R selects f_0 and measures the SINR. If the measured SINR meets the required quality, the selected channel is allocated and its priority value is increased. Otherwise, the priority value of that channel is decreased and the procedure goes to the next step.

Step 2: The WP_R selects the channel with the highest priority among the unchecked channels.

Step 3: If the channel is used for transmitting data in another call or is used by the WP_T for reception, the WP_R decreases the priority value of the selected channel and goes back to Step 2. Otherwise, the procedure goes to the next step.

Step 4: The WP_R measures the SINR of the selected channel. If the measured SINR is larger than the required SINR Λ_{req}, the channel is allocated and its priority value is increased. Otherwise, the priority value of the selected channel is decreased and the procedure goes back to Step 2.

If the channel allocations for links over the multi-hop route from the MT to the CP are successful, the call is established. Otherwise, the call is blocked.

14.4.2 Computer Simulation

In this section, the blocking probability is evaluated by computer simulation. Comparison between the multi-hop VCN and the present cellular network is given. The impact of different parameters on the blocking probability is also discussed.

A total of 19 VCs of hexagonal layout (the center VC is the cell of interest) are considered. For a fair comparison, the CP of each VC is set in the middle of the cell. We assume that the overall network traffic arrival follows a Poisson distribution with a mean arrival rate of v, the holding time of each call is exponentially distributed with a mean of μ [22]. The offered load G per cell is defined as

$$G = v\mu. \tag{14.19}$$

The call arrival events are generated before the start of the simulation and are known *a priori* on the time scale.

In CS-DCA, the measurement of the SINR is necessary. The SINR is affected by distance dependent path loss, shadowing loss, and multi-path fading. We assume L-path Rayleigh fading with uniform power delay profile. Assuming quadrature phase shift keying (QPSK) data modulation and an ideal SNR-based fast TPC, the required SINR Λ_{req} for a required bit error rate (BER) of $10^{-2}(10^{-3})$ is given by 7.3 dB (9.8 dB) [23].

The propagation channel can be modeled as the product of distance dependent path loss, log-normally distributed shadowing loss, and multi-path fading [7]. Assuming an L-path fading channel with uniform power delay profile, the received power $P_r(i, j)$ of the signal transmitted from the WP $\sharp i$ and received at the WP $\sharp j$ is given by

$$P_r(i, j) = P_t(i) d_{i,j}^{-\alpha} 10^{-\frac{\eta_{i,j}}{10}} \sum_{l=0}^{L-1} |\xi_{i,j}(l)|^2, \tag{14.20}$$

where $P_t(i)$ is the transmit power of the WP $\sharp i$, α is the path loss exponent, and $d_{i,j}$, $\eta_{i,j}$, and $\xi_{i,j}(l)$ are, respectively, the distance, log normally distributed shadowing loss with the standard deviation σ in dB, and the l-th path's complex path gain between the WPs $\sharp i$ and $\sharp j$. Also, $\{\xi_{i,j}(l); i, j, l\}$ are characterized by time-invariant independent (but location dependent) complex Gaussian variables with zero mean and a variance of $1/L$. SNR-based ideal TPC is assumed. The transmit power $P_t(i)$ is determined as

$$\frac{P_t(i)}{P_{noise}} = \Lambda_{target} d_{i,j}^{\alpha} 10^{\frac{\eta_{i,j}}{10}} \left(\sum_{l=0}^{L-1} |\xi_{i,j}(l)|^2 \right)^{-1}, \tag{14.21}$$

where Λ_{target} is the target SNR and P_{noise} is the noise power.

Using (14.20) and (14.21) and assuming ideal L-finger coherent RAKE combining based on MRC, the SINR after RAKE combining at WP $\sharp i$ is given by [9]

$$\lambda = \Lambda_{target} \sum_{l=0}^{L-1} \lambda_l, \tag{14.22}$$

where

$$\lambda_l = \cfrac{\cfrac{|\xi_{i,j}(l)|^2}{\sum_{l=0}^{L-1} |\xi_{i,j}(l)|^2}}{1 + \cfrac{\Lambda_{target}}{SF} \left\{ (m+1) \cdot \left[1 - \cfrac{|\xi_{i,j}(l)|^2}{\sum_{l=0}^{L-1} |\xi_{i,j}(l)|^2} \right] + \sum_{\substack{k \\ ((k,q(k)) \neq (i,j))}} \left(\cfrac{r_{k,j}}{r_{k,q(k)}} \right)^{-\alpha} 10^{\frac{-(\eta_{k,j} - \eta_{k,q(k)})}{10}} \cfrac{\sum_{l=0}^{L-1} |\xi_{k,j}(l)|^2}{\sum_{l=0}^{L-1} |\xi_{k,q(k)}(l)|^2} \right\}}. \tag{14.23}$$

In (14.23), SF is the spreading factor and m is the number of times the same link ($\sharp i$, $\sharp j$) is being used for different calls. Also, $\sharp q(k)$ is the index of the receiver WP of the interfering link ($\sharp k$, $\sharp q(k)$). The second term of the denominator is its own interpath interference (IPI) and the third term is the interference from other transmitters in other links. Because the orthogonal spreading code is assumed, the same path interference is suppressed. The expression derived above is used for computing the SINR in the computer simulation.

First, we show in Figure 14.15 an example of the distribution of channels allocated by the CS-DCA (the number indicates the channel index) where the number C of available channels is $C = 4$, $G = 4$, $N = 5$, $SF = 16$, $\alpha = 3.5$, $\sigma = 6$ dB, and $L = 2$. The CP is located in the middle of the VC. The other 19 WPs, each having omnidirectional transmit/receive antenna, are randomly located. We can see that the same channel (e.g., channel $\sharp 3$), is reused for many links in the communications resulting in an efficient frequency usage. Also we can see that the same channel (e.g., channel $\sharp 1$ and $\sharp 2$) is

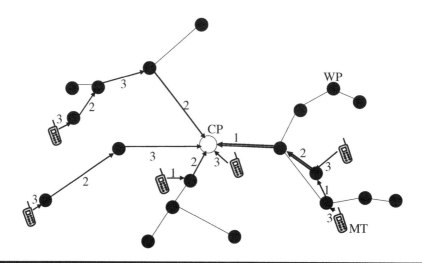

Figure 14.15   Example of Multi-Hop Calls Channel Allocation.

used for the same link (WP-WP) for two different users resulting in efficient frequency usage.

Next, the simulation results for the blocking probability in both the multi-hop VCN and the present DS-CDMA cellular network are presented and discussed. These results are evaluated as a function of the average offered cell load.

1. Impact of the number of available channels.

A comparison of the blocking probabilities of the multi-hop VCN and the present cellular network is shown in Figure 14.16 with C as a parameter, for $SF = 16$, $N = 4$, $\alpha = 3.5$, $\sigma = 6$ dB, and $L = 2$. In Figure 14.16(a), the required SINR Λ_{req} is assumed to be 7.3 dB, while in Figure 14.16(b) it is 9.8 dB. In both cases, we see the same trend of blocking probabilities of the two networks; the performance difference between the two networks depends on C. In high-traffic load, the multi-hop VCN performance overcomes the present cellular network. This is because in high-traffic load, the large intercell interference (from neighboring cells) affects the performance of the present cellular network. However, with using the multi-hop communication, the total transmit power needed for each communication is decreased leading to less interference to neighboring cells. If frequency channels are few, like in the case of $C = 2$, then the impact of channel exhaustion results in almost equal blocking probability in the multi-hop VCN and the present cellular network. However, as more frequency channels are available, the blocking reduction from multi-hopping is larger as shown from the curves of $C = 4$ and 8. As a consequence, the multi-hop VCN can provide smaller blocking probability while significantly reducing the MT transmit power. The detailed blocking performance in the multi-hop VCN is discussed below.

The blocking in the multi-hop VCN can occur because of two major contributing factors: poor coverage ($SINR < SINR_{req}$) and unavailability of free channels (because they are used in the adjacent links). In Figure 14.17, we evaluate the contribution of these two factors in the multi-hop VCN performances of Figure 14.16(a), for $C = 2$, 4, and 8 when $\Lambda_{req} = 7.3$ dB, $SF = 16$, $N = 4$, $\alpha = 3.5$, $\sigma = 6$ dB, and $L = 2$. Figure 14.17 gives the contribution percentage of the total blocking calls, where the blocking occurs because n channels are used in the adjacent links and m channels have poor coverage. In Figure 14.17 the n and m explained above are expressed in the (n, m) shown on the

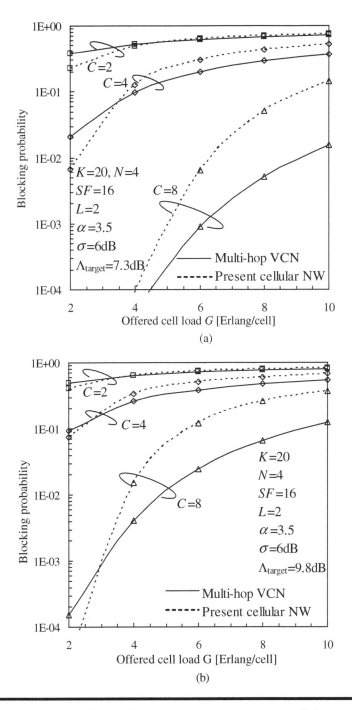

Figure 14.16 Blocking Probabilities of the VCN and the Present Cellular Network (CN) for Different Number *C* of Available Channels. (a) $\Lambda_{target} = 7.3\,dB$, (b) $\Lambda_{target} = 9.8\,dB$.

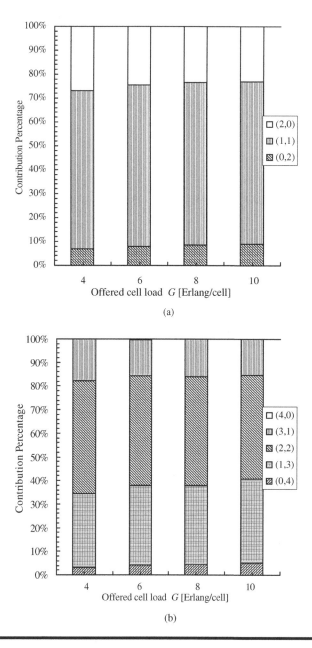

Figure 14.17 Contribution of Different Factors in Blockings of the Multi-Hop VCN for Different C, (a) $C = 2$, (b) $C = 4$.

right of the graphs. It can be seen that the unavailability of many channels, because they are used in the adjacent links, causes a high blocking probability. This leads to a need to several channels in the multi-hop VCN.

2. Impact of number K of WPs per VC.

The impact of number K of WPs on the blocking probability is shown in Figure 14.18 for $C = 8$, $N = 4$, $\alpha = 3.5$, $\sigma = 6$ dB, $L = 2$, and $SF = 16$. We can see that as K increases, the blocking probability reduces. This is due to a further decrease of the transmit power

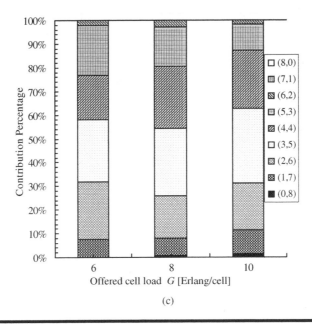

Figure 14.17   (Continued) (c) $C = 8$.

of the multi-hop links, and hence less interference on the hops. As K increases, the probability to find a route with smaller transmit power increases.

3. Impact of propagation parameters.

As understood from (14.23), the SINR of each port is affected by the propagation parameters (α, σ, and L). Below, we evaluate the impact of α, σ, and L on the blocking probability.

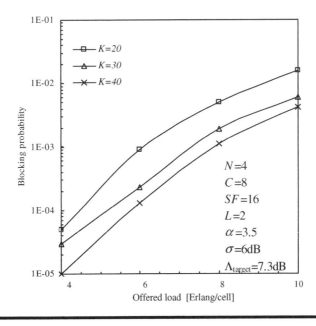

Figure 14.18   Impact of Number K of WPs Per VC on Blocking Probability.

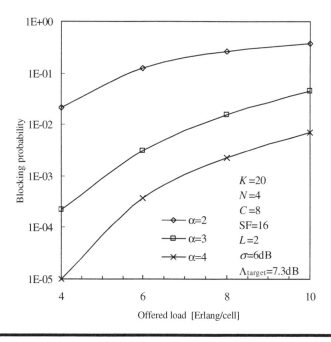

Figure 14.19   Impact of α on Blocking Probability.

The impact of α is shown in Figure 14.19 for $K = 20$, $N = 4$, $C = 8$, $\sigma = 6$ dB, $L = 2$, and $SF = 16$. We can see that as α increases, the blocking probability decreases. This is due to the decrease of the interference when α increases.

The impact of σ on the blocking probability is shown in Figure 14.20 for $K = 20$, $N = 4$, $C = 8$, $\alpha = 3.5$, $L = 2$, and $SF = 16$. The blocking probability is almost insensitive to σ. Possible reasons for this are discussed below. Increasing σ means large

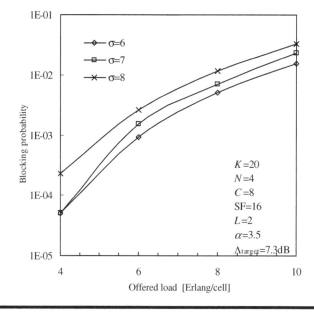

Figure 14.20   Impact of σ on the Blocking Probability in the Multi-Hop VCN.

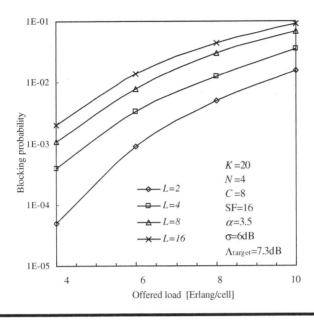

Figure 14.21   Impact of _L_ on the Blocking Probability in the Multi-Hop VCN.

variations in the shadowing losses between different links. This can decrease the transmit power of the multi-hop routes from the WPs to the CP, and hence can decrease the interference to other links. Therefore, the blocking probability can be decreased. On the other hand, increasing σ may lead to an increase in the interference power, thereby increasing the blocking probability. As a consequence, the blocking probability becomes almost insensitive to σ.

The impact of the number L of paths on the blocking probability is shown in Figure 14.21 for $K = 20$, $N = 4$, $C = 8$, $SF = 16$, $\alpha = 3.5$, and $\sigma = 6$ dB. We can see that the trend of the blocking probability changes at a certain value of L; in the used conditions this happens when $L = 2$. This is because increasing L can increase the path diversity effect obtained by RAKE combining and hence the probability of the SINR falling below the required value can decrease. Therefore, the blocking probability can decrease. However, as L increases, the blocking probability starts to increase due to a high interpath interference (the first term inside the brackets of the denominator of (14.23)).

14.5   CONCLUSIONS

A wireless multi-hop VCN concept for high-speed packet data services in future mobile communication was introduced. In VCN, each VC consists of many distributed WPs, and a group of distributed WPs acts as one base station in the present cellular network. Because an MT communicates with its surrounding WPs (the end WPs) simultaneously, then each end WP can act as a site diversity branch and therefore, the VCN can significantly reduce the MT transmit power. For data relaying between the CP and the end WP, a wireless multi-hop technique is used. The total transmit power minimization route construction algorithm was introduced to reduce the interference power and increase the frequency efficiency. The channel segregation dynamic channel allocation CS-DCA algorithm was introduced as an efficient channel allocation technique. However, before the realization of the multi-hop VCN, there are many technical issues such as packet scheduling, resource management, etc.

REFERENCES

[1] F. Adachi, M. Sawahashi, and H. Suda, "Wideband DS-CDMA for next generation mobile communication systems," *IEEE Commn. Mag.*, vol. 36, no. 9, pp. 55–69, Sept. 1998.

[2] 3GPP, TP25.848, "Physical layer aspects of UTRA high speed downlink packet access," v. 4.0.0, March 2006.

[3] K. Imai, W. Takita, S. Kano, and A. Kodate, "An extension of 4G mobile networks towards the ubiquitous real space," *IEICE Trans. Commn.*, vol. E88-B, no. 7, pp. 2700–2708, July 2005.

[4] F. Adachi, "Wireless past and future-evolving mobile communication systems," *IEICE Trans. Fundamentals*, vol. E84-A, no. 1, pp. 55–60, Jan. 2001.

[5] F. Adachi, D. Garg, S. Takaoka, and K. Takeda, "Broadband CDMA techniques," *IEEE Wireless Commn. Mag.*, vol. 12, no. 2, pp. 8–18, April 2005.

[6] ITU-R Recommendation, Future public land mobile telecommunication systems (FPLMTS), Rec. ITU-R M.687-2, 1997.

[7] W.C. Jakes, Jr., Ed., *Microwave Mobile Communication*, John Wiley & Sons, New York, 1974.

[8] E. Kudoh and F. Adachi, "Power and frequency efficient multi-hop virtual cellular concept," *IEICE Trans. Commn.*, vol. E88-B, no. 4, pp. 1613–1621, April 2005.

[9] E. Kudoh and F. Adachi, "Distributed dynamic channel assignment for a multi-hop DS-CDMA virtual cellular network," *IEICE Trans. Commn.*, vol. E88-B, no. 6, pp. 2525–2531, June 2005.

[10] T. Otsu, Y. Aburakawa, and Y. Yamao, "Multi-hop wireless link system for new generation mobile radio access networks," *IEICE Trans. Commn.*, vol. E85-B, no. 8, pp. 1542–1551, Aug. 2002.

[11] R. Bruno, M. Conti, and E. Gregori, "Mesh networks: commodity multi-hop ad hoc networks," *IEEE Commn. Mag.* vol. 43, no. 3, pp. 123-131, March 2005.

[12] E.M. Royer and C.K. Toh, "A review of current routing protocols for ad hoc mobile wireless networks," *IEEE Personal Commn.*, vol. 6, no. 9, pp. 46–55, April 1999.

[13] T. Mukai, H. Murata, and S. Yoshida, "Study on channel selection algorithm and number of established routes of multi-hop autonomous distributed radio network," (in Japanese) *IEICE Trans. Commn.*, vol. J85-B, no. 12, pp. 2080–2086, Dec. 2002.

[14] C.E. Perkins, E.M. Belding-Royer, and S. Das. "Ad hoc on demand distance vector (AODV) routing," *IETF RFC* 3561, July 2003.

[15] A. Fujiwara, S. Takeda, H. Yoshino, T. Otsu, and Y. Yamao, "Capacity improvement with a multi-hop access scheme in broadband CDMA cellular system," (in Japanese) *IEICE. Trans. Commn.*, vol. J85-B, no. 12, pp. 2073–2079, Dec. 2002.

[16] I. Daou, E. Kudoh, and F. Adachi, "Transmit Power Efficiency of Multi-Hop MRC Diversity for a DS-CDMA Virtual Cellular Network," *IEICE Trans. Commn.*, vol. E88-B, no. 9, pp. 2525–2531, Sept. 2005.

[17] I. Daou, E. Kudoh, and F. Adachi, "Transmit power efficiency of multi-hop hybrid selection/MRC diversity for a DS-CDMA virtual cellular network," *Proc. IEEE VTC'2005 Fall*, Sept. 2004.

[18] A. Katoh, E. Kudoh, and F. Adachi, "A study on optimum weights for delay transmit diversity for DS-CDMA in a frequency non selective fading channel," *IEICE Trans. Commn.*, vol. E87-B, no. 4, pp. 838–848, April 2004.

[19] I. Katzela and M. Naghshineh, "Channel assignment schemes for cellular mobile telecommunication systems: a comprehensive survey," *IEEE Personal Commn.*, vol. 3, no. 3, pp. 10–31, June 1996.

[20] Y. Furuya and Y. Akaiwa, "Channel segregation, a distributed adaptive channel allocation scheme for mobile communication systems," *IEICE Trans.*, vol. E74, no. 6, pp. 1531–1537, June 1991.

[21] L. Soundous, E. Kudoh, and F. Adachi, "Blocking probability of a DS-CDMA multi-hop virtual cellular network," *IEICE Trans. Commn.*, vol. E89-A, no. 7, pp. 1875–1883, July 2006.

[22] M. Naghshineh and M. Schwartz, "Distributed call admission control in mobile/wireless network," *IEEE J. Select. Areas Commn.*, vol. 14, pp. 711–717, May 1996.

[23] J.G. Proakis, *Digital Communications*, 3rd ed., New York, McGraw-Hill, 1995.

15

DAS FOR DVB-H NETWORKS

Xiaodong Yang

Contents

The distributed antenna system (DAS) is a transport infrastructure that distributes wireless signals to remote locations from a central point. Digital video broadcasting (DVB) is the emerging broadcast technology. When mobile TV becomes reality, the technology supporting it, DVB for handheld (DVB-H), becomes a focus. The use of repeaters in DVB-H networks is a typical application example of a DAS. This chapter first provides the state-of-the-art introduction of DAS in DVB-H networks, including the application of DAS in DVB-H networks, repeaters, which includes both passive and active repeaters. To examine the application and performance of the DAS system in DVB-H networks, a handover algorithm, called RA_handover, is introduced as a case study to show the application of the DAS in DVB-H. RA_handover is designed to aid the handover process using DAS, namely, repeaters. Simulation and analysis are presented. It is shown that the repeaters in DVB-H networks not only provide better radio signal coverage to the "dead spot" in the DVB-H radio coverage area, but also provide other functions like better quality of service with the virtual cell concept, and provide soft handover support in multifrequency DVB-H networks. Furthermore, the strengths and weaknesses of the DAS for handover in DVB-H networks are discussed. This chapter is concluded by summarizing the current research status and identifying some open research issues on DAS applications in DVB-H networks and a short conclusion.[1]

15.1 DAS AND DVB-H

Digital video broadcasting for handhelds (DVB-H) is a standard for broadcasting internet protocol (IP) data services to portable devices. Handover in unidirectional broadcast networks is a novel issue introduced by this technology. The distributed antenna system (DAS) is a transport infrastructure that distributes wireless signals to remote locations from a central point. One typical example of a DAS in DVB-H is the use of repeaters.

Repeaters provide an efficient way to increase the coverage of broadcasting networks [1] like DVB-H. In DVB-H, the network operators usually put high power transmitters at strategic points first to quickly ensure an attractive coverage, and then in a second step, increase their coverage by placing low power repeaters in the dead spot or shadow areas, such as a tunnel, valley, or indoor area. A repeater is simply a device that receives an analogue or digital signal and regenerates the signal along the next leg of the medium.

Handover in unidirectional broadcast networks is a novel issue introduced by DVB-H. In the last few years since the birth of DVB-H technology, great attention has been given to the performance analysis of DVB-H mobile terminals. Handover is one of the main topics in mobile terminal performance analysis. Better reception quality and greater power efficiency are the key targets of the handover research in DVB-H. Considering the benefits of the DAS, repeaters are used to aid the handover process in DVB-H. Repeater aided handover (RA)_handover is such an algorithm, which will be introduced in detail in the following sections of the chapter.

15.1.1 Introduction of DVB-H

With multimedia broadcast getting more attention from both broadcast and telecommunications operators [2,3], DVB-H [4], as a "one-to-many" broadcast system targeting

---

[1] Inderscience Publishers, 2006. This is a revision of the work published in *International Journal of Services and Standards*, vol. 2, no. 3, pp. 238–256.

personal digital assistants (PDAs), mobile phones, and laptop computers, is now being rolled out.

Digital video broadcasting-Terrestrial (DVB-T) [5] is not suitable for handheld devices partly because it would drain the batteries of a handheld terminal too fast for effective use [6]. DVB-H is designed to overcome this limitation and offers a new outlet for content providers. DVB-H also provides much better service quality for mobile receivers than DVB-T.

The service contents in DVB-H networks will be delivered in the form of IP-packets using IPv4/v6 protocol. Low power DVB-H transmitters offer the possibility of multifrequency cellular DVB-H networks for the broadcast of localized services. With decreasing cell sizes, handover in DVB-H becomes a critical issue. There is already some work that has been reported on handover in multifrequency cellular DVB-H networks and a few papers are available.

Time slicing, multiprotocol encapsulation-forward error correction (MPE-FEC), 4K mode, in-depth interleavers, and DVB-H signaling are the essential elements that are introduced by DVB-H [4]. These features are located in the data link layer and the physical layer of the DVB-H protocol stack. Time slicing (in the data link layer) and DVB-H signaling (in the data link layer and the physical layer) are the two features that are directly related to DVB-H handover.

15.1.1.1 Protocol Stack

The DVB-H protocol stack is shown in Figure 15.1. The newly introduced DVB-H technical features are in the data link layer and the physical layer. The application services can be sent via RTP (real-time protocol) [7] for real-time content (for example, a TV program). Nonreal-time data can be sent via a FLUTE/ALC (file delivery over unidirectional transport/asynchronous layered coding) [8] data carousel (for example, for file downloads). The ESG (electronic service guide) is also broadcast using a FLUTE/ALC.

15.1.1.2 Time Slicing

Time slicing is used in DVB-H to transmit data in periodic bursts. For the same service, the burst bit rates are significantly higher compared with that of DVB-T. Time slicing enables the tuner in a receiver to stay active only a fraction of the time, while receiving

Application Layer	Real Time Content	File Based Content	ESG
Presentation Layer	Source Coding H.264 (Mpeg4)	Source Coding H.264 (Mpeg4)	Coding Encapsulation (XML)
Session Layer	RTP	FLUTE/ALC	
Transport Layer	UDP		
Network Layer	IP (IPv4/IPv6)		
Data Link Layer	MPE (MPE-FEC/Time Slicing)		
	MPEG-2 Transport Stream		
Physical Layer	TPS	DVB-T (4K Mode, In-depth Interleaver)	

Figure 15.1 DVB-H Protocol Stack.

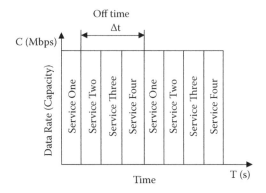

Figure 15.2   Time Slicing Illustration.

bursts of a requested service; this saves battery power. It is claimed that up to a 95% power savings can be achieved compared with conventional and continuously operating DVB-T tuners [9]. The high bit rate signals will be buffered in the receiver memory. A brief performance analysis of a time slicing scheme in DVB-H is done by simulation in paper [10]. Time slicing offers, as an extra benefit, the possibility to use the same front end to monitor neighboring cells between bursts, making seamless soft handover possible. May [11] showed how the off times between the transmissions bursts can be used to perform handovers, how they have to be synchronized, and what boundary conditions exist. A technology called "phase shifting" is proposed as a solution. Time slicing is illustrated in Figure 15.2. It is possible to use a combination of DVB-H (time sliced) and DVB-T (not time sliced) services in a single multiplex shown in Figure 15.3 [12]. However, the power saving is decreased in this case due to a smaller data rate being

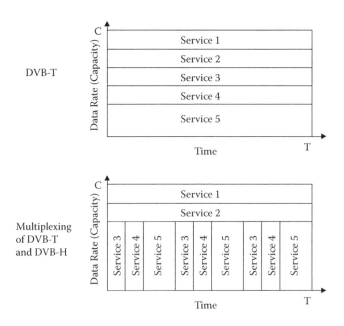

Figure 15.3   DVB-T and Multiplexing of DVB-T and DVB-H.

available for time sliced services [12]. Another benefit of time slicing in DVB-H is that it is unique in terms of the power saving achieved. This means that the same amount of power savings achieved by time slicing in DVB-H could not be obtained when time slicing is used in digital audio broadcasting (DAB) or digital multimedia broadcasting (DMB) [13]. Depending on the transmission bit rate, burst size, and burst duration, the off time Δt in the transmission stream can vary [14,15].

15.1.1.3 MPE-FEC

Multi-Protocol Encapsulation (MPE) is a method to transmit IP data over DVB networks [16]. It specifies the carriage of IP packets within motion picture experts group (MPEG) private data sections. The mobile reception of DVB-H is characterized by nonline of sight, multipath, Doppler impairments, strong propagation loss (especially for indoor reception), poor receiving antenna gain [17], and mobile channel interferences from adjacent TV, global system for mobile communication (GSM) channels and environmental factors like weather and traffic. As a result, accessing a downstream high bitrate service with a handheld terminal is very demanding. The objective of the MPE-FEC is to improve the carrier/noise (C/N) ratio and Doppler tolerance in mobile channels and to improve the tolerance to impulse interference [4]. But MPE-FEC only works within individual time slices [18] and the size of one time slicing burst exactly corresponds to the content of one MPE-FEC frame [19]. So if a single transmission error cannot be corrected, the service drops out not only for the duration of the burst, but also during the time until the next burst is transmitted.

15.1.1.4 4K Mode and In-Depth Interleavers

The 4K mode and the in-depth interleavers affect the physical layer of DVB-H, but do not affect the soft handover directly. However, their objectives are to improve the single frequency network (SFN) planning flexibility and to protect against short noise impulses caused by, e.g., ignition interference and interference from various electrical appliances [4,20]. In this case, they affect the mobile reception of DVB-H signals. The 4K mode offers a tradeoff between mobility and SFN size in the network planning [4]. Because DVB-T does not include this mode, it is an option only in dedicated DVB-H networks [19]. For the 2K and 4K modes, the in-depth interleaver increases the flexibility of the symbol interleaving by decoupling the choice of the inner interleaver from the transmission mode [4,15].

15.1.1.5 DVB-H Signaling

The objective of DVB-H signaling is to provide robust and easy-to-access signaling to DVB-H receivers, thus enhancing and speeding up service discovery [4]. DVB-H signaling is also the main feature of DVB-H that can be directly utilized by the DAS systems. It should be noted that DVB-H is based on DVB-T, and most of the DVB-H specifications in the physical layers are the same as those of DVB-T that can be found in [6]. Besides the specifications in common with DVB-T, DVB-H has unique physical specifications.

There are two important signaling parameters in DVB-H. One is transmission parameter signaling (TPS) signaling bits in the physical layer. The other is program specific information (PSI)/service information (SI) [16,21,22]. PSI/SI is the core signaling for enabling service discovery within DVB-T and also within other DVB systems. Because the PSI/SI used within DVB-H is different to that of other DVB systems, a subset of PSI/SI for IP datacast (IPDC) over DVB-H is defined in [23]. The PSI/SI data enables a DVB-H receiver to discover IPDC over DVB-H specific services from the transport stream and it also provides essential information for enabling handover. The TPS is defined over 68 consecutive orthogonal frequency division multiplex (OFDM) symbols, referred to as one OFDM frame. Each OFDM symbol conveys one TPS bit, so each TPS block contains 68 bits [6]. The TPS bits are located within the physical layer [20] so the signal for synchronization in the handover process is first obtained by utilizing the information contained in the TPS bits [24]. The synchronization word bits aid the receiver in synchronizing with the target frequency. The cell identifier conveys unique cell identification information to the receiver. The PSI/SI provides information on the DVB-H services carried by the different transport streams.

15.1.2 Application of DAS in DVB-H

15.1.2.1 Passive and Active Repeaters

In DVB-H networks, there are two different kinds of repeaters. They are passive repeaters, which are also called gap fillers, and active repeaters that are called regenerative repeaters. A passive repeater receives and retransmits a DVB-H signal without changing the signaling information bits. The signal is only boosted. An active repeater can demodulate the incoming signal, perform error recovery, and then remodulate the bit stream. The output of the error recovery can even be connected to a local remultiplexer to enable insertion of local programs. This means that the entire signal is regenerated. The building blocks of the passive and active repeater configurations are shown in Figure 15.4.

15.1.2.2 Application of Passive and Active Repeaters in DVB-H Networks

The passive repeaters are mostly used as "gap fillers" to cover the area where the main transmitters are not able to provide signals with better quality. On the other hand, because the active repeaters can add extra signaling information to the boosted signals, they are often used to facilitate the handover decision-making process of DVB-H. However, passive repeaters only relay signals without adding extra signaling information, thus it is not possible at the moment for passive repeaters to be used in, for example, DVB-H handovers.

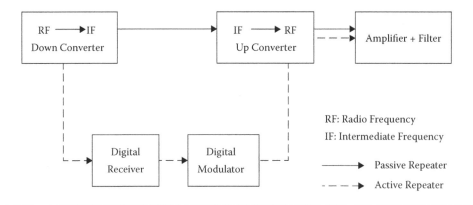

Figure 15.4 Building Blocks of Passive and Active Repeater Configurations.

In spite of the fact that active repeaters have more functionalities, most of the repeaters in DVB-H networks are passive repeaters, partly because the price of active repeaters is much higher, partly because active repeaters are still not mature enough to be widely used in the DVB-H handover process.

15.2 APPLICATIONS OF ACTIVE REPEATERS IN DVB-H NETWORKS

This chapter will only investigate the application of active repeaters in DVB-H, as active repeaters provide additional benefits other than what the common passive repeaters can provide. This novel application of DAS, using repeaters for DVB-H handover, is introduced here as the so-called RA_handover.

RA_handover is designed to decide when soft handover should occur by incorporating active digital repeaters into DVB-H networks. The DVB-H terminal considered in this approach is a DVB-H receiving-capability-only receiver with multiple-input multiple-output (MIMO) antennas [26], so that the receiver can receive and process signals from different transmitters and repeaters at the same time. In this handover approach, an intelligent active repeater structure is proposed where each repeater can add repeater identification bits to the received DVB-H signal and retransmit it to the repeater-covered area. Such an algorithm will greatly improve the quality of service of the received signals and reduce the receiver battery power consumption without considerably increasing the overall cost.

The novel active repeater structure used in the RA_handover algorithm is shown in Figure 15.5, where the TPS adapter adds unique repeater specific information to TPS bits in the transport stream.

In the RA_handover approach, the active repeaters are located in the cell border area. Each repeater-covered area is defined as one subcell. Unlike a passive repeater

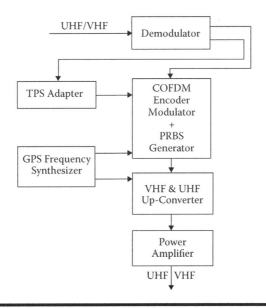

Figure 15.5   Active Repeater Structure in RA Handover.

that simply amplifies and relays an incoming signal, an intelligent active repeater can demodulate the incoming transport stream, add handover scheme information and subcell ID information to the TPS bits, and add subcell ID information to the SI bits in the transport stream.

In the RA_handover approach, the DVB-H receivers are assumed to have MIMO antennas that can provide receiving and decoding capability from different transmitters and repeaters at the same time.

For the proposed RA_handover approach, intelligent active repeaters are put uniformly around the cell borders in a cellular DVB-H network. In fact, there should already be repeaters installed in the broadcast cell borders. A repeater-covered area defines a subcell and when a mobile device moves into such a subcell it receives the subcell's unique repeater identification information from the repeater-transmitted signals, so the mobile device will know in which specific subcell it is located. When the device is in the repeater covered subcell, it will begin to measure the signal strength using the off-burst time. Otherwise, the receiver is in idle mode in the off-burst time. In this way, the measurement frequency in the off burst time is greatly reduced, saving battery power and improving quality of service. In addition, it also reduces the ping-pong effect.

15.2.1 Soft Handover in DVB-H

DVB-H is intended to provide IP data services to mobile handheld devices. To provide diverse IP data services it is expected that a DVB-H cell will usually be smaller than a DVB-T cell, and the multifrequency cellular network structure will be a typical DVB-H network structure. Thus, low power transmitters serving a dense multi-frequency cellular network are expected to be one of the main network structures for DVB-H. Because roaming will be a quite common scenario in future DVB-H networks, handover becomes a critical issue.

15.2.1.1 Introduction of Soft Handover in DVB-H Networks

Handover means the switching of a mobile signal from one channel or cell to another. This chapter defines handover in DVB-H as a change of transport stream and frequency when the receiver moves from one DVB-H cell to another.

Soft handover means that radio links are added and removed in such a way that the device always keeps at least one radio link to a base station [27]. In DVB-H, this means that the received frequency or transport stream is changed without interruption of the ongoing service. When the DVB-H terminal moves from one cell to another, it will try to synchronize with the new frequency of the target cell. This chapter only considers the soft handover process in DVB-H. Wherever the word handover appears in this chapter, it will mean soft handover.

The handover process in cellular telecommunications networks always involves the participation of both the base stations and the handheld terminals. However, in the handover process of DVB-H networks the DVB-H transmitter cannot get a measurement report from a DVB-H receiver. Thus, the handover process will be initialized and completed by the handheld device only. In the case of cellular telecommunications networks, the handover is called active handover because of the involvement of the base station, while the handover in DVB-H networks is called passive handover because the transmitter is not involved in the handover process. Portable devices that can receive only DVB-H services will always use passive handover. However, portable devices that can receive DVB-H services and at the same time have cellular telecommunications capabilities can use both active handover and passive handover. So handover in DVB-H networks can also be divided into active handover and passive handover [28]. This is dependent on whether the network base stations and transmitters control the handover process or not. This chapter will only consider passive soft handover where the receiver is a pure DVB-H receiver.

Handover in DVB-H consists of three stages: handover measurement, handover decision-making, and handover execution [29]. The handover measurement process provides the required measurement parameters, such as RSSI (received signal strength indicator) or SNR (signal noise ratio) to facilitate the handover decision-making process. Parameters such as RSSI or SNR are required for any handover decision-making algorithm. The handover decision-making stage is the stage where the handover decision is made according to predefined handover criteria and the obtained measurement parameters from both the handover measurement stage and the handover decision-making stage. The handover execution process performs the work of synchronizing to signals of the targeted handover cell after the targeted handover cell is chosen in the handover decision-making stage. The RA_handover algorithm presented in this chapter focuses on the handover decision-making process.

15.2.1.2 Time Slicing and Soft Handover

Without time slicing, soft handover in DVB-H would not be possible. Time slicing, which is shown in Figure 15.2, enables DVB-H to transmit data in burst mode. The receiver will only stay active for a fraction of the time and then switch to idle mode to save battery power. In the off-burst duration, the receiver will measure the received signal strength from different cells and perform handover to the targeted cell. Because the receiver only makes handover measurements in the off-burst time, the service will not be interrupted and therefore is a soft handover. An illustration of soft handover with time slicing mode in DVB-H is shown in Figure 15.6. Because the receiver can only receive the data passively without interactive communication with the transmitters, it has to make handover measurements during the off-burst time. The more handover measurements made, the more battery power will be consumed and the more quality

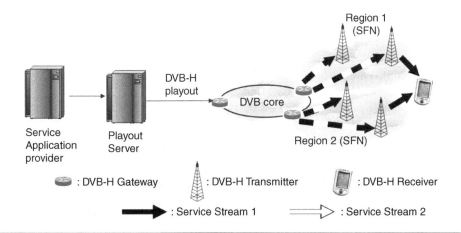

Figure 15.6 Soft Handover Illustration.

Table 15.1 TPS Signaling Information for Handover

Bit Number	Purpose/Content
$S_1 - S_{16}$	Synchronization word
$S_{40} - S_{47}$	Cell identifier
$S_{48} - S_{49}$	DVB-H signaling
$S_{50} - S_{51}$	Handover types
$S_{51} - S_{53}$	Reserved for future use

of service decreases. So finding a way to reduce the handover measurement frequency becomes an important issue in the soft handover process of DVB-H.

15.2.1.3 DVB-H Signaling for Soft Handover

To implement handover in DVB-H, the receiver needs to receive signaling information from the network. There are two kinds of signaling information the DVB-H receiver can use. One is TPS bits in the physical layer. The other is service information (SI) description data that forms part of the DVB-H transport streams [21]. In this chapter, new signaling information for TPS and SI in DVB-H soft handover is proposed and described briefly below.

TPS is defined over 68 consecutive OFDM symbols referred to as one OFDM frame. Each OFDM symbol conveys one TPS bit so each TPS block contains 68 bits [6]. The TPS bits needed for handover are derived from [6] and listed in Table 15.1.

The synchronization word bits in Table 15.1 aid the receiver in synchronizing with the target transport stream or frequency. The cell identifier in Table 15.1 conveys unique cell identification information to the receiver. Bits numbered S51–S53 in Table 15.1 were originally defined as "reserved for future use." All the other TPS bits are already used for certain functions in the DVB standard [6]. Some of these "reserved for future use" bits could be used to identify the RA_handover approach.

The SI data provides information on the DVB-H services carried by the different transport streams. Handover related information in SI is contained in the (network information table) NIT, which is derived from [6] and defined in Table 15.2.

Table 15.2 Handover Related Information in NIT According to [16,21]

Descriptor	Purpose/Content
Network_name_descriptor	Contains network name information
Service_list_descriptor	Contains services listings
Linkage_descriptor	Contains information accessing INT
Frequency_list_descriptor	Contains a list of frequencies for a transport stream
Cell_list_descriptor	Contains a list of cells and subcells including their coverage areas
Cell_frequency_link_descriptor	Contains a list of cells and frequencies used for the transport streams
Terrestrial_delivery_system_descriptor	Contains information about the center frequency, bandwidth, code rate, etc.
Time_slice_FEC_identifier_descriptor	Contains information about the time slicing and MPE-FEC being used

If the cell ID information is announced in the TPS bits, the NIT in SI data will contain both a cell_frequency_link_descriptor and a cell_list_descriptor announcing all cells and subcells within the DVB-H network.

Using the SI and TPS information, the receiver can initialize and decide when handover should take place.

15.2.1.4 Related Work on Soft Handover for DVB-H

Handover in DVB-H networks is a new topic. However, some research work has already been published. An instantaneous (Received Signal Strength Indication) RSSI value-based handover scheme was proposed in [14]. This handover scheme is, to the author's knowledge, the first for DVB-H published in the literature. This scheme uses the off-burst time to measure the RSSI value. After comparing the current RSSI value with the RSSI values of adjacent cells, it hands over to the cell with the strongest RSSI value. Because the RSSI value can vary, due to multipath interference or other environmental effects, it may not give a true indication of communication performance or range, and mistakenly measuring the RSSI value would result in unnecessarily consuming battery power because more off-burst time would be used in handover measurement. It is possible, in a worst-case scenario, that the RSSI value could end up being measured at least once every off-burst period. Such an algorithm cannot avoid the ping-pong effect. The ping-pong effect is unnecessary frequent handover caused by signal fluctuation. This scheme cannot effectively eliminate the possibility of receiving "fake signals" either [14]. A fake signal is a signal that has a similar frequency to that of an adjacent cell, but is from a faraway cell. Constant measuring of the adjacent cell's signal level without any handover prediction leads to unnecessary battery power consumption, receiving "fake signals," and to a degraded quality of service. To overcome these shortcomings, a handover prediction algorithm has to be developed. Yang et al. [27] proposed a handover decision-making approach based on postprocessing of the measured SNR value to avoid the ping-pong effect and to get rid of the received "fake signals." May [11] proposed a technology called "phase shifting" to show how the off-burst times can be used to perform soft handovers. Vare et al. [14] proposed a cell description table (CDT) based method to improve the performance of soft handover for a DVB-H terminal with global positioning system (GPS) support. In a recent paper [29], different handover decision-making algorithms are presented and a novel hybrid handover decision-making algorithm is proposed. Research on the handover issue in DVB-H was being conducted in the IST-INSTINCT project (http://www.ist-instinct.org).

In this chapter, intelligent active repeaters are presented to provide location information to mobile receivers. In this way, a receiver does not need to measure the handover parameters before it reaches the handover location, reducing the ping-pong effect and consequently battery power consumption. On the other hand, "fake signals" will be completely eliminated because all the repeaters provide their unique identification information to the receivers.

15.2.2 Using Repeaters Forming Virtual Cells in DVB-H

Virtual cells in DVB-H refer to the cells that are formed by repeaters and main transmitters. The typical virtual cell structure is shown in Figure 15.7. In the virtual cell strucutre, the receiver will always feel it is at the center of the cell no matter where the receiver moves within the cell.

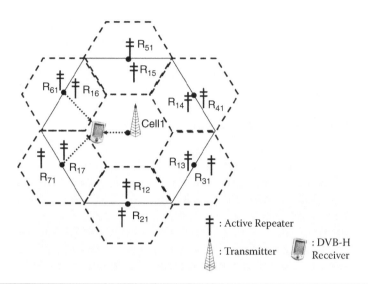

Figure 15.7   Virtual Cell Structure.

15.2.3   Repeater Aided Soft Handover Algorithm

This section provides the details of the introduced RA_handover algorithm.

Figure 15.8 illustrates the relationship between the different components and their operations in RA_handover, and can be taken as a general illustrative case of the RA_handover algorithm. The RA_handover algorithm's job is located in the handover decision-making stage of the DVB-H handover algorithm. In the handover measurement stage, the receiver receives the services from both the main transmitter and the repeater. Meanwhile, the signal RSSI or SNR is obtained constantly. In the handover decision-making stage, the signaling bits are obtained from the repeaters; this triggers the handover decision-making process. According to the RSSI or SNR value obtained from the handover measurement stage and the signaling bits from the repeaters, the handover decision is made based on predefined handover criteria. The result of the handover decision-making stage is a decision on whether or not handover will take place and if so, the handover target cell is chosen. In the handover execution stage, synchronization to the appropriate signal of the target handover cell is performed.

The cellular network structure for RA_handover is illustrated in Figure 15.9. It shows a seven-cell DVB-H network topology. Each cell contains six repeaters, i.e., six subcells, allocated uniformly around the border of the cell. R_{12} is one subcell in cell 1 at the border between cell 1 and cell 2. R_{21} is one subcell in cell 2 at the border between cell 2 and cell 1. Correspondingly, R_{ij} and R_{ji} are the subcells at the border between cell i and cell $j (i, j = 1, 2, 3, 4, 5, 6,$ and $7)$, respectively. Suppose the repeaters are using directional antennas and each repeater can only cover the subcell area where it is located. When the mobile receiver moves into any subcell area covered by a repeater, it will get the corresponding repeater information from the signaling bits it receives within the on-burst time. At this location the receiver will begin to carry out the handover measurement in the off-burst time. This means that the receiver will not measure the signal strength using the off-burst time until it reaches one of the subcell areas covered by a repeater. In this way the receiver does not need to measure the signal strength level constantly, thus saving battery power. With the installation of the repeaters in the

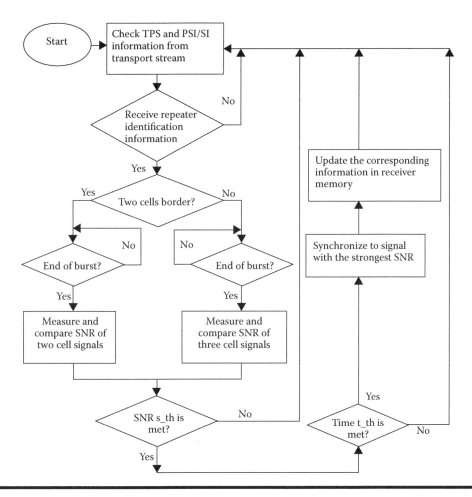

Figure 15.8   RA_handover Algorithm.

cell border area, the quality of service will also be increased compared with having no repeaters installed in the border. Because broadcast cells are usually very large, some specific service information may need to be broadcast in part of the cell. The active repeaters proposed can be used to do this job. This provides an extra tool and outlet for the service management.

With MIMO antennas on the DVB-H receiver, the receiver can receive signals from different directions and combine them into a better quality signal, thus improving the quality of service. Take cell 1 of Figure 15.7, for example, it is easy to see that another advantage of the RA_handover algorithm is that the receiver will always feel it is at the center of the cell, no matter where the receiver moves within the cell. In this way, the RA_handover algorithm not only improves the quality of the service that the receiver receives, but also keeps the quality of the service consistent over all the cells.

With the addition of repeaters, the main transmitter power can be reduced, thus reducing the cost of the main transmitter. Although the addition of the repeaters will add to the cost of the network equipment, the overall cost of the system will not necessarily be increased when the increased quality of service and the savings on the terminal side (because of reduced power consumption) are considered.

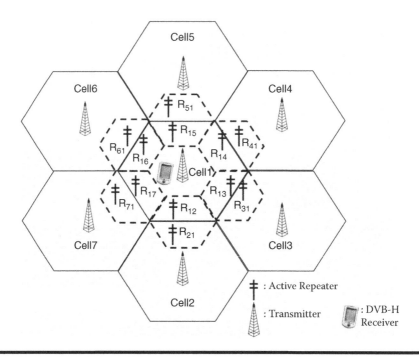

Figure 15.9   RA_handover Algorithm Cellular Structure.

Furthermore, the standardization of the handover algorithm in the DVB-H standard is not yet finalized. The RA_handover algorithm will provide a very competitive candidate for the selection of the handover algorithm to be incorporated into the standards of DVB-H. This also brings DAS into a novel application-field handover in DVB-H.

15.2.4   Simulation and Analysis

The performance of the RA_handover algorithm is analyzed in this section with respect to the front-end battery power consumption, the received quality of service, and the cost of the overall network system. The approach of this section is to build a simulation model in MATLAB to identify the relationship between the received signal strength and the battery power consumption. The received signal strength is related to the repeater-covered area. The simulation scenario is that the repeater-covered area is changed as the received signal strength from the repeaters is changed. Cell radius, antenna height, transmitter power, transmission frequency, and time percentage are the common parameters on which received signal strength depends. Time percentage is a term widely used in propagation modeling; it accounts for variations in hourly median values of attenuation due to, for example, slow changes in atmospheric refraction or in the intensity of atmospheric turbulence. The value of time percentage gives the fraction of time during which the actual received field strength is expected to be equal to or higher than the hourly median field strength. This variable allows the time variability of changing atmospheric (and other) effects to be specified. As the received signal strength can be thought of as proportional to the received quality of service, the relationship between the quality of service and the battery power consumption can be obtained. The simulation parameter data was derived from the DVB-H standards [6,21,30] and International Telecommunication Union (ITU) standards [31].

Research has shown that human factors are an essential component of a successful service delivery system for wireless telecommunications [32]. Because cost is one of the human factors that is a very important issue in business and standardization processes [33], the cost issue is described analytically in the last part of this section.

First, the percentage of battery power consumed using the RA_handover algorithm is compared with that of an algorithm in which every off-burst time is used to make handover measurements, as may happen without repeaters. From the network in Figure 15.7, it is easy to see that with the use of the repeaters, the receiver will only make handover measurements in the six subcell areas instead of the whole area of cell 1. More handover measurements mean more battery power consumption. The handover probability can be obtained from the area where the handover will happen and the whole service area [34]. By using the same methods, the saved power consumption can be calculated from the difference of the repeater covered area and the whole cell area. Suppose that the whole cell area and the repeater-covered area are ideal hexagonal shapes as shown in Figure 15.7, and the DVB-H receiver is uniformly distributed in both time and location in the cell. Figure 15.7 shows the maximum area that the repeaters are able to cover, and it can be seen that the following equation holds:

$$S = \frac{A_c - A_r}{A_c} = 25\%. \tag{15.1}$$

In equation (15.1), A_c is the area of the whole cell, A_r is the whole area covered by the six repeaters, and S is the saved battery power compared with the handover algorithm utilizing every off-burst time. Thus, it can be seen that at least 25% of the battery power consumption on the handover decision-making stage can be saved.

It needs to be noted that in the network topology shown in Figure 15.7, the receiver will receive the best quality of service because it always receives as if it is near the center of the cell. On the other hand, Figure 15.7 shows the maximum area that the repeaters cover. In this case, the saved battery power consumption S is minimum. If the repeater covered area A_r is decreased, the saved battery power consumption S will be increased, but the quality of service will be decreased too. Because the receiver will not receive as if it is near the center of the cell again when A_r is decreased, the quality of the service will not be coherent all over the cell.

To determine the relationship between the battery power consumed by the handover decision-making algorithm and the received quality of service, a model is built up for simulation. Without losing generality, suppose the received quality of service is directly related to the received signal strength. Although the received quality of service and the received signal strength are a nonlinear relationship, it is simpler and does not affect the purpose of the simulation to assume that there is a fixed linear relationship between the received quality of service Q and the received signal strength E_b:

$$Q = \alpha E_b, \tag{15.2}$$

where coefficient α is a constant parameter that links the Q and the E_b together.

Correspondingly, for a simpler simulation and without affecting the purpose of the simulation, suppose that there is a fixed linear relationship between the battery power consumed by the handover decision-making algorithm C and the size of the repeater covered area A_r.

$$C = \beta A_r, \tag{15.3}$$

where coefficient β is also a constant parameter that connects the C and the A_r together.

The repeater covered area or the range of the repeaters depends on several things, such as antennas and their heights, expected reception quality, the propagation paths of signals, geographical location and terrain, presence of interference, receiver sensitivity, and transmitter power. Given a receiver and a fixed location, the adjustable parameters are the antennas and their height and the transmitter power. In this case, ITU-R P.1546-1 provides easy-to-follow procedures to calculate the field strength given antenna height and transmitter power [31].

ITU-R P.1546-1 is the ITU Recommendation for point-to-area field strength predictions for terrestrial services in the frequency range 30 MHz to 3000 MHz. Here, land use only is considered. Based on recommendation ITU-R P.1546-1, a simulation is built up in the following way:

Step 1: The dimensionless parameter k is calculated using the transmitter or repeater height h, as follows:

$$k = \frac{\log\left[\frac{h}{9.375}\right]}{\log(2)}, \tag{15.4}$$

where h is in the range of 9.375 and 1200 m; k is an integer in the range between 0 and 7.

Step 2: An intermediate field strength E_u at the distance d for transmitter height h is calculated as follows:

$$E_u = p_b \times \log \frac{10^{\frac{E1+E2}{p_b}}}{10^{\frac{E1}{p_b}} + 10^{\frac{E2}{p_b}}}, \tag{15.5}$$

where

$$p_b = d_0 + d_1 \cdot \sqrt{k}, \tag{15.6}$$

$$E1 = (a_0 \cdot k^2 + a_1 \cdot k + a_2) \cdot \log(d) + 0.1995 \cdot k^2 + 1.8671 \cdot k + a_3, \tag{15.7}$$

$$E_2 = E_{off} + E_{ref}, \tag{15.8}$$

where

$$E_{off} = \frac{C_0}{2} \cdot k \cdot k \left[1 - tgh\left[c_1 \cdot \left[\log(d) - c_2 - \frac{c_3^k}{c_4}\right]\right]\right] + c_5 \cdot k^{c_6}, \tag{15.9}$$

$$E_{ref} = b_0[exp[-b_4 \cdot 10^\xi] - 1] + b_1 \cdot exp\left[-\left(\frac{\log(d) - b_2}{b_3}\right)^2\right] - b_6 \cdot \log(d) + b_7, \tag{15.10}$$

where

$$\xi = \log(d)^{b_5}. \tag{15.11}$$

In the equations in Step 2 above, a_0 to a_3, b_0 to b_7, c_0 to c_6, and d_0 to d_1 are parameters given in Table 15.3. Because a DVB-H is most likely to be used in UHF band

Table 15.3   Coefficients for the Generation of the Land Tabulations

Frequency	600 MHz			2000 MHz		
Time (%)	50	10	1	50	10	1
a_0	0.0946	0.0913	0.0870	0.0946	0.0941	0.0918
a_1	0.8849	0.8539	0.8141	0.8849	0.8805	0.8584
a_2	−35.399	−34.160	−32.567	−35.399	−35.222	−34.337
a_3	92.778	92.778	92.778	94.493	94.493	94.493
b_0	51.6386	35.3453	36.8836	30.0051	25.0641	31.3878
b_1	10.9877	15.7595	13.8843	15.4202	22.1011	15.6683
b_2	2.2113	2.2252	2.3469	2.2978	2.3183	2.3941
b_3	0.5384	0.5285	0.5246	0.4971	0.5636	0.5633
b_4	4.323 $\times 10^{-6}$	1.704 $\times 10^{-7}$	5.169 $\times 10^{-7}$	1.677 $\times 10^{-7}$	3.126 $\times 10^{-8}$	1.439 $\times 10^{-7}$
b_5	1.52	1.76	1.69	1.762	1.86	1.77
b_6	49.52	49.06	46.5	55.21	54.39	49.18
b_7	97.28	98.93	101.59	101.89	101.39	100.39
c_0	6.4701	5.8636	4.7453	6.9657	6.5809	6.0398
c_1	2.9820	3.0122	2.9581	3.6532	3.547	2.5951
c_2	1.7604	1.7335	1.9286	1.7658	1.7750	1.9153
c_3	1.7508	1.7452	1.7378	1.6268	1.7321	1.6542
c_4	198.33	216.91	247.68	114.39	219.54	186.67
c_5	0.1432	0.1690	0.1842	0.1309	0.1704	0.1019
c_6	2.2690	2.1985	2.0873	2.3286	2.1977	2.3954
d_0	5	5	8	8	8	8
d_1	1.2	1.2	0	0	0	0

(470–838 MHz) and L band (1440–1790 MHz) [35], only the transmitting frequencies 600 MHz (in UHF band) and 2000 MHz (adjacent to L band for convenience) are used and different time percentages (50%, 10% and 1%) for land area are used.

Step 3: The final field strength E_b at the distance d for transmitter height b is:

$$E_b = p_b b \cdot \log \left[\frac{10^{\frac{E_u + E_{fs}}{P_{bb}}}}{10^{\frac{E_u}{P_{bb}}} + 10^{\frac{E_{fs}}{P_{bb}}}} \right]. \tag{15.12}$$

In the above equation, E_{fs} is the free space field strength assuming that the transmitter *ERP* (effective radiated power) is 1 KW and E_{fs} is given by:

$$E_{fs} = 106.9 - 20 \log(d) \quad \text{dB}(\mu V/m), \tag{15.13}$$

and P_{bb} in Equation (15.12) is the blend coefficient set to value 8 according to ITU-R P.1546-1.

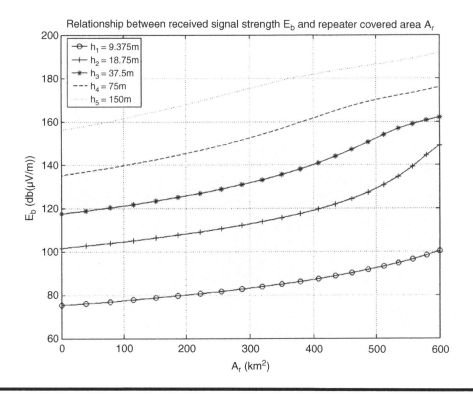

Figure 15.10 The Relationship Between Received Signal Strength and Repeater Covered Area.

The relationship between the receiver-received quality of service Q and the consumed battery power C can be analytically expressed as Equation (15.14) and Equation (15.15) below:

$$Q = \alpha \cdot f(h_1, h_2, h_3, l_1, l_2, l_3), \tag{15.14}$$

$$C = \beta \cdot g(l_1, l_2, l_3), \tag{15.15}$$

where $Q = \alpha \cdot f(h_1, h_2, h_3, l_1, l_2, l_3)$ and $C = \beta \cdot g(l_1, l_2, l_3)$ are the abstracted functions obtained from Equation (15.2) and Equation (15.3), while l_1, l_2, and l_3 are the corresponding distances from the DVB-H receiver to the central main transmitter and the nearest two repeaters, and h_1, h_2, and h_3 are the corresponding antenna height of the central main transmitter and the nearest two repeaters.

Based on the field strength prediction procedures above, a simulation model is built in MATLAB. The simulation parameters are: DVB-H cell radius is 30 km; antenna height is between 9.375 and 1200 m; 600 MHz, land path, 50% time. After simulation, the relationship obtained between E_b and A_r is shown in Figure 15.10. In Figure 15.10, $h_i(i = 1, 2, \ldots, 5)$ is the antenna height of the main transmitter and the repeaters.

For simplicity it is supposed that α and β are both equal to 1, then the receiver-received quality of service Q and the consumed battery power C is shown in Figure 15.11. It is easy to see that the received quality of service Q is increased with the increasing battery power consumption C, and given a fixed value of cell radius, antenna height,

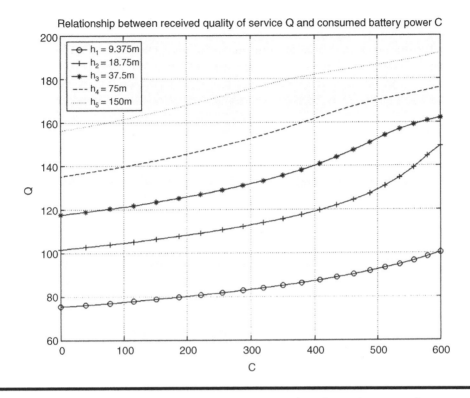

Figure 15.11   The Relationship Between Received Quality of Service and Consumed Battery Power.

transmitter power, transmission frequency, and time percent a fixed relationship between Q and C is able to be obtained.

Now the cost of the RA_handover scheme is considered. Active repeaters are expensive. On the other hand, the cost of the repeaters is connected with the cost of the main transmitter. Because low power repeaters cover small areas, to provide the same quality of service for the users, even in the border area of the cell, it is necessary to install very costly high power main transmitters. Without repeaters, the main transmitter must use high power to provide an acceptable quality of service in the cell border area. The more transmission power the main transmitters and repeaters have, the more costly they will be. However, it is not very easy to get the exact cost of installing the repeaters and the main transmitters. Though it is hard to compare the exact cost of the RA_handover algorithm and the algorithm without the repeaters, it can be seen that by implementing the RA_handover algorithm, the handheld DVB-H receiver can save considerable battery power consumption and improve the quality of service. This will drive the consumers' desire to use the DVB-H service, which equates to profit making for the whole system. The exact cost comparison will be done in the future.

15.3   CURRENT RESEARCH STATUS AND OPEN ISSUES

This section presents the current research status of DAS for handover in DVB-H and some open research issues to encourage the research on DAS in DVB-H networks.

15.3.1 Current Research Status

Handover for unidirectional broadcasting networks like DVB-H is a novel issue and a new challenge. Low power transmitters constituting dense multifrequency cellular DVB-H networks will be a typical network structure of DVB-H, which makes handover a very important issue in DVB-H network planning and optimization. The charachteristics of DAS make it very suitable to be used in addressing the handover problem of DVB-H.

Currently, the only approach, called RA_handover, has been proposed for DVB-H receivers with MIMO antennas to decide when soft handover should occur, and is based on a proposed intelligent repeater structure. Simple mathematical calculation showed that the RA_handover scheme could save 25% of the battery power consumed by handover measurements compared with a handover algorithm utilizing every off-burst time for handover measurement, as may happen without repeaters. A simulation model has been developed to show the performance of the RA_handover approach. Simulation results show that the receiver-received quality of service increases as the repeater-covered area is increased, and that the maximum quality of service happens when the receiver always feels it is located in the center of the cell. The cost issues introduced by the RA_handover algorithm were also analyzed. Although it is still difficult to determine the cost of introducing the RA_handover algorithm, it has been shown that the cost will not be an obstacle to the implementation of the RA_handover algorithm when overall system costs and revenues are considered. In the RA_handover algorithm, the repeaters are active repeaters. These repeaters can improve the quality of services in the repeater-covered area as described in the chapter. The active repeaters can also add extra signaling information to the received signals. For the service providers, the extra signaling information can be used to signal the services or even additional localized services in the repeater covered area. This will provide an extra tool for the management of the provided services. Because RA_handover is a feasible handover algorithm for DVB-H, as demonstrated through simulation results reported in this chapter, it is very promising to be considered by the DVB group for eventual incorporation into the soft handover standard for DVB-H.

15.3.2 Open Issues

The handover in DVB-H networks is the subject of ongoing research and the application of DAS for addressing the handover issues in DVB-H will be a hot topic for some time in the future as mobile TVs are rolling out. Besides the ongoing research for the RA_handover algorithm itself, in this section, some key points are presented as criteria for designing an efficient handover algorithm using DAS in DVB-H networks identified from the research conducted by the author. And these criteria should be seriously considered when applying DAS to the handover research in DVB-H [36].

15.3.2.1 Handover Decision-Making Stage

One of the key concerns in designing an efficient handover algorithm for DVB-H is reducing battery power consumption. The handover decision-making stage is the handover phase where the battery power consumption reduction can be fully exploited. The main objective in the handover decision-making stage is to try to predict the handover moment to reduce the number of off-burst time intervals that are used for handover measurement.

15.3.2.2 The Complexity of Handover Algorithms

The handover algorithm using DAS for DVB-H should be simple. Complex handover algorithms mean more central processing unit (CPU) power consumption on the handheld terminal. If a complex handover algorithm using DAS has to be introduced, the tradeoff between the CPU power consumption of the algorithm itself and the power savings due to the reduced measurement frequency resulting from introducing DAS to the handover algorithm has to be evaluated first.

15.3.2.3 Additional Signaling Information

Additional signaling information introduced by DAS should always be fully exploited by the handover algorithm. Handover in dedicated DVB-H networks is a passive handover. If additional signaling information is available, DAS should likely be used to help the handover process. Take the converged terminal as an example; converged DVB-H and mobile cellular general packet radio service/universal mobile telecommunications system (GPRS/UMTS) terminals have the advantage of having an interactive uplink channel. In this case, the uplink channel can be utilized to aid the handover process and this UMTS aided handover is a kind of active handover. The network parameters transmitted from transmitters and repeaters can also be fully utilized by introducing DAS to aid the handover process.

15.3.2.4 Additional Cost Introduced by Handover Algorithms

DVB-H terminals should be affordable to the consumers. An additional attachment such as a GPS receiver can improve the handover efficiency, but it can also increase the terminal prices. Introducing DAS to the network faces the same problem as introducing the GPS receivers to the terminals. The DVB-H handover algorithms should focus on making full use of the limited DAS to exploit the signaling information available and avoid the extra cost of introducing new network equipment (such as expensive repeaters) or terminal attachments (such as GPS receivers) purely for handover purposes.

15.4 CONCLUSION

This chapter described how DAS systems can be used in the DVB-H networks and presented an example of its application by applying active repeaters to aid the handover process of DVB-H. To give a detailed analysis of this kind of application, a handover algorithm called RA_handover is presented. The RA_handover algorithm is analyzed by simulation and mathematical analysis; results showed that using DAS for the handover process of DVB-H brings many benefits, including reduced battery power consumption for the terminals and improved receiving service quality.

To encourage the research on DAS for handover algorithms in DVB-H, the current status of DAS research in DVB-H and some open issues for future research have been presented.

REFERENCES

[1] C. Trolet, *SPOT: Filling Gaps in DVB-T Networks With Digital Repeaters*, presented by G. Faria, at BroadcastAsia2002 International Conference, available: www.broadcast.harris.com.

[2] J. Cosmas, T. Itagaki, K. Krishnapillai, and A. Lucas, "Multimedia broadcast and internet satellite system design and user trial results," *Int. J. Serv. Stand.*, vol. 1, 2005, pp. 336–357.

[3] C. Shchiglik, S. J. Barnes, and E. Scornavacca, "Mobile entertainment services: a study of consumer perceptions towards games delivered via the wireless application protocol," *Int. J. Serv. Stand.*, vol. 1, 2004, pp. 155–171.

[4] ETSI EN 302 304 V1.1.1, "Digital Video Broadcasting (DVB); Transmission System for Handheld Terminals (DVB-H)." Nov. 2004.

[5] B. Heidkamp, A. Pohl, U. Schiek, F. Klinkenberg, J. Hynynen, A. Sieber, P. Christ, T. Owens, J. Cosmas, T. Itagaki, and F. Sun, "Demonstrating the feasibility of standardised application programme interfaces that will allow mobile/portable terminals to receive services combining UMTS and DVB-T," *Int. J. Serv. Stand.*, vol. 1, 2004, pp. 228–242.

[6] ETSI, EN 300 744, "Digital Video Broadcasting (DVB): Framing Structure, Channel Coding and Modulation for Digital Terrestrial Television." V1.5.1 (2004–11).

[7] H. Schulzrinne, S. Casner, R. Frederick, and V. Jacobson, "RTP: a transport protocol for real-time applications," IETF RFC 3550, 2003.

[8] T. Paila, M. Luby, R. Lehtonen, V. Roca, and R. Walsh, "FLUTE — file delivery over unidirectional transport," IETF RFC 3926, 2004.

[9] DigiTAG, "Television on a handheld receiver," v. 1.1, 2005: www.digitag.org.

[10] X.D. Yang, Y.H. Song, T.J. Owens, J. Cosmas, and T. Itagaki, "Performance analysis of time slicing in DVB-H," *Sympotic2004*, 2004, pp. 183–186.

[11] G. May, "Loss-free handover for IP datacast over DVB-H networks," *IEEE ISCE 2005*, Macau, June 2005.

[12] G. May, "The IP datacast system — overview and mobility aspects," *2004 IEEE International Symposium on Consumer Electronics*, Sept. 2004, pp. 509–514.

[13] www.digitalradiotech.co.uk/dvb-h_dab_dmb.htm, Sept. 2005.

[14] J. Väre and M. Puputti, "Soft handover in terrestrial broadcast networks," *MDM2004 Proceedings*, 2004, pp. 236–242.

[15] ETSI, TR 102 377, "Digital Video Broadcasting (DVB); DVB-H Implementation Guidelines," v1.2.1 (2005–11).

[16] ETSI, EN 301 192, "Digital video broadcasting (DVB); DVB Specification for Data Broadcasting," v1.4.1 (2004–11).

[17] A. Bria and D.-G. Barquero, "Scalability of DVB-H deployment on existing wireless infrastructure," *IEEE PIMRC*, Sept. 2005.

[18] A. Sieber and C. Weck, "What's the difference between DVB-H and DAB — in the mobile environment?" *Technical Review, EBU*, July 2004.

[19] M. Kornfeld, "DVB-H — the Emerging Standard for Mobile Data Communication," *2004 IEEE International Symposium on Consumer Electronics*, Sept. 2004, pp. 193–198.

[20] G. Faria, "DVB-H to deliver digital TV to hand-held terminals," 2004: www.broadcastpapers.com/BcastAsia04/BAsia04TeamcastDVBH.pdf.

[21] ETSI, EN 300 468, "Digital Video Broadcasting (DVB); Specification for Service Information (SI) in DVB Systems," v1.6.1 (2004–11).

[22] ETSI, TR 101 211, "Digital Video Broadcasting (DVB); Guidelines on Implementation and Usage of Service Information (SI)," v1.6.1 (2004–05).

[23] DVB Document A079 Rev.1, "IP datacast over DVB-H: PSI/SI," Nov. 2005, available: www.dvb-h-online.org.

[24] L. Schwoerer and J. Vesma, "Fast scattered pilot synchronization for DVB-T and DVB-H." *Proc. 8th International OFDM Workshop*, Hamburg, Germany, Sept. 2003.

[25] C. Trolet, "SPOT: filling gaps in DVB-T networks with digital repeaters," presented by G. Faria, at BroadcastAsia2002 International Conference, available: www.broadcast.harris.com.

[26] E. Telatar, "Capacity of multi-antenna Gaussian channels," *Eur. Trans. Telecomm. ETT*, 10(6): 585–596, Nov. 1999.

[27] X.D. Yang, Y.H. Song, T.J. Owens, J. Cosmas, and T. Itagaki, "Seamless soft handover in DVB-H networks," *Softcom2004*, Split (Croatia), Dubrovnik (Croatia), Venice (Italy), Oct. 2004.

[28] Digital Video Broadcasting, "IPDC in DVB-H: Technical Requirements," CBMS 1026 v.1.0.0 rev. 1/TM 3095 rev. 2, 2004.

[29] X.D. Yang, Y.H. Song, T.J. Owens, J. Cosmas, and T. Itagaki, "An investigation and a proposal for handover decision-making in DVB-H," *14th IST Mobile and Wireless Communications Summit*, Dresden, Germany, June 2005.

[30] ETSI, TR 102 401, "Digital Video Broadcasting (DVB): Transmission to Handheld Terminals (DVB-H); Validation Task Force Report," v.1.1.1 (2005–05).

[31] ITU Recommendation ITU-R P.1546-1: "Method for point-to-area predictions for terrestrial services in the frequency range 30 MHz to 3000 MHz," *ITU*, 2001–2003.

[32] Y. B. Choi, J. S. Krause, M. A. Imperio, S. P. Macchio, and K. A. Rill, "Applications of 'human factors' in wireless telecommunications service delivery," *Int. J. Serv. Stand.*, vol. 1, 2005, pp. 287–298.

[33] S. K. Herath and A. Gupta, "A framework for analysing cost structures in business process reengineering (BPR)," *Int. J. Serv. Stand.*, vol. 1, 2005, pp. 494–511.

[34] S.N.P. Van Cauwenberge, "Study of Soft Handover in UMTS, " Master's thesis, 2003, Danmarks Tekniske Universitet, Copenhagen.

[35] B. Tyler, "New Demands of Broadcasting for Spectrum," The IEE seminar on broadcasting spectrum: the issues, London, 2005.

[36] X.D. Yang, J. Väre, and T.J. Owens, "A survey of handover algorithms in DVB-H," *IEEE Commni. Surv. Tutor.*, Dec. 2006.

INDEX

T - #0313 - 101024 - C0 - 254/178/26 [28] - CB - 9781420042887 - Gloss Lamination